Special Distributions

Binomial $b(n, p)$

$$f(x) = \frac{n!}{x!\,(n-x)!} p^x (1-p)^{n-x},$$

$x = 0, 1, 2, \ldots, n$

$M(t) = (1 - p + pe^t)^n$

$\mu = np, \qquad \sigma^2 = np(1-p)$

Poisson

$$f(x) = \frac{\mu^x e^{-\mu}}{x!}, \qquad x = 0, 1, 2, \ldots$$

$M(t) = e^{\mu(e^t - 1)}, \qquad E(X) = \mu = \sigma^2$

Gamma

$$f(x) = \frac{1}{\Gamma(\alpha)\beta^\alpha} x^{\alpha-1} e^{-x/\beta}, \qquad 0 < x < \infty$$

$M(t) = (1 - \beta t)^{-\alpha}, \qquad t < 1/\beta$

$\mu = \alpha\beta, \qquad \sigma^2 = \alpha\beta^2$

Exponential Special gamma with $\alpha = 1$

Chi-square $\chi^2(r)$ Special gamma with $\alpha = r/2$, $\beta = 2$

Normal $N(\mu, \sigma^2)$

$$f(x) = \frac{1}{\sigma\sqrt{2\pi}} \exp\left[-\frac{(x-\mu)^2}{2\sigma^2}\right], \qquad -\infty < x < \infty$$

$M(t) = \exp(\mu t + \sigma^2 t^2/2)$

$E(X) = \mu, \qquad \text{var}(X) = \sigma^2$

Beta

$$f(x) = \frac{\Gamma(\alpha+\beta)}{\Gamma(\alpha)\Gamma(\beta)} x^{\alpha-1}(1-x)^{\beta-1}, \qquad 0 < x < 1$$

$$\mu = \frac{\alpha}{\alpha+\beta}, \qquad \sigma^2 = \frac{\alpha\beta}{(\alpha+\beta+1)(\alpha+\beta)^2}$$

Student's t

$$T = \frac{W}{\sqrt{V/r}}, \qquad E(T) = 0, \qquad \text{var}(T) = \frac{r}{r-2},$$

provided that $r > 2$, where W is $N(0, 1)$, V is $\chi^2(r)$, and W and V are independent

Fisher's F

$$F = \frac{U/r_1}{V/r_2}, \qquad E(F) = \frac{r_2}{r_2 - 2},$$

$$\text{var}(F) = \frac{2r_2^2(r_1 + r_2 - 2)}{r_1(r_2-2)^2(r_2-4)}, \qquad r_2 > 4,$$

where U is $\chi^2(r_1)$, V is $\chi^2(r_2)$, and U and V are independent

Introduction to Mathematical Statistics

Introduction to Mathematical Statistics

FIFTH EDITION

ROBERT V. HOGG
University of Iowa

ALLEN T. CRAIG
Late Professor of Statistics,
University of Iowa

PRENTICE HALL, Upper Saddle River, New Jersey 07458

Library of Congress Cataloging-in-Publication Data

Hogg, Robert V.
Introduction to mathematical statistics/Robert V. Hogg, Allen T. Craig.—5th ed.
p. cm.
Includes bibliographical references and index.
ISBN 0-02-355722-2
1. Mathematical statistics. I. Craig, A. T. (Allen Thornton)
II. Title.
QA276.H59 1995
519.5—dc20

Editor: Robert W. Pirtle
Production Supervisor: Elaine W. Wetterau
Production Manager: Francesca Drago
Text and Cover Designer: Robert Freese

Upper Saddle River, New Jersey 07458

PRINTED IN THE UNITED STATES OF AMERICA

10

Prentice-Hall International (UK) Limited, *London*
Prentice-Hall of Australia Pty. Limited, *Sydney*
Prentice-Hall Canada Inc., *Toronto*
Prentice-Hall Hispanoamericana, S.A., *Mexico*
Prentice-Hall of India Private Limited, *New Delhi*
Prentice-Hall of Japan, Inc., *Tokyo*
Prentice-Hall Asia Pte. Ltd., *Singapore*
Editora Prentice-Hall do Brasil, Ltda., *Rio de Janeiro*

Contents

CHAPTER 3 116

Some Special Distributions

CHAPTER 4 155

Distributions of Functions of Random Variables

CHAPTER 5 233

Limiting Distributions

CHAPTER 6 259

Introduction to Statistical Inference

Preface

When Allen T. Craig died in late November 1978, I lost my advisor, mentor, colleague, and very dear friend. Due to his health, Allen did nothing on the fourth edition and, of course, this revision is mine alone. There is, however, a great deal of Craig's influence in this book. As a matter of fact, when I would debate with myself whether or not to change something, I could hear Allen saying, "It's very good now, Bob; don't mess it up." Often, I would follow that advice.

Nevertheless, there were a number of things that needed to be done. I have had many suggestions from my colleagues at the University of Iowa; in particular, Jim Broffitt, Jon Cryer, Dick Dykstra, Subhash Kochar (a visitor), Joe Lang, Russ Lenth, and Tim Robertson provided me with a great deal of constructive criticism. In addition, three reviewers suggested a number of other topics to include. I have also had statisticians and students from around the world write to me about possible improvements. Elliot Tanis, my good friend and co-author of our *Probability and Statistical Inference*, gave me permission to use a few of the figures, examples, and exercises used in that book. I truly thank these people, who have been so helpful and generous.

Clearly, I could not use all of these ideas. As a matter of fact, I resisted adding "real" problems, although a few slipped into the exercises. Allen and I wanted to write about the mathematics of statistics, and I have followed that guideline. Hopefully, without those problems, there is still enough motivation to study mathematical statistics in this book. In addition, there are a number of excellent books on applied statistics, and most students have had a little exposure to applications before studying this book.

The major differences between this edition and the preceding one are the following:

- There is a better discussion of assigning probabilities to events, including introducing independent events and Bayes' theorem in the text.
- The consideration of random variables and their expectations is greatly improved.
- Sufficient statistics are presented earlier (as was true in the very early editions of the book), and minimal sufficient statistics are introduced.
- Invariance of the maximum likelihood estimators and invariant location- and scale-statistics are considered.
- The expressions "convergence in distribution" and "convergence in probability" are used, and the delta method for finding asymptotic distributions is spelled out.
- Fisher information is given, and the Rao–Cramér lower bound is presented for an estimator of a function of a parameter, not just for an unbiased estimator.
- The asymptotic distribution of the maximum likelihood estimator is included.
- The discussion of Bayesian procedures has been improved and expanded somewhat.

There are also a number of little items that should improve the understanding of the text: the expressions var and cov are used; the convolution formula is in the text; there is more explanation of p-values; the relationship between two-sided tests and confidence intervals is noted; the indicator function is used when helpful; the multivariate normal distribution is given earlier (for those with an appropriate background in matrices, although this is still not necessary in the use of this book); and there is more on conditioning.

I believe that the order of presentation has been improved; in particular, sufficient statistics are presented earlier. More exercises have been introduced; and at the end of each chapter, there are several additional exercises that have not been ordered by section or by difficulty (several students had suggested this). Moreover, answers have not been given for any of these additional exercises because I thought some instructors might want to use them for questions on examinations. Finally, the index has been improved greatly, another suggestion of students as well as of some of my colleagues at Iowa.

There is really enough material in this book for a three-semester sequence. However, most instructors find that selections from the first five chapters provide a good one-semester background in the probability needed for the mathematical statistics based on selections from the remainder of the book, which certainly would include most of Chapters 6 and 7.

I am obligated to Catherine M. Thompson and Maxine Merrington and to Professor E. S. Pearson for permission to include Tables II and V, which are abridgments and adaptations of tables published in *Biometrika*. I wish to thank Oliver & Boyd Ltd., Edinburgh, for permission to include Table IV, which is an abridgment and adaptation of Table III from the book *Statistical Tables for Biological, Agricultural, and Medical Research* by the late Professor Sir Ronald A. Fisher, Cambridge, and Dr. Frank Yates, Rothamsted.

Finally, I would like to dedicate this edition to the memory of Allen Craig and my wife, Carolyn, who died June 25, 1990. Without the love and support of these two caring persons, I could not have done as much professionally as I have. My friends in Iowa City and my children (Mary, Barbara, Allen, and Robert) have given me the strength to continue. After four previous efforts, I really hope that I have come close to "getting it right this fifth time." I will let the readers be the judge.

R. V. H.

CHAPTER 1

Probability and Distributions

1.1 Introduction

Many kinds of investigations may be characterized in part by the fact that repeated experimentation, under essentially the same conditions, is more or less standard procedure. For instance, in medical research, interest may center on the effect of a drug that is to be administered; or an economist may be concerned with the prices of three specified commodities at various time intervals; or the agronomist may wish to study the effect that a chemical fertilizer has on the yield of a cereal grain. The only way in which an investigator can elicit information about any such phenomenon is to perform his experiment. Each experiment terminates with an *outcome*. But it is characteristic of these experiments that the outcome cannot be predicted with certainty prior to the performance of the experiment.

Suppose that we have such an experiment, the outcome of which cannot be predicted with certainty, but the experiment is of such a nature that a collection of every possible outcome can be described prior to its performance. If this kind of experiment can be repeated under the same conditions, it is called a *random experiment*, and the collection of every possible outcome is called the experimental space or the *sample space*.

Example 1. In the toss of a coin, let the outcome tails be denoted by T and let the outcome heads be denoted by H. If we assume that the coin may be repeatedly tossed under the same conditions, then the toss of this coin is an example of a random experiment in which the outcome is one of the two symbols T and H; that is, the sample space is the collection of these two symbols.

Example 2. In the cast of one red die and one white die, let the outcome be the ordered pair (number of spots up on the red die, number of spots up on the white die). If we assume that these two dice may be repeatedly cast under the same conditions, then the cast of this pair of dice is a random experiment and the sample space consists of the following 36 ordered pairs: $(1, 1), \ldots, (1, 6), (2, 1), \ldots, (2, 6), \ldots, (6, 6)$.

Let $\mathscr{C}$ denote a sample space, and let C represent a part of $\mathscr{C}$. If, upon the performance of the experiment, the outcome is in C, we shall say that the *event* C has occurred. Now conceive of our having made N repeated performances of the random experiment. Then we can count the number f of times (the frequency) that the event C actually occurred throughout the N performances. The ratio f/N is called the *relative frequency* of the event C in these N experiments. A relative frequency is usually quite erratic for small values of N, as you can discover by tossing a coin. But as N increases, experience indicates that we associate with the event C a number, say p, that is equal or approximately equal to that number about which the relative frequency seems to stabilize. If we do this, then the number p can be interpreted as that number which, in future performances of the experiment, the relative frequency of the event C will either equal or approximate. Thus, although we *cannot* predict the outcome of a random experiment, we *can*, for a large value of N, predict approximately the relative frequency with which the outcome will be in C. The number p associated with the event C is given various names. Sometimes it is called the *probability* that the outcome of the random experiment is in C; sometimes it is called the *probability* of the event C; and sometimes it is called the *probability measure* of C. The context usually suggests an appropriate choice of terminology.

Example 3. Let $\mathscr{C}$ denote the sample space of Example 2 and let C be the collection of every ordered pair of $\mathscr{C}$ for which the sum of the pair is equal to seven. Thus C is the collection (1, 6), (2, 5), (3, 4), (4, 3), (5, 2), and (6, 1). Suppose that the dice are cast $N = 400$ times and let f, the frequency of a sum of seven, be $f = 60$. Then the relative frequency with which the outcome was in C is $f/N = \frac{60}{400} = 0.15$. Thus we might associate with C a number p that is close to 0.15, and p would be called the probability of the event C.

Remark. The preceding interpretation of probability is sometimes referred to as the *relative frequency approach*, and it obviously depends upon the fact that an experiment can be repeated under essentially identical conditions. However, many persons extend probability to other situations by treating it as a rational measure of belief. For example, the statement $p = \frac{2}{5}$ would mean to them that their *personal* or *subjective* probability of the event C is equal to $\frac{2}{5}$. Hence, if they are not opposed to gambling, this could be interpreted as a willingness on their part to bet on the outcome of C so that the two possible payoffs are in the ratio $p/(1 - p) = \frac{2}{5}/\frac{3}{5} = \frac{2}{3}$. Moreover, if they truly believe that $p = \frac{2}{5}$ is correct, they would be willing to accept either side of the bet: (a) win 3 units if C occurs and lose 2 if it does not occur, or (b) win 2 units if C does not occur and lose 3 if it does. However, since the mathematical properties of probability given in Section 1.3 are consistent with either of these interpretations, the subsequent mathematical development does not depend upon which approach is used.

The primary purpose of having a mathematical theory of statistics is to provide mathematical models for random experiments. Once a model for such an experiment has been provided and the theory worked out in detail, the statistician may, within this framework, make inferences (that is, draw conclusions) about the random experiment. The construction of such a model requires a theory of probability. One of the more logically satisfying theories of probability is that based on the concepts of sets and functions of sets. These concepts are introduced in Section 1.2.

1.2 Set Theory

The concept of a *set* or a *collection* of objects is usually left undefined. However, a particular set can be described so that there is no misunderstanding as to what collection of objects is under consideration. For example, the set of the first 10 positive integers is sufficiently well described to make clear that the numbers $\frac{3}{4}$ and 14 are not in the set, while the number 3 is in the set. If an object belongs to a set, it is said to be an *element* of the set. For example, if A denotes the set of real numbers x for which $0 \le x \le 1$, then $\frac{3}{4}$ is an element of the set A. The fact that $\frac{3}{4}$ is an element of the set A is indicated by writing $\frac{3}{4} \in A$. More generally, $a \in A$ means that a is an element of the set A.

The sets that concern us will frequently be *sets of numbers*. However, the language of sets of *points* proves somewhat more convenient than that of sets of numbers. Accordingly, we briefly indicate how we use this terminology. In analytic geometry considerable emphasis is placed on the fact that to each point on a line (on which an origin and a unit point have been selected) there corresponds one and only one number, say x; and that to each number x there corresponds one and only one point on the line. This one-to-one correspondence between the numbers and points on a line enables us to speak, without misunderstanding, of the "point x" instead of the "number x." Furthermore, with a plane rectangular coordinate system and with x and y numbers, to each symbol (x, y) there corresponds one and only one point in the plane; and to each point in the plane there corresponds but one such symbol. Here again, we may speak of the "point (x, y)," meaning the "ordered number pair x and y." This convenient language can be used when we have a rectangular coordinate system in a space of three or more dimensions. Thus the "point $(x_1, x_2, \ldots, x_n)$" means the numbers $x_1, x_2, \ldots, x_n$ in the order stated. Accordingly, in describing our sets, we frequently speak of a set of points (a set whose elements are points), being careful, of course, to describe the set so as to avoid any ambiguity. The notation $A = \{x : 0 \le x \le 1\}$ is read "A is the one-dimensional set of points x for which $0 \le x \le 1$." Similarly, $A = \{(x, y) : 0 \le x \le 1, 0 \le y \le 1\}$ can be read "A is the two-dimensional set of points (x, y) that are interior to, or on the boundary of, a square with opposite vertices at $(0, 0)$ and $(1, 1)$." We now give some definitions (together with illustrative examples) that lead to an elementary algebra of sets adequate for our purposes.

Definition 1. If each element of a set A_1 is also an element of set A_2, the set A_1 is called a *subset* of the set A_2. This is indicated by writing $A_1 \subset A_2$. If $A_1 \subset A_2$ and also $A_2 \subset A_1$, the two sets have the same elements, and this is indicated by writing $A_1 = A_2$.

Example 1. Let $A_1 = \{x : 0 \le x \le 1\}$ and $A_2 = \{x : -1 \le x \le 2\}$. Here the one-dimensional set A_1 is seen to be a subset of the one-dimensional set A_2; that is, $A_1 \subset A_2$. Subsequently, when the dimensionality of the set is clear, we shall not make specific reference to it.

Example 2. Let $A_1 = \{(x, y) : 0 \leq x = y \leq 1\}$ and $A_2 = \{(x, y) : 0 \leq x \leq 1, 0 \leq y \leq 1\}$. Since the elements of A_1 are the points on one diagonal of the square, then $A_1 \subset A_2$.

Definition 2. If a set A has no elements, A is called the *null set*. This is indicated by writing $A = \varnothing$.

Definition 3. The set of all elements that belong to at least one of the sets A_1 and A_2 is called the *union* of A_1 and A_2. The union of A_1 and A_2 is indicated by writing $A_1 \cup A_2$. The union of several sets $A_1, A_2, A_3, \ldots$ is the set of all elements that belong to at least one of the several sets. This union is denoted by $A_1 \cup A_2 \cup A_3 \cup \cdots$ or by $A_1 \cup A_2 \cup \cdots \cup A_k$ if a finite number k of sets is involved.

Example 3. Let $A_1 = \{x : x = 0, 1, \ldots, 10\}$ and $A_2 = \{x : x = 8, 9, 10, 11,$ or $11 < x \leq 12\}$. Then $A_1 \cup A_2 = \{x : x = 0, 1, \ldots, 8, 9, 10, 11,$ or $11 < x \leq 12\} = \{x : x = 0, 1, \ldots, 8, 9, 10,$ or $11 \leq x \leq 12\}$.

Example 4. Let A_1 and A_2 be defined as in Example 1. Then $A_1 \cup A_2 = A_2$.

Example 5. Let $A_2 = \varnothing$. Then $A_1 \cup A_2 = A_1$ for every set A_1.

Example 6. For every set A, $A \cup A = A$.

Example 7. Let

$$A_k = \left\{x : \frac{1}{k+1} \leq x \leq 1\right\}, \qquad k = 1, 2, 3, \ldots.$$

Then $A_1 \cup A_2 \cup A_3 \cup \cdots = \{x : 0 < x \leq 1\}$. Note that the number zero is not in this set, since it is not in one of the sets $A_1, A_2, A_3, \ldots$.

Definition 4. The set of all elements that belong to each of the sets A_1 and A_2 is called the *intersection* of A_1 and A_2. The intersection of A_1 and A_2 is indicated by writing $A_1 \cap A_2$. The intersection of several sets $A_1, A_2, A_3, \ldots$ is the set of all elements that belong to each of the sets $A_1, A_2, A_3, \ldots$. This intersection is denoted by $A_1 \cap A_2 \cap A_3 \cap \cdots$ or by $A_1 \cap A_2 \cap \cdots \cap A_k$ if a finite number k of sets is involved.

Example 8. Let $A_1 = \{(0, 0), (0, 1), (1, 1)\}$ and $A_2 = \{(1, 1), (1, 2), (2, 1)\}$. Then $A_1 \cap A_2 = \{(1, 1)\}$.

Example 9. Let $A_1 = \{(x, y) : 0 \leq x + y \leq 1\}$ and $A_2 = \{(x, y) : 1 < x + y\}$. Then A_1 and A_2 have no points in common and $A_1 \cap A_2 = \varnothing$.

Example 10. For every set A, $A \cap A = A$ and $A \cap \varnothing = \varnothing$.

Example 11. Let

$$A_k = \left\{x : 0 < x < \frac{1}{k}\right\}, \qquad k = 1, 2, 3, \ldots.$$

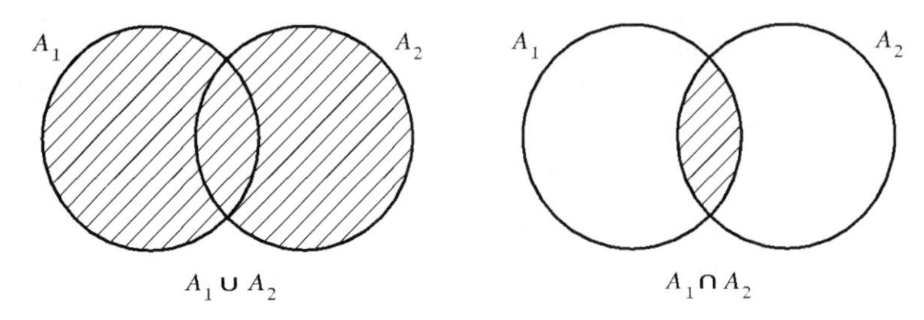

FIGURE 1.1

Then $A_1 \cap A_2 \cap A_3 \cdots$ is the null set, since there is no point that belongs to each of the sets $A_1, A_2, A_3, \ldots$.

Example 12. Let A_1 and A_2 represent the sets of points enclosed, respectively, by two intersecting circles. Then the sets $A_1 \cup A_2$ and $A_1 \cap A_2$ are represented, respectively, by the shaded regions in the *Venn diagrams* in Figure 1.1.

Example 13. Let A_1, A_2, and A_3 represent the sets of points enclosed, respectively, by three intersecting circles. Then the sets $(A_1 \cup A_2) \cap A_3$ and $(A_1 \cap A_2) \cup A_3$ are depicted in Figure 1.2.

Definition 5. In certain discussions or considerations, the totality of all elements that pertain to the discussion can be described. This set of all elements under consideration is given a special name. It is called the *space*. We shall often denote spaces by capital script letters such as $\mathscr{A}$, $\mathscr{B}$, and $\mathscr{C}$.

Example 14. Let the number of heads, in tossing a coin four times, be denoted by x. Of necessity, the number of heads will be one of the numbers 0, 1, 2, 3, 4. Here, then, the space is the set $\mathscr{A} = \{0, 1, 2, 3, 4\}$.

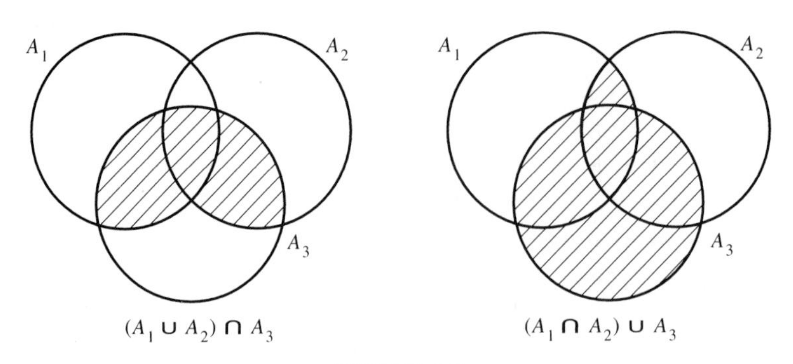

FIGURE 1.2

Example 15. Consider all nondegenerate rectangles of base x and height y. To be meaningful, both x and y must be positive. Thus the space is the set $\mathscr{A} = \{(x, y) : x > 0, y > 0\}$.

Definition 6. Let $\mathscr{A}$ denote a space and let A be a subset of the set $\mathscr{A}$. The set that consists of all elements of $\mathscr{A}$ that are not elements of A is called the *complement* of A (actually, with respect to $\mathscr{A}$). The complement of A is denoted by A^*. In particular, $\mathscr{A}^* = \varnothing$.

Example 16. Let $\mathscr{A}$ be defined as in Example 14, and let the set $A = \{0, 1\}$. The complement of A (with respect to $\mathscr{A}$) is $A^* = \{2, 3, 4\}$.

Example 17. Given $A \subset \mathscr{A}$. Then $A \cup A^* = \mathscr{A}$, $A \cap A^* = \varnothing$, $A \cup \mathscr{A} = \mathscr{A}$, $A \cap \mathscr{A} = A$, and $(A^*)^* = A$.

In the calculus, functions such as

$$f(x) = 2x, \qquad -\infty < x < \infty,$$

or

$$\begin{aligned} g(x, y) &= e^{-x-y}, \qquad 0 < x < \infty, \quad 0 < y < \infty, \\ &= 0 \qquad \text{elsewhere,} \end{aligned}$$

or possibly

$$\begin{aligned} h(x_1, x_2, \ldots, x_n) &= 3x_1x_2 \cdots x_n, \qquad 0 \le x_i \le 1, \quad i = 1, 2, \ldots, n, \\ &= 0 \qquad \text{elsewhere,} \end{aligned}$$

were of common occurrence. The value of $f(x)$ at the "point $x = 1$" is $f(1) = 2$; the value of $g(x, y)$ at the "point $(-1, 3)$" is $g(-1, 3) = 0$; the value of $h(x_1, x_2, \ldots, x_n)$ at the "point $(1, 1, \ldots, 1)$" is 3. Functions such as these are called functions of a point or, more simply, *point functions* because they are evaluated (if they have a value) at a point in a space of indicated dimension.

There is no reason why, if they prove useful, we should not have functions that can be evaluated, not necessarily at a point, but for an entire set of points. Such functions are naturally called functions of a set or, more simply, *set functions*. We shall give some examples of set functions and evaluate them for certain simple sets.

Example 18. Let A be a set in one-dimensional space and let $Q(A)$ be equal to the number of points in A which correspond to positive integers. Then $Q(A)$ is a function of the set A. Thus, if $A = \{x : 0 < x < 5\}$, then $Q(A) = 4$; if $A = \{-2, -1\}$, then $Q(A) = 0$; if $A = \{x : -\infty < x < 6\}$, then $Q(A) = 5$.

Example 19. Let A be a set in two-dimensional space and let $Q(A)$ be the area of A, if A has a finite area; otherwise, let $Q(A)$ be undefined. Thus, if

$A = \{(x, y) : x^2 + y^2 \le 1\}$, then $Q(A) = \pi$; if $A = \{(0, 0), (1, 1), (0, 1)\}$, then $Q(A) = 0$; if $A = \{(x, y) : 0 \le x, 0 \le y, x + y \le 1\}$, then $Q(A) = \frac{1}{2}$.

Example 20. Let A be a set in three-dimensional space and let $Q(A)$ be the volume of A, if A has a finite volume; otherwise, let $Q(A)$ be undefined. Thus, if $A = \{(x, y, z) : 0 \le x \le 2, 0 \le y \le 1, 0 \le z \le 3\}$, then $Q(A) = 6$; if $A = \{(x, y, z) : x^2 + y^2 + z^2 \ge 1\}$, then $Q(A)$ is undefined.

At this point we introduce the following notations. The symbol

$$\int_A f(x)\, dx$$

will mean the ordinary (Riemann) integral of $f(x)$ over a prescribed one-dimensional set A; the symbol

$$\int_A\int g(x, y)\, dx\, dy$$

will mean the Riemann integral of $g(x, y)$ over a prescribed two-dimensional set A; and so on. To be sure, unless these sets A and these functions $f(x)$ and $g(x, y)$ are chosen with care, the integrals will frequently fail to exist. Similarly, the symbol

$$\sum_A f(x)$$

will mean the sum extended over all $x \in A$; the symbol

$$\sum_A\sum g(x, y)$$

will mean the sum extended over all $(x, y) \in A$; and so on.

Example 21. Let A be a set in one-dimensional space and let $Q(A) = \sum_A f(x)$, where

$$f(x) = (\tfrac{1}{2})^x, \qquad x = 1, 2, 3, \ldots,$$
$$= 0 \qquad \text{elsewhere.}$$

If $A = \{x : 0 \le x \le 3\}$, then

$$Q(A) = \tfrac{1}{2} + (\tfrac{1}{2})^2 + (\tfrac{1}{2})^3 = \tfrac{7}{8}.$$

Example 22. Let $Q(A) = \sum_A f(x)$, where

$$f(x) = p^x(1 - p)^{1-x}, \qquad x = 0, 1,$$
$$= 0 \qquad \text{elsewhere.}$$

If $A = \{0\}$, then

$$Q(A) = \sum_{x=0}^{0} p^x(1 - p)^{1-x} = 1 - p;$$

if $A = \{x : 1 \le x \le 2\}$, then $Q(A) = f(1) = p$.

Example 23. Let A be a one-dimensional set and let

$$Q(A) = \int_A e^{-x}\,dx.$$

Thus, if $A = \{x : 0 \le x < \infty\}$, then

$$Q(A) = \int_0^\infty e^{-x}\,dx = 1;$$

if $A = \{x : 1 \le x \le 2\}$, then

$$Q(A) = \int_1^2 e^{-x}\,dx = e^{-1} - e^{-2};$$

if $A_1 = \{x : 0 \le x \le 1\}$ and $A_2 = \{x : 1 < x \le 3\}$, then

$$\begin{aligned} Q(A_1 \cup A_2) &= \int_0^3 e^{-x}\,dx \\ &= \int_0^1 e^{-x}\,dx + \int_1^3 e^{-x}\,dx \\ &= Q(A_1) + Q(A_2); \end{aligned}$$

if $A = A_1 \cup A_2$, where $A_1 = \{x : 0 \le x \le 2\}$ and $A_2 = \{x : 1 \le x \le 3\}$, then

$$\begin{aligned} Q(A) = Q(A_1 \cup A_2) &= \int_0^3 e^{-x}\,dx \\ &= \int_0^2 e^{-x}\,dx + \int_1^3 e^{-x}\,dx - \int_1^2 e^{-x}\,dx \\ &= Q(A_1) + Q(A_2) - Q(A_1 \cap A_2). \end{aligned}$$

Example 24. Let A be a set in n-dimensional space and let

$$Q(A) = \int \cdots \int_A dx_1\,dx_2 \cdots dx_n.$$

If $A = \{(x_1, x_2, \ldots, x_n) : 0 \le x_1 \le x_2 \le \cdots \le x_n \le 1\}$, then

$$\begin{aligned} Q(A) &= \int_0^1 \int_0^{x_n} \cdots \int_0^{x_3} \int_0^{x_2} dx_1\,dx_2 \cdots dx_{n-1}\,dx_n \\ &= \frac{1}{n!}, \qquad \text{where } n! = n(n-1) \cdots 3 \cdot 2 \cdot 1. \end{aligned}$$

EXERCISES

1.1. Find the union $A_1 \cup A_2$ and the intersection $A_1 \cap A_2$ of the two sets A_1 and A_2, where:
(a) $A_1 = \{0, 1, 2\}$, $A_2 = \{2, 3, 4\}$.
(b) $A_1 = \{x : 0 < x < 2\}$, $A_2 = \{x : 1 \le x < 3\}$.
(c) $A_1 = \{(x, y) : 0 < x < 2, 0 < y < 2\}$,
$A_2 = \{(x, y) : 1 < x < 3, 1 < y < 3\}$.

1.2. Find the complement A^* of the set A with respect to the space $\mathscr{A}$ if:
(a) $\mathscr{A} = \{x : 0 < x < 1\}$, $A = \{x : \frac{5}{8} \le x < 1\}$.
(b) $\mathscr{A} = \{(x, y, z) : x^2 + y^2 + z^2 \le 1\}$, $A = \{(x, y, z) : x^2 + y^2 + z^2 = 1\}$.
(c) $\mathscr{A} = \{(x, y) : |x| + |y| \le 2\}$, $A = \{(x, y) : x^2 + y^2 < 2\}$.

1.3. List all possible arrangements of the four letters m, a, r, and y. Let A_1 be the collection of the arrangements in which y is in the last position. Let A_2 be the collection of the arrangements in which m is in the first position. Find the union and intersection of A_1 and A_2.

1.4. By use of Venn diagrams, in which the space $\mathscr{A}$ is the set of points enclosed by a rectangle containing the circles, compare the following sets:
(a) $A_1 \cap (A_2 \cup A_3)$ and $(A_1 \cap A_2) \cup (A_1 \cap A_3)$.
(b) $A_1 \cup (A_2 \cap A_3)$ and $(A_1 \cup A_2) \cap (A_1 \cup A_3)$.
(c) $(A_1 \cup A_2)^*$ and $A_1^* \cap A_2^*$.
(d) $(A_1 \cap A_2)^*$ and $A_1^* \cup A_2^*$.

1.5. If a sequence of sets $A_1, A_2, A_3, \ldots$ is such that $A_k \subset A_{k+1}$, $k = 1, 2, 3, \ldots$, the sequence is said to be a *nondecreasing sequence*. Give an example of this kind of sequence of sets.

1.6. If a sequence of sets $A_1, A_2, A_3, \ldots$ is such that $A_k \supset A_{k+1}$, $k = 1, 2, 3, \ldots$, the sequence is said to be a *nonincreasing sequence*. Give an example of this kind of sequence of sets.

1.7. If $A_1, A_2, A_3, \ldots$ are sets such that $A_k \subset A_{k+1}$, $k = 1, 2, 3, \ldots$, $\lim_{k\to\infty} A_k$ is defined as the union $A_1 \cup A_2 \cup A_3 \cup \cdots$. Find $\lim_{k\to\infty} A_k$ if:

(a) $A_k = \{x : 1/k \le x \le 3 - 1/k\}$, $k = 1, 2, 3, \ldots$.
(b) $A_k = \{(x, y) : 1/k \le x^2 + y^2 \le 4 - 1/k\}$, $k = 1, 2, 3, \ldots$.

1.8. If $A_1, A_2, A_3, \ldots$ are sets such that $A_k \supset A_{k+1}$, $k = 1, 2, 3, \ldots$, $\lim_{k\to\infty} A_k$ is defined as the intersection $A_1 \cap A_2 \cap A_3 \cap \cdots$. Find $\lim_{k\to\infty} A_k$ if:

(a) $A_k = \{x : 2 - 1/k < x \le 2\}$, $k = 1, 2, 3, \ldots$.
(b) $A_k = \{x : 2 < x \le 2 + 1/k\}$, $k = 1, 2, 3, \ldots$.
(c) $A_k = \{(x, y) : 0 \le x^2 + y^2 \le 1/k\}$, $k = 1, 2, 3, \ldots$.

1.9. For every one-dimensional set A, let $Q(A) = \sum_A f(x)$, where $f(x) = (\frac{2}{3})(\frac{1}{3})^x$, $x = 0, 1, 2, \ldots,$ zero elsewhere. If $A_1 = \{x : x = 0, 1, 2, 3\}$ and $A_2 = \{x : x = 0, 1, 2, \ldots\}$, find $Q(A_1)$ and $Q(A_2)$.

Hint: Recall that $S_n = a + ar + \cdots + ar^{n-1} = a(1 - r^n)/(1 - r)$ and $\lim_{n \to \infty} S_n = a/(1 - r)$ provided that $|r| < 1$.

1.10. For every one-dimensional set A for which the integral exists, let $Q(A) = \int_A f(x)\, dx$, where $f(x) = 6x(1 - x)$, $0 < x < 1$, zero elsewhere; otherwise, let $Q(A)$ be undefined. If $A_1 = \{x : \frac{1}{4} < x < \frac{3}{4}\}$, $A_2 = \{\frac{1}{2}\}$, and $A_3 = \{x : 0 < x < 10\}$, find $Q(A_1)$, $Q(A_2)$, and $Q(A_3)$.

1.11. Let $Q(A) = \int_A \int (x^2 + y^2)\, dx\, dy$ for every two-dimensional set A for which the integral exists; otherwise, let $Q(A)$ be undefined. If $A_1 = \{(x, y) : -1 \le x \le 1, -1 \le y \le 1\}$, $A_2 = \{(x, y) : -1 \le x = y \le 1\}$, and $A_3 = \{(x, y) : x^2 + y^2 \le 1\}$, find $Q(A_1)$, $Q(A_2)$, and $Q(A_3)$.

Hint: In evaluating $Q(A_2)$, recall the definition of the double integral (or consider the volume under the surface $z = x^2 + y^2$ above the line segment $-1 \le x = y \le 1$ in the xy-plane). Use polar coordinates in the calculation of $Q(A_3)$.

1.12. Let $\mathscr{A}$ denote the set of points that are interior to, or on the boundary of, a square with opposite vertices at the points $(0, 0)$ and $(1, 1)$. Let $Q(A) = \int_A \int dy\, dx$.

(a) If $A \subset \mathscr{A}$ is the set $\{(x, y) : 0 < x < y < 1\}$, compute $Q(A)$.
(b) If $A \subset \mathscr{A}$ is the set $\{(x, y) : 0 < x = y < 1\}$, compute $Q(A)$.
(c) If $A \subset \mathscr{A}$ is the set $\{(x, y) : 0 < x/2 \le y \le 3x/2 < 1\}$, compute $Q(A)$.

1.13. Let $\mathscr{A}$ be the set of points interior to or on the boundary of a cube with edge of length 1. Moreover, say that the cube is in the first octant with one vertex at the point $(0, 0, 0)$ and an opposite vertex at the point $(1, 1, 1)$. Let $Q(A) = \iiint_A dx\, dy\, dz$.

(a) If $A \subset \mathscr{A}$ is the set $\{(x, y, z) : 0 < x < y < z < 1\}$, compute $Q(A)$.
(b) If A is the subset $\{(x, y, z) : 0 < x = y = z < 1\}$, compute $Q(A)$.

1.14. Let A denote the set $\{(x, y, z) : x^2 + y^2 + z^2 \le 1\}$. Evaluate $Q(A) = \iiint_A \sqrt{x^2 + y^2 + z^2}\, dx\, dy\, dz$.

Hint: Use spherical coordinates.

1.15. To join a certain club, a person must be either a statistician or a mathematician or both. Of the 25 members in this club, 19 are statisticians and 16 are mathematicians. How many persons in the club are both a statistician and a mathematician?

1.16. After a hard-fought football game, it was reported that, of the 11 starting players, 8 hurt a hip, 6 hurt an arm, 5 hurt a knee, 3 hurt both a hip and an arm, 2 hurt both a hip and a knee, 1 hurt both an arm and a knee, and no one hurt all three. Comment on the accuracy of the report.

1.3. The Probability Set Function

Let $\mathscr{C}$ denote the set of every possible outcome of a random experiment; that is, $\mathscr{C}$ is the sample space. It is our purpose to define a set function $P(C)$ such that if C is a subset of $\mathscr{C}$, then $P(C)$ is the probability that the outcome of the random experiment is an element of C. Henceforth it will be tacitly assumed that the structure of each set C is sufficiently simple to allow the computation. We have already seen that advantages accrue if we take $P(C)$ to be that number about which the relative frequency f/N of the event C tends to stabilize after a long series of experiments. This important fact suggests some of the properties that we would surely want the set function $P(C)$ to possess. For example, no relative frequency is ever negative; accordingly, we would want $P(C)$ to be a nonnegative set function. Again, the relative frequency of the whole sample space $\mathscr{C}$ is always 1. Thus we would want $P(\mathscr{C}) = 1$. Finally, if $C_1, C_2, C_3, \ldots$ are subsets of $\mathscr{C}$ such that no two of these subsets have a point in common, the relative frequency of the union of these sets is the sum of the relative frequencies of the sets, and we would want the set function $P(C)$ to reflect this additive property. We now formally define a probability set function.

Definition 7. If $P(C)$ is defined for a type of subset of the space $\mathscr{C}$, and if

(a) $P(C) \geq 0$,
(b) $P(C_1 \cup C_2 \cup C_3 \cup \cdots) = P(C_1) + P(C_2) + P(C_3) + \cdots$, where the sets C_i, $i = 1, 2, 3, \ldots$, are such that no two have a point in common (that is, where $C_i \cap C_j = \varnothing$, $i \neq j$),
(c) $P(\mathscr{C}) = 1$,

then P is called the *probability set function* of the outcome of the random experiment. For each subset C of $\mathscr{C}$, the number $P(C)$ is called the probability that the outcome of the random experiment is an element of the set C, or the probability of the event C, or the probability measure of the set C.

A probability set function tells us how the probability is distributed over various subsets C of a sample space $\mathscr{C}$. In this sense we speak of a distribution of probability.

Remark. In the definition, the phrase "a type of subset of the space $\mathscr{C}$" refers to the fact that P is a probability measure on a sigma field of subsets of $\mathscr{C}$ and would be explained more fully in a more advanced course. Nevertheless, a few observations can be made about the collection of subsets that are of the type. From condition (c) of the definition, we see that the space $\mathscr{C}$ must be in the collection. Condition (b) implies that if the sets $C_1, C_2, C_3, \ldots$ are in the collection, their union is also one of that type. Finally, we observe from the following theorems and their proofs that if the set C is in the collection, its complement must be one of those subsets. In particular, the null set, which is the complement of $\mathscr{C}$, must be in the collection.

The following theorems give us some other properties of a probability set function. In the statement of each of these theorems, $P(C)$ is taken, tacitly, to be a probability set function defined for a certain type of subset of the sample space $\mathscr{C}$.

Theorem 1. *For each $C \subset \mathscr{C}$, $P(C) = 1 - P(C^*)$.*

Proof. We have $\mathscr{C} = C \cup C^*$ and $C \cap C^* = \varnothing$. Thus, from (c) and (b) of Definition 7, it follows that

$$1 = P(C) + P(C^*),$$

which is the desired result.

Theorem 2. *The probability of the null set is zero; that is, $P(\varnothing) = 0$.*

Proof. In Theorem 1, take $C = \varnothing$ so that $C^* = \mathscr{C}$. Accordingly, we have

$$P(\varnothing) = 1 - P(\mathscr{C}) = 1 - 1 = 0,$$

and the theorem is proved.

Theorem 3. *If C_1 and C_2 are subsets of $\mathscr{C}$ such that $C_1 \subset C_2$, then $P(C_1) \le P(C_2)$.*

Proof. Now $C_2 = C_1 \cup (C_1^* \cap C_2)$ and $C_1 \cap (C_1^* \cap C_2) = \varnothing$. Hence, from (b) of Definition 7,

$$P(C_2) = P(C_1) + P(C_1^* \cap C_2).$$

However, from (a) of Definition 7, $P(C_1^* \cap C_2) \geq 0$; accordingly, $P(C_2) \geq P(C_1)$.

Theorem 4. *For each* $C \subset \mathscr{C}$, $0 \leq P(C) \leq 1$.

Proof. Since $\varnothing \subset C \subset \mathscr{C}$, we have by Theorem 3 that

$$P(\varnothing) \leq P(C) \leq P(\mathscr{C}) \qquad \text{or} \qquad 0 \leq P(C) \leq 1,$$

the desired result.

Theorem 5. *If* C_1 *and* C_2 *are subsets of* $\mathscr{C}$, *then*

$$P(C_1 \cup C_2) = P(C_1) + P(C_2) - P(C_1 \cap C_2).$$

Proof. Each of the sets $C_1 \cup C_2$ and C_2 can be represented, respectively, as a union of nonintersecting sets as follows:

$$C_1 \cup C_2 = C_1 \cup (C_1^* \cap C_2) \qquad \text{and} \qquad C_2 = (C_1 \cap C_2) \cup (C_1^* \cap C_2).$$

Thus, from (b) of Definition 7,

$$P(C_1 \cup C_2) = P(C_1) + P(C_1^* \cap C_2)$$

and

$$P(C_2) = P(C_1 \cap C_2) + P(C_1^* \cap C_2).$$

If the second of these equations is solved for $P(C_1^* \cap C_2)$ and this result substituted in the first equation, we obtain

$$P(C_1 \cup C_2) = P(C_1) + P(C_2) - P(C_1 \cap C_2).$$

This completes the proof.

Example 1. Let $\mathscr{C}$ denote the sample space of Example 2 of Section 1.1. Let the probability set function assign a probability of $\frac{1}{36}$ to each of the 36 points in $\mathscr{C}$. If $C_1 = \{(1, 1), (2, 1), (3, 1), (4, 1), (5, 1)\}$ and $C_2 = \{(1, 2), (2, 2), (3, 2)\}$, then $P(C_1) = \frac{5}{36}$, $P(C_2) = \frac{3}{36}$, $P(C_1 \cup C_2) = \frac{8}{36}$, and $P(C_1 \cap C_2) = 0$.

Example 2. Two coins are to be tossed and the outcome is the ordered pair (face on the first coin, face on the second coin). Thus the sample space may be represented as $\mathscr{C} = \{(\text{H}, \text{H}), (\text{H}, \text{T}), (\text{T}, \text{H}), (\text{T}, \text{T})\}$. Let the probability set function assign a probability of $\frac{1}{4}$ to each element of $\mathscr{C}$. Let $C_1 = \{(\text{H}, \text{H}), (\text{H}, \text{T})\}$ and $C_2 = \{(\text{H}, \text{H}), (\text{T}, \text{H})\}$. Then $P(C_1) = P(C_2) = \frac{1}{2}$, $P(C_1 \cap C_2) = \frac{1}{4}$, and, in accordance with Theorem 5, $P(C_1 \cup C_2) = \frac{1}{2} + \frac{1}{2} - \frac{1}{4} = \frac{3}{4}$.

Let $\mathscr{C}$ denote a sample space and let $C_1, C_2, C_3, \ldots$ denote subsets of $\mathscr{C}$. If these subsets are such that no two have an element in common, they are called mutually disjoint sets and the corresponding events $C_1, C_2, C_3, \ldots$ are said to be *mutually exclusive events*. Then, for example, $P(C_1 \cup C_2 \cup C_3 \cup \cdots) = P(C_1) + P(C_2) + P(C_3) + \cdots$, in accordance with (b) of Definition 7. Moreover, if $\mathscr{C} = C_1 \cup C_2 \cup C_3 \cup \cdots$, the mutually exclusive events are further characterized as being *exhaustive* and the probability of their union is obviously equal to 1.

Let $\mathscr{C}$ be partitioned into k mutually disjoint subsets $C_1, C_2, \ldots, C_k$ in such a way that the union of these k mutually disjoint subsets is the sample space $\mathscr{C}$. Thus the events $C_1, C_2, \ldots, C_k$ are mutually exclusive and exhaustive. Suppose that the random experiment is of such a character that it is reasonable to *assume* that each of the mutually exclusive and exhaustive events C_i, $i = 1, 2, \ldots, k$, has the same probability. It is necessary, then, that $P(C_i) = 1/k$, $i = 1, 2, \ldots, k$; and we often say that the events $C_1, C_2, \ldots, C_k$ are *equally likely*. Let the event E be the union of r of these mutually exclusive events, say

$$E = C_1 \cup C_2 \cup \cdots \cup C_r, \qquad r \le k.$$

Then

$$P(E) = P(C_1) + P(C_2) + \cdots + P(C_r) = \frac{r}{k}.$$

Frequently, the integer k is called the total number of ways (for this particular partition of $\mathscr{C}$) in which the random experiment can terminate and the integer r is called the number of ways that are favorable to the event E. So, in this terminology, $P(E)$ is equal to the number of ways favorable to the event E divided by the total number of ways in which the experiment can terminate. It should be emphasized that in order to assign, *in this manner*, the probability r/k to the event E, we must assume that each of the mutually exclusive and exhaustive events $C_1, C_2, \ldots, C_k$ has the same probability $1/k$. This assumption of equally likely events then becomes a *part* of our probability model. Obviously, if this assumption is not realistic in an application, the probability of the event E cannot be computed in this way.

We next present an example that is illustrative of this model.

Example 3. Let a card be drawn at random from an ordinary deck of

52 playing cards. The sample space $\mathscr{C}$ is the union of $k = 52$ outcomes, and it is reasonable to assume that each of these outcomes has the same probability $\frac{1}{52}$. Accordingly, if E_1 is the set of outcomes that are spades, $P(E_1) = \frac{13}{52} = \frac{1}{4}$ because there are $r_1 = 13$ spades in the deck; that is, $\frac{1}{4}$ is the probability of drawing a card that is a spade. If E_2 is the set of outcomes that are kings, $P(E_2) = \frac{4}{52} = \frac{1}{13}$ because there are $r_2 = 4$ kings in the deck; that is, $\frac{1}{13}$ is the probability of drawing a card that is a king. These computations are very easy because there are no difficulties in the determination of the appropriate values of r and k. However, instead of drawing only one card, suppose that five cards are taken, at random and without replacement, from this deck. We can think of each five-card hand as being an outcome in a sample space. It is reasonable to assume that each of these outcomes has the same probability. Now if E_1 is the set of outcomes in which each card of the hand is a spade, $P(E_1)$ is equal to the number r_1 of all spade hands divided by the total number, say k, of five-card hands. It is shown in many books on algebra that

$$r_1 = \binom{13}{5} = \frac{13!}{5!\,8!} \qquad \text{and} \qquad k = \binom{52}{5} = \frac{52!}{5!\,47!}.$$

In general, if n is a positive integer and if x is a nonnegative integer with $x \le n$, then the binomial coefficient

$$\binom{n}{x} = \frac{n!}{x!\,(n-x)!}$$

is equal to the number of combinations of n things taken x at a time. If $x = 0$, $0! = 1$, so that $\binom{n}{0} = 1$. Thus, in the special case involving E_1,

$$P(E_1) = \frac{\binom{13}{5}}{\binom{52}{5}} = \frac{(13)(12)(11)(10)(9)}{(52)(51)(50)(49)(48)} = 0.0005,$$

approximately. Next, let E_2 be the set of outcomes in which at least one card is a spade. Then E_2^* is the set of outcomes in which no card is a spade. There are $r_2^* = \binom{39}{5}$ such outcomes. Hence

$$P(E_2^*) = \frac{\binom{39}{5}}{\binom{52}{5}} \qquad \text{and} \qquad P(E_2) = 1 - P(E_2^*).$$

Now suppose that E_3 is the set of outcomes in which exactly three cards are kings and exactly two cards are queens. We can select the three kings in any one of $\binom{4}{3}$ ways and the two queens in any one of $\binom{4}{2}$ ways. By a well-known counting principle, the number of outcomes in E_3 is $r_3 = \binom{4}{3}\binom{4}{2}$. Thus $P(E_3) = \binom{4}{3}\binom{4}{2}\Big/\binom{52}{5}$. Finally, let E_4 be the set of outcomes in which there are exactly two kings, two queens, and one jack. Then

$$P(E_4) = \frac{\binom{4}{2}\binom{4}{2}\binom{4}{1}}{\binom{52}{5}},$$

because the numerator of this fraction is the number of outcomes in E_4.

Example 3 and the previous discussion allow us to see one way in which we can define a probability set function, that is, a set function that satisfies the requirements of Definition 7. Suppose that our space $\mathscr{C}$ consists of k distinct points, which, for this discussion, we take to be in a one-dimensional space. If the random experiment that ends in one of those k points is such that it is reasonable to assume that these points are equally likely, we could assign $1/k$ to each point and let, for $C \subset \mathscr{C}$,

$$\begin{aligned} P(C) &= \frac{\text{number of points in } C}{k} \\ &= \sum_{x \in C} f(x), \qquad \text{where } f(x) = \frac{1}{k}, \quad x \in \mathscr{C}. \end{aligned}$$

For illustration, in the cast of a die, we could take $\mathscr{C} = \{1, 2, 3, 4, 5, 6\}$ and $f(x) = \frac{1}{6}$, $x \in \mathscr{C}$, if we believe the die to be unbiased. Clearly, such a set function satisfies Definition 7.

The word *unbiased* in this illustration suggests the possibility that all six points might *not*, in all such cases, be equally likely. As a matter of fact, *loaded* dice do exist. In the case of a loaded die, some numbers occur more frequently than others in a sequence of casts of that die. For example, suppose that a die has been loaded so that the relative frequencies of the numbers in $\mathscr{C}$ *seem to stabilize* proportional to the number of spots that are on the *up* side. Thus we might assign $f(x) = x/21$, $x \in \mathscr{C}$, and the corresponding

$$P(C) = \sum_{x \in C} f(x)$$

would satisfy Definition 7. For illustration, this means that if $C = \{1, 2, 3\}$, then

$$P(C) = \sum_{x=1}^{3} f(x) = \frac{1}{21} + \frac{2}{21} + \frac{3}{21} = \frac{6}{21} = \frac{2}{7}.$$

Whether this probability set function is realistic can only be checked by performing the random experiment a large number of times.

EXERCISES

1.17. A positive integer from one to six is to be chosen by casting a die. Thus the elements c of the sample space $\mathscr{C}$ are 1, 2, 3, 4, 5, 6. Let $C_1 = \{1, 2, 3, 4\}$, $C_2 = \{3, 4, 5, 6\}$. If the probability set function P assigns a probability of $\frac{1}{6}$ to each of the elements of $\mathscr{C}$, compute $P(C_1)$, $P(C_2)$, $P(C_1 \cap C_2)$, and $P(C_1 \cup C_2)$.

1.18. A random experiment consists of drawing a card from an ordinary deck of 52 playing cards. Let the probability set function P assign a probability of $\frac{1}{52}$ to each of the 52 possible outcomes. Let C_1 denote the collection of the 13 hearts and let C_2 denote the collection of the 4 kings. Compute $P(C_1)$, $P(C_2)$, $P(C_1 \cap C_2)$, and $P(C_1 \cup C_2)$.

1.19. A coin is to be tossed as many times as necessary to turn up one head. Thus the elements c of the sample space $\mathscr{C}$ are H, TH, TTH, TTTH, and so forth. Let the probability set function P assign to these elements the respective probabilities $\frac{1}{2}, \frac{1}{4}, \frac{1}{8}, \frac{1}{16}$, and so forth. Show that $P(\mathscr{C}) = 1$. Let $C_1 = \{c : c \text{ is H, TH, TTH, TTTH, or TTTTH}\}$. Compute $P(C_1)$. Let $C_2 = \{c : c \text{ is TTTTH or TTTTTH}\}$. Compute $P(C_2)$, $P(C_1 \cap C_2)$, and $P(C_1 \cup C_2)$.

1.20. If the sample space is $\mathscr{C} = C_1 \cup C_2$ and if $P(C_1) = 0.8$ and $P(C_2) = 0.5$, find $P(C_1 \cap C_2)$.

1.21. Let the sample space be $\mathscr{C} = \{c : 0 < c < \infty\}$. Let $C \subset \mathscr{C}$ be defined by $C = \{c : 4 < c < \infty\}$ and take $P(C) = \int_C e^{-x}\, dx$. Evaluate $P(C)$, $P(C^*)$, and $P(C \cup C^*)$.

1.22. If the sample space is $\mathscr{C} = \{c : -\infty < c < \infty\}$ and if $C \subset \mathscr{C}$ is a set for which the integral $\int_C e^{-|x|}\, dx$ exists, show that this set function is not a probability set function. What constant do we multiply the integral by to make it a probability set function?

1.23. If C_1 and C_2 are subsets of the sample space $\mathscr{C}$, show that

$$P(C_1 \cap C_2) \le P(C_1) \le P(C_1 \cup C_2) \le P(C_1) + P(C_2).$$

1.24. Let C_1, C_2, and C_3 be three mutually disjoint subsets of the sample space $\mathscr{C}$. Find $P[(C_1 \cup C_2) \cap C_3]$ and $P(C_1^* \cup C_2^*)$.

1.25. If C_1, C_2, and C_3 are subsets of $\mathscr{C}$, show that

$$P(C_1 \cup C_2 \cup C_3) = P(C_1) + P(C_2) + P(C_3) - P(C_1 \cap C_2)$$
$$- P(C_1 \cap C_3) - P(C_2 \cap C_3) + P(C_1 \cap C_2 \cap C_3).$$

What is the generalization of this result to four or more subsets of $\mathscr{C}$?

Hint: Write $P(C_1 \cup C_2 \cup C_3) = P[C_1 \cup (C_2 \cup C_3)]$ and use Theorem 5.

Remark. In order to solve a number of exercises, like 1.26–1.31, certain reasonable assumptions must be made.

1.26. A bowl contains 16 chips, of which 6 are red, 7 are white, and 3 are blue. If four chips are taken at random and without replacement, find the probability that: (a) each of the 4 chips is red; (b) none of the 4 chips is red; (c) there is at least 1 chip of each color.

1.27. A person has purchased 10 of 1000 tickets sold in a certain raffle. To determine the five prize winners, 5 tickets are to be drawn at random and without replacement. Compute the probability that this person will win at least one prize.

Hint: First compute the probability that the person does not win a prize.

1.28. Compute the probability of being dealt at random and without replacement a 13-card bridge hand consisting of: (a) 6 spades, 4 hearts, 2 diamonds, and 1 club; (b) 13 cards of the same suit.

1.29. Three distinct integers are chosen at random from the first 20 positive integers. Compute the probability that: (a) their sum is even; (b) their product is even.

1.30. There are 5 red chips and 3 blue chips in a bowl. The red chips are numbered 1, 2, 3, 4, 5, respectively, and the blue chips are numbered 1, 2, 3, respectively. If 2 chips are to be drawn at random and without replacement, find the probability that these chips have either the same number or the same color.

1.31. In a lot of 50 light bulbs, there are 2 bad bulbs. An inspector examines 5 bulbs, which are selected at random and without replacement.

(a) Find the probability of at least 1 defective bulb among the 5.

(b) How many bulbs should he examine so that the probability of finding at least 1 bad bulb exceeds $\frac{1}{2}$?

1.4 Conditional Probability and Independence

In some random experiments, we are interested only in those outcomes that are elements of a subset C_1 of the sample space $\mathscr{C}$. This means, for our purposes, that the sample space is effectively the subset C_1. We are now confronted with the problem of defining a probability set function with C_1 as the "new" sample space.

Let the probability set function $P(C)$ be defined on the sample space $\mathscr{C}$ and let C_1 be a subset of $\mathscr{C}$ such that $P(C_1) > 0$. We agree to consider only those outcomes of the random experiment that are elements of C_1; in essence, then, we take C_1 to be a sample space. Let C_2 be another subset of $\mathscr{C}$. How, relative to the new sample space C_1, do we want to define the probability of the event C_2? Once defined, this probability is called the *conditional probability* of the event C_2, relative to the hypothesis of the event C_1; or, more briefly, the conditional probability of C_2, given C_1. Such a conditional probability is denoted by the symbol $P(C_2|C_1)$. We now return to the question that was raised about the definition of this symbol. Since C_1 is now the sample space, the only elements of C_2 that concern us are those, if any, that are also elements of C_1, that is, the elements of $C_1 \cap C_2$. It seems desirable, then, to define the symbol $P(C_2|C_1)$ in such a way that

$$P(C_1|C_1) = 1 \qquad \text{and} \qquad P(C_2|C_1) = P(C_1 \cap C_2|C_1).$$

Moreover, from a relative frequency point of view, it would seem logically inconsistent if we did not require that the ratio of the probabilities of the events $C_1 \cap C_2$ and C_1, relative to the space C_1, be the same as the ratio of the probabilities of these events relative to the space $\mathscr{C}$; that is, we should have

$$\frac{P(C_1 \cap C_2|C_1)}{P(C_1|C_1)} = \frac{P(C_1 \cap C_2)}{P(C_1)}.$$

These three desirable conditions imply that the relation

$$P(C_2|C_1) = \frac{P(C_1 \cap C_2)}{P(C_1)}$$

is a suitable *definition* of the conditional probability of the event C_2, given the event C_1, provided that $P(C_1) > 0$. Moreover, we have

1. $P(C_2|C_1) \geq 0$.
2. $P(C_2 \cup C_3 \cup \cdots |C_1) = P(C_2|C_1) + P(C_3|C_1) + \cdots$, provided that $C_2, C_3, \ldots$ are mutually disjoint sets.
3. $P(C_1|C_1) = 1$.

Properties (1) and (3) are evident; proof of property (2) is left as an exercise (1.32). But these are precisely the conditions that a probability set function must satisfy. Accordingly, $P(C_2|C_1)$ is a probability set function, defined for subsets of C_1. It may be called the conditional probability set function, relative to the hypothesis C_1; or the conditional probability set function, given C_1. It should be noted that this conditional probability set function, given C_1, is defined at this time only when $P(C_1) > 0$.

Example 1. A hand of 5 cards is to be dealt at random without replacement from an ordinary deck of 52 playing cards. The conditional probability of an all-spade hand (C_2), relative to the hypothesis that there are at least 4 spades in the hand (C_1), is, since $C_1 \cap C_2 = C_2$,

$$P(C_2|C_1) = \frac{P(C_2)}{P(C_1)} = \frac{\binom{13}{5}\Big/\binom{52}{5}}{\left[\binom{13}{4}\binom{39}{1} + \binom{13}{5}\right]\Big/\binom{52}{5}}$$

$$= \frac{\binom{13}{5}}{\binom{13}{4}\binom{39}{1} + \binom{13}{5}}.$$

From the definition of the conditional probability set function, we observe that

$$P(C_1 \cap C_2) = P(C_1)P(C_2|C_1).$$

This relation is frequently called the *multiplication rule* for probabilities. Sometimes, after considering the nature of the random experiment, it is possible to make reasonable assumptions so that both $P(C_1)$ and $P(C_2|C_1)$ can be assigned. Then $P(C_1 \cap C_2)$ can be computed under these assumptions. This will be illustrated in Examples 2 and 3.

Example 2. A bowl contains eight chips. Three of the chips are red and the remaining five are blue. Two chips are to be drawn successively, at random and without replacement. We want to compute the probability that the first draw results in a red chip (C_1) and that the second draw results in a blue chip (C_2). It is reasonable to assign the following probabilities:

$$P(C_1) = \tfrac{3}{8} \quad \text{and} \quad P(C_2|C_1) = \tfrac{5}{7}.$$

Thus, under these assignments, we have $P(C_1 \cap C_2) = (\frac{3}{8})(\frac{5}{7}) = \frac{15}{56}$.

Example 3. From an ordinary deck of playing cards, cards are to be drawn successively, at random and without replacement. The probability that the third spade appears on the sixth draw is computed as follows. Let C_1 be the event of two spades in the first five draws and let C_2 be the event of a spade on the sixth draw. Thus the probability that we wish to compute is $P(C_1 \cap C_2)$. It is reasonable to take

$$P(C_1) = \frac{\binom{13}{2}\binom{39}{3}}{\binom{52}{5}} \qquad \text{and} \qquad P(C_2|C_1) = \frac{11}{47}.$$

The desired probability $P(C_1 \cap C_2)$ is then the product of these two numbers.

The multiplication rule can be extended to three or more events. In the case of three events, we have, by using the multiplication rule for two events,

$$\begin{aligned} P(C_1 \cap C_2 \cap C_3) &= P[(C_1 \cap C_2) \cap C_3] \\ &= P(C_1 \cap C_2)P(C_3|C_1 \cap C_2). \end{aligned}$$

But $P(C_1 \cap C_2) = P(C_1)P(C_2|C_1)$. Hence

$$P(C_1 \cap C_2 \cap C_3) = P(C_1)P(C_2|C_1)P(C_3|C_1 \cap C_2).$$

This procedure can be used to extend the multiplication rule to four or more events. The general formula for k events can be proved by mathematical induction.

Example 4. Four cards are to be dealt successively, at random and without replacement, from an ordinary deck of playing cards. The probability of receiving a spade, a heart, a diamond, and a club, in that order, is $(\frac{13}{52})(\frac{13}{51})(\frac{13}{50})(\frac{13}{49})$. This follows from the extension of the multiplication rule. In this computation, the assumptions that are involved seem clear.

Let the space $\mathscr{C}$ be partitioned into k mutually exclusive and exhaustive events $C_1, C_2, \ldots, C_k$ such that $P(C_i) > 0$, $i = 1, 2, \ldots, k$. Here the events $C_1, C_2, \ldots, C_k$ do *not* need to be equally likely. Let C be another event such that $P(C) > 0$. Thus C occurs with one and only one of the events $C_1, C_2, \ldots, C_k$; that is,

$$\begin{aligned} C &= C \cap (C_1 \cup C_2 \cup \cdots \cup C_k) \\ &= (C \cap C_1) \cup (C \cap C_2) \cup \cdots \cup (C \cap C_k). \end{aligned}$$

Since $C \cap C_i$, $i = 1, 2, \ldots, k$, are mutually exclusive, we have

$$P(C) = P(C \cap C_1) + P(C \cap C_2) + \cdots + P(C \cap C_k).$$

However, $P(C \cap C_i) = P(C_i)P(C|C_i)$, $i = 1, 2, \ldots, k$; so

$$P(C) = P(C_1)P(C|C_1) + P(C_2)P(C|C_2) + \cdots + P(C_k)P(C|C_k)$$
$$= \sum_{i=1}^{k} P(C_i)P(C|C_i).$$

This result is sometimes called the *law of total probability*.

From the definition of conditional probability, we have, using the law of total probability, that

$$P(C_j|C) = \frac{P(C \cap C_j)}{P(C)} = \frac{P(C_j)P(C|C_j)}{\sum_{i=1}^{k} P(C_i)P(C|C_i)},$$

which is the well-known *Bayes' theorem*. This permits us to calculate the conditional probability of C_j, given C, from the probabilities of $C_1, C_2, \ldots, C_k$ and the conditional probabilities of C, given C_i, $i = 1, 2, \ldots, k$.

Example 5. Say it is known that bowl C_1 contains 3 red and 7 blue chips and bowl C_2 contains 8 red and 2 blue chips. All chips are identical in size and shape. A die is cast and bowl C_1 is selected if five or six spots show on the side that is up; otherwise, bowl C_2 is selected. In a notation that is fairly obvious, it seems reasonable to assign $P(C_1) = \frac{2}{6}$ and $P(C_2) = \frac{4}{6}$. The selected bowl is handed to another person and one chip is taken at random. Say that this chip is red, an event which we denote by C. By considering the contents of the bowls, it is reasonable to assign the conditional probabilities $P(C|C_1) = \frac{3}{10}$ and $P(C|C_2) = \frac{8}{10}$. Thus the conditional probability of bowl C_1, given that a red chip is drawn, is

$$P(C_1|C) = \frac{P(C_1)P(C|C_1)}{P(C_1)P(C|C_1) + P(C_2)P(C|C_2)}$$
$$= \frac{(\frac{2}{6})(\frac{3}{10})}{(\frac{2}{6})(\frac{3}{10}) + (\frac{4}{6})(\frac{8}{10})} = \frac{3}{19}.$$

In a similar manner, we have $P(C_2|C) = \frac{16}{19}$.

In Example 5, the probabilities $P(C_1) = \frac{2}{6}$ and $P(C_2) = \frac{4}{6}$ are called *prior probabilities* of C_1 and C_2, respectively, because they are known to be due to the random mechanism used to select the bowls. After the chip is taken and observed to be red, the conditional probabilities $P(C_1|C) = \frac{3}{19}$ and $P(C_2|C) = \frac{16}{19}$ are called *posterior probabilities*. Since C_2 has a larger proportion of red chips than does C_1, it appeals to one's intuition that $P(C_2|C)$ should be larger than $P(C_2)$ and, of course, $P(C_1|C)$ should be smaller than $P(C_1)$. That is, intuitively the chances of having bowl C_2 are better once that a red chip is observed than before a chip is taken. Bayes' theorem provides a method of determining exactly what those probabilities are.

Example 6. Three plants, C_1, C_2, and C_3, produce respectively, 10, 50, and 40 percent of a company's output. Although plant C_1 is a small plant, its manager believes in high quality and only 1 percent of its products are defective. The other two, C_2 and C_3, are worse and produce items that are 3 and 4 percent defective, respectively. All products are sent to a central warehouse. One item is selected at random and observed to be defective, say event C. The conditional probability that it comes from plant C_1 is found as follows. It is natural to assign the respective prior probabilities of getting an item from the plants as $P(C_1) = 0.1$, $P(C_2) = 0.5$, and $P(C_3) = 0.4$, while the conditional probabilities of defective are $P(C|C_1) = 0.01$, $P(C|C_2) = 0.03$, and $P(C|C_3) = 0.04$. Thus the posterior probability of C_1, given a defective, is

$$P(C_1|C) = \frac{P(C_1 \cap C)}{P(C)} = \frac{(0.10)(0.01)}{(0.10)(0.01) + (0.50)(0.03) + (0.40)(0.04)},$$

which equals $\frac{1}{32}$; this is much smaller than the prior probability $P(C_1) = \frac{1}{10}$. This is as it should be because the fact that the item is defective decreases the chances that it comes from the high-quality plant C_1.

Sometimes it happens that the occurrence of event C_1 does not change the probability of event C_2; that is, when $P(C_1) > 0$,

$$P(C_2|C_1) = P(C_2).$$

In this case, we say that the events C_1 and C_2 are *independent*. Moreover, the multiplication rule becomes

$$P(C_1 \cap C_2) = P(C_1)P(C_2|C_1) = P(C_1)P(C_2).$$

This, in turn, implies, when $P(C_2) > 0$, that

$$P(C_1|C_2) = \frac{P(C_1 \cap C_2)}{P(C_2)} = \frac{P(C_1)P(C_2)}{P(C_2)} = P(C_1).$$

Remark. Events that are *independent* are sometimes called *statistically independent*, *stochastically independent*, or *independent in a probability sense*. In most instances, we use independent without a modifier if there is no possibility of misunderstanding.

It is interesting to note that C_1 and C_2 are independent if $P(C_1) = 0$ or $P(C_2) = 0$ because then $P(C_1 \cap C_2) = 0$ since $(C_1 \cap C_2) \subset C_1$ and $(C_1 \cap C_2) \subset C_2$. Thus the left- and right-hand members of

$$P(C_1 \cap C_2) = P(C_1)P(C_2)$$

are both equal to zero and are, of course, equal to each other. Also, if C_1 and C_2 are independent events, so are the three pairs: C_1 and C_2^*, C_1^* and C_2, and C_1^* and C_2^* (see Exercise 1.41).

Example 7. A red die and a white die are cast in such a way that the number of spots on the two sides that are up are independent events. If C_1 represents a four on the red die and C_2 represents a three on the white die, with an equally likely assumption for each side, we assign $P(C_1) = \frac{1}{6}$ and $P(C_2) = \frac{1}{6}$. Thus, from independence, the probability of the ordered pair (red = 4, white = 3) is

$$P[(4, 3)] = (\tfrac{1}{6})(\tfrac{1}{6}) = \tfrac{1}{36}.$$

The probability that the sum of the up spots of the two dice equals seven is

$$\begin{aligned} &P[(1, 6), (2, 5), (3, 4), (4, 3), (5, 2), (6, 1)] \\ &\quad = (\tfrac{1}{6})(\tfrac{1}{6}) + (\tfrac{1}{6})(\tfrac{1}{6}) + (\tfrac{1}{6})(\tfrac{1}{6}) + (\tfrac{1}{6})(\tfrac{1}{6}) + (\tfrac{1}{6})(\tfrac{1}{6}) + (\tfrac{1}{6})(\tfrac{1}{6}) = \tfrac{6}{36}. \end{aligned}$$

In a similar manner, it is easy to show that the probabilities of the sums of 2, 3, 4, 5, 6, 7, 8, 9, 10, 11, 12 are, respectively,

$$\tfrac{1}{36}, \tfrac{2}{36}, \tfrac{3}{36}, \tfrac{4}{36}, \tfrac{5}{36}, \tfrac{6}{36}, \tfrac{5}{36}, \tfrac{4}{36}, \tfrac{3}{36}, \tfrac{2}{36}, \tfrac{1}{36}.$$

Suppose now that we have three events, C_1, C_2, and C_3. We say that they are *mutually independent* if and only if they are *pairwise independent*:

$$P(C_1 \cap C_3) = P(C_1)P(C_3), \qquad P(C_1 \cap C_2) = P(C_1)P(C_2),$$

$$P(C_2 \cap C_3) = P(C_2)P(C_3)$$

and

$$P(C_1 \cap C_2 \cap C_3) = P(C_1)P(C_2)P(C_3).$$

More generally, the n events $C_1, C_2, \ldots, C_n$ are *mutually independent* if and only if for every collection of k of these events, $2 \le k \le n$, the following is true:

Say that $d_1, d_2, \ldots, d_k$ are k distinct integers from $1, 2, \ldots, n$; then

$$P(C_{d_1} \cap C_{d_2} \cap \cdots \cap C_{d_k}) = P(C_{d_1})P(C_{d_2}) \cdots P(C_{d_k}).$$

In particular, if $C_1, C_2, \ldots, C_n$ are mutually independent, then

$$P(C_1 \cap C_2 \cap \cdots \cap C_n) = P(C_1)P(C_2) \cdots P(C_n).$$

Also, as with two sets, many combinations of these events and their complements are independent, such as

C_1^* and $(C_2 \cup C_3^* \cup C_4)$ are independent;

$C_1 \cup C_2^*$, C_3^*, and $C_4 \cap C_5^*$ are mutually independent.

If there is no possibility of misunderstanding, *independent* is often used without the modifier *mutually* when considering more than two events.

We often perform a sequence of random experiments in such a way that the events associated with one of them are independent of the events associated with the others. For convenience, we refer to these events as *independent experiments*, meaning that the respective events are independent. Thus we often refer to independent flips of a coin or independent casts of a die or—more generally—independent trials of some given random experiment.

Example 8. A coin is flipped independently several times. Let the event C_i represent a head (H) on the ith toss; thus C_i^* represents a tail (T). Assume that C_i and C_i^* are equally likely; that is, $P(C_i) = P(C_i^*) = \frac{1}{2}$. Thus the probability of an ordered sequence like HHTH is, from independence,

$$P(C_1 \cap C_2 \cap C_3^* \cap C_4) = P(C_1)P(C_2)P(C_3^*)P(C_4) = (\tfrac{1}{2})^4 = \tfrac{1}{16}.$$

Similarly, the probability of observing the first head on the third flip is

$$P(C_1^* \cap C_2^* \cap C_3) = P(C_1^*)P(C_2^*)P(C_3) = (\tfrac{1}{2})^3 = \tfrac{1}{8}.$$

Also, the probability of getting at least one head on four flips is

$$\begin{aligned} P(C_1 \cup C_2 \cup C_3 \cup C_4) &= 1 - P[(C_1 \cup C_2 \cup C_3 \cup C_4)^*] \\ &= 1 - P(C_1^* \cap C_2^* \cap C_3^* \cap C_4^*) \\ &= 1 - (\tfrac{1}{2})^4 = \tfrac{15}{16}. \end{aligned}$$

See Exercise 1.43 to justify this last probability.

EXERCISES

1.32. If $P(C_1) > 0$ and if $C_2, C_3, C_4, \ldots$ are mutually disjoint sets, show that $P(C_2 \cup C_3 \cup \cdots | C_1) = P(C_2|C_1) + P(C_3|C_1) + \cdots$.

1.33. Prove that

$$P(C_1 \cap C_2 \cap C_3 \cap C_4) = P(C_1)P(C_2|C_1)P(C_3|C_1 \cap C_2)P(C_4|C_1 \cap C_2 \cap C_3).$$

1.34. A bowl contains 8 chips. Three of the chips are red and 5 are blue. Four chips are to be drawn successively at random and without replacement.
(a) Compute the probability that the colors alternate.

(b) Compute the probability that the first blue chip appears on the third draw.

1.35. A hand of 13 cards is to be dealt at random and without replacement from an ordinary deck of playing cards. Find the conditional probability that there are at least three kings in the hand relative to the hypothesis that the hand contains at least two kings.

1.36. A drawer contains eight pairs of socks. If six socks are taken at random and without replacement, compute the probability that there is at least one matching pair among these six socks.

Hint: Compute the probability that there is not a matching pair.

1.37. A bowl contains 10 chips. Four of the chips are red, 5 are white, and 1 is blue. If 3 chips are taken at random and without replacement, compute the conditional probability that there is 1 chip of each color relative to the hypothesis that there is exactly 1 red chip among the 3.

1.38. Bowl I contains 3 red chips and 7 blue chips. Bowl II contains 6 red chips and 4 blue chips. A bowl is selected at random and then 1 chip is drawn from this bowl.

(a) Compute the probability that this chip is red.

(b) Relative to the hypothesis that the chip is red, find the conditional probability that it is drawn from bowl II.

1.39. Bowl I contains 6 red chips and 4 blue chips. Five of these 10 chips are selected at random and without replacement and put in bowl II, which was originally empty. One chip is then drawn at random from bowl II. Relative to the hypothesis that this chip is blue, find the conditional probability that 2 red chips and 3 blue chips are transferred from bowl I to bowl II.

1.40. A professor of statistics has two boxes of computer disks: box C_1 contains seven Verbatim disks and three Control Data disks and box C_2 contains two Verbatim disks and eight Control Data disks. She selects a box at random with probabilities $P(C_1) = \frac{2}{3}$ and $P(C_2) = \frac{1}{3}$ because of their respective locations. A disk is then selected at random and the event C occurs if it is from Control Data. Using an equally likely assumption for each disk in the selected box, compute $P(C_1|C)$ and $P(C_2|C)$.

1.41. If C_1 and C_2 are independent events, show that the following pairs of events are also independent: (a) C_1 and C_2^*, (b) C_1^* and C_2, and (c) C_1^* and C_2^*.

Hint: In (a), write $P(C_1 \cap C_2^*) = P(C_1)P(C_2^*|C_1) = P(C_1)[1 - P(C_2|C_1)]$. From independence of C_1 and C_2, $P(C_2|C_1) = P(C_2)$.

1.42. Let C_1 and C_2 be independent events with $P(C_1) = 0.6$ and $P(C_2) = 0.3$. Compute (a) $P(C_1 \cap C_2)$; (b) $P(C_1 \cup C_2)$; (c) $P(C_1 \cup C_2^*)$.

1.43. Generalize Exercise 1.4 to obtain

$$(C_1 \cup C_2 \cup \cdots \cup C_k)^* = C_1^* \cap C_2^* \cap \cdots \cap C_k^*.$$

Say that $C_1, C_2, \ldots, C_k$ are independent events that have respective probabilities $p_1, p_2, \ldots, p_k$. Argue that the probability of at least one of $C_1, C_2, \ldots, C_k$ is equal to

$$1 - (1 - p_1)(1 - p_2) \cdots (1 - p_k).$$

1.44. Each of four persons fires one shot at a target. Let C_k denote the event that the target is hit by person k, $k = 1, 2, 3, 4$. If C_1, C_2, C_3, C_4 are independent and if $P(C_1) = P(C_2) = 0.7$, $P(C_3) = 0.9$, and $P(C_4) = 0.4$, compute the probability that (a) all of them hit the target; (b) exactly one hits the target; (c) no one hits the target; (d) at least one hits the target.

1.45. A bowl contains three red (R) balls and seven white (W) balls of exactly the same size and shape. Select balls successively at random and with replacement so that the events of white on the first trial, white on the second, and so on, can be assumed to be independent. In four trials, make certain assumptions and compute the probabilities of the following ordered sequences: (a) WWRW; (b) RWWW; (c) WWWR; and (d) WRWW. (*e*) Compute the probability of exactly one red ball in the four trials.

1.46. A coin is tossed two independent times, each resulting in a tail (T) or a head (H). The sample space consists of four ordered pairs: TT, TH, HT, HH. Making certain assumptions, compute the probability of each of these ordered pairs. What is the probability of at least one head?

1.5 Random Variables of the Discrete Type

The reader will perceive that a sample space $\mathscr{C}$ may be tedious to describe if the elements of $\mathscr{C}$ are not numbers. We shall now discuss how we may formulate a rule, or a set of rules, by which the elements c of $\mathscr{C}$ may be represented by numbers. We begin the discussion with a very simple example. Let the random experiment be the toss of a coin and let the sample space associated with the experiment be $\mathscr{C} = \{c : \text{where } c \text{ is T or } c \text{ is H}\}$ and T and H represent, respectively, tails and heads. Let X be a function such that $X(c) = 0$ if c is T and let $X(c) = 1$ if c is H. Thus X is a real-valued function defined on the sample space $\mathscr{C}$ which takes us from the sample space $\mathscr{C}$ to a space of real numbers $\mathscr{A} = \{0, 1\}$. We call X a random variable and, in this example, the space associated with X is $\mathscr{A} = \{0, 1\}$. We now formulate the definition of a random variable and its space.

Definition 8. Consider a random experiment with a sample space $\mathscr{C}$. A function X, which assigns to each element $c \in \mathscr{C}$ one and only one real number $X(c) = x$, is called a *random variable*. The *space* of X is the set of real numbers $\mathscr{A} = \{x : x = X(c), c \in \mathscr{C}\}$.

It may be that the set $\mathscr{C}$ has elements which are themselves real numbers. In such an instance we could write $X(c) = c$ so that $\mathscr{A} = \mathscr{C}$.

Let X be a random variable that is defined on a sample space $\mathscr{C}$, and let $\mathscr{A}$ be the space of X. Further, let A be a subset of $\mathscr{A}$. Just as we used the terminology "the event C," with $C \subset \mathscr{C}$, we shall now speak of "the event A." The probability $P(C)$ of the event C has been defined. We wish now to define the probability of the event A. This probability will be denoted by $\Pr(X \in A)$, where Pr is an abbreviation for "the probability that." With A a subset of $\mathscr{A}$, let C be that subset of $\mathscr{C}$ such that $C = \{c : c \in \mathscr{C} \text{ and } X(c) \in A\}$. Thus C has as its elements all outcomes in $\mathscr{C}$ for which the random variable X has a value that is in A. This prompts us to define, as we now do, $\Pr(X \in A)$ to be equal to $P(C)$, where $C = \{c : c \in \mathscr{C} \text{ and } X(c) \in A\}$. Thus $\Pr(X \in A)$ is an assignment of probability to a set A, which is a subset of the space $\mathscr{A}$ associated with the random variable X. This assignment is determined by the probability set function P and the random variable X and is sometimes denoted by $P_X(A)$. That is,

$$\Pr(X \in A) = P_X(A) = P(C),$$

where $C = \{c : c \in \mathscr{C} \text{ and } X(c) \in A\}$. Thus a random variable X is a function that carries the probability from a sample space $\mathscr{C}$ to a space $\mathscr{A}$ of real numbers. In this sense, with $A \subset \mathscr{A}$, the probability $P_X(A)$ is often called an *induced probability*.

Remark. In a more advanced course, it would be noted that the random variable X is a Borel measurable function. This is needed to assure that we can find the induced probabilities on the sigma field of the subsets of $\mathscr{A}$. We need this requirement throughout this book for every function that is a random variable, but no further mention of it is made.

The function $P_X(A)$ satisfies the conditions (a), (b), and (c) of the definition of a probability set function (Section 1.3). That is, $P_X(A)$ is also a probability set function. Conditions (a) and (c) are easily verified by observing, for an appropriate C, that

$$P_X(A) = P(C) \geq 0,$$

and that $\mathscr{C} = \{c : c \in \mathscr{C} \text{ and } X(c) \in \mathscr{A}\}$ requires

$$P_X(\mathscr{A}) = P(\mathscr{C}) = 1.$$

In discussing the condition (b), let us restrict our attention to the two mutually exclusive events A_1 and A_2. Here $P_X(A_1 \cup A_2) = P(C)$, where $C = \{c : c \in \mathscr{C} \text{ and } X(c) \in A_1 \cup A_2\}$. However,

$$C = \{c : c \in \mathscr{C} \text{ and } X(c) \in A_1\} \cup \{c : c \in \mathscr{C} \text{ and } X(c) \in A_2\},$$

or, for brevity, $C = C_1 \cup C_2$. But C_1 and C_2 are disjoint sets. This must be so, for if some c were common, say c_i, then $X(c_i) \in A_1$ and $X(c_i) \in A_2$. That is, the same number $X(c_i)$ belongs to both A_1 and A_2. This is a contradiction because A_1 and A_2 are disjoint sets. Accordingly,

$$P(C) = P(C_1) + P(C_2).$$

However, by definition, $P(C_1)$ is $P_X(A_1)$ and $P(C_2)$ is $P_X(A_2)$ and thus

$$P_X(A_1 \cup A_2) = P_X(A_1) + P_X(A_2).$$

This is condition (b) for two disjoint sets.

Thus each of $P_X(A)$ and $P(C)$ is a probability set function. But the reader should fully recognize that the probability set function P is defined for subsets C of $\mathscr{C}$, whereas P_X is defined for subsets A of $\mathscr{A}$, and, in general, they are not the same set function. Nevertheless, they are closely related and some authors even drop the index X and write $P(A)$ for $P_X(A)$. They think it is quite clear that $P(A)$ means the probability of A, a subset of $\mathscr{A}$, and $P(C)$ means the probability of C, a subset of $\mathscr{C}$. From this point on, we shall adopt this convention and simply write $P(A)$.

Perhaps an additional example will be helpful. Let a coin be tossed two independent times and let our interest be in the *number* of heads to be observed. Thus the sample space is $\mathscr{C} = \{c : \text{where } c \text{ is TT or TH or HT or HH}\}$. Let $X(c) = 0$ if c is TT; let $X(c) = 1$ if c is either TH or HT; and let $X(c) = 2$ if c is HH. Thus the space of the random variable X is $\mathscr{A} = \{0, 1, 2\}$. Consider the subset A of the space $\mathscr{A}$, where $A = \{1\}$. How is the probability of the event A defined? We take the subset C of $\mathscr{C}$ to have as its elements all outcomes in $\mathscr{C}$ for which the random variable X has a value that is an element of A. Because $X(c) = 1$ if c is either TH or HT, then $C = \{c : \text{where } c \text{ is TH or HT}\}$. Thus $P(A) = \Pr(X \in A) = P(C)$. Since $A = \{1\}$, then $P(A) = \Pr(X \in A)$ can be written more simply as $\Pr(X = 1)$. Let $C_1 = \{c : c \text{ is TT}\}$, $C_2 = \{c : c \text{ is TH}\}$, $C_3 = \{c : c \text{ is HT}\}$, and $C_4 = \{c : c \text{ is HH}\}$ denote subsets of $\mathscr{C}$. From independence and equally likely assumptions (see Exercise 1.46), our probability set function $P(C)$ assigns a probability of $\frac{1}{4}$ to each of the sets C_i, $i = 1, 2, 3, 4$. Then $P(C_1) = \frac{1}{4}$, $P(C_2 \cup C_3) = \frac{1}{4} + \frac{1}{4} = \frac{1}{2}$, and $P(C_4) = \frac{1}{4}$. Let us now point out how much simpler it is to couch these statements in a language that involves the random variable X. Because X is the number of heads to be observed in tossing a coin two times, we have

$$\Pr(X = 0) = \tfrac{1}{4}, \qquad \text{since } P(C_1) = \tfrac{1}{4};$$

$$\Pr(X = 1) = \tfrac{1}{2}, \qquad \text{since } P(C_2 \cup C_3) = \tfrac{1}{2};$$

and

$$\Pr(X = 2) = \tfrac{1}{4}, \qquad \text{since } P(C_4) = \tfrac{1}{4}.$$

This may be further condensed in the following table:

x	0	1	2
$\Pr(X = x)$	$\frac{1}{4}$	$\frac{1}{2}$	$\frac{1}{4}$

This table depicts the distribution of probability over the elements of $\mathscr{A}$, the space of the random variable X. This can be written more simply as

$$\Pr(X = x) = \binom{2}{x}\left(\frac{1}{2}\right)^2, \qquad x \in \mathscr{A}.$$

Example 1. Consider a sequence of independent flips of a coin, each resulting in a head (H) or a tail (T). Moreover, on each flip, we assume that H and T are equally likely, that is, $P(\text{H}) = P(\text{T}) = \frac{1}{2}$. The sample space $\mathscr{C}$ consists of sequences like TTHTHHT $\cdots$. Let the random variable X equal the number of flips needed to obtain the first head. For this given sequence, $X = 3$. Clearly, the space of X is $\mathscr{A} = \{1, 2, 3, 4, \ldots\}$. We see that $X = 1$ when the sequence begins with an H and thus $\Pr(X = 1) = \frac{1}{2}$. Likewise, $X = 2$ when the sequence begins with TH, which has probability $\Pr(X = 2) = (\frac{1}{2})(\frac{1}{2}) = \frac{1}{4}$ from the independence. More generally, if $X = x$, where $x = 1, 2, 3, 4, \ldots,$ there must be a string of $x - 1$ tails followed by a head, that is, TT $\cdots$ TH, where there are $x - 1$ tails in TT $\cdots$ T. Thus, from independence, we have

$$\Pr(X = x) = \left(\frac{1}{2}\right)^{x-1}\left(\frac{1}{2}\right) = \left(\frac{1}{2}\right)^x, \qquad x = 1, 2, 3, \ldots.$$

Let us make some observations about these three illustrations of a random variable. In each case the number of points in the space $\mathscr{A}$ was finite, as with $\{0, 1\}$ and $\{0, 1, 2\}$, or countable, as with $\{1, 2, 3, \ldots\}$. There was a function, say $f(x) = \Pr(X = x)$, that described how the probability is distributed over the space $\mathscr{A}$. In each of these illustrations, there is a simple formula (although that is not necessary in general) for that function, namely:

$$f(x) = \frac{1}{2}, \qquad x \in \{0, 1\},$$

$$f(x) = \binom{2}{x}\left(\frac{1}{2}\right)^2, \qquad x \in \{0, 1, 2\},$$

and

$$f(x) = \left(\frac{1}{2}\right)^x, \qquad x \in \{1, 2, 3, \ldots\}.$$

Moreover, the sum of $f(x)$ over all $x \in \mathscr{A}$ equals 1:

$$\sum_{x=0}^{1} \left(\frac{1}{2}\right) = \frac{1}{2} + \frac{1}{2} = 1,$$

$$\sum_{x=0}^{2} \binom{2}{x}\left(\frac{1}{2}\right)^2 = \frac{1}{4} + \frac{1}{2} + \frac{1}{4} = 1,$$

$$\sum_{x=1}^{\infty} \left(\frac{1}{2}\right)^x = \frac{1}{2} + \left(\frac{1}{2}\right)^2 + \left(\frac{1}{2}\right)^3 + \cdots = \frac{\frac{1}{2}}{1 - \frac{1}{2}} = 1.$$

Finally, if $A \subset \mathscr{A}$, we can compute the probability of $X \in A$ by the summation

$$\Pr(X \in A) = \sum_A f(x).$$

For illustrations, using the random variable of Example 1,

$$\Pr(X = 1, 2, 3) = \sum_{x=1}^{3} \left(\frac{1}{2}\right)^x = \frac{1}{2} + \frac{1}{4} + \frac{1}{8} = \frac{7}{8}$$

and

$$\Pr(X = 1, 3, 5, \ldots) = \left(\frac{1}{2}\right) + \left(\frac{1}{2}\right)^3 + \left(\frac{1}{2}\right)^5 + \cdots$$

$$= \frac{\frac{1}{2}}{1 - \frac{1}{4}} = \frac{2}{3}.$$

We have special names for this type of random variable X and for a function $f(x)$ like that in each of these three illustrations, which we now give.

Let X denote a random variable with a one-dimensional space $\mathscr{A}$. Suppose that $\mathscr{A}$ consists of a countable number of points; that is, $\mathscr{A}$ contains a finite number of points or the points of $\mathscr{A}$ can be put into a one-to-one correspondence with the positive integers. Such a space

$\mathscr{A}$ is called a *discrete set* of points. Let a function $f(x)$ be such that $f(x) > 0$, $x \in \mathscr{A}$, and

$$\sum_{\mathscr{A}} f(x) = 1.$$

Whenever a probability set function $P(A)$, $A \subset \mathscr{A}$, can be expressed in terms of such an $f(x)$ by

$$P(A) = \Pr(X \in A) = \sum_{A} f(x),$$

then X is called a *random variable of the discrete type* and $f(x)$ is called the *probability density function* of X. Hereafter the *probability density function* is abbreviated p.d.f.

Our notation can be simplified somewhat so that we do not need to spell out the space in each instance. For illustration, let the random variable be the number of flips necessary to obtain the first head. We now extend the definition of the p.d.f. from on $\mathscr{A} = \{1, 2, 3, \ldots\}$ to all the real numbers by writing

$$\begin{aligned} f(x) &= \left(\frac{1}{2}\right)^x, \qquad x = 1, 2, 3, \ldots, \\ &= 0 \qquad \text{elsewhere.} \end{aligned}$$

From such a function, we see that the space $\mathscr{A}$ is clearly the set of positive integers which is a discrete set of points. Thus the corresponding random variable is one of the discrete type.

Example 2. A lot, consisting of 100 fuses, is inspected by the following procedure. Five of these fuses are chosen at random and tested; if all 5 "blow" at the correct amperage, the lot is accepted. If, in fact, there are 20 defective fuses in the lot, the probability of accepting the lot is, under appropriate assumptions,

$$\frac{\binom{80}{5}}{\binom{100}{5}} = 0.32,$$

approximately. More generally, let the random variable X be the number of defective fuses among the 5 that are inspected. The p.d.f. of X is given by

$$\begin{aligned} f(x) = \Pr(X = x) &= \frac{\binom{20}{x}\binom{80}{5-x}}{\binom{100}{5}}, \qquad x = 0, 1, 2, 3, 4, 5, \\ &= 0 \qquad \text{elsewhere.} \end{aligned}$$

Clearly, the space of X is $\mathscr{A} = \{0, 1, 2, 3, 4, 5\}$. Thus this is an example of a random variable of the discrete type whose distribution is an illustration of a *hypergeometric distribution*.

Let the random variable X have the probability set function $P(A)$, where A is a one-dimensional set. Take x to be a real number and consider the set A which is an unbounded set from $-\infty$ to x, including the point x itself. For all such sets A we have $P(A) = \Pr(X \in A) = \Pr(X \le x)$. This probability depends on the point x; that is, this probability is a function of the point x. This point function is denoted by the symbol $F(x) = \Pr(X \le x)$. The function $F(x)$ is called the *distribution function* (sometimes, *cumulative distribution function*) of the random variable X. Since $F(x) = \Pr(X \le x)$, then, with $f(x)$ the p.d.f., we have

$$F(x) = \sum_{w \le x} f(w),$$

for the discrete type.

Example 3. Let the random variable X of the discrete type have the p.d.f. $f(x) = x/6$, $x = 1, 2, 3$, zero elsewhere. The distribution function of X is

$$\begin{aligned} F(x) &= 0, & x &< 1, \\ &= \tfrac{1}{6}, & 1 \le x &< 2, \\ &= \tfrac{3}{6}, & 2 \le x &< 3, \\ &= 1, & 3 &\le x. \end{aligned}$$

Here, as depicted in Figure 1.3, $F(x)$ is a step function that is constant in every interval not containing 1, 2, or 3, but has steps of heights $\frac{1}{6}$, $\frac{2}{6}$, and $\frac{3}{6}$, which are the probabilities at those respective points. It is also seen that $F(x)$ is everywhere continuous from the right. The p.d.f. of X is displayed as a bar

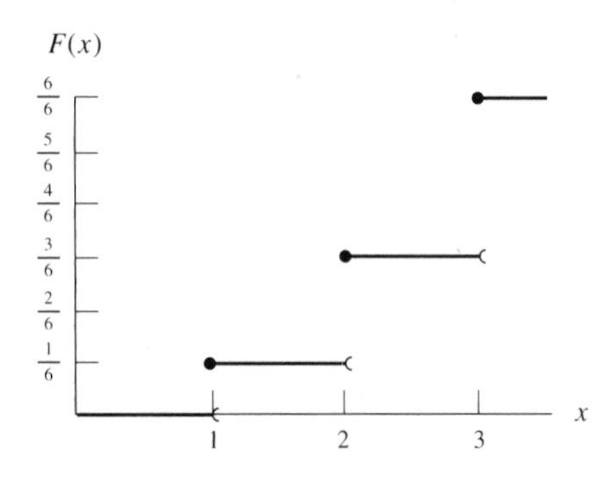

FIGURE 1.3

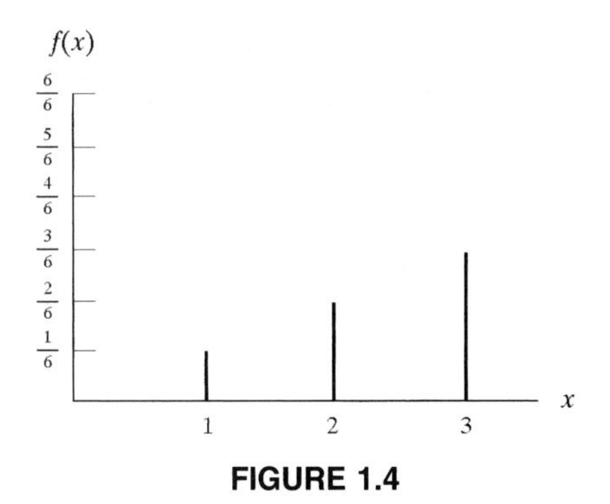

FIGURE 1.4

graph in Figure 1.4. We see that $f(x)$ represents the probability at each x while $F(x)$ cumulates all the probability of points that are less than or equal to x. Thus we can compute a probability like

$$\Pr(1.5 < X \le 4.5) = F(4.5) - F(1.5) = 1 - \tfrac{1}{6} = \tfrac{5}{6}$$

or as

$$\Pr(1.5 < X \le 4.5) = f(2) + f(3) = \tfrac{2}{6} + \tfrac{3}{6} = \tfrac{5}{6}.$$

While the properties of a *distribution function* $F(x) = \Pr(X \le x)$ are discussed in more detail in Section 1.7, we can make a few observations now since $F(x)$ is a probability.

1. $0 \le F(x) \le 1$.
2. $F(x)$ is a nondecreasing function as it cumulates probability as x increases.
3. $F(y) = 0$ for every point y that is less than the smallest value in the space of X.
4. $F(z) = 1$ for every point z that is greater than the largest value in the space of X.
5. If X is a random variable of the discrete type, then $F(x)$ is a step function and the height of the step at x in the space of X is equal to the probability $f(x) = \Pr(X = x)$.

EXERCISES

1.47. Let a card be selected from an ordinary deck of playing cards. The outcome c is one of these 52 cards. Let $X(c) = 4$ if c is an ace, let $X(c) = 3$ if c is a king, let $X(c) = 2$ if c is a queen, let $X(c) = 1$ if c is a jack, and let $X(c) = 0$ otherwise. Suppose that P assigns a probability of $\frac{1}{52}$ to each outcome c. Describe the induced probability $P_X(A)$ on the space $\mathscr{A} = \{0, 1, 2, 3, 4\}$ of the random variable X.

1.48. For each of the following, find the constant c so that $f(x)$ satisfies the condition of being a p.d.f. of one random variable X.
(a) $f(x) = c(\frac{2}{3})^x$, $x = 1, 2, 3, \ldots$, zero elsewhere.
(b) $f(x) = cx$, $x = 1, 2, 3, 4, 5, 6$, zero elsewhere.

1.49. Let $f(x) = x/15$, $x = 1, 2, 3, 4, 5$, zero elsewhere, be the p.d.f. of X. Find $\Pr(X = 1 \text{ or } 2)$, $\Pr(\frac{1}{2} < X < \frac{5}{2})$, and $\Pr(1 \le X \le 2)$.

1.50. Let $f(x)$ be the p.d.f. of a random variable X. Find the distribution function $F(x)$ of X and sketch its graph along with that of $f(x)$ if:
(a) $f(x) = 1$, $x = 0$, zero elsewhere.
(b) $f(x) = \frac{1}{3}$, $x = -1, 0, 1$, zero elsewhere.
(c) $f(x) = x/15$, $x = 1, 2, 3, 4, 5$, zero elsewhere.

1.51. Let us select five cards at random and without replacement from an ordinary deck of playing cards.
(a) Find the p.d.f. of X, the number of hearts in the five cards.
(b) Determine $\Pr(X \le 1)$.

1.52. Let X equal the number of heads in four independent flips of a coin. Using certain assumptions, determine the p.d.f. of X and compute the probability that X is equal to an odd number.

1.53. Let X have the p.d.f. $f(x) = x/5050$, $x = 1, 2, 3, \ldots, 100$, zero elsewhere.
(a) Compute $\Pr(X \le 50)$.
(b) Show that the distribution function of X is $F(x) = [x]([x] + 1)/10100$, for $1 \le x \le 100$, where $[x]$ is the greatest integer in x.

1.54. Let a bowl contain 10 chips of the same size and shape. One and only one of these chips is red. Continue to draw chips from the bowl, one at a time and at random and without replacement, until the red chip is drawn.
(a) Find the p.d.f. of X, the number of trials needed to draw the red chip.
(b) Compute $\Pr(X \le 4)$.

1.55. Cast a die a number of independent times until a six appears on the up side of the die.
(a) Find the p.d.f. $f(x)$ of X, the number of casts needed to obtain that first six.
(b) Show that $\sum_{x=1}^{\infty} f(x) = 1$.
(c) Determine $\Pr(X = 1, 3, 5, 7, \ldots)$.
(d) Find the distribution function $F(x) = \Pr(X \le x)$.

1.56. Cast a die two independent times and let X equal the absolute value of the difference of the two resulting values (the numbers on the up sides). Find the p.d.f. of X.

Hint: It is not necessary to find a formula for the p.d.f.

1.6 Random Variables of the Continuous Type

A random variable was defined in Section 1.5, and only those of the discrete type were considered there. Let us begin the discussion of random variables of the continuous type with an example.

Let a random experiment be a selection of a point that is interior to a circle of radius 1 that has center at the origin of a two-dimensional space. We call this space $\mathscr{C}$ and the area of this circle is π. The random selection is in such a way that the probability of being in a certain set C interior to $\mathscr{C}$ is proportional to the area of C; in particular, if $C \subset \mathscr{C}$,

$$P(C) = \frac{\text{area of } C}{\pi}.$$

First we observe that $P(\mathscr{C}) = 1$. In addition, if C_1 is that subset of $\mathscr{C}$ that is in the first quadrant, $P(C_1) = (\pi/4)/\pi = \frac{1}{4}$. If C_2 is the interior of a circle of radius $\frac{1}{3}$ such that $C_2 \subset \mathscr{C}$, then $P(C_2) = \pi(\frac{1}{3})^2/\pi = \frac{1}{9}$. It is interesting to note that the probability of a point, a line segment, or any curve in $\mathscr{C}$ is equal to zero because those areas would be zero. In particular, if C_3 is the boundary of the set C_2 (that is, C_3 is the actual circle of radius $\frac{1}{3}$), then $P(C_3) = 0$.

We define a random variable X, associated with this random experiment, as the distance of the selected point from the origin. The space of X is $\mathscr{A} = \{x : 0 \le x < 1\}$. Of course, for any $x \in \mathscr{A}$, $\Pr(X = x) = 0$, because $X = x$ is the event that the random point falls on a circle, symmetric with respect to the origin, of radius x and the associated area equals zero. However, it does make sense to consider the induced probability of the event $X \le x$, namely the distribution function of X. If $x \in \mathscr{A}$, then

$$F(x) = \Pr(X \le x) = \frac{\text{area of a certain circle of radius } x}{\pi}$$

$$= \frac{\pi x^2}{\pi} = x^2, \qquad 0 \le x < 1.$$

Clearly, if $x < 0$, then $F(x) = 0$; and if $x > 1$, then $F(x) = 1$. Thus we can write

$$\begin{aligned} F(x) &= 0, & x &< 0, \\ &= x^2, & 0 &\le x < 1, \\ &= 1, & 1 &\le x. \end{aligned}$$

Recall, in the discrete case, we had a function f that was associated with F through the equation

$$F(x) = \sum_{w \le x} f(w).$$

Either F or f could be used to compute probabilities like

$$\Pr(a < X \le b) = F(b) - F(a) = \sum_{w \in A} f(w),$$

where $A = \{w : a < w \le b\}$. We have observed, in this continuous case, that $\Pr(X = x) = 0$, so a summation of such probabilities is no longer appropriate. However, it is easy to find an integral that relates F to f through

$$F(x) = \int_{w \le x} f(w)\, dw.$$

Since $\mathscr{A} = \{x : 0 \le x < 1\}$, this can be written as

$$F(x) = x^2 = \int_0^x f(w)\, dw, \qquad x \in \mathscr{A}.$$

By one form of the fundamental theorem of calculus, we know that the derivative of the right-hand member of this equation is $f(x)$. Thus taking derivatives of each member of the equation, we obtain

$$2x = f(x), \qquad 0 \le x < 1$$

Of course, at $x = 0$, this is only a right-hand derivative. We observe that $f(x) \ge 0$, $x \in \mathscr{A}$, and

$$\int_0^1 2x\, dx = 1.$$

Probabilities can now be computed through

$$\Pr(X \in A) = \int_A f(w)\, dw.$$

For illustration,

$$\Pr\left(\tfrac{1}{4} < X \le \tfrac{1}{2}\right) = \int_{1/4}^{1/2} 2w\,dw = \left[w^2\right]_{1/4}^{1/2}$$

$$= F\left(\tfrac{1}{2}\right) - F\left(\tfrac{1}{4}\right) = \tfrac{1}{4} - \tfrac{1}{16} = \tfrac{3}{16}.$$

With the background of this example, we give the definition of a random variable of the continuous type.

Let X denote a random variable with a one-dimensional space $\mathscr{A}$, which consists of an interval or a union of intervals. Let a function $f(x)$ be nonnegative such that

$$\int_{\mathscr{A}} f(x)\,dx = 1.$$

Whenever a probability set function $P(A)$, $A \subset \mathscr{A}$, can be expressed in terms of such an $f(x)$ by

$$P(A) = \Pr(X \in A) = \int_A f(x)\,dx,$$

then X is said to be a *random variable of the continuous type* and $f(x)$ is called the *probability density function* (p.d.f.) of X.

Example 1. Let the random variable of the continuous type X equal the distance in feet between bad records of a used computer tape. Say that the space of X is $\mathscr{A} = \{x : 0 < x < \infty\}$. Suppose that a reasonable probability model for X is given by the p.d.f.

$$f(x) = \tfrac{1}{40} e^{-x/40}, \qquad x \in \mathscr{A}.$$

Here $f(x) \ge 0$ for $x \in \mathscr{A}$, and

$$\int_0^\infty \tfrac{1}{40} e^{-x/40}\,dx = \left[-e^{-x/40}\right]_0^\infty = 1.$$

If we are interested in the probability that the distance between bad records is greater than 40 feet, then $A = \{x : 40 < x < \infty\}$ and

$$\Pr(X \in A) = \int_{40}^\infty \tfrac{1}{40} e^{-x/40}\,dx = e^{-1}.$$

The p.d.f. and the probability of interest are depicted in Figure 1.5.

If we restrict ourselves to random variables of either the discrete type or the continuous type, we may work exclusively with the p.d.f. $f(x)$. This affords an enormous simplification; but it should be recognized that this simplification is obtained at considerable cost from a mathematical point of view. Not only shall we exclude from

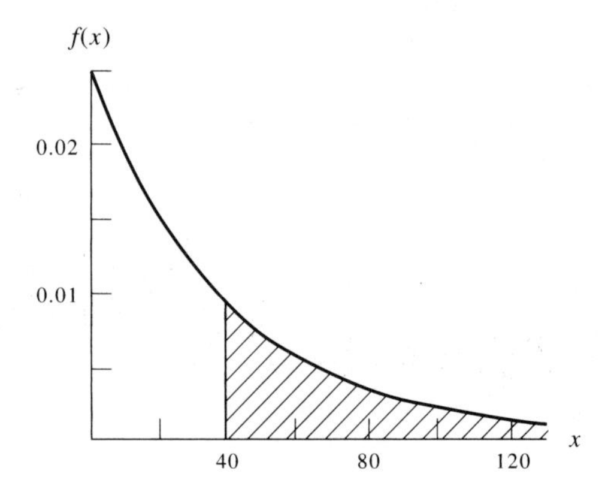

FIGURE 1.5

consideration many random variables that do not have these types of distributions, but we shall also exclude many interesting subsets of the space. In this book, however, we shall in general restrict ourselves to these simple types of random variables.

Remarks. Let X denote the number of spots that show when a die is cast. We can assume that X is a random variable with $\mathscr{A} = \{1, 2, \ldots, 6\}$ and with a p.d.f. $f(x) = \frac{1}{6}$, $x \in \mathscr{A}$. Other assumptions can be made to provide different mathematical models for this experiment. Experimental evidence can be used to help one decide which model is the more realistic. Next, let X denote the point at which a balanced pointer comes to rest. If the circumference is graduated $0 \le x < 1$, a reasonable mathematical model for this experiment is to take X to be a random variable with $\mathscr{A} = \{x : 0 \le x < 1\}$ and with a p.d.f. $f(x) = 1$, $x \in \mathscr{A}$.

Both types of probability density functions can be used as distributional models for many random variables found in real situations. For illustrations consider the following. If X is the number of automobile accidents during a given day, then $f(0), f(1), f(2), \ldots$ represent the probabilities of 0, 1, 2, ... accidents. On the other hand, if X is length of life of a female born in a certain community, the integral [area under the graph of $f(x)$ that lies above the x-axis and between the vertical lines $x = 40$ and $x = 50$]

$$\int_{40}^{50} f(x)\, dx$$

represents the probability that she dies between 40 and 50 (or the percentage of those females dying between 40 and 50). A particular $f(x)$ will be suggested later for each of these situations, but again experimental evidence must be used to decide whether we have realistic models.

Our notation can be considerably simplified when we restrict ourselves to random variables of the continuous or discrete types. Suppose that the space of a continuous type of random variable X is $\mathscr{A} = \{x : 0 < x < \infty\}$ and that the p.d.f. of X is e^{-x}, $x \in \mathscr{A}$. We shall in no manner alter the distribution of X [that is, alter any $P(A)$, $A \subset \mathscr{A}$] if we extend the definition of the p.d.f. of X by writing

$$\begin{aligned} f(x) &= e^{-x}, \qquad 0 < x < \infty, \\ &= 0 \qquad \text{elsewhere,} \end{aligned}$$

and then refer to $f(x)$ as the p.d.f. of X. We have

$$\int_{-\infty}^{\infty} f(x)\,dx = \int_{-\infty}^{0} 0\,dx + \int_{0}^{\infty} e^{-x}\,dx = 1.$$

Thus we may treat the entire axis of reals as though it were the space of X. Accordingly, we now replace

$$\int_{\mathscr{A}} f(x)\,dx \qquad \text{by} \qquad \int_{-\infty}^{\infty} f(x)\,dx.$$

If $f(x)$ is the p.d.f. of a continuous type of random variable X and if A is the set $\{x : a < x < b\}$, then $P(A) = \Pr(X \in A)$ can be written as

$$\Pr(a < X < b) = \int_{a}^{b} f(x)\,dx.$$

Moreover, if $A = \{a\}$, then

$$P(A) = \Pr(X \in A) = \Pr(X = a) = \int_{a}^{a} f(x)\,dx = 0,$$

since the integral $\int_a^a f(x)\,dx$ is defined in calculus to be zero. That is, if X is a random variable of the continuous type, the probability of every set consisting of a single point is zero. This fact enables us to write, say,

$$\Pr(a < X < b) = \Pr(a \le X \le b).$$

More important, this fact allows us to change the value of the p.d.f. of a continuous type of random variable X at a single point without altering the distribution of X. For instance, the p.d.f.

$$\begin{aligned} f(x) &= e^{-x}, \qquad 0 < x < \infty, \\ &= 0 \qquad \text{elsewhere,} \end{aligned}$$

can be written as

$$\begin{aligned} f(x) &= e^{-x}, \qquad 0 \le x < \infty, \\ &= 0, \qquad \text{elsewhere,} \end{aligned}$$

without changing any $P(A)$. We observe that these two functions differ only at $x = 0$ and $\Pr(X = 0) = 0$. More generally, if two probability density functions of random variables of the continuous type differ only on a set having probability zero, the two corresponding probability set functions are exactly the same. Unlike the continuous type, the p.d.f. of a discrete type of random variable may not be changed at any point, since a change in such a p.d.f. alters the distribution of probability.

Example 2. Let the random variable X of the continuous type have the p.d.f. $f(x) = 2/x^3$, $1 < x < \infty$, zero elsewhere. The distribution function of X is

$$\begin{aligned} F(x) &= \int_{-\infty}^{x} 0\, dw = 0, \qquad x < 1, \\ &= \int_{1}^{x} \frac{2}{w^3}\, dw = 1 - \frac{1}{x^2}, \qquad 1 \le x. \end{aligned}$$

The graph of this distribution function is depicted in Figure 1.6. Here $F(x)$ is a continuous function for all real numbers x; in particular, $F(x)$ is everywhere continuous from the right. Moreover, the derivative of $F(x)$ with respect to x exists at all points except at $x = 1$. Thus the p.d.f. of X is defined by this derivative except at $x = 1$. Since the set $A = \{1\}$ is a set of probability measure zero [that is, $P(A) = 0$], we are free to define the p.d.f. at $x = 1$ in any manner we please. One way to do this is to write $f(x) = 2/x^3$, $1 < x < \infty$, zero elsewhere.

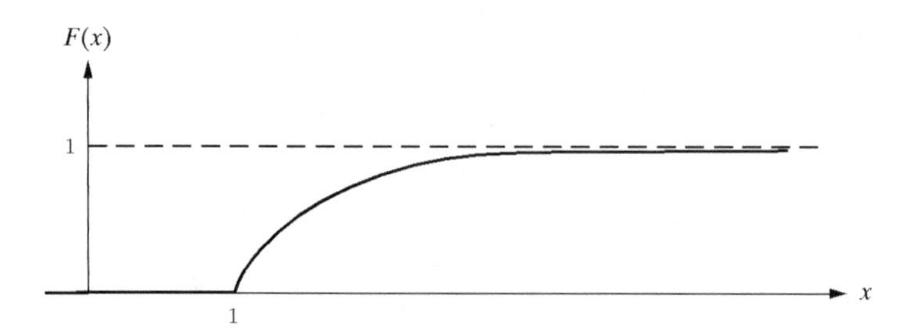

FIGURE 1.6

EXERCISES

1.57. Let a point be selected from the sample space $\mathscr{C} = \{c : 0 < c < 10\}$. Let $C \subset \mathscr{C}$ and let the probability set function be $P(C) = \int_C \frac{1}{10}\,dz$. Define the random variable X to be $X(c) = c^2$. Find the distribution function and the p.d.f. of X.

1.58. Let the probability set function $P(A)$ of the random variable X be $P(A) = \int_A f(x)\,dx$, where $f(x) = 2x/9$, $x \in \mathscr{A} = \{x : 0 < x < 3\}$. Let $A_1 = \{x : 0 < x < 1\}$, $A_2 = \{x : 2 < x < 3\}$. Compute $P(A_1) = \Pr[X \in A_1]$, $P(A_2) = \Pr(X \in A_2)$, and $P(A_1 \cup A_2) = \Pr(X \in A_1 \cup A_2)$.

1.59. Let the space of the random variable X be $\mathscr{A} = \{x : 0 < x < 1\}$. If $A_1 = \{x : 0 < x < \frac{1}{2}\}$ and $A_2 = \{x : \frac{1}{2} \le x < 1\}$, find $P(A_2)$ if $P(A_1) = \frac{1}{4}$.

1.60. Let the space of the random variable X be $\mathscr{A} = \{x : 0 < x < 10\}$ and let $P(A_1) = \frac{3}{8}$, where $A_1 = \{x : 1 < x < 5\}$. Show that $P(A_2) \le \frac{5}{8}$, where $A_2 = \{x : 5 \le x < 10\}$.

1.61. Let the subsets $A_1 = \{x : \frac{1}{4} < x < \frac{1}{2}\}$ and $A_2 = \{x : \frac{1}{2} \le x < 1\}$ of the space $\mathscr{A} = \{x : 0 < x < 1\}$ of the random variable X be such that $P(A_1) = \frac{1}{8}$ and $P(A_2) = \frac{1}{2}$. Find $P(A_1 \cup A_2)$, $P(A_1^*)$, and $P(A_1^* \cap A_2^*)$.

1.62. Given $\int_A [1/\pi(1 + x^2)]\,dx$, where $A \subset \mathscr{A} = \{x : -\infty < x < \infty\}$. Show that the integral could serve as a probability set function of a random variable X whose space is $\mathscr{A}$.

1.63. Let the probability set function of the random variable X be

$$P(A) = \int_A e^{-x}\,dx, \qquad \text{where } \mathscr{A} = \{x : 0 < x < \infty\}.$$

Let $A_k = \{x : 2 - 1/k < x \le 3\}$, $k = 1, 2, 3, \ldots$. Find $\lim_{k \to \infty} A_k$ and $P\left(\lim_{k\to\infty} A_k\right)$.

Find $P(A_k)$ and $\lim_{k\to\infty} P(A_k)$. Note that $\lim_{k\to\infty} P(A_k) = P\left(\lim_{k\to\infty} A_k\right)$.

1.64. For each of the following probability density functions of X, compute $\Pr(|X| < 1)$ and $\Pr(X^2 < 9)$.

(a) $f(x) = x^2/18$, $-3 < x < 3$, zero elsewhere.

(b) $f(x) = (x + 2)/18$, $-2 < x < 4$, zero elsewhere.

1.65. Let $f(x) = 1/x^2$, $1 < x < \infty$, zero elsewhere, be the p.d.f. of X. If $A_1 = \{x : 1 < x < 2\}$ and $A_2 = \{x : 4 < x < 5\}$, find $P(A_1 \cup A_2)$ and $P(A_1 \cap A_2)$.

1.66. A *mode* of a distribution of one random variable X is a value of x that maximizes the p.d.f. $f(x)$. For X of the continuous type, $f(x)$ must be continuous. If there is only one such x, it is called the *mode of the distribution*. Find the mode of each of the following distributions:

(a) $f(x) = (\frac{1}{2})^x$, $x = 1, 2, 3, \ldots$, zero elsewhere.

(b) $f(x) = 12x^2(1 - x)$, $0 < x < 1$, zero elsewhere.
(c) $f(x) = (\frac{1}{2})x^2e^{-x}$, $0 < x < \infty$, zero elsewhere.

1.67. A *median* of a distribution of one random variable X of the discrete or continuous type is a value of x such that $\Pr(X < x) \le \frac{1}{2}$ and $\Pr(X \le x) \ge \frac{1}{2}$. If there is only one such x, it is called the *median of the distribution*. Find the median of each of the following distributions:

(a) $f(x) = \dfrac{4!}{x!\,(4-x)!}\left(\dfrac{1}{4}\right)^x\left(\dfrac{3}{4}\right)^{4-x}$, $x = 0, 1, 2, 3, 4$, zero elsewhere.

(b) $f(x) = 3x^2$, $0 < x < 1$, zero elsewhere.

(c) $f(x) = \dfrac{1}{\pi(1 + x^2)}$, $-\infty < x < \infty$.

Hint: In parts (b) and (c), $\Pr(X < x) = \Pr(X \le x)$ and thus that common value must equal $\frac{1}{2}$ if x is to be the median of the distribution.

1.68. Let $0 < p < 1$. A $(100p)$th *percentile* (*quantile* of order p) of the distribution of a random variable X is a value ξ_p such that $\Pr(X < \xi_p) \le p$ and $\Pr(X \le \xi_p) \ge p$. Find the twentieth percentile of the distribution that has p.d.f. $f(x) = 4x^3$, $0 < x < 1$, zero elsewhere.

Hint: With a continuous-type random variable X, $\Pr(X < \xi_p) = \Pr(X \le \xi_p)$ and hence that common value must equal p.

1.69. Find the distribution function $F(x)$ associated with each of the following probability density functions. Sketch the graphs of $f(x)$ and $F(x)$.
(a) $f(x) = 3(1 - x)^2$, $0 < x < 1$, zero elsewhere.
(b) $f(x) = 1/x^2$, $1 < x < \infty$, zero elsewhere.
(c) $f(x) = \frac{1}{3}$, $0 < x < 1$ or $2 < x < 4$, zero elsewhere.
Also find the median and 25th percentile of each of these distributions.

1.70. Consider the distribution function $F(x) = 1 - e^{-x} - xe^{-x}$, $0 \le x < \infty$, zero elsewhere. Find the p.d.f., the mode, and the median (by numerical methods) of this distribution.

1.7 Properties of the Distribution Function

In Section 1.5 we defined the distribution function of a random variable X as $F(x) = \Pr(X \le x)$. This concept was used in Section 1.6 to find the probability distribution of a random variable of the continuous type. So, in terms of the p.d.f. $f(x)$, we know that

$$F(x) = \sum_{w \le x} f(w),$$

for the discrete type of random variable, and

$$F(x) = \int_{-\infty}^{x} f(w)\, dw,$$

for the continuous type of random variable. We speak of a distribution function $F(x)$ as being of the continuous or discrete type, depending on whether the random variable is of the continuous or discrete type.

Remark. If X is a random variable of the continuous type, the p.d.f. $f(x)$ has at most a finite number of discontinuities in every finite interval. This means (1) that the distribution function $F(x)$ is everywhere continuous and (2) that the derivative of $F(x)$ with respect to x exists and is equal to $f(x)$ at each point of continuity of $f(x)$. That is, $F'(x) = f(x)$ at each point of continuity of $f(x)$. If the random variable X is of the discrete type, most surely the p.d.f. $f(x)$ is *not* the derivative of $F(x)$ with respect to x (that is, with respect to Lebesgue measure); but $f(x)$ *is* the (Radon–Nikodym) derivative of $F(x)$ with respect to a counting measure. A derivative is often called a *density*. Accordingly, we call these derivatives *probability density functions*.

There are several properties of a distribution function $F(x)$ that can be listed as a consequence of the properties of the probability set function. Some of these are the following. In listing these properties, we shall not restrict X to be a random variable of the discrete or continuous type. We shall use the symbols $F(\infty)$ and $F(-\infty)$ to mean $\lim_{x\to\infty} F(x)$ and $\lim_{x\to-\infty} F(x)$, respectively. In like manner, the symbols $\{x : x < \infty\}$ and $\{x : x < -\infty\}$ represent, respectively, the limits of the sets $\{x : x \le b\}$ and $\{x : x \le -b\}$ as $b \to \infty$.

1. $0 \le F(x) \le 1$ because $0 \le \Pr(X \le x) \le 1$.
2. $F(x)$ is a nondecreasing function of x. For, if $x' < x''$, then

$$\{x : x \le x''\} = \{x : x \le x'\} \cup \{x : x' < x \le x''\}$$

and

$$\Pr(X \le x'') = \Pr(X \le x') + \Pr(x' < X \le x'').$$

That is,

$$F(x'') - F(x') = \Pr(x' < X \le x'') \ge 0.$$

3. $F(\infty) = 1$ and $F(-\infty) = 0$ because the set $\{x : x \le \infty\}$ is the entire one-dimensional space and the set $\{x : x \le -\infty\}$ is the null set.

From the proof of property 2, it is observed that, if $a < b$, then

$$\Pr(a < X \leq b) = F(b) - F(a).$$

Suppose that we want to use $F(x)$ to compute the probability $\Pr(X = b)$. To do this, consider, with $h > 0$,

$$\lim_{h \to 0} \Pr(b - h < X \leq b) = \lim_{h \to 0} [F(b) - F(b - h)].$$

Intuitively, it seems that $\lim_{h \to 0} \Pr(b - h < X \leq b)$ should exist and be equal to $\Pr(X = b)$ because, as h tends to zero, the limit of the set $\{x : b - h < x \leq b\}$ is the set that contains the single point $x = b$. The fact that this limit is $\Pr(X = b)$ is a theorem that we accept without proof. Accordingly, we have

$$\Pr(X = b) = F(b) - F(b-),$$

where $F(b-)$ is the left-hand limit of $F(x)$ at $x = b$. That is, the probability that $X = b$ is the height of the step that $F(x)$ has at $x = b$. Hence, if the distribution function $F(x)$ is continuous at $x = b$, then $\Pr(X = b) = 0$.

There is a fourth property of $F(x)$ that is now listed.

4. $F(x)$ is continuous from the right, that is, right-continuous.

To prove this property, consider, with $h > 0$,

$$\lim_{h \to 0} \Pr(a < X \leq a + h) = \lim_{h \to 0} [F(a + h) - F(a)].$$

We accept without proof a theorem which states, with $h > 0$, that

$$\lim_{h \to 0} \Pr(a < X \leq a + h) = P(\varnothing) = 0.$$

Here also, the theorem is intuitively appealing because, as h tends to zero, the limit of the set $\{x : a < x \leq a + h\}$ is the null set. Accordingly, we write

$$0 = F(a+) - F(a),$$

where $F(a+)$ is the right-hand limit of $F(x)$ at $x = a$. Hence $F(x)$ is continuous from the right at every point $x = a$.

Remark. In the arguments concerning several of these properties, we appeal to the reader's intuition. However, most of these properties can be proved in formal ways using the definition of $\lim_{k \to \infty} A_k$, given in Exercises 1.7

and 1.8, and the fact that the probability set function P is countably additive; that is, P enjoys (b) of Definition 7.

The preceding discussion may be summarized in the following manner: A distribution function $F(x)$ is a nondecreasing function of x, which is everywhere continuous from the right and has $F(-\infty) = 0$, $F(\infty) = 1$. The probability $\Pr(a < X \le b)$ is equal to the difference $F(b) - F(a)$. If x is a discontinuity point of $F(x)$, then the probability $\Pr(X = x)$ is equal to the jump which the distribution function has at the point x. If x is a continuity point of $F(x)$, then $\Pr(X = x) = 0$.

Remark. The definition of the distribution function makes it clear that the probability set function P determines the distribution function F. It is true, although not so obvious, that a probability set function P can be found from a distribution function F. That is, P and F give the same information about the distribution of probability, and which function is used is a matter of convenience.

Often, probability models can be constructed that make reasonable assumptions about the probability set function and thus the distribution function. For a simple illustration, consider an experiment in which one chooses at random a point from the closed interval $[a, b]$, $a < b$, that is on the real line. Thus the sample space $\mathscr{C}$ is $[a, b]$. Let the random variable X be the identity function defined on $\mathscr{C}$. Thus the space $\mathscr{A}$ of X is $\mathscr{A} = \mathscr{C}$. Suppose that it is reasonable to *assume*, from the nature of the experiment, that if an interval A is a subset of $\mathscr{A}$, the probability of the event A is proportional to the length of A. Hence, if A is the interval $[a, x]$, $x \le b$, then

$$P(A) = \Pr(X \in A) = \Pr(a \le X \le x) = c(x - a),$$

where c is the constant of proportionality.

In the expression above, if we take $x = b$, we have

$$1 = \Pr(a \le X \le b) = c(b - a),$$

so $c = 1/(b - a)$. Thus we will have an appropriate probability model if we take the distribution function of X, $F(x) = \Pr(X \le x)$, to be

$$\begin{aligned} F(x) &= 0, & x &< a, \\ &= \frac{x-a}{b-a}, & a &\le x \le b, \\ &= 1, & b &< x. \end{aligned}$$

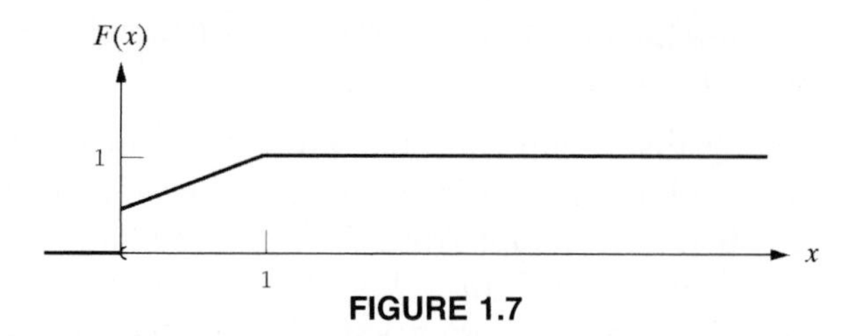

FIGURE 1.7

Accordingly, the p.d.f. of X, $f(x) = F'(x)$, may be written

$$f(x) = \frac{1}{b-a}, \qquad a \le x \le b,$$
$$= 0 \qquad \text{elsewhere.}$$

The derivative of $F(x)$ does not exist at $x = a$ nor at $x = b$; but the set $\{x : x = a, b\}$ is a set of probability measure zero, and we elect to define $f(x)$ to be equal to $1/(b - a)$ at those two points, just as a matter of convenience. We observe that this p.d.f. is a constant on $\mathscr{A}$. If the p.d.f. of one or more variables of the continuous type or of the discrete type is a constant on the space $\mathscr{A}$, we say that the probability is distributed *uniformly* over $\mathscr{A}$. Thus, in the example above, we say that X has a *uniform distribution* over the interval $[a, b]$.

We now give an illustrative example of a distribution that is neither of the discrete nor continuous type.

Example 1. Let a distribution function be given by

$$F(x) = 0, \qquad x < 0,$$
$$= \frac{x+1}{2}, \qquad 0 \le x < 1,$$
$$= 1, \qquad 1 \le x.$$

Then, for instance,

$$\Pr(-3 < X \le \tfrac{1}{2}) = F(\tfrac{1}{2}) - F(-3) = \tfrac{3}{4} - 0 = \tfrac{3}{4}$$

and

$$\Pr(X = 0) = F(0) - F(0-) = \tfrac{1}{2} - 0 = \tfrac{1}{2}.$$

The graph of $F(x)$ is shown in Figure 1.7. We see that $F(x)$ is not always continuous, nor is it a step function. Accordingly, the corresponding distribution is neither of the continuous type nor of the discrete type. It may be described as a mixture of those types.

Distributions that are mixtures of the continuous and discrete types do, in fact, occur frequently in practice. For illustration, in life testing, suppose we know that the length of life, say X, exceeds the number b, but the exact value is unknown. This is called *censoring*. For instance, this can happen when a subject in a cancer study simply disappears; the investigator knows that the subject has lived a certain number of months, but the exact length of life is unknown. Or it might happen when an investigator does not have enough time in an investigation to observe the moments of deaths of all the animals, say rats, in some study. Censoring can also occur in the insurance industry; in particular, consider a loss with a limited-pay policy in which the top amount is exceeded but it is not known by how much.

Example 2. Reinsurance companies are concerned with large losses because they might agree, for illustration, to cover losses due to wind damages that are between \$2,000,000 and \$10,000,000. Say that X equals the size of a wind loss in millions of dollars, and suppose that it has the distribution function

$$\begin{aligned} F(x) &= 0, & -\infty < x < 0, \\ &= 1 - \left(\frac{10}{10+x}\right)^3, & 0 \le x < \infty. \end{aligned}$$

If losses beyond \$10,000,000 are reported only as 10, then the distribution function of this censored distribution is

$$\begin{aligned} F(x) &= 0, & -\infty < x < 0, \\ &= 1 - \left(\frac{10}{10+x}\right)^3, & 0 \le x < 10, \\ &= 1, & 10 \le x < \infty, \end{aligned}$$

which has a jump of $[10/(10+10)]^3 = \frac{1}{8}$ at $x = 10$.

We shall now point out an important fact about a function of a random variable. Let X denote a random variable with space $\mathscr{A}$. Consider the function $Y = u(X)$ of the random variable X. Since X is a function defined on a sample space $\mathscr{C}$, then $Y = u(X)$ is a composite function defined on $\mathscr{C}$. That is, $Y = u(X)$ is itself a random variable which has its own space $\mathscr{B} = \{y : y = u(x), x \in \mathscr{A}\}$ and its own probability set function. If $y \in \mathscr{B}$, the event $Y = u(X) \le y$ occurs when, and only when, the event $X \in A \subset \mathscr{A}$ occurs, where $A = \{x : u(x) \le y\}$. That is, the distribution function of Y is

$$G(y) = \Pr(Y \le y) = \Pr[u(X) \le y] = P(A).$$

The following example illustrates a method of finding the distribution function and the p.d.f. of a function of a random variable. This method is called the *distribution-function technique*.

Example 3. Let $f(x) = \frac{1}{2}$, $-1 < x < 1$, zero elsewhere, be the p.d.f. of the random variable X. Define the random variable Y by $Y = X^2$. We wish to find the p.d.f. of Y. If $y \geq 0$, the probability $\Pr(Y \leq y)$ is equivalent to

$$\Pr(X^2 \leq y) = \Pr(-\sqrt{y} \leq X \leq \sqrt{y}).$$

Accordingly, the distribution function of Y, $G(y) = \Pr(Y \leq y)$, is given by

$$\begin{aligned} G(y) &= 0, \qquad y < 0, \\ &= \int_{-\sqrt{y}}^{\sqrt{y}} \tfrac{1}{2}\,dx = \sqrt{y}, \qquad 0 \leq y < 1, \\ &= 1, \qquad 1 \leq y. \end{aligned}$$

Since Y is a random variable of the continuous type, the p.d.f. of Y is $g(y) = G'(y)$ at all points of continuity of $g(y)$. Thus we may write

$$\begin{aligned} g(y) &= \frac{1}{2\sqrt{y}}, \qquad 0 < y < 1, \\ &= 0 \qquad \text{elsewhere.} \end{aligned}$$

Remarks. Many authors use f_X and f_Y to denote the respective probability density functions of the random variables X and Y. Here we use f and g because we can avoid the use of subscripts. However, at other times, we will use subscripts as in f_X and f_Y or even f_1 and f_2, depending upon the circumstances. In a given example, we do not use the same symbol, without subscripts, to represent different functions. That is, in Example 2, we do not use $f(x)$ and $f(y)$ to represent different probability density functions.

In addition, while we ordinarily use the letter x in the description of the p.d.f. of X, this is not necessary at all because it is unimportant which letter we use in describing a function. For illustration, in Example 3, we could say that the random variable Y has the p.d.f. $g(w) = 1/2\sqrt{w}, 0 < w < 1$, zero elsewhere, and it would have exactly the same meaning as Y has the p.d.f. $g(y) = 1/2\sqrt{y}$, $0 < y < 1$, zero elsewhere.

These remarks apply to other functions too, such as distribution functions. In Example 3, we could have written the distribution function of Y, where $0 \leq w < 1$, as

$$F_Y(w) = \Pr(Y \leq w) = \sqrt{w}.$$

EXERCISES

1.71. Given the distribution function

$$F(x)=0, \qquad x<-1,$$
$$=\frac{x+2}{4}, \qquad -1\le x<1,$$
$$=1, \qquad 1\le x.$$

Sketch the graph of $F(x)$ and then compute: (a) $\Pr(-\frac{1}{2}<X\le\frac{1}{2})$; (b) $\Pr(X=0)$; (c) $\Pr(X=1)$; (d) $\Pr(2<X\le 3)$.

1.72. Let $f(x)=1$, $0<x<1$, zero elsewhere, be the p.d.f. of X. Find the distribution function and the p.d.f. of $Y=\sqrt{X}$.

Hint: $\Pr(Y\le y)=\Pr(\sqrt{X}\le y)=\Pr(X\le y^2)$, $0<y<1$.

1.73. Let $f(x)=x/6$, $x=1, 2, 3$, zero elsewhere, be the p.d.f. of X. Find the distribution function and the p.d.f. of $Y=X^2$.

Hint: Note that X is a random variable of the discrete type.

1.74. Let $f(x)=(4-x)/16$, $-2<x<2$, zero elsewhere, be the p.d.f. of X.

(a) Sketch the distribution function and the p.d.f. of X on the same set of axes.

(b) If $Y=|X|$, compute $\Pr(Y\le 1)$.

(c) If $Z=X^2$, compute $\Pr(Z\le\frac{1}{4})$.

1.75. Let X have the p.d.f. $f(x)=2x$, $0<x<1$, zero elsewhere. Find the distribution function and p.d.f. of $Y=X^2$.

1.76. Let X have the p.d.f. $f(x)=4x^3$, $0<x<1$, zero elsewhere. Find the distribution function and p.d.f. of $Y=-2\ln X^4$.

1.77. Explain why, with $h>0$, the two limits, $\lim_{h\to 0}\Pr(b-h<X\le b)$ and $\lim_{h\to 0}F(b-h)$, exist.

Hint: Note that $\Pr(b-h<X\le b)$ is bounded below by zero and $F(b-h)$ is bounded above by both $F(b)$ and 1.

1.78. Let $F(x)$ be the distribution function of the random variable X. If m is a number such that $F(m)=\frac{1}{2}$, show that m is a median of the distribution.

1.79. Let $f(x)=\frac{1}{3}$, $-1<x<2$, zero elsewhere, be the p.d.f. of X. Find the distribution function and the p.d.f. of $Y=X^2$.

Hint: Consider $\Pr(X^2\le y)$ for two cases: $0\le y<1$ and $1\le y<4$.

1.8 Expectation of a Random Variable

Let X be a random variable having a p.d.f. $f(x)$ such that we have certain absolute convergence; namely, in the discrete case,

$$\sum_x |x| f(x) \qquad \text{converges to a finite limit,}$$

or, in the continuous case,

$$\int_{-\infty}^{\infty} |x| f(x)\, dx \qquad \text{converges to a finite limit.}$$

The *expectation of a random variable* is

$$E(X) = \sum_x x f(x), \qquad \text{in the discrete case,}$$

or

$$E(X) = \int_{-\infty}^{\infty} x f(x)\, dx, \qquad \text{in the continuous case.}$$

Sometimes the expectation $E(X)$ is called the *mathematical expectation of X* or the *expected value of X*.

Remark. The terminology of expectation or expected value has its origin in games of chance. This can be illustrated as follows: Four small similar chips, numbered 1, 1, 1, and 2, respectively, are placed in a bowl and are mixed. A player is blindfolded and is to draw a chip from the bowl. If she draws one of the three chips numbered 1, she will receive one dollar. If she draws the chip numbered 2, she will receive two dollars. It seems reasonable to assume that the player has a "$\frac{3}{4}$ claim" on the \$1 and a "$\frac{1}{4}$ claim" on the \$2. Her "total claim" is $(1)(\frac{3}{4}) + 2(\frac{1}{4}) = \frac{5}{4}$, that is, \$1.25. Thus the expectation of X is precisely the player's claim in this game.

Example 1. Let the random variable X of the discrete type have the p.d.f. given by the table

x	1	2	3	4
$f(x)$	$\frac{4}{10}$	$\frac{1}{10}$	$\frac{3}{10}$	$\frac{2}{10}$

Here $f(x) = 0$ if x is not equal to one of the first four positive integers. This illustrates the fact that there is no need to have a formula to describe a p.d.f. We have

$$E(X) = (1)(\tfrac{4}{10}) + 2(\tfrac{1}{10}) + 3(\tfrac{3}{10}) + 4(\tfrac{2}{10}) = \tfrac{23}{10} = 2.3.$$

Example 2. Let X have the p.d.f.

$$f(x) = 4x^3, \qquad 0 < x < 1,$$
$$= 0 \qquad \text{elsewhere.}$$

Then

$$E(X) = \int_0^1 x(4x^3)\,dx = \int_0^1 4x^4\,dx = \left[\frac{4x^5}{5}\right]_0^1 = \frac{4}{5}.$$

Let us consider a function of a random variable X with space $\mathscr{A}$. Call this function $Y = u(X)$. For convenience, let X be of the continuous type and $y = u(x)$ be a continuous increasing function of X with an inverse function $x = w(y)$, which, of course, is also increasing. So Y is a random variable and its distribution function is

$$G(y) = \Pr(Y \le y) = \Pr[u(X) \le y] = \Pr[X \le w(y)]$$
$$= \int_{-\infty}^{w(y)} f(x)\,dx,$$

where $f(x)$ is the p.d.f. of X. By one form of the fundamental theorem of calculus,

$$g(y) = G'(y) = f[w(y)]w'(y), \qquad y \in \mathscr{B},$$
$$= 0 \qquad \text{elsewhere,}$$

where

$$\mathscr{B} = \{y : y = u(x), \quad x \in \mathscr{A}\}.$$

By definition, given absolute convergence, the expected value of Y is

$$E(Y) = \int_{-\infty}^{\infty} yg(y)\,dy.$$

Since $y = u(x)$, we might ask how $E(Y)$ compares to the integral

$$I = \int_{-\infty}^{\infty} u(x)f(x)\,dx.$$

To answer this, change the variable of integration through $y = u(x)$ or, equivalently, $x = w(y)$. Since

$$\frac{dx}{dy} = w'(y) > 0,$$

we have

$$I = \int_{-\infty}^{\infty} yf[w(y)]w'(y)\,dy = \int_{-\infty}^{\infty} yg(y)\,dy.$$

That is, in this special case,

$$E(Y) = \int_{-\infty}^{\infty} yg(y)\,dy = \int_{-\infty}^{\infty} u(x)f(x)\,dx.$$

However, this is true more generally and it also makes no difference whether X is of the discrete or continuous type and $Y = u(X)$ need not be an increasing function of X (Exercise 1.87 illustrates this).

So if $Y = u(X)$ has an expectation, we can find it from

$$E[u(X)] = \int_{-\infty}^{\infty} u(x)f(x)\,dx, \tag{1}$$

in the continuous case, and

$$E[u(X)] = \sum_{x} u(x)f(x), \tag{2}$$

in the discrete case. Accordingly, we say that $E[u(X)]$ is the expectation (mathematical expectation or expected value) of $u(X)$.

Remark. If the mathematical expectation of Y exists, recall that the integral (or sum)

$$\int_{-\infty}^{\infty} |y|g(y)\,dy \qquad \left[\text{or} \quad \sum_{y} |y|g(y)\right]$$

exists. Hence the existence of $E[u(X)]$ implies that the corresponding integral (or sum) converges absolutely.

Next, we shall point out some fairly obvious but useful facts about expectations when they exist.

1. If k is a constant, then $E(k) = k$. This follows from expression (1) [or (2)] upon setting $u = k$ and recalling that an integral (or sum) of a constant times a function is the constant times the integral (or sum) of the function. Of course, the integral (or sum) of the function f is 1.
2. If k is a constant and v is a function, then $E(kv) = kE(v)$. This follows from expression (1) [or (2)] upon setting $u = kv$ and rewriting expression (1) [or (2)] as k times the integral (or sum) of the product vf.
3. If k_1 and k_2 are constants and v_1 and v_2 are functions, then $E(k_1v_1 + k_2v_2) = k_1E(v_1) + k_2E(v_2)$. This, too, follows from ex-

pression (1) [or (2)] upon setting $u = k_1v_1 + k_2v_2$ because the integral (or sum) of $(k_1v_1 + k_2v_2)f$ is equal to the integral (or sum) of k_1v_1f plus the integral (or sum) of k_2v_2f. Repeated application of this property shows that if $k_1, k_2, \ldots, k_m$ are constants and $v_1, v_2, \ldots, v_m$ are functions, then

$$E(k_1v_1 + k_2v_2 + \cdots + k_mv_m) = k_1E(v_1) + k_2E(v_2) + \cdots + k_mE(v_m).$$

This property of expectation leads us to characterize the symbol E as a linear operator.

Example 3. Let X have the p.d.f.

$$f(x) = 2(1 - x), \qquad 0 < x < 1,$$
$$= 0 \qquad \text{elsewhere.}$$

Then

$$E(X) = \int_{-\infty}^{\infty} xf(x)\,dx = \int_0^1 (x)2(1 - x)\,dx = \tfrac{1}{3},$$

$$E(X^2) = \int_{-\infty}^{\infty} x^2f(x)\,dx = \int_0^1 (x^2)2(1 - x)\,dx = \tfrac{1}{6},$$

and, of course,

$$E(6X + 3X^2) = 6(\tfrac{1}{3}) + 3(\tfrac{1}{6}) = \tfrac{5}{2}.$$

Example 4. Let X have the p.d.f.

$$f(x) = \frac{x}{6}, \qquad x = 1, 2, 3,$$
$$= 0 \qquad \text{elsewhere.}$$

Then

$$E(X^3) = \sum_x x^3f(x) = \sum_{x=1}^{3} x^3\frac{x}{6}$$

$$= \tfrac{1}{6} + \tfrac{16}{6} + \tfrac{81}{6} = \tfrac{98}{6}.$$

Example 5. Let us divide, at random, a horizontal line segment of length 5 into two parts. If X is the length of the left-hand part, it is reasonable to assume that X has the p.d.f.

$$f(x) = \tfrac{1}{5}, \qquad 0 < x < 5,$$
$$= 0 \qquad \text{elsewhere.}$$

The expected value of the length X is $E(X) = \frac{5}{2}$ and the expected value of the length $5 - X$ is $E(5 - X) = \frac{5}{2}$. But the expected value of the product of the two lengths is equal to

$$E[X(5 - X)] = \int_0^5 x(5 - x)(\tfrac{1}{5})\, dx = \tfrac{25}{6} \neq (\tfrac{5}{2})^2.$$

That is, in general, the expected value of a product is not equal to the product of the expected values.

Example 6. A bowl contains five chips, which cannot be distinguished by a sense of touch alone. Three of the chips are marked \$1 each and the remaining two are marked \$4 each. A player is blindfolded and draws, at random and without replacement, two chips from the bowl. The player is paid an amount equal to the sum of the values of the two chips that he draws and the game is over. If it costs \$4.75 to play this game, would we care to participate for any protracted period of time? Because we are unable to distinguish the chips by sense of touch, we assume that each of the 10 pairs that can be drawn has the same probability of being drawn. Let the random variable X be the number of chips, of the two to be chosen, that are marked \$1. Then, under our assumption, X has the hypergeometric p.d.f.

$$f(x) = \frac{\binom{3}{x}\binom{2}{2-x}}{\binom{5}{2}}, \qquad x = 0, 1, 2,$$

$$= 0 \qquad \text{elsewhere.}$$

If $X = x$, the player receives $u(x) = x + 4(2 - x) = 8 - 3x$ dollars. Hence his mathematical expectation is equal to

$$E[8 - 3X] = \sum_{x=0}^{2} (8 - 3x)f(x) = \tfrac{44}{10},$$

or \$4.40.

EXERCISES

1.80. Let X have the p.d.f. $f(x) = (x + 2)/18$, $-2 < x < 4$, zero elsewhere. Find $E(X)$, $E[(X + 2)^3]$, and $E[6X - 2(X + 2)^3]$.

1.81. Suppose that $f(x) = \frac{1}{5}$, $x = 1, 2, 3, 4, 5$, zero elsewhere, is the p.d.f. of the discrete type of random variable X. Compute $E(X)$ and $E(X^2)$. Use these two results to find $E[(X + 2)^2]$ by writing $(X + 2)^2 = X^2 + 4X + 4$.

1.82. Let X be a number selected at random from a set of numbers $\{51, 52, 53, \ldots, 100\}$. Approximate $E(1/X)$.

Hint: Find reasonable upper and lower bounds by finding integrals bounding $E(1/X)$.

1.83. Let the p.d.f. $f(x)$ be positive at $x = -1, 0, 1$ and zero elsewhere.
(a) If $f(0) = \frac{1}{4}$, find $E(X^2)$.
(b) If $f(0) = \frac{1}{4}$ and if $E(X) = \frac{1}{4}$, determine $f(-1)$ and $f(1)$.

1.84. Let X have the p.d.f. $f(x) = 3x^2$, $0 < x < 1$, zero elsewhere. Consider a random rectangle whose sides are X and $(1 - X)$. Determine the expected value of the area of the rectangle.

1.85. A bowl contains 10 chips, of which 8 are marked $2 each and 2 are marked $5 each. Let a person choose, at random and without replacement, 3 chips from this bowl. If the person is to receive the sum of the resulting amounts, find his expectation.

1.86. Let X be a random variable of the continuous type that has p.d.f. $f(x)$. If m is the unique median of the distribution of X and b is a real constant, show that

$$E(|X - b|) = E(|X - m|) + 2\int_m^b (b - x)f(x)\,dx,$$

provided that the expectations exist. For what value of b is $E(|X - b|)$ a minimum?

1.87. Let $f(x) = 2x$, $0 < x < 1$, zero elsewhere, be the p.d.f. of X.
(a) Compute $E(1/X)$.
(b) Find the distribution function and the p.d.f. of $Y = 1/X$.
(c) Compute $E(Y)$ and compare this result with the answer obtained in part (a).
Hint: Here $\mathscr{A} = \{x : 0 < x < 1\}$, find $\mathscr{B}$.

1.88. Two distinct integers are chosen at random and without replacement from the first six positive integers. Compute the expected value of the absolute value of the difference of these two numbers.

1.9 Some Special Expectations

Certain expectations, if they exist, have special names and symbols to represent them. First, let X be a random variable of the discrete type having a p.d.f. $f(x)$. Then

$$E(X) = \sum_x x f(x).$$

If the discrete points of the space of positive probability density are $a_1, a_2, a_3, \ldots$, then

$$E(X) = a_1 f(a_1) + a_2 f(a_2) + a_3 f(a_3) + \cdots.$$

This sum of products is seen to be a "weighted average" of the values $a_1, a_2, a_3, \ldots$, the "weight" associated with each a_i being $f(a_i)$. This suggests that we call $E(X)$ the arithmetic mean of the values of X, or, more simply, the *mean value* of X (or the mean value of the distribution).

The mean value μ of a random variable X is defined, when it exists, to be $\mu = E(X)$, where X is a random variable of the discrete or of the continuous type.

Another special expectation is obtained by taking $u(X) = (X - \mu)^2$. If, initially, X is a random variable of the discrete type having a p.d.f. $f(x)$, then

$$\begin{aligned} E[(X - \mu)^2] &= \sum_x (x - \mu)^2 f(x) \\ &= (a_1 - \mu)^2 f(a_1) + (a_2 - \mu)^2 f(a_2) + \cdots, \end{aligned}$$

if $a_1, a_2, \ldots$ are the discrete points of the space of positive probability density. This sum of products may be interpreted as a "weighted average" of the squares of the deviations of the numbers $a_1, a_2, \ldots$ from the mean value μ of those numbers where the "weight" associated with each $(a_i - \mu)^2$ is $f(a_i)$. This mean value of the square of the deviation of X from its mean value μ is called the *variance* of X (or the variance of the distribution).

The variance of X will be denoted by σ^2, and we define σ^2, if it exists, by $\sigma^2 = E[(X - \mu)^2]$, whether X is a discrete or a continuous type of random variable. Sometimes the variance of X is written var (X).

It is worthwhile to observe that var (X) equals

$$\sigma^2 = E[(X - \mu)^2] = E(X^2 - 2\mu X + \mu^2);$$

and since E is a linear operator,

$$\begin{aligned} \sigma^2 &= E(X^2) - 2\mu E(X) + \mu^2 \\ &= E(X^2) - 2\mu^2 + \mu^2 \\ &= E(X^2) - \mu^2. \end{aligned}$$

This frequency affords an easier way of computing the variance of X.

It is customary to call σ (the positive square root of the variance) the *standard deviation* of X (or the standard deviation of the distribution). The number σ is sometimes interpreted as a measure of the dispersion of the points of the space relative to the mean value μ. We note that if the space contains only one point x for which $f(x) > 0$, then $\sigma = 0$.

Remark. Let the random variable X of the continuous type have the p.d.f. $f(x) = 1/2a$, $-a < x < a$, zero elsewhere, so that $\sigma = a/\sqrt{3}$ is the standard deviation of the distribution of X. Next, let the random variable Y of the continuous type have the p.d.f. $g(y) = 1/4a$, $-2a < y < 2a$, zero elsewhere, so that $\sigma = 2a/\sqrt{3}$ is the standard deviation of the distribution of Y. Here the standard deviation of Y is greater than that of X; this reflects the fact that the probability for Y is more widely distributed (relative to the mean zero) than is the probability for X.

We next define a third special mathematical expectation, called the *moment-generating function* (abbreviated m.g.f.) of a random variable X. Suppose that there is a positive number h such that for $-h < t < h$ the mathematical expectation $E(e^{tX})$ exists. Thus

$$E(e^{tX}) = \int_{-\infty}^{\infty} e^{tx} f(x)\, dx,$$

if X is a continuous type of random variable, or

$$E(e^{tX}) = \sum_x e^{tx} f(x),$$

if X is a discrete type of random variable. This expectation is called the moment-generating function (m.g.f.) of X (or of the distribution) and is denoted by $M(t)$. That is,

$$M(t) = E(e^{tX}).$$

It is evident that if we set $t = 0$, we have $M(0) = 1$. As will be seen by example, not every distribution has an m.g.f., but it is difficult to overemphasize the importance of an m.g.f., when it does exist. This importance stems from the fact that the m.g.f. is unique and completely determines the distribution of the random variable; thus, if two random variables have the same m.g.f., they have the same distribution. This property of an m.g.f. will be very useful in subsequent chapters. Proof of the uniqueness of the m.g.f. is based on the theory of transforms in analysis, and therefore we merely assert this uniqueness.

Although the fact that an m.g.f. (when it exists) completely determines the distribution of one random variable will not be proved, it does seem desirable to try to make the assertion plausible. This can be done if the random variable is of the discrete type. For example, let it be given that

$$M(t) = \tfrac{1}{10}e^t + \tfrac{2}{10}e^{2t} + \tfrac{3}{10}e^{3t} + \tfrac{4}{10}e^{4t}$$

is, for all real values of t, the m.g.f. of a random variable X of the discrete type. If we let $f(x)$ be the p.d.f. of X and let $a, b, c, d, \ldots$ be the discrete points in the space of X at which $f(x) > 0$, then

$$M(t) = \sum_x e^{tx} f(x),$$

or

$$\tfrac{1}{10}e^t + \tfrac{2}{10}e^{2t} + \tfrac{3}{10}e^{3t} + \tfrac{4}{10}e^{4t} = f(a)e^{at} + f(b)e^{bt} + \cdots.$$

Because this is an identity for all real values of t, it seems that the right-hand member should consist of but four terms and that each of the four should equal, respectively, one of those in the left-hand member; hence we may take $a = 1, f(a) = \tfrac{1}{10}$; $b = 2, f(b) = \tfrac{2}{10}$; $c = 3$, $f(c) = \tfrac{3}{10}$; $d = 4, f(d) = \tfrac{4}{10}$. Or, more simply, the p.d.f. of X is

$$\begin{aligned} f(x) &= \frac{x}{10}, \qquad x = 1, 2, 3, 4, \\ &= 0 \qquad \text{elsewhere.} \end{aligned}$$

On the other hand, let X be a random variable of the continuous type and let it be given that

$$M(t) = \frac{1}{1-t}, \qquad t < 1,$$

is the m.g.f. of X. That is, we are given

$$\frac{1}{1-t} = \int_{-\infty}^{\infty} e^{tx} f(x)\, dx, \qquad t < 1.$$

It is not at all obvious how $f(x)$ is found. However, it is easy to see that a distribution with p.d.f.

$$\begin{aligned} f(x) &= e^{-x}, \qquad 0 < x < \infty, \\ &= 0 \qquad \text{elsewhere} \end{aligned}$$

has the m.g.f. $M(t) = (1-t)^{-1}$, $t < 1$. Thus the random variable X

has a distribution with this p.d.f. in accordance with the assertion of the uniqueness of the m.g.f.

Since a distribution that has an m.g.f. $M(t)$ is completely determined by $M(t)$, it would not be surprising if we could obtain some properties of the distribution directly from $M(t)$. For example, the existence of $M(t)$ for $-h < t < h$ implies that derivatives of all order exist at $t = 0$. Thus, using a theorem in analysis that allows us to change the order of differentiation and integration, we have

$$\frac{dM(t)}{dt} = M'(t) = \int_{-\infty}^{\infty} xe^{tx}f(x)\,dx,$$

if X is of the continuous type, or

$$\frac{dM(t)}{dt} = M'(t) = \sum_x xe^{tx}f(x),$$

if X is of the discrete type. Upon setting $t = 0$, we have in either case

$$M'(0) = E(X) = \mu.$$

The second derivative of $M(t)$ is

$$M''(t) = \int_{-\infty}^{\infty} x^2e^{tx}f(x)\,dx \qquad \text{or} \qquad \sum_x x^2e^{tx}f(x),$$

so that $M''(0) = E(X^2)$. Accordingly, the var (X) equals

$$\sigma^2 = E(X^2) - \mu^2 = M''(0) - [M'(0)]^2.$$

For example, if $M(t) = (1 - t)^{-1}$, $t < 1$, as in the illustration above, then

$$M'(t) = (1 - t)^{-2} \qquad \text{and} \qquad M''(t) = 2(1 - t)^{-3}.$$

Hence

$$\mu = M'(0) = 1$$

and

$$\sigma^2 = M''(0) - \mu^2 = 2 - 1 = 1.$$

Of course, we could have computed μ and σ^2 from the p.d.f. by

$$\mu = \int_{-\infty}^{\infty} xf(x)\,dx \qquad \text{and} \qquad \sigma^2 = \int_{-\infty}^{\infty} x^2f(x)\,dx - \mu^2,$$

respectively. Sometimes one way is easier than the other.

In general, if m is a positive integer and if $M^{(m)}(t)$ means the mth derivative of $M(t)$, we have, by repeated differentiation with respect to t,

$$M^{(m)}(0) = E(X^m).$$

Now

$$E(X^m) = \int_{-\infty}^{\infty} x^m f(x)\,dx \qquad \text{or} \qquad \sum_x x^m f(x),$$

and integrals (or sums) of this sort are, in mechanics, called *moments*. Since $M(t)$ generates the values of $E(X^m)$, $m = 1, 2, 3, \ldots$, it is called the moment-generating function (m.g.f.). In fact, we shall sometimes call $E(X^m)$ the mth moment of the distribution, or the mth moment of X.

Example 1. Let X have the p.d.f.

$$\begin{aligned} f(x) &= \tfrac{1}{2}(x + 1), \qquad -1 < x < 1, \\ &= 0 \qquad \text{elsewhere.} \end{aligned}$$

Then the mean value of X is

$$\mu = \int_{-\infty}^{\infty} x f(x)\,dx = \int_{-1}^{1} x\,\frac{x+1}{2}\,dx = \frac{1}{3}$$

while the variance of X is

$$\sigma^2 = \int_{-\infty}^{\infty} x^2 f(x)\,dx - \mu^2 = \int_{-1}^{1} x^2\,\frac{x+1}{2}\,dx - (\tfrac{1}{3})^2 = \frac{2}{9}.$$

Example 2. If X has the p.d.f.

$$\begin{aligned} f(x) &= \frac{1}{x^2}, \qquad 1 < x < \infty, \\ &= 0 \qquad \text{elsewhere,} \end{aligned}$$

then the mean value of X does not exist, since

$$\begin{aligned} \int_{1}^{\infty} |x|\,\frac{1}{x^2}\,dx &= \lim_{b\to\infty} \int_{1}^{b} \frac{1}{x}\,dx \\ &= \lim_{b\to\infty} (\ln b - \ln 1) \end{aligned}$$

does not exist.

Example 3. It is known that the series

$$\frac{1}{1^2} + \frac{1}{2^2} + \frac{1}{3^2} + \cdots$$

converges to $\pi^2/6$. Then

$$f(x) = \frac{6}{\pi^2 x^2}, \qquad x = 1, 2, 3, \ldots,$$
$$= 0 \qquad \text{elsewhere},$$

is the p.d.f. of a discrete type of random variable X. The m.g.f. of this distribution, if it exists, is given by

$$M(t) = E(e^{tX}) = \sum_x e^{tx} f(x)$$
$$= \sum_{x=1}^{\infty} \frac{6e^{tx}}{\pi^2 x^2}.$$

The ratio test may be used to show that this series diverges if $t > 0$. Thus there does not exist a positive number h such that $M(t)$ exists for $-h < t < h$. Accordingly, the distribution having the p.d.f. $f(x)$ of this example does not have an m.g.f.

Example 4. Let X have the m.g.f. $M(t) = e^{t^2/2}$, $-\infty < t < \infty$. We can differentiate $M(t)$ any number of times to find the moments of X. However, it is instructive to consider this alternative method. The function $M(t)$ is represented by the following MacLaurin's series.

$$e^{t^2/2} = 1 + \frac{1}{1!}\left(\frac{t^2}{2}\right) + \frac{1}{2!}\left(\frac{t^2}{2}\right)^2 + \cdots + \frac{1}{k!}\left(\frac{t^2}{2}\right)^k + \cdots$$

$$= 1 + \frac{1}{2!}t^2 + \frac{(3)(1)}{4!}t^4 + \cdots + \frac{(2k-1)\cdots(3)(1)}{(2k)!}t^{2k} + \cdots.$$

In general, the MacLaurin's series for $M(t)$ is

$$M(t) = M(0) + \frac{M'(0)}{1!}t + \frac{M''(0)}{2!}t^2 + \cdots + \frac{M^{(m)}(0)}{m!}t^m + \cdots$$

$$= 1 + \frac{E(X)}{1!}t + \frac{E(X^2)}{2!}t^2 + \cdots + \frac{E(X^m)}{m!}t^m + \cdots.$$

Thus the coefficient of $(t^m/m!)$ in the MacLaurin's series representation of $M(t)$ is $E(X^m)$. So, for our particular $M(t)$, we have

$$E(X^{2k}) = (2k-1)(2k-3)\cdots(3)(1) = \frac{(2k)!}{2^k k!},$$

$k = 1, 2, 3, \ldots$, and $E(X^{2k-1}) = 0$, $k = 1, 2, 3, \ldots$.

Remarks. In a more advanced course, we would not work with the m.g.f. because so many distributions do not have moment-generating functions. Instead, we would let i denote the imaginary unit, t an arbitrary real, and we would define $\varphi(t) = E(e^{itX})$. This expectation exists for *every* distribution and it is called the *characteristic function* of the distribution. To see why $\varphi(t)$ exists for all real t, we note, in the continuous case, that its absolute value

$$|\varphi(t)| = \left| \int_{-\infty}^{\infty} e^{itx} f(x)\, dx \right| \le \int_{-\infty}^{\infty} |e^{itx} f(x)|\, dx.$$

However, $|f(x)| = f(x)$ since $f(x)$ is nonnegative and

$$|e^{itx}| = |\cos tx + i \sin tx| = \sqrt{\cos^2 tx + \sin^2 tx} = 1.$$

Thus

$$|\varphi(t)| \le \int_{-\infty}^{\infty} f(x)\, dx = 1.$$

Accordingly, the integral for $\varphi(t)$ exists for all real values of t. In the discrete case, a summation would replace the integral.

Every distribution has a unique characteristic function; and to each characteristic function there corresponds a unique distribution of probability. If X has a distribution with characteristic function $\varphi(t)$, then, for instance, if $E(X)$ and $E(X^2)$ exist, they are given, respectively, by $iE(X) = \varphi'(0)$ and $i^2E(X^2) = \varphi''(0)$. Readers who are familiar with complex-valued functions may write $\varphi(t) = M(it)$ and, throughout this book, may prove certain theorems in complete generality.

Those who have studied Laplace and Fourier transforms will note a similarity between these transforms and $M(t)$ and $\varphi(t)$; it is the uniqueness of these transforms that allows us to assert the uniqueness of each of the moment-generating and characteristic functions.

EXERCISES

1.89. Find the mean and variance, if they exist, of each of the following distributions.

(a) $f(x) = \dfrac{3!}{x!\,(3-x)!}\left(\dfrac{1}{2}\right)^3$, $x = 0, 1, 2, 3$, zero elsewhere.

(b) $f(x) = 6x(1 - x)$, $0 < x < 1$, zero elsewhere.

(c) $f(x) = 2/x^3$, $1 < x < \infty$, zero elsewhere.

1.90. Let $f(x) = (\frac{1}{2})^x$, $x = 1, 2, 3, \ldots$, zero elsewhere, be the p.d.f. of the random variable X. Find the m.g.f., the mean, and the variance of X.

1.91. For each of the following probability density functions, compute $\Pr(\mu - 2\sigma < X < \mu + 2\sigma)$.

(a) $f(x) = 6x(1 - x)$, $0 < x < 1$, zero elsewhere.
(b) $f(x) = (\frac{1}{2})^x$, $x = 1, 2, 3, \ldots$, zero elsewhere.

1.92. If the variance of the random variable X exists, show that

$$E(X^2) \geq [E(X)]^2.$$

1.93. Let a random variable X of the continuous type have a p.d.f. $f(x)$ whose graph is symmetric with respect to $x = c$. If the mean value of X exists, show that $E(X) = c$.

Hint: Show that $E(X - c)$ equals zero by writing $E(X - c)$ as the sum of two integrals: one from $-\infty$ to c and the other from c to ∞. In the first, let $y = c - x$; and, in the second, $z = x - c$. Finally, use the symmetry condition $f(c - y) = f(c + y)$ in the first.

1.94. Let the random variable X have mean μ, standard deviation σ, and m.g.f. $M(t)$, $-h < t < h$. Show that

$$E\left(\frac{X - \mu}{\sigma}\right) = 0, \qquad E\left[\left(\frac{X - \mu}{\sigma}\right)^2\right] = 1,$$

and

$$E\left\{\exp\left[t\left(\frac{X - \mu}{\sigma}\right)\right]\right\} = e^{-\mu t/\sigma} M\left(\frac{t}{\sigma}\right), \qquad -h\sigma < t < h\sigma.$$

1.95. Show that the m.g.f. of the random variable X having the p.d.f. $f(x) = \frac{1}{3}$, $-1 < x < 2$, zero elsewhere, is

$$M(t) = \frac{e^{2t} - e^{-t}}{3t}, \qquad t \neq 0,$$

$$= 1, \qquad t = 0.$$

1.96. Let X be a random variable such that $E[(X - b)^2]$ exists for all real b. Show that $E[(X - b)^2]$ is a minimum when $b = E(X)$.

1.97. Let X denote a random variable for which $E[(X - a)^2]$ exists. Give an example of a distribution of a discrete type such that this expectation is zero. Such a distribution is called a *degenerate distribution*.

1.98. Let X be a random variable such that $K(t) = E(t^X)$ exists for all real values of t in a certain open interval that includes the point $t = 1$. Show that $K^{(m)}(1)$ is equal to the mth *factorial moment* $E[X(X - 1) \cdots (X - m + 1)]$.

1.99. Let X be a random variable. If m is a positive integer, the expectation $E[(X - b)^m]$, if it exists, is called the mth moment of the distribution about the point b. Let the first, second, and third moments of the distribution about the point 7 be 3, 11, and 15, respectively. Determine the mean μ of X, and then find the first, second, and third moments of the distribution about the point μ.

1.100. Let X be a random variable such that $R(t) = E(e^{t(X-b)})$ exists for $-h < t < h$. If m is a positive integer, show that $R^{(m)}(0)$ is equal to the mth moment of the distribution about the point b.

1.101. Let X be a random variable with mean μ and variance σ^2 such that the third moment $E[(X-\mu)^3]$ about the vertical line through μ exists. The value of the ratio $E[(X-\mu)^3]/\sigma^3$ is often used as a measure of *skewness*. Graph each of the following probability density functions and show that this measure is negative, zero, and positive for these respective distributions (which are said to be skewed to the left, not skewed, and skewed to the right, respectively).
(a) $f(x) = (x+1)/2$, $-1 < x < 1$, zero elsewhere.
(b) $f(x) = \frac{1}{2}$, $-1 < x < 1$, zero elsewhere.
(c) $f(x) = (1-x)/2$, $-1 < x < 1$, zero elsewhere.

1.102. Let X be a random variable with mean μ and variance σ^2 such that the fourth moment $E[(X-\mu)^4]$ about the vertical line through μ exists. The value of the ratio $E[(X-\mu)^4]/\sigma^4$ is often used as a measure of *kurtosis*. Graph each of the following probability density functions and show that this measure is smaller for the first distribution.
(a) $f(x) = \frac{1}{2}$, $-1 < x < 1$, zero elsewhere.
(b) $f(x) = 3(1-x^2)/4$, $-1 < x < 1$, zero elsewhere.

1.103. Let the random variable X have p.d.f.

$$\begin{aligned} f(x) &= p, & x &= -1, 1, \\ &= 1-2p, & x &= 0, \\ &= 0 & &\text{elsewhere,} \end{aligned}$$

where $0 < p < \frac{1}{2}$. Find the measure of kurtosis as a function of p. Determine its value when $p = \frac{1}{3}$, $p = \frac{1}{5}$, $p = \frac{1}{10}$, and $p = \frac{1}{100}$. Note that the kurtosis increases as p decreases.

1.104. Let $\psi(t) = \ln M(t)$, where $M(t)$ is the m.g.f. of a distribution. Prove that $\psi'(0) = \mu$ and $\psi''(0) = \sigma^2$.

1.105. Find the mean and the variance of the distribution that has the distribution function

$$\begin{aligned} F(x) &= 0, & x &< 0, \\ &= \frac{x}{8}, & 0 &\le x < 2, \\ &= \frac{x^2}{16}, & 2 &\le x < 4, \\ &= 1, & 4 &\le x. \end{aligned}$$

1.106. Find the moments of the distribution that has m.g.f. $M(t) = (1 - t)^{-3}$, $t < 1$.

Hint: Find the MacLaurin's series for $M(t)$.

1.107. Let X be a random variable of the continuous type with p.d.f. $f(x)$, which is positive provided $0 < x < b < \infty$, and is equal to zero elsewhere. Show that

$$E(X) = \int_0^b [1 - F(x)]\, dx,$$

where $F(x)$ is the distribution function of X.

1.108. Let X be a random variable of the discrete type with p.d.f. $f(x)$ that is positive on the nonnegative integers and is equal to zero elsewhere. Show that

$$E(X) = \sum_{x=0}^{\infty} [1 - F(x)],$$

where $F(x)$ is the distribution function of X.

1.109. Let X have the p.d.f. $f(x) = 1/k$, $x = 1, 2, \ldots, k$, zero elsewhere. Show that the m.g.f. is

$$M(t) = \frac{e^t(1 - e^{kt})}{k(1 - e^t)}, \qquad t \neq 0,$$
$$= 1, \qquad t = 0.$$

1.110. Let X have the distribution function $F(x)$ that is a mixture of the continuous and discrete types, namely

$$F(x) = 0, \qquad x < 0,$$
$$= \frac{x + 1}{4}, \qquad 0 \le x < 1,$$
$$= 1, \qquad 1 \le x.$$

Find $\mu = E(X)$ and $\sigma^2 = \text{var}(X)$.

Hint: Determine that part of the p.d.f. associated with each of the discrete and continuous types, and then sum for the discrete part and integrate for the continuous part.

1.111. Consider k continuous-type distributions with the following characteristics: p.d.f. $f_i(x)$, mean μ_i, and variance σ_i^2, $i = 1, 2, \ldots, k$. If $c_i \ge 0$, $i = 1, 2, \ldots, k$, and $c_1 + c_2 + \cdots + c_k = 1$, show that the mean and the variance of the distribution having p.d.f. $c_1 f_1(x) + \cdots + c_k f_k(x)$ are $\mu = \sum_{i=1}^{k} c_i \mu_i$ and $\sigma^2 = \sum_{i=1}^{k} c_i[\sigma_i^2 + (\mu_i - \mu)^2]$, respectively.

1.10 Chebyshev's Inequality

In this section we prove a theorem that enables us to find upper (or lower) bounds for certain probabilities. These bounds, however, are not necessarily close to the exact probabilities and, accordingly, we ordinarily do not use the theorem to approximate a probability. The principal uses of the theorem and a special case of it are in theoretical discussions in other chapters.

Theorem 6. *Let $u(X)$ be a nonnegative function of the random variable X. If $E[u(X)]$ exists, then, for every positive constant c,*

$$\Pr[u(X) \geq c] \leq \frac{E[u(X)]}{c}.$$

Proof. The proof is given when the random variable X is of the continuous type; but the proof can be adapted to the discrete case if we replace integrals by sums. Let $A = \{x : u(x) \geq c\}$ and let $f(x)$ denote the p.d.f. of X. Then

$$E[u(X)] = \int_{-\infty}^{\infty} u(x)f(x)\,dx = \int_A u(x)f(x)\,dx + \int_{A^*} u(x)f(x)\,dx.$$

Since each of the integrals in the extreme right-hand member of the preceding equation is nonnegative, the left-hand member is greater than or equal to either of them. In particular,

$$E[u(X)] \geq \int_A u(x)f(x)\,dx.$$

However, if $x \in A$, then $u(x) \geq c$; accordingly, the right-hand member of the preceding inequality is not increased if we replace $u(x)$ by c. Thus

$$E[u(X)] \geq c\int_A f(x)\,dx.$$

Since

$$\int_A f(x)\,dx = \Pr(X \in A) = \Pr[u(X) \geq c],$$

it follows that

$$E[u(X)] \geq c\Pr[u(X) \geq c],$$

which is the desired result.

The preceding theorem is a generalization of an inequality that is often called *Chebyshev's inequality*. This inequality will now be established.

Theorem 7. Chebyshev's Inequality. *Let the random variable X have a distribution of probability about which we assume only that there is a finite variance σ^2. This, of course, implies that there is a mean μ. Then for every $k > 0$,*

$$\Pr(|X - \mu| \geq k\sigma) \leq \frac{1}{k^2},$$

or, equivalently,

$$\Pr(|X - \mu| < k\sigma) \geq 1 - \frac{1}{k^2}.$$

Proof. In Theorem 6 take $u(X) = (X - \mu)^2$ and $c = k^2\sigma^2$. Then we have

$$\Pr[(X - \mu)^2 \geq k^2\sigma^2] \leq \frac{E[(X - \mu)^2]}{k^2\sigma^2}.$$

Since the numerator of the right-hand member of the preceding inequality is σ^2, the inequality may be written

$$\Pr(|X - \mu| \geq k\sigma) \leq \frac{1}{k^2},$$

which is the desired result. Naturally, we would take the positive number k to be greater than 1 to have an inequality of interest.

It is seen that the number $1/k^2$ is an upper bound for the probability $\Pr(|X - \mu| \geq k\sigma)$. In the following example this upper bound and the exact value of the probability are compared in special instances.

Example 1. Let X have the p.d.f.

$$\begin{aligned} f(x) &= \frac{1}{2\sqrt{3}}, \qquad -\sqrt{3} < x < \sqrt{3}, \\ &= 0 \qquad \text{elsewhere.} \end{aligned}$$

Here $\mu = 0$ and $\sigma^2 = 1$. If $k = \frac{3}{2}$, we have the exact probability

$$\Pr(|X - \mu| \geq k\sigma) = \Pr\left(|X| \geq \frac{3}{2}\right) = 1 - \int_{-3/2}^{3/2} \frac{1}{2\sqrt{3}}\, dx = 1 - \frac{\sqrt{3}}{2}.$$

By Chebyshev's inequality, the preceding probability has the upper bound $1/k^2 = \frac{4}{9}$. Since $1 - \sqrt{3}/2 = 0.134$, approximately, the exact probability in this case is considerably less than the upper bound $\frac{4}{9}$. If we take $k = 2$, we have the exact probability $\Pr(|X - \mu| \geq 2\sigma) = \Pr(|X| \geq 2) = 0$. This again is considerably less than the upper bound $1/k^2 = \frac{1}{4}$ provided by Chebyshev's inequality.

In each of the instances in the preceding example, the probability $\Pr(|X - \mu| \geq k\sigma)$ and its upper bound $1/k^2$ differ considerably. This suggests that this inequality might be made sharper. However, if we want an inequality that holds for every $k > 0$ and holds for all random variables having finite variance, such an improvement is impossible, as is shown by the following example.

Example 2. Let the random variable X of the discrete type have probabilities $\frac{1}{8}, \frac{6}{8}, \frac{1}{8}$ at the points $x = -1, 0, 1$, respectively. Here $\mu = 0$ and $\sigma^2 = \frac{1}{4}$. If $k = 2$, then $1/k^2 = \frac{1}{4}$ and $\Pr(|X - \mu| \geq k\sigma) = \Pr(|X| \geq 1) = \frac{1}{4}$. That is, the probability $\Pr(|X - \mu| \geq k\sigma)$ here attains the upper bound $1/k^2 = \frac{1}{4}$. Hence the inequality cannot be improved without further assumptions about the distribution of X.

EXERCISES

1.112. Let X be a random variable with mean μ and let $E[(X - \mu)^{2k}]$ exist. Show, with $d > 0$, that $\Pr(|X - \mu| \geq d) \leq E[(X - \mu)^{2k}]/d^{2k}$. This is essentially Chebyshev's inequality when $k = 1$. The fact that this holds for all $k = 1, 2, 3, \ldots$, when those $(2k)$th moments exist, usually provides a much smaller upper bound for $\Pr(|X - \mu| \geq d)$ than does Chebyshev's result.

1.113. Let X be a random variable such that $\Pr(X \leq 0) = 0$ and let $\mu = E(X)$ exist. Show that $\Pr(X \geq 2\mu) \leq \frac{1}{2}$.

1.114. If X is a random variable such that $E(X) = 3$ and $E(X^2) = 13$, use Chebyshev's inequality to determine a lower bound for the probability $\Pr(-2 < X < 8)$.

1.115. Let X be a random variable with m.g.f. $M(t)$, $-h < t < h$. Prove that

$$\Pr(X \geq a) \leq e^{-at}M(t), \qquad 0 < t < h,$$

and that

$$\Pr(X \leq a) \leq e^{-at}M(t), \qquad -h < t < 0.$$

Hint: Let $u(x) = e^{tx}$ and $c = e^{ta}$ in Theorem 6. *Note.* These results imply that $\Pr(X \geq a)$ and $\Pr(X \leq a)$ are less than the respective greatest lower bounds for $e^{-at}M(t)$ when $0 < t < h$ and when $-h < t < 0$.

1.116. The m.g.f. of X exists for all real values of t and is given by

$$M(t) = \frac{e^t - e^{-t}}{2t}, \qquad t \neq 0, \quad M(0) = 1.$$

Use the results of the preceding exercise to show that $\Pr(X \geq 1) = 0$ and $\Pr(X \leq -1) = 0$. Note that here h is infinite.

ADDITIONAL EXERCISES

1.117. Players A and B play a sequence of independent games. Player A throws a die first and wins on a "six." If he fails, B throws and wins on a "five" or "six." If he fails, A throws again and wins on a "four," "five," or "six." And so on. Find the probability of each player winning the sequence.

1.118. Let X be the number of gallons of ice cream that is requested at a certain store on a hot summer day. Let us assume that the p.d.f. of X is $f(x) = 12x(1000 - x)^2/10^{12}$, $0 < x < 1000$, zero elsewhere. How many gallons of ice cream should the store have on hand each of these days, so that the probability of exhausting its supply on a particular day is 0.05?

1.119. Find the 25th percentile of the distribution having p.d.f. $f(x) = |x|/4$, $-2 < x < 2$, zero elsewhere.

1.120. Let A_1, A_2, A_3 be independent events with probabilities $\frac{1}{2}$, $\frac{1}{3}$, $\frac{1}{4}$, respectively. Compute $\Pr(A_1 \cup A_2 \cup A_3)$.

1.121. From a bowl containing 5 red, 3 white, and 7 blue chips, select 4 at random and without replacement. Compute the conditional probability of 1 red, 0 white, and 3 blue chips, given that there are at least 3 blue chips in this sample of 4 chips.

1.122. Let the three independent events A, B, and C be such that $P(A) = P(B) = P(C) = \frac{1}{4}$. Find $P[(A^* \cap B^*) \cup C]$.

1.123. Person A tosses a coin and then person B rolls a die. This is repeated independently until a head or one of the numbers 1, 2, 3, 4 appears, at which time the game is stopped. Person A wins with the head and B wins with one of the numbers 1, 2, 3, 4. Compute the probability that A wins the game.

1.124. Find the mean and variance of the random variable X having distribution function

$$\begin{aligned} F(x) &= 0, && x < 0, \\ &= \frac{x}{4}, && 0 \leq x < 1, \end{aligned}$$

$$= \frac{x^2}{4}, \qquad 1 \le x < 2,$$

$$= 1, \qquad 2 \le x.$$

1.125. Let X be a random variable having distribution function

$$\begin{aligned} F(x) &= 0, && x < 0, \\ &= 2x^2, && 0 \le x < \tfrac{1}{2}, \\ &= 1 - 2(1-x)^2, && \tfrac{1}{2} \le x < \tfrac{3}{4}, \\ &= 1, && \tfrac{3}{4} \le x. \end{aligned}$$

Find $\Pr(\frac{1}{4} < X < \frac{5}{8})$ and the variance of the distribution.
Hint: Note that there is a step in $F(x)$.

1.126. Bowl I contains 7 red and 3 white chips and bowl II has 4 red and 6 white chips. Two chips are selected at random and without replacement from I and transferred to II. Three chips are then selected at random and without replacement from II.
(a) What is the probability that all three are white?
(b) Given that three white chips are selected from II, what is the conditional probability that two white chips were transferred from I?

1.127. A bowl contains ten chips numbered 1, 2, . . . , 10, respectively. Five chips are drawn at random, one at a time, and without replacement. What is the probability that exactly two even-numbered chips are drawn and they occur on even-numbered draws?

1.128. Let $E(X^r) = \dfrac{1}{r+1}$, $r = 1, 2, 3, \ldots$. Find the series representation for the m.g.f. of X. Sum this series.

1.129. Let X have the p.d.f. $f(x) = 2x$, $0 < x < 1$, zero elsewhere. Compute the probability that X is at least $\frac{3}{4}$ given that X is at least $\frac{1}{2}$.

1.130. Divide a line segment into two parts by selecting a point at random. Find the probability that the larger segment is at least three times the shorter. Assume a uniform distribution.

1.131. Three chips are selected at random and without replacement from a bowl containing 5 white, 4 black, and 7 red chips. Find the probability that these three chips are alike in color.

1.132. Factories A, B, and C produce, respectively, 20, 30, and 50% of a certain company's output. The items produced in A, B, and C are 1, 2, and 3 percent defective, respectively. We observe one item from the company's output at random and find it defective. What is the conditional probability that the item was from A?

1.133. The probabilities that the independent events A, B, and C will occur are $\frac{3}{4}$, $\frac{1}{2}$, and $\frac{1}{4}$. What is the probability that at least one of the three events will happen?

1.134. A person bets 1 dollar to b dollars that he can draw two cards from an ordinary deck without replacement and that they will be of the same suit. Find b so that the bet will be fair.

1.135. A bowl contains 6 chips: 4 are red and 2 are white. Three chips are selected at random and without replacement; then a coin is tossed a number of independent times that is equal to the number of red chips in this sample of 3. For example, if we have 2 red and 1 white, the coin is tossed twice. Given that one head results, compute the conditional probability that the sample contains 1 red and 2 white.

CHAPTER 2

Multivariate Distributions

2.1 Distributions of Two Random Variables

We begin the discussion of two random variables with the following example. A coin is to be tossed three times and our interest is in the ordered number pair (number of H's on first two tosses, number of H's on all three tosses), where H and T represent, respectively, heads and tails. Thus the sample space is $\mathscr{C} = \{c : c = c_i, i = 1, 2, \ldots, 8\}$, where c_1 is TTT, c_2 is TTH, c_3 is THT, c_4 is HTT, c_5 is THH, c_6 is HTH, c_7 is HHT, and c_8 is HHH. Let X_1 and X_2 be two functions such that $X_1(c_1) = X_1(c_2) = 0$, $X_1(c_3) = X_1(c_4) = X_1(c_5) = X_1(c_6) = 1$, $X_1(c_7) = X_1(c_8) = 2$; and $X_2(c_1) = 0$, $X_2(c_2) = X_2(c_3) = X_2(c_4) = 1$, $X_2(c_5) = X_2(c_6) = X_2(c_7) = 2$, $X_2(c_8) = 3$. Thus X_1 and X_2 are real-valued functions defined on the sample space $\mathscr{C}$, which take us from that sample space to the space of ordered number pairs

$$\mathscr{A} = \{(0, 0), (0, 1), (1, 1), (1, 2), (2, 2), (2, 3)\}.$$

Thus X_1 and X_2 are two random variables defined on the space $\mathscr{C}$, and, in this example, the space of these random variables is the two-dimensional set $\mathscr{A}$ given immediately above. We now formulate the definition of the space of two random variables.

Definition 1. Given a random experiment with a sample space $\mathscr{C}$. Consider two random variables X_1 and X_2, which assign to each element c of $\mathscr{C}$ one and only one ordered pair of numbers $X_1(c) = x_1$, $X_2(c) = x_2$. The *space* of X_1 and X_2 is the set of ordered pairs $\mathscr{A} = \{(x_1, x_2) : x_1 = X_1(c), x_2 = X_2(c), c \in \mathscr{C}\}$.

Let $\mathscr{A}$ be the space associated with the two random variables X_1 and X_2 and let A be a subset of $\mathscr{A}$. As in the case of one random variable, we shall speak of the event A. We wish to define the probability of the event A, which we denote by $\Pr[(X_1, X_2) \in A]$. Take $C = \{c : c \in \mathscr{C}$ and $[X_1(c), X_2(c)] \in A\}$, where $\mathscr{C}$ is the sample space. We then define $\Pr[(X_1, X_2) \in A] = P(C)$, where P is the probability set function defined for subsets C of $\mathscr{C}$. Here again we could denote $\Pr[(X_1, X_2) \in A]$ by the probability set function $P_{X_1,X_2}(A)$; but, with our previous convention, we simply write

$$P(A) = \Pr[(X_1, X_2) \in A].$$

Again it is important to observe that this function is a probability set function defined for subsets A of the space $\mathscr{A}$.

Let us return to the example in our discussion of two random variables. Consider the subset A of $\mathscr{A}$, where $A = \{(1, 1), (1, 2)\}$. To compute $\Pr[(X_1, X_2) \in A] = P(A)$, we must include as elements of C all outcomes in $\mathscr{C}$ for which the random variables X_1 and X_2 take values (x_1, x_2) which are elements of A. Now $X_1(c_3) = 1$, $X_2(c_3) = 1$, $X_1(c_4) = 1$, and $X_2(c_4) = 1$. Also, $X_1(c_5) = 1$, $X_2(c_5) = 2$, $X_1(c_6) = 1$, and $X_2(c_6) = 2$. Thus $P(A) = \Pr[(X_1, X_2) \in A] = P(C)$, where $C = \{c_3, c_4, c_5, \text{or } c_6\}$. Suppose that our probability set function $P(C)$ assigns a probability of $\frac{1}{8}$ to each of the eight elements of $\mathscr{C}$. This assignment seems reasonable if $P(\text{T}) = P(\text{H}) = \frac{1}{2}$ and the tosses are independent. For illustration,

$$P(\{c_1\}) = \Pr(\text{TTT}) = (\tfrac{1}{2})(\tfrac{1}{2})(\tfrac{1}{2}) = \tfrac{1}{8}.$$

Then $P(A)$, which can be written as $\Pr(X_1 = 1, X_2 = 1 \text{ or } 2)$, is equal to $\frac{4}{8} = \frac{1}{2}$. It is left for the reader to show that we can tabulate the probability, which is then assigned to each of the elements of $\mathscr{A}$, with the following result:

(x_1, x_2)	$(0, 0)$	$(0, 1)$	$(1, 1)$	$(1, 2)$	$(2, 2)$	$(2, 3)$
$\Pr[(X_1, X_2) = (x_1, x_2)]$	$\frac{1}{8}$	$\frac{1}{8}$	$\frac{2}{8}$	$\frac{2}{8}$	$\frac{1}{8}$	$\frac{1}{8}$

This table depicts the distribution of probability over the elements of $\mathscr{A}$, the space of the random variables X_1 and X_2.

Again in statistics we are more interested in the space $\mathscr{A}$ of two random variables, say X and Y, than that of $\mathscr{C}$. Moreover, the notion of the p.d.f. of one random variable X can be extended to the notion of the p.d.f. of two or more random variables. Under certain restrictions on the space $\mathscr{A}$ and the function $f > 0$ on $\mathscr{A}$ (restrictions that will not be enumerated here), we say that the two random variables X and Y are of the discrete type or of the continuous type, and have a distribution of that type, according as the probability set function $P(A)$, $A \subset \mathscr{A}$, can be expressed as

$$P(A) = \Pr[(X, Y) \in A] = \sum_A\sum f(x, y),$$

or as

$$P(A) = \Pr[(X, Y) \in A] = \int_A\int f(x, y)\,dx\,dy.$$

In either case f is called the p.d.f. of the two random variables X and Y. Of necessity, $P(\mathscr{A}) = 1$ in each case.

We may extend the definition of a p.d.f. $f(x, y)$ over the entire xy-plane by using zero elsewhere. We shall do this consistently so that tedious, repetitious references to the space $\mathscr{A}$ can be avoided. Once this is done, we replace

$$\int_{\mathscr{A}}\int f(x, y)\,dx\,dy \qquad \text{by} \qquad \int_{-\infty}^{\infty}\int_{-\infty}^{\infty} f(x, y)\,dx\,dy.$$

Similarly, after extending the definition of a p.d.f. of the discrete type, we replace

$$\sum_{\mathscr{A}}\sum f(x, y) \qquad \text{by} \qquad \sum_y\sum_x f(x, y).$$

In accordance with this convention (of extending the definition of a p.d.f.), it is seen that a point function f, whether in one or two variables, essentially satisfies the conditions of being a p.d.f. if (a) f

is defined and is nonnegative for all real values of its argument(s) and if (b) its integral [for the continuous type of random variable(s)], or its sum [for the discrete type of random variable(s)] over all real values of its arguments(s) is 1.

Finally, if a p.d.f. in one or more variables is explicitly defined, we can see by inspection whether the random variables are of the continuous or discrete type. For example, it seems obvious that the p.d.f.

$$f(x, y) = \frac{9}{4^{x+y}}, \qquad x = 1, 2, 3, \ldots, \quad y = 1, 2, 3, \ldots,$$

$$= 0 \qquad \text{elsewhere,}$$

is a p.d.f. of two discrete-type random variables X and Y, whereas the p.d.f.

$$f(x, y) = 4xye^{-x^2-y^2}, \qquad 0 < x < \infty, \quad 0 < y < \infty,$$

$$= 0 \qquad \text{elsewhere,}$$

is clearly a p.d.f. of two continuous-type random variables X and Y. In such cases it seems unnecessary to specify which of the two simpler types of random variables is under consideration.

Example 1. Let

$$f(x, y) = 6x^2y, \qquad 0 < x < 1, \quad 0 < y < 1,$$

$$= 0 \qquad \text{elsewhere,}$$

be the p.d.f. of two random variables X and Y, which must be of the continuous type. We have, for instance,

$$\Pr\left(0 < X < \tfrac{3}{4}, \tfrac{1}{3} < Y < 2\right) = \int_{1/3}^{2} \int_{0}^{3/4} f(x, y)\, dx\, dy$$

$$= \int_{1/3}^{1} \int_{0}^{3/4} 6x^2y\, dx\, dy + \int_{1}^{2} \int_{0}^{3/4} 0\, dx\, dy$$

$$= \tfrac{3}{8} + 0 = \tfrac{3}{8}.$$

Note that this probability is the volume under the surface $f(x, y) = 6x^2y$ and above the rectangular set $\{(x, y) : 0 < x < \frac{3}{4}, \frac{1}{3} < y < 1\}$ in the xy-plane.

Let the random variables X and Y have the probability set function $P(A)$, where A is a two-dimensional set. If A is the unbounded set $\{(u, v) : u \le x, v \le y\}$, where x and y are real numbers, we have

$$P(A) = \Pr[(X, Y) \in A] = \Pr(X \le x, Y \le y).$$

This function of the point (x, y) is called the *distribution function* of X and Y and is denoted by

$$F(x, y) = \Pr(X \le x, Y \le y).$$

If X and Y are random variables of the continuous type that have p.d.f. $f(x, y)$, then

$$F(x, y) = \int_{-\infty}^{y} \int_{-\infty}^{x} f(u, v)\, du\, dv.$$

Accordingly, at points of continuity of $f(x, y)$, we have

$$\frac{\partial^2 F(x, y)}{\partial x\, \partial y} = f(x, y).$$

It is left as an exercise to show, in every case, that

$$\Pr(a < X \le b, c < Y \le d) = F(b, d) - F(b, c) - F(a, d) + F(a, c),$$

for all real constants $a < b$, $c < d$.

Consider next an experiment in which a person chooses at random a point (X, Y) from the unit square $\mathscr{C} = \mathscr{A} = \{(x, y) : 0 < x < 1, 0 < y < 1\}$. Suppose that our interest is not in X or in Y but in $Z = X + Y$. Once a suitable probability model has been adopted, we shall see how to find the p.d.f. of Z. To be specific, let the nature of the random experiment be such that it is reasonable to *assume* that the distribution of probability over the unit square is uniform. Then the p.d.f. of X and Y may be written

$$\begin{aligned} f(x, y) &= 1, \qquad 0 < x < 1, \quad 0 < y < 1, \\ &= 0 \qquad \text{elsewhere,} \end{aligned}$$

and this describes the probability model. Now let the distribution function of Z be denoted by $G(z) = \Pr(X + Y \le z)$. Then

$$\begin{aligned} G(z) &= 0, \qquad z < 0, \\ &= \int_0^z \int_0^{z-x} dy\, dx = \frac{z^2}{2}, \qquad 0 \le z < 1, \\ &= 1 - \int_{z-1}^1 \int_{z-x}^1 dy\, dx = 1 - \frac{(2 - z)^2}{2}, \qquad 1 \le z < 2, \\ &= 1, \qquad 2 \le z. \end{aligned}$$

Since $G'(z)$ exists for all values of z, the p.d.f. of Z may then be written

$$
\begin{aligned}
g(z) &= z, && 0 < z < 1, \\
&= 2 - z, && 1 \le z < 2, \\
&= 0 && \text{elsewhere.}
\end{aligned}
$$

It is clear that a different choice of the p.d.f. $f(x, y)$ that describes the probability model will, in general, lead to a different p.d.f. of Z.

Let $f(x_1, x_2)$ be the p.d.f. of two random variables X_1 and X_2. From this point on, for emphasis and clarity, we shall call a p.d.f. or a distribution function a *joint* p.d.f. or a *joint* distribution function when more than one random variable is involved. Thus $f(x_1, x_2)$ is the joint p.d.f. of the random variables X_1 and X_2. Consider the event $a < X_1 < b$, $a < b$. This event can occur when and only when the event $a < X_1 < b$, $-\infty < X_2 < \infty$ occurs; that is, the two events are equivalent, so that they have the same probability. But the probability of the latter event has been defined and is given by

$$\Pr(a < X_1 < b, -\infty < X_2 < \infty) = \int_a^b \int_{-\infty}^{\infty} f(x_1, x_2)\, dx_2\, dx_1$$

for the continuous case, and by

$$\Pr(a < X_1 < b_1, -\infty < X_2 < \infty) = \sum_{a < x_1 < b} \sum_{x_2} f(x_1, x_2)$$

for the discrete case. Now each of

$$\int_{-\infty}^{\infty} f(x_1, x_2)\, dx_2 \qquad \text{and} \qquad \sum_{x_2} f(x_1, x_2)$$

is a function of x_1 alone, say $f_1(x_1)$. Thus, for every $a < b$, we have

$$
\begin{aligned}
\Pr(a < X_1 < b) &= \int_a^b f_1(x_1)\, dx_1 && \text{(continuous case),} \\
&= \sum_{a < x_1 < b} f_1(x_1) && \text{(discrete case),}
\end{aligned}
$$

so that $f_1(x_1)$ is the p.d.f. of X_1 alone. Since $f_1(x_1)$ is found by summing (or integrating) the joint p.d.f. $f(x_1, x_2)$ over all x_2 for a fixed x_1, we can think of recording this sum in the "margin" of the

x_1x_2-plane. Accordingly, $f_1(x_1)$ is called the marginal p.d.f. of X_1. In like manner

$$f_2(x_2) = \int_{-\infty}^{\infty} f(x_1, x_2)\, dx_1 \qquad \text{(continuous case)},$$

$$= \sum_{x_1} f(x_1, x_2) \qquad \text{(discrete case)},$$

is called the marginal p.d.f. of X_2.

Example 2. Consider a random experiment that consists of drawing at random one chip from a bowl containing 10 chips of the same shape and size. Each chip has an ordered pair of numbers on it: one with (1, 1), one with (2, 1), two with (3, 1), one with (1, 2), two with (2, 2), and three with (3, 2). Let the random variables X_1 and X_2 be defined as the respective first and second values of the ordered pair. Thus the joint p.d.f. $f(x_1, x_2)$ of X_1 and X_2 can be given by the following table, with $f(x_1, x_2)$ equal to zero elsewhere.

	x_1			
x_2	1	2	3	$f_2(x_2)$
1	$\frac{1}{10}$	$\frac{1}{10}$	$\frac{2}{10}$	$\frac{4}{10}$
2	$\frac{1}{10}$	$\frac{2}{10}$	$\frac{3}{10}$	$\frac{6}{10}$
$f_1(x_1)$	$\frac{2}{10}$	$\frac{3}{10}$	$\frac{5}{10}$	

The joint probabilities have been summed in each row and each column and these sums recorded in the margins to give the marginal probability density functions of X_1 and X_2, respectively. Note that it is not necessary to have a formula for $f(x_1, x_2)$ to do this.

Example 3. Let X_1 and X_2 have the joint p.d.f.

$$f(x_1, x_2) = x_1 + x_2, \qquad 0 < x_1 < 1, \quad 0 < x_2 < 1,$$
$$= 0 \qquad \text{elsewhere}.$$

The marginal p.d.f. of X_1 is

$$f_1(x_1) = \int_0^1 (x_1 + x_2)\, dx_2 = x_1 + \tfrac{1}{2}, \qquad 0 < x_1 < 1,$$

zero elsewhere, and the marginal p.d.f. of X_2 is

$$f_2(x_2) = \int_0^1 (x_1 + x_2)\, dx_1 = \tfrac{1}{2} + x_2, \qquad 0 < x_2 < 1,$$

zero elsewhere. A probability like $\Pr(X_1 \le \frac{1}{2})$ can be computed from either $f_1(x_1)$ or $f(x_1, x_2)$ because

$$\int_0^{1/2}\int_0^1 f(x_1, x_2)\,dx_2\,dx_1 = \int_0^{1/2} f_1(x_1)\,dx_1 = \tfrac{3}{8}.$$

However to find a probability like $\Pr(X_1 + X_2 \le 1)$, we must use the joint p.d.f. $f(x_1, x_2)$ as follows:

$$\int_0^1\int_0^{1-x_1} (x_1 + x_2)\,dx_2\,dx_1 = \int_0^1 \left[x_1(1 - x_1) + \frac{(1 - x_1)^2}{2}\right] dx_1$$

$$= \int_0^1 \left(\frac{1}{2} - \frac{1}{2}x_1^2\right) dx_1 = \frac{1}{3}.$$

This latter probability is the volume under the surface $f(x_1, x_2) = x_1 + x_2$ above the set $\{(x_1, x_2) : 0 < x_1, 0 < x_2, x_1 + x_2 \le 1\}$.

EXERCISES

2.1. Let $f(x_1, x_2) = 4x_1x_2$, $0 < x_1 < 1$, $0 < x_2 < 1$, zero elsewhere, be the p.d.f. of X_1 and X_2. Find $\Pr(0 < X_1 < \frac{1}{2}, \frac{1}{4} < X_2 < 1)$, $\Pr(X_1 = X_2)$, $\Pr(X_1 < X_2)$, and $\Pr(X_1 \le X_2)$.

Hint: Recall that $\Pr(X_1 = X_2)$ would be the volume under the surface $f(x_1, x_2) = 4x_1x_2$ and above the line segment $0 < x_1 = x_2 < 1$ in the x_1x_2-plane.

2.2. Let $A_1 = \{(x, y) : x \le 2, y \le 4\}$, $A_2 = \{(x, y) : x \le 2, y \le 1\}$, $A_3 = \{(x, y) : x \le 0, y \le 4\}$, and $A_4 = \{(x, y) : x \le 0, y \le 1\}$ be subsets of the space $\mathscr{A}$ of two random variables X and Y, which is the entire two-dimensional plane. If $P(A_1) = \frac{7}{8}$, $P(A_2) = \frac{4}{8}$, $P(A_3) = \frac{3}{8}$, and $P(A_4) = \frac{2}{8}$, find $P(A_5)$, where $A_5 = \{(x, y) : 0 < x \le 2, 1 < y \le 4\}$.

2.3. Let $F(x, y)$ be the distribution function of X and Y. Show that $\Pr(a < X \le b, c < Y \le d) = F(b, d) - F(b, c) - F(a, d) + F(a, c)$, for all real constants $a < b$, $c < d$.

2.4. Show that the function $F(x, y)$ that is equal to 1 provided that $x + 2y \ge 1$, and that is equal to zero provided that $x + 2y < 1$, cannot be a distribution function of two random variables.

Hint: Find four numbers $a < b$, $c < d$, so that

$$F(b, d) - F(a, d) - F(b, c) + F(a, c)$$

is less than zero.

2.5. Given that the nonnegative function $g(x)$ has the property that

$$\int_0^\infty g(x)\,dx = 1.$$

Show that

$$f(x_1, x_2) = [2g(\sqrt{x_1^2 + x_2^2})]/(\pi\sqrt{x_1^2 + x_2^2}),\ 0 < x_1 < \infty,\ 0 < x_2 < \infty,$$

zero elsewhere, satisfies the conditions of being a p.d.f. of two continuous-type random variables X_1 and X_2.

Hint: Use polar coordinates.

2.6. Let $f(x, y) = e^{-x-y}$, $0 < x < \infty$, $0 < y < \infty$, zero elsewhere, be the p.d.f. of X and Y. Then if $Z = X + Y$, compute $\Pr(Z \le 0)$, $\Pr(Z \le 6)$, and, more generally, $\Pr(Z \le z)$, for $0 < z < \infty$. What is the p.d.f. of Z?

2.7. Let X and Y have the p.d.f. $f(x, y) = 1$, $0 < x < 1$, $0 < y < 1$, zero elsewhere. Find the p.d.f. of the product $Z = XY$.

2.8. Let 13 cards be taken, at random and without replacement, from an ordinary deck of playing cards. If X is the number of spades in these 13 cards, find the p.d.f. of X. If, in addition, Y is the number of hearts in these 13 cards, find the probability $\Pr(X = 2, Y = 5)$. What is the joint p.d.f. of X and Y?

2.9. Let the random variables X_1 and X_2 have the joint p.d.f. described as follows:

(x_1, x_2)	$(0, 0)$	$(0, 1)$	$(0, 2)$	$(1, 0)$	$(1, 1)$	$(1, 2)$
$f(x_1, x_2)$	$\frac{2}{12}$	$\frac{3}{12}$	$\frac{2}{12}$	$\frac{2}{12}$	$\frac{2}{12}$	$\frac{1}{12}$

and $f(x_1, x_2)$ is equal to zero elsewhere.

(a) Write these probabilities in a rectangular array as in Example 2, recording each marginal p.d.f. in the "margins."

(b) What is $\Pr(X_1 + X_2 = 1)$?

2.10. Let X_1 and X_2 have the joint p.d.f. $f(x_1, x_2) = 15x_1^2x_2$, $0 < x_1 < x_2 < 1$, zero elsewhere. Find each marginal p.d.f. and compute $\Pr(X_1 + X_2 \le 1)$.

Hint: Graph the space of X_1 and X_2 and carefully choose the limits of integration in determining each marginal p.d.f.

2.2 Conditional Distributions and Expectations

We shall now discuss the notion of a conditional p.d.f. Let X_1 and X_2 denote random variables of the discrete type which have the joint p.d.f. $f(x_1, x_2)$ which is positive on $\mathscr{A}$ and is zero elsewhere. Let $f_1(x_1)$ and $f_2(x_2)$ denote, respectively, the marginal probability density functions of X_1 and X_2. Take A_1 to be the set $A_1 = \{(x_1, x_2) : x_1 = x_1', -\infty < x_2 < \infty\}$, where x_1' is such that $P(A_1) = \Pr(X_1 = x_1') = f_1(x_1') > 0$, and take A_2 to be the set

$A_2 = \{(x_1, x_2) : -\infty < x_1 < \infty, x_2 = x_2'\}$. Then, by definition, the conditional probability of the event A_2, given the event A_1, is

$$P(A_2|A_1) = \frac{P(A_1 \cap A_2)}{P(A_1)} = \frac{\Pr(X_1 = x_1', X_2 = x_2')}{\Pr(X_1 = x_1')} = \frac{f(x_1', x_2')}{f_1(x_1')}.$$

That is, if (x_1, x_2) is any point at which $f_1(x_1) > 0$, the conditional probability that $X_2 = x_2$, given that $X_1 = x_1$, is $f(x_1, x_2)/f_1(x_1)$. With x_1 held fast, and with $f_1(x_1) > 0$, this function of x_2 satisfies the conditions of being a p.d.f. of a discrete type of random variable X_2 because $f(x_1, x_2)/f_1(x_1)$ is nonnegative and

$$\sum_{x_2} \frac{f(x_1, x_2)}{f_1(x_1)} = \frac{1}{f_1(x_1)} \sum_{x_2} f(x_1, x_2) = \frac{f_1(x_1)}{f_1(x_1)} = 1.$$

We now define the symbol $f_{2|1}(x_2|x_1)$ by the relation

$$f_{2|1}(x_2|x_1) = \frac{f(x_1, x_2)}{f_1(x_1)}, \qquad f_1(x_1) > 0,$$

and we call $f_{2|1}(x_2|x_1)$ the *conditional p.d.f.* of the discrete type of random variable X_2, given that the discrete type of random variable $X_1 = x_1$. In a similar manner we define the symbol $f_{1|2}(x_1|x_2)$ by the relation

$$f_{1|2}(x_1|x_2) = \frac{f(x_1, x_2)}{f_2(x_2)}, \qquad f_2(x_2) > 0,$$

and we call $f_{1|2}(x_1|x_2)$ the conditional p.d.f. of the discrete type of random variable X_1, given that the discrete type of random variable $X_2 = x_2$.

Now let X_1 and X_2 denote random variables of the continuous type that have the joint p.d.f. $f(x_1, x_2)$ and the marginal probability density functions $f_1(x_1)$ and $f_2(x_2)$, respectively. We shall use the results of the preceding paragraph to motivate a definition of a conditional p.d.f. of a continuous type of random variable. When $f_1(x_1) > 0$, we define the symbol $f_{2|1}(x_2|x_1)$ by the relation

$$f_{2|1}(x_2|x_1) = \frac{f(x_1, x_2)}{f_1(x_1)}.$$

In this relation, x_1 is to be thought of as having a fixed (but any fixed) value for which $f_1(x_1) > 0$. It is evident that $f_{2|1}(x_2|x_1)$ is nonnegative and that

$$\int_{-\infty}^{\infty} f_{2|1}(x_2|x_1)\,dx_2 = \int_{-\infty}^{\infty} \frac{f(x_1, x_2)}{f_1(x_1)}\,dx_2$$
$$= \frac{1}{f_1(x_1)} \int_{-\infty}^{\infty} f(x_1, x_2)\,dx_2$$
$$= \frac{1}{f_1(x_1)} f_1(x_1) = 1.$$

That is, $f_{2|1}(x_2|x_1)$ has the properties of a p.d.f. of one continuous type of random variable. It is called the *conditional p.d.f.* of the continuous type of random variable X_2, given that the continuous type of random variable X_1 has the value x_1. When $f_2(x_2) > 0$, the conditional p.d.f. of the continuous type of random variable X_1, given that the continuous type of random variable X_2 has the value x_2, is defined by

$$f_{1|2}(x_1|x_2) = \frac{f(x_1, x_2)}{f_2(x_2)}, \qquad f_2(x_2) > 0.$$

Since each of $f_{2|1}(x_2|x_1)$ and $f_{1|2}(x_1|x_2)$ is a p.d.f. of one random variable (whether of the discrete or the continuous type), each has all the properties of such a p.d.f. Thus we can compute probabilities and mathematical expectations. If the random variables are of the continuous type, the probability

$$\Pr(a < X_2 < b|X_1 = x_1) = \int_a^b f_{2|1}(x_2|x_1)\,dx_2$$

is called "the conditional probability that $a < X_2 < b$, given that $X_1 = x_1$." If there is no ambiguity, this may be written in the form $\Pr(a < X_2 < b|x_1)$. Similarly, the conditional probability that $c < X_1 < d$, given $X_2 = x_2$, is

$$\Pr(c < X_1 < d|X_2 = x_2) = \int_c^d f_{1|2}(x_1|x_2)\,dx_1.$$

If $u(X_2)$ is a function of X_2, the expectation

$$E[u(X_2)|x_1] = \int_{-\infty}^{\infty} u(x_2) f_{2|1}(x_2|x_1)\,dx_2$$

is called the conditional expectation of $u(X_2)$, given that $X_1 = x_1$. In particular, if they do exist, then $E(X_2|x_1)$ is the mean and

$E\{[X_2 - E(X_2|x_1)]^2|x_1\}$ is the variance of the conditional distribution of X_2, given $X_1 = x_1$, which can be written more simply as var $(X_2|x_1)$. It is convenient to refer to these as the "conditional mean" and the "conditional variance" of X_2, given $X_1 = x_1$. Of course, we have

$$\text{var}(X_2|x_1) = E(X_2^2|x_1) - [E(X_2|x_1)]^2$$

from an earlier result. In like manner, the conditional expectation of $u(X_1)$, given $X_2 = x_2$, is given by

$$E[u(X_1)|x_2] = \int_{-\infty}^{\infty} u(x_1)f_{1|2}(x_1|x_2)\,dx_1.$$

With random variables of the discrete type, these conditional probabilities and conditional expectations are computed by using summation instead of integration. An illustrative example follows.

Example 1. Let X_1 and X_2 have the joint p.d.f.

$$\begin{aligned} f(x_1, x_2) &= 2, \qquad 0 < x_1 < x_2 < 1, \\ &= 0 \qquad \text{elsewhere.} \end{aligned}$$

Then the marginal probability density functions are, respectively,

$$\begin{aligned} f_1(x_1) &= \int_{x_1}^{1} 2\,dx_2 = 2(1 - x_1), \qquad 0 < x_1 < 1, \\ &= 0 \qquad \text{elsewhere,} \end{aligned}$$

and

$$\begin{aligned} f_2(x_2) &= \int_0^{x_2} 2\,dx_1 = 2x_2, \qquad 0 < x_2 < 1, \\ &= 0 \qquad \text{elsewhere.} \end{aligned}$$

The conditional p.d.f. of X_1, given $X_2 = x_2$, $0 < x_2 < 1$, is

$$\begin{aligned} f_{1|2}(x_1|x_2) &= \frac{2}{2x_2} = \frac{1}{x_2}, \qquad 0 < x_1 < x_2, \\ &= 0 \qquad \text{elsewhere.} \end{aligned}$$

Here the conditional mean and conditional variance of X_1, given $X_2 = x_2$, are, respectively,

$$\begin{aligned} E(X_1|x_2) &= \int_{-\infty}^{\infty} x_1 f_{1|2}(x_1|x_2)\,dx_1 \\ &= \int_0^{x_2} x_1\left(\frac{1}{x_2}\right) dx_1 \\ &= \frac{x_2}{2}, \qquad 0 < x_2 < 1, \end{aligned}$$

and

$$\text{var}(X_1|x_2) = \int_0^{x_2} \left(x_1 - \frac{x_2}{2}\right)^2 \left(\frac{1}{x_2}\right) dx_1$$
$$= \frac{x_2^2}{12}, \qquad 0 < x_2 < 1.$$

Finally, we shall compare the values of

$$\Pr(0 < X_1 < \tfrac{1}{2}|X_2 = \tfrac{3}{4}) \qquad \text{and} \qquad \Pr(0 < X_1 < \tfrac{1}{2}).$$

We have

$$\Pr(0 < X_1 < \tfrac{1}{2}|X_2 = \tfrac{3}{4}) = \int_0^{1/2} f_{1|2}(x_1|\tfrac{3}{4})\, dx_1 = \int_0^{1/2} (\tfrac{4}{3})\, dx_1 = \tfrac{2}{3},$$

but

$$\Pr(0 < X_1 < \tfrac{1}{2}) = \int_0^{1/2} f_1(x_1)\, dx_1 = \int_0^{1/2} 2(1 - x_1)\, dx_1 = \tfrac{3}{4}.$$

Since $E(X_2|x_1)$ is a function of x_1, then $E(X_2|X_1)$ is a random variable with its own distribution, mean, and variance. Let us consider the following illustration of this.

Example 2. Let X_1 and X_2 have the joint p.d.f.

$$f(x_1, x_2) = 6x_2, \qquad 0 < x_2 < x_1 < 1,$$
$$= 0 \qquad \text{elsewhere.}$$

Then the marginal p.d.f. of X_1 is

$$f_1(x_1) = \int_0^{x_1} 6x_2\, dx_2 = 3x_1^2, \qquad 0 < x_1 < 1,$$

zero elsewhere. The conditional p.d.f. of X_2, given $X_1 = x_1$, is

$$f_{2|1}(x_2|x_1) = \frac{6x_2}{3x_1^2} = \frac{2x_2}{x_1^2}, \qquad 0 < x_2 < x_1,$$

zero elsewhere, where $0 < x_1 < 1$. The conditional mean of X_2, given $X_1 = x_1$, is

$$E(X_2|x_1) = \int_0^{x_1} x_2\left(\frac{2x_2}{x_1^2}\right) dx_2 = \frac{2}{3}x_1, \qquad 0 < x_1 < 1.$$

Now $E(X_2|X_1) = 2X_1/3$ is a random variable, say Y. The distribution function of $Y = 2X_1/3$ is

$$G(y) = \Pr(Y \le y) = \Pr\left(X_1 \le \frac{3y}{2}\right), \qquad 0 \le y < \frac{2}{3}.$$

From the p.d.f. $f_1(x_1)$, we have

$$G(y) = \int_0^{3y/2} 3x_1^2\, dx_1 = \frac{27y^3}{8}, \qquad 0 \le y < \frac{2}{3}.$$

Of course, $G(y) = 0$, if $y < 0$, and $G(y) = 1$, if $\frac{2}{3} < y$. The p.d.f., mean, and variance of $Y = 2X_1/3$ are

$$g(y) = \frac{81y^2}{8}, \qquad 0 \le y < \frac{2}{3},$$

zero elsewhere,

$$E(Y) = \int_0^{2/3} y\left(\frac{81y^2}{8}\right) dy = \frac{1}{2},$$

and

$$\text{var}(Y) = \int_0^{2/3} y^2\left(\frac{81y^2}{8}\right) dy - \frac{1}{4} = \frac{1}{60}.$$

Since the marginal p.d.f. of X_2 is

$$f_2(x_2) = \int_{x_2}^{1} 6x_2\, dx_1 = 6x_2(1 - x_2), \qquad 0 < x_2 < 1,$$

zero elsewhere, it is easy to show that $E(X_2) = \frac{1}{2}$ and $\text{var}(X_2) = \frac{1}{20}$. That is, here

$$E(Y) = E[E(X_2|X_1)] = E(X_2)$$

and

$$\text{var}(Y) = \text{var}[E(X_2|X_1)] \le \text{var}(X_2).$$

Example 2 is excellent, as it provides us with the opportunity to apply many of these new definitions as well as review the distribution function technique for finding the distribution of a function of a random variable, namely $Y = 2X_1/3$. Moreover, the two observations at the end of Example 2 are no accident because it is true, in general, that

$$E[E(X_2|X_1)] = E(X_2) \quad \text{and} \quad \text{var}[E(X_2|X_1)] \le \text{var}(X_2).$$

To prove these two facts, we must first comment on the expectation of a function of two random variables, say $u(X_1, X_2)$. We do this for the continuous case, but the argument holds in the discrete case with summations replacing integrals. Of course, $Y = u(X_1, X_2)$ is a random variable and has a p.d.f., say $g(y)$, and

$$E(Y) = \int_{-\infty}^{\infty} yg(y)\, dy.$$

However, as before, it can be proved (Section 4.7) that $E(Y)$ equals

$$E[u(X_1, X_2)] = \int_{-\infty}^{\infty}\int_{-\infty}^{\infty} u(x_1, x_2)f(x_1, x_2)\, dx_1\, dx_2.$$

We call $E[u(X_1, X_2)]$ the expectation (mathematical expectation or expected value) of $u(X_1, X_2)$, and it can be shown to be a linear operator as in the one-variable case. We also note that the expected value of X_2 can be found in two ways:

$$E(X_2) = \int_{-\infty}^{\infty} \int_{-\infty}^{\infty} x_2 f(x_1, x_2)\, dx_1\, dx_2 = \int_{-\infty}^{\infty} x_2 f_2(x_2)\, dx_2,$$

the latter single integral being obtained from the double integral by integrating on x_1 first.

Example 3. Let X_1 and X_2 have the p.d.f.

$$\begin{aligned} f(x_1, x_2) &= 8x_1x_2, \qquad 0 < x_1 < x_2 < 1, \\ &= 0 \qquad \text{elsewhere.} \end{aligned}$$

Then

$$\begin{aligned} E(X_1X_2^2) &= \int_{-\infty}^{\infty} \int_{-\infty}^{\infty} x_1x_2^2 f(x_1, x_2)\, dx_1\, dx_2 \\ &= \int_0^1 \int_0^{x_2} 8x_1^2x_2^3\, dx_1\, dx_2 \\ &= \int_0^1 \tfrac{8}{3}x_2^6\, dx_2 = \tfrac{8}{21}. \end{aligned}$$

In addition,

$$E(X_2) = \int_0^1 \int_0^{x_2} x_2(8x_1x_2)\, dx_1\, dx_2 = \tfrac{4}{5}.$$

Since X_2 has the p.d.f. $f_2(x_2) = 4x_2^3$, $0 < x_2 < 1$, zero elsewhere, the latter expectation can be found by

$$E(X_2) = \int_0^1 x_2(4x_2^3)\, dx_2 = \tfrac{4}{5}.$$

Finally,

$$\begin{aligned} E(7X_1X_2^2 + 5X_2) &= 7E(X_1X_2^2) + 5E(X_2) \\ &= (7)(\tfrac{8}{21}) + (5)(\tfrac{4}{5}) = \tfrac{20}{3}. \end{aligned}$$

We begin the proof of $E[E(X_2|X_1)] = E(X_2)$ and $\text{var}\,[E(X_2|X_1)] \le \text{var}\,(X_2)$ by noting that

$$E(X_2) = \int_{-\infty}^{\infty} \int_{-\infty}^{\infty} x_2 f(x_1, x_2)\, dx_2\, dx_1$$

$$= \int_{-\infty}^{\infty} \left[\int_{-\infty}^{\infty} x_2 \frac{f(x_1, x_2)}{f_1(x_1)} \, dx_2 \right] f_1(x_1) \, dx_1$$

$$= \int_{-\infty}^{\infty} E(X_2|x_1) f_1(x_1) \, dx_1$$

$$= E[E(X_2|X_1)],$$

which is the first result. Consider next, with $\mu_2 = E(X_2)$,

$$\begin{aligned} \text{var}(X_2) &= E[(X_2 - \mu_2)^2] \\ &= E\{[X_2 - E(X_2|X_1) + E(X_2|X_1) - \mu_2]^2\} \\ &= E\{[X_2 - E(X_2|X_1)]^2\} + E\{[E(X_2|X_1) - \mu_2]^2\} \\ &\quad + 2E\{[X_2 - E(X_2|X_1)][E(X_2|X_1) - \mu_2]\}. \end{aligned}$$

We shall show that the last term of the right-hand member of the immediately preceding equation is zero. It is equal to

$$2 \int_{-\infty}^{\infty} \int_{-\infty}^{\infty} [x_2 - E(X_2|x_1)][E(X_2|x_1) - \mu_2] f(x_1, x_2) \, dx_2 \, dx_1$$

$$= 2 \int_{-\infty}^{\infty} [E(X_2|x_1) - \mu_2]$$

$$\times \left\{ \int_{-\infty}^{\infty} [x_2 - E(X_2|x_1)] \frac{f(x_1, x_2)}{f_1(x_1)} \, dx_2 \right\} f_1(x_1) \, dx_1.$$

But $E(X_2|x_1)$ is the conditional mean of X_2, given $X_1 = x_1$. Since the expression in the inner braces is equal to

$$E(X_2|x_1) - E(X_2|x_1) = 0,$$

the double integral is equal to zero. Accordingly, we have

$$\text{var}(X_2) = E\{[X_2 - E(X_2|X_1)]^2\} + E\{[E(X_2|X_1) - \mu_2]^2\}.$$

The first term in the right-hand member of this equation is nonnegative because it is the expected value of a nonnegative function, namely $[X_2 - E(X_2|X_1)]^2$. Since $E[E(X_2|X_1)] = \mu_2$, the second term will be the var $[E(X_2|X_1)]$. Hence we have

$$\text{var}(X_2) \geq \text{var}[E(X_2|X_1)],$$

which completes the proof.

Intuitively, this result could have this useful interpretation. Both the random variables X_2 and $E(X_2|X_1)$ have the same mean μ_2. If we did not know μ_2, we could use either of the two random variables to guess at the unknown μ_2. Since, however, var $(X_2) \geq$ var $[E(X_2|X_1)]$ we would put more reliance in $E(X_2|X_1)$ as a guess. That is, if we observe the pair (X_1, X_2) to be (x_1, x_2), we would prefer to use $E(X_2|x_1)$ to x_2 as a guess at the unknown μ_2. When studying the use of sufficient statistics in estimation in Chapter 7, we make use of this famous result, attributed to C. R. Rao and David Blackwell.

EXERCISES

2.11. Let X_1 and X_2 have the joint p.d.f. $f(x_1, x_2) = x_1 + x_2$, $0 < x_1 < 1$, $0 < x_2 < 1$, zero elsewhere. Find the conditional mean and variance of X_2, given $X_1 = x_1$, $0 < x_1 < 1$.

2.12. Let $f_{1|2}(x_1|x_2) = c_1 x_1/x_2^2$, $0 < x_1 < x_2$, $0 < x_2 < 1$, zero elsewhere, and $f_2(x_2) = c_2 x_2^4$, $0 < x_2 < 1$, zero elsewhere, denote, respectively, the conditional p.d.f. of X_1, given $X_2 = x_2$, and the marginal p.d.f. of X_2. Determine:
(a) The constants c_1 and c_2.
(b) The joint p.d.f. of X_1 and X_2.
(c) Pr $(\frac{1}{4} < X_1 < \frac{1}{2}|X_2 = \frac{5}{8})$.
(d) Pr $(\frac{1}{4} < X_1 < \frac{1}{2})$.

2.13. Let $f(x_1, x_2) = 21x_1^2x_2^3$, $0 < x_1 < x_2 < 1$, zero elsewhere, be the joint p.d.f. of X_1 and X_2.
(a) Find the conditional mean and variance of X_1, given $X_2 = x_2$, $0 < x_2 < 1$.
(b) Find the distribution of $Y = E(X_1|X_2)$.
(c) Determine $E(Y)$ and var (Y) and compare these to $E(X_1)$ and var (X_1), respectively.

2.14. If X_1 and X_2 are random variables of the discrete type having p.d.f. $f(x_1, x_2) = (x_1 + 2x_2)/18$, $(x_1, x_2) = (1, 1), (1, 2), (2, 1), (2, 2)$, zero elsewhere, determine the conditional mean and variance of X_2, given $X_1 = x_1$, for $x_1 = 1$ or 2. Also compute $E(3X_1 - 2X_2)$.

2.15. Five cards are drawn at random and without replacement from a bridge deck. Let the random variables X_1, X_2, and X_3 denote, respectively, the number of spades, the number of hearts, and the number of diamonds that appear among the five cards.
(a) Determine the joint p.d.f. of X_1, X_2, and X_3.
(b) Find the marginal probability density functions of X_1, X_2, and X_3.
(c) What is the joint conditional p.d.f. of X_2 and X_3, given that $X_1 = 3$?

2.16. Let X_1 and X_2 have the joint p.d.f. $f(x_1, x_2)$ described as follows:

(x_1, x_2)	(0, 0)	(0, 1)	(1, 0)	(1, 1)	(2, 0)	(2, 1)
$f(x_1, x_2)$	$\frac{1}{18}$	$\frac{3}{18}$	$\frac{4}{18}$	$\frac{3}{18}$	$\frac{6}{18}$	$\frac{1}{18}$

and $f(x_1, x_2)$ is equal to zero elsewhere. Find the two marginal probability density functions and the two conditional means.

Hint: Write the probabilities in a rectangular array.

2.17. Let us choose at random a point from the interval (0, 1) and let the random variable X_1 be equal to the number which corresponds to that point. Then choose a point at random from the interval $(0, x_1)$, where x_1 is the experimental value of X_1; and let the random variable X_2 be equal to the number which corresponds to this point.

(a) Make assumptions about the marginal p.d.f. $f_1(x_1)$, and the conditional p.d.f. $f_{2|1}(x_2|x_1)$.

(b) Compute $\Pr(X_1 + X_2 \geq 1)$.

(c) Find the conditional mean $E(X_1|x_2)$.

2.18. Let $f(x)$ and $F(x)$ denote, respectively, the p.d.f. and the distribution function of the random variable X. The conditional p.d.f. of X, given $X > x_0$, x_0 a fixed number, is defined by $f(x|X > x_0) = f(x)/[1 - F(x_0)]$, $x_0 < x$, zero elsewhere. This kind of conditional p.d.f. finds application in a problem of time until death, given survival until time x_0.

(a) Show that $f(x|X > x_0)$ is a p.d.f.

(b) Let $f(x) = e^{-x}$, $0 < x < \infty$, and zero elsewhere. Compute $\Pr(X > 2|X > 1)$.

2.19. Let X and Y have the joint p.d.f. $f(x, y) = 6(1 - x - y)$, $0 < x$, $0 < y$, $x + y < 1$, and zero elsewhere. Compute $\Pr(2X + 3Y < 1)$ and $E(XY + 2X^2)$.

2.3 The Correlation Coefficient

Because the result that we obtain in this section is more familiar in terms of X and Y, we use X and Y rather than X_1 and X_2 as symbols for our two random variables. Let X and Y have joint p.d.f. $f(x, y)$. If $u(x, y)$ is a function of x and y, then $E[u(X, Y)]$ was defined, subject to its existence, in Section 2.2. The existence of all mathematical expectations will be assumed in this discussion. The means of X and Y, say μ_1 and μ_2, are obtained by taking $u(x, y)$ to be x and y, respectively; and the variances of X and Y, say σ_1^2 and σ_2^2, are obtained by setting the function $u(x, y)$ equal to $(x - \mu_1)^2$ and $(y - \mu_2)^2$, respectively. Consider the mathematical expectation

$$\begin{aligned} E[(X - \mu_1)(Y - \mu_2)] &= E(XY - \mu_2 X - \mu_1 Y + \mu_1\mu_2) \\ &= E(XY) - \mu_2 E(X) - \mu_1 E(Y) + \mu_1\mu_2 \\ &= E(XY) - \mu_1\mu_2. \end{aligned}$$

This number is called the *covariance* of X and Y and is often denoted by cov (X, Y). If each of σ_1 and σ_2 is positive, the number

$$\rho = \frac{E[(X - \mu_1)(Y - \mu_2)]}{\sigma_1\sigma_2} = \frac{\text{cov}\,(X, Y)}{\sigma_1\sigma_2}$$

is called the *correlation coefficient* of X and Y. If the standard deviations are positive, the correlation coefficient of any two random variables is defined to be the covariance of the two random variables divided by the product of the standard deviations of the two random variables. It should be noted that the expected value of the product of two random variables is equal to the product of their expectations plus their covariance; that is, $E(XY) = \mu_1\mu_2 + \rho\sigma_1\sigma_2 = \mu_1\mu_2 + \text{cov}\,(X, Y)$.

Example 1. Let the random variables X and Y have the joint p.d.f.

$$\begin{aligned} f(x, y) &= x + y, \qquad 0 < x < 1, \quad 0 < y < 1, \\ &= 0 \qquad \text{elsewhere.} \end{aligned}$$

We shall compute the correlation coefficient of X and Y. When only two variables are under consideration, we shall denote the correlation coefficient by ρ. Now

$$\mu_1 = E(X) = \int_0^1 \int_0^1 x(x + y)\,dx\,dy = \frac{7}{12}$$

and

$$\sigma_1^2 = E(X^2) - \mu_1^2 = \int_0^1 \int_0^1 x^2(x + y)\,dx\,dy - \left(\frac{7}{12}\right)^2 = \frac{11}{144}.$$

Similarly,

$$\mu_2 = E(Y) = \frac{7}{12} \qquad \text{and} \qquad \sigma_2^2 = E(Y^2) - \mu_2^2 = \frac{11}{144}.$$

The covariance of X and Y is

$$E(XY) - \mu_1\mu_2 = \int_0^1 \int_0^1 xy(x + y)\,dx\,dy - \left(\frac{7}{12}\right)^2 = \frac{-1}{144}.$$

Accordingly, the correlation coefficient of X and Y is

$$\rho = \frac{-\frac{1}{144}}{\sqrt{(\frac{11}{144})(\frac{11}{144})}} = -\frac{1}{11}.$$

Remark. For certain kinds of distributions of two random variables, say X and Y, the correlation coefficient ρ proves to be a very useful characteristic of the distribution. Unfortunately, the formal definition of ρ does not reveal this fact. At this time we make some observations about ρ, some of which will be explored more fully at a later stage. It will soon be seen that if a joint distribution of two variables has a correlation coefficient (that is, if both of the variances are positive), then ρ satisfies $-1 \le \rho \le 1$. If $\rho = 1$, there is a line with equation $y = a + bx$, $b > 0$, the graph of which contains all of the probability of the distribution of X and Y. In this extreme case, we have $\Pr(Y = a + bX) = 1$. If $\rho = -1$, we have the same state of affairs except that $b < 0$. This suggests the following interesting question: When ρ does not have one of its extreme values, is there a line in the xy-plane such that the probability for X and Y tends to be concentrated in a band about this line? Under certain restrictive conditions this is in fact the case, and under those conditions we can look upon ρ as a measure of the intensity of the concentration of the probability for X and Y about that line.

Next, let $f(x, y)$ denote the joint p.d.f. of two random variables X and Y and let $f_1(x)$ denote the marginal p.d.f. of X. The conditional p.d.f. of Y, given $X = x$, is

$$f_{2|1}(y|x) = \frac{f(x, y)}{f_1(x)}$$

at points where $f_1(x) > 0$. Then the conditional mean of Y, given $X = x$, is given by

$$E(Y|x) = \int_{-\infty}^{\infty} y f_{2|1}(y|x)\, dy = \frac{\displaystyle\int_{-\infty}^{\infty} y f(x, y)\, dy}{f_1(x)},$$

when dealing with random variables of the continuous type. This conditional mean of Y, given $X = x$, is, of course, a function of x alone, say $u(x)$. In like vein, the conditional mean of X, given $Y = y$, is a function of y alone, say $v(y)$.

In case $u(x)$ is a linear function of x, say $u(x) = a + bx$, we say the conditional mean of Y is linear in x; or that Y has a linear conditional mean. When $u(x) = a + bx$, the constants a and b have simple values which will now be determined.

It will be assumed that neither σ_1^2 nor σ_2^2, the variances of X and Y, is zero. From

$$E(Y|x) = \frac{\int_{-\infty}^{\infty} yf(x, y)\, dy}{f_1(x)} = a + bx,$$

we have

$$\int_{-\infty}^{\infty} yf(x, y)\, dy = (a + bx)f_1(x). \tag{1}$$

If both members of Equation (1) are integrated on x, it is seen that

$$E(Y) = a + bE(X),$$

or

$$\mu_2 = a + b\mu_1, \tag{2}$$

where $\mu_1 = E(X)$ and $\mu_2 = E(Y)$. If both members of Equation (1) are first multiplied by x and then integrated on x, we have

$$E(XY) = aE(X) + bE(X^2),$$

or

$$\rho\sigma_1\sigma_2 + \mu_1\mu_2 = a\mu_1 + b(\sigma_1^2 + \mu_1^2), \tag{3}$$

where $\rho\sigma_1\sigma_2$ is the covariance of X and Y. The simultaneous solution of Equations (2) and (3) yields

$$a = \mu_2 - \rho\frac{\sigma_2}{\sigma_1}\mu_1 \qquad \text{and} \qquad b = \rho\frac{\sigma_2}{\sigma_1}.$$

That is,

$$u(x) = E(Y|x) = \mu_2 + \rho\frac{\sigma_2}{\sigma_1}(x - \mu_1)$$

is the conditional mean of Y, given $X = x$, when the conditional mean of Y is linear in x. If the conditional mean of X, given $Y = y$, is linear in y, then that conditional mean is given by

$$v(y) = E(X|y) = \mu_1 + \rho\frac{\sigma_1}{\sigma_2}(y - \mu_2).$$

We shall next investigate the variance of a conditional distribution under the assumption that the conditional mean is linear. The conditional variance of Y is given by

$$\operatorname{var}(Y|x) = \int_{-\infty}^{\infty} \left[y - \mu_2 - \rho \frac{\sigma_2}{\sigma_1}(x - \mu_1) \right]^2 f_{2|1}(y|x)\, dy$$

$$= \frac{\displaystyle\int_{-\infty}^{\infty} \left[(y - \mu_2) - \rho \frac{\sigma_2}{\sigma_1}(x - \mu_1) \right]^2 f(x, y)\, dy}{f_1(x)} \tag{4}$$

when the random variables are of the continuous type. This variance is nonnegative and is at most a function of x alone. If then, it is multiplied by $f_1(x)$ and integrated on x, the result obtained will be nonnegative. This result is

$$\int_{-\infty}^{\infty} \int_{-\infty}^{\infty} \left[(y - \mu_2) - \rho \frac{\sigma_2}{\sigma_1}(x - \mu_1) \right]^2 f(x, y)\, dy\, dx$$

$$= \int_{-\infty}^{\infty} \int_{-\infty}^{\infty} \left[(y - \mu_2)^2 - 2\rho \frac{\sigma_2}{\sigma_1}(y - \mu_2)(x - \mu_1) \right.$$

$$\left. + \rho^2 \frac{\sigma_2^2}{\sigma_1^2}(x - \mu_1)^2 \right] f(x, y)\, dy\, dx$$

$$= E[(Y - \mu_2)^2] - 2\rho \frac{\sigma_2}{\sigma_1} E[(X - \mu_1)(Y - \mu_2)]$$

$$+ \rho^2 \frac{\sigma_2^2}{\sigma_1^2} E[(X - \mu_1)^2]$$

$$= \sigma_2^2 - 2\rho \frac{\sigma_2}{\sigma_1} \rho\sigma_1\sigma_2 + \rho^2 \frac{\sigma_2^2}{\sigma_1^2} \sigma_1^2$$

$$= \sigma_2^2 - 2\rho^2\sigma_2^2 + \rho^2\sigma_2^2 = \sigma_2^2(1 - \rho^2) \ge 0.$$

That is, if the variance, Equation (4), is denoted by $k(x)$, then $E[k(X)] = \sigma_2^2(1 - \rho^2) \ge 0$. Accordingly, $\rho^2 \le 1$, or $-1 \le \rho \le 1$. It is left as an exercise to prove that $-1 \le \rho \le 1$ whether the conditional mean is or is not linear.

Suppose that the variance, Equation (4), is positive but not a function of x; that is, the variance is a constant $k > 0$. Now if k is multiplied by $f_1(x)$ and integrated on x, the result is k, so that $k = \sigma_2^2(1 - \rho^2)$. Thus, in this case, the variance of each conditional distribution of Y, given $X = x$, is $\sigma_2^2(1 - \rho^2)$. If $\rho = 0$, the variance of each conditional distribution of Y, given $X = x$, is σ_2^2, the variance of the marginal distribution of Y. On the other hand, if ρ^2 is near one, the variance of each conditional distribution of Y, given $X = x$, is relatively small, and there is a high concentration of the probability for this conditional distribution near the mean $E(Y|x) = \mu_2 + \rho(\sigma_2/\sigma_1)(x - \mu_1)$.

It should be pointed out that if the random variables X and Y in the preceding discussion are taken to be of the discrete type, the results just obtained are valid.

Example 2. Let the random variables X and Y have the linear conditional means $E(Y|x) = 4x + 3$ and $E(X|y) = \frac{1}{16}y - 3$. In accordance with the general formulas for the linear conditional means, we see that $E(Y|x) = \mu_2$ if $x = \mu_1$ and $E(X|y) = \mu_1$ if $y = \mu_2$. Accordingly, in this special case, we have $\mu_2 = 4\mu_1 + 3$ and $\mu_1 = \frac{1}{16}\mu_2 - 3$ so that $\mu_1 = -\frac{15}{4}$ and $\mu_2 = -12$. The general formulas for the linear conditional means also show that the product of the coefficients of x and y, respectively, is equal to ρ^2 and that the quotient of these coefficients is equal to σ_2^2/σ_1^2. Here $\rho^2 = 4(\frac{1}{16}) = \frac{1}{4}$ with $\rho = \frac{1}{2}$ (not $-\frac{1}{2}$), and $\sigma_2^2/\sigma_1^2 = 64$. Thus, from the two linear conditional means, we are able to find the values of μ_1, μ_2, ρ, and σ_2/σ_1, but not the values of σ_1 and σ_2.

Example 3. To illustrate how the correlation coefficient measures the intensity of the concentration of the probability for X and Y about a line, let these random variables have a distribution that is uniform over the area depicted in Figure 2.1. That is, the joint p.d.f. of X and Y is

$$\begin{aligned} f(x, y) &= \frac{1}{4ah}, \qquad -a + bx < y < a + bx, \quad -h < x < h, \\ &= 0 \qquad \text{elsewhere.} \end{aligned}$$

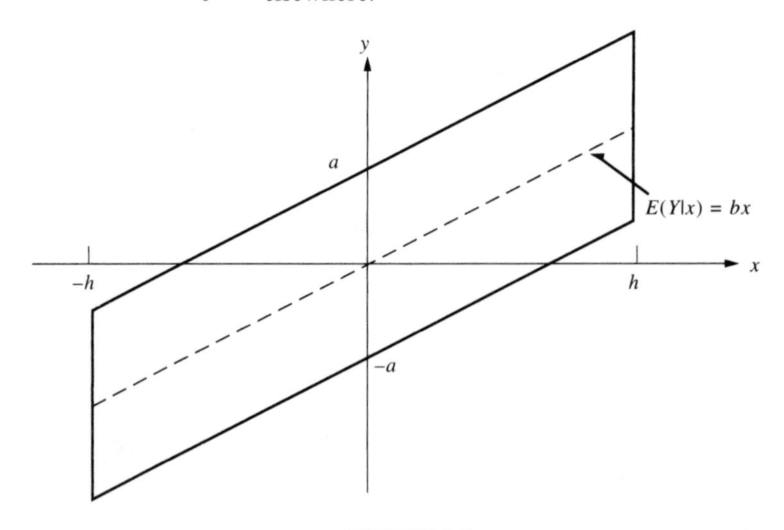

FIGURE 2.1

We assume here that $b \geq 0$, but the argument can be modified for $b \leq 0$. It is easy to show that the p.d.f. of X is uniform, namely

$$f_1(x) = \int_{-a+bx}^{a+bx} \frac{1}{4ah}\,dy = \frac{1}{2h}, \qquad -h < x < h,$$

$$= 0 \qquad \text{elsewhere.}$$

Thus the conditional p.d.f. of Y, given $X = x$, is uniform:

$$f_{2|1}(y|x) = \frac{1/4ah}{1/2h} = \frac{1}{2a}, \qquad -a + bx < y < a + bx,$$

$$= 0 \qquad \text{elsewhere.}$$

The conditional mean and variance are

$$E(Y|x) = bx \qquad \text{and} \qquad \text{var}\,(Y|x) = \frac{a^2}{3}.$$

From the general expressions for those characteristics we know that

$$b = \rho\frac{\sigma_2}{\sigma_1} \qquad \text{and} \qquad \frac{a^2}{3} = \sigma_2^2(1 - \rho^2).$$

In addition, we know that $\sigma_1^2 = h^2/3$. If we solve these three equations, we obtain an expression for the correlation coefficient, namely

$$\rho = \frac{bh}{\sqrt{a^2 + b^2h^2}}.$$

Referring to Figure 2.1, we note:

1. As a gets small (large), the straight line effect is more (less) intense and ρ is closer to 1 (zero).
2. As h gets large (small), the straight line effect is more (less) intense and ρ is closer to 1 (zero).
3. As b gets large (small), the straight line effect is more (less) intense and ρ is closer to 1 (zero).

This section will conclude with a definition and an illustrative example. Let $f(x, y)$ denote the joint p.d.f. of the two random variables X and Y. If $E(e^{t_1X + t_2Y})$ exists for $-h_1 < t_1 < h_1$, $-h_2 < t_2 < h_2$, where h_1 and h_2 are positive, it is denoted by $M(t_1, t_2)$ and is called the *moment-generating function* (m.g.f.) of the joint distribution of X and Y. As in the case of one random variable, the m.g.f. $M(t_1, t_2)$ completely determines the joint distribution of X and Y, and hence the marginal distributions of X and Y. In fact, the m.g.f. $M_1(t_1)$ of X is

$$M_1(t_1) = E(e^{t_1X}) = M(t_1, 0)$$

and the m.g.f. $M_2(t_2)$ of Y is

$$M_2(t_2) = E(e^{t_2 Y}) = M(0, t_2).$$

In addition, in the case of random variables of the continuous type,

$$\frac{\partial^{k+m} M(t_1, t_2)}{\partial t_1^k \, \partial t_2^m} = \int_{-\infty}^{\infty} \int_{-\infty}^{\infty} x^k y^m e^{t_1 x + t_2 y} f(x, y) \, dx \, dy,$$

so that

$$\left. \frac{\partial^{k+m} M(t_1, t_2)}{\partial t_1^k \, \partial t_2^m} \right|_{t_1 = t_2 = 0} = \int_{-\infty}^{\infty} \int_{-\infty}^{\infty} x^k y^m f(x, y) \, dx \, dy = E(X^k Y^m).$$

For instance, in a simplified notation which appears to be clear,

$$\mu_1 = E(X) = \frac{\partial M(0, 0)}{\partial t_1}, \qquad \mu_2 = E(Y) = \frac{\partial M(0, 0)}{\partial t_2},$$

$$\sigma_1^2 = E(X^2) - \mu_1^2 = \frac{\partial^2 M(0, 0)}{\partial t_1^2} - \mu_1^2,$$

$$\sigma_2^2 = E(Y^2) - \mu_2^2 = \frac{\partial^2 M(0, 0)}{\partial t_2^2} - \mu_2^2, \tag{5}$$

$$E[(X - \mu_1)(Y - \mu_2)] = \frac{\partial^2 M(0, 0)}{\partial t_1 \, \partial t_2} - \mu_1 \mu_2,$$

and from these we can compute the correlation coefficient ρ.

It is fairly obvious that the results of Equations (5) hold if X and Y are random variables of the discrete type. Thus the correlation coefficients may be computed by using the m.g.f. of the joint distribution if that function is readily available. An illustrative example follows. In this, we let $e^w = \exp(w)$.

Example 4. Let the continuous-type random variables X and Y have the joint p.d.f.

$$f(x, y) = e^{-y}, \qquad 0 < x < y < \infty,$$
$$= 0 \qquad \text{elsewhere.}$$

The m.g.f. of this joint distribution is

$$M(t_1, t_2) = \int_0^{\infty} \int_x^{\infty} \exp(t_1 x + t_2 y - y) \, dy \, dx$$

$$= \frac{1}{(1 - t_1 - t_2)(1 - t_2)},$$

provided that $t_1 + t_2 < 1$ and $t_2 < 1$. For this distribution, Equations (5) become

$$\mu_1 = 1, \qquad \mu_2 = 2,$$
$$\sigma_1^2 = 1, \qquad \sigma_2^2 = 2, \tag{6}$$
$$E[(X - \mu_1)(Y - \mu_2)] = 1.$$

Verification of results of Equations (6) is left as an exercise. If, momentarily, we accept these results, the correlation coefficient of X and Y is $\rho = 1/\sqrt{2}$. Furthermore, the moment-generating functions of the marginal distributions of X and Y are, respectively,

$$M(t_1, 0) = \frac{1}{1 - t_1}, \qquad t_1 < 1,$$

$$M(0, t_2) = \frac{1}{(1 - t_2)^2}, \qquad t_2 < 1.$$

These moment-generating functions are, of course, respectively, those of the marginal probability density functions,

$$f_1(x) = \int_x^\infty e^{-y}\,dy = e^{-x}, \qquad 0 < x < \infty,$$

zero elsewhere, and

$$f_2(y) = e^{-y} \int_0^y dx = ye^{-y}, \qquad 0 < y < \infty,$$

zero elsewhere.

EXERCISES

2.20. Let the random variables X and Y have the joint p.d.f.
(a) $f(x, y) = \frac{1}{3}$, $(x, y) = (0, 0), (1, 1), (2, 2)$, zero elsewhere.
(b) $f(x, y) = \frac{1}{3}$, $(x, y) = (0, 2), (1, 1), (2, 0)$, zero elsewhere.
(c) $f(x, y) = \frac{1}{3}$, $(x, y) = (0, 0), (1, 1), (2, 0)$, zero elsewhere.
In each case compute the correlation coefficient of X and Y.

2.21. Let X and Y have the joint p.d.f. described as follows:

(x, y)	$(1, 1)$	$(1, 2)$	$(1, 3)$	$(2, 1)$	$(2, 2)$	$(2, 3)$
$f(x, y)$	$\frac{2}{15}$	$\frac{4}{15}$	$\frac{3}{15}$	$\frac{1}{15}$	$\frac{1}{15}$	$\frac{4}{15}$

and $f(x, y)$ is equal to zero elsewhere. (a) Find the means μ_1 and μ_2, the variances σ_1^2 and σ_2^2, and the correlation coefficient ρ. (b) Compute $E(Y|X = 1)$, $E(Y|X = 2)$, and the line $\mu_2 + \rho(\sigma_2/\sigma_1)(x - \mu_1)$. Do the points $[k, E(Y|X = k)]$, $k = 1, 2$, lie on this line?

2.22. Let $f(x, y) = 2$, $0 < x < y$, $0 < y < 1$, zero elsewhere, be the joint p.d.f. of X and Y. Show that the conditional means are, respectively, $(1 + x)/2$, $0 < x < 1$, and $y/2$, $0 < y < 1$. Show that the correlation coefficient of X and Y is $\rho = \frac{1}{2}$.

2.23. Show that the variance of the conditional distribution of Y, given $X = x$, in Exercise 2.22, is $(1 - x)^2/12$, $0 < x < 1$, and that the variance of the conditional distribution of X, given $Y = y$, is $y^2/12$, $0 < y < 1$.

2.24. Verify the results of Equations (6) of this section.

2.25. Let X and Y have the joint p.d.f. $f(x, y) = 1$, $-x < y < x$, $0 < x < 1$, zero elsewhere. Show that, on the set of positive probability density, the graph of $E(Y|x)$ is a straight line, whereas that of $E(X|y)$ is not a straight line.

2.26. If the correlation coefficient ρ of X and Y exists, show that $-1 \le \rho \le 1$.
Hint: Consider the discriminant of the nonnegative quadratic function $h(v) = E\{[(X - \mu_1) + v(Y - \mu_2)]^2\}$, where v is real and is not a function of X nor of Y.

2.27. Let $\psi(t_1, t_2) = \ln M(t_1, t_2)$, where $M(t_1, t_2)$ is the m.g.f. of X and Y. Show that

$$\frac{\partial \psi(0, 0)}{\partial t_i}, \qquad \frac{\partial^2 \psi(0, 0)}{\partial t_i^2}, \qquad i = 1, 2,$$

and

$$\frac{\partial^2 \psi(0, 0)}{\partial t_1 \, \partial t_2}$$

yield the means, the variances, and the covariance of the two random variables. Use this result to find the means, the variances, and the covariance of X and Y of Example 4.

2.4 Independent Random Variables

Let X_1 and X_2 denote random variables of either the continuous or the discrete type which have the joint p.d.f. $f(x_1, x_2)$ and marginal probability density functions $f_1(x_1)$ and $f_2(x_2)$, respectively. In accordance with the definition of the conditional p.d.f. $f_{2|1}(x_2|x_1)$, we may write the joint p.d.f. $f(x_1, x_2)$ as

$$f(x_1, x_2) = f_{2|1}(x_2|x_1) f_1(x_1).$$

Suppose that we have an instance where $f_{2|1}(x_2|x_1)$ does not depend upon x_1. Then the marginal p.d.f. of X_2 is, for random variables of the continuous type,

$$f_2(x_2) = \int_{-\infty}^{\infty} f_{2|1}(x_2|x_1)f_1(x_1)\,dx_1$$

$$= f_{2|1}(x_2|x_1)\int_{-\infty}^{\infty} f_1(x_1)\,dx_1$$

$$= f_{2|1}(x_2|x_1).$$

Accordingly,

$$f_2(x_2) = f_{2|1}(x_2|x_1) \qquad \text{and} \qquad f(x_1, x_2) = f_1(x_1)f_2(x_2),$$

when $f_{2|1}(x_2|x_1)$ does not depend upon x_1. That is, if the conditional distribution of X_2, given $X_1 = x_1$, is independent of any assumption about x_1, then $f(x_1, x_2) = f_1(x_1)f_2(x_2)$. These considerations motivate the following definition.

Definition 2. Let the random variables X_1 and X_2 have the joint p.d.f. $f(x_1, x_2)$ and the marginal probability density functions $f_1(x_1)$ and $f_2(x_2)$, respectively. The random variables X_1 and X_2 are said to be independent if, and only if, $f(x_1, x_2) \equiv f_1(x_1)f_2(x_2)$. Random variables that are not independent are said to be dependent.

Remarks. Two comments should be made about the preceding definition. First, the product of two positive functions $f_1(x_1)f_2(x_2)$ means a function that is positive on a product space. That is, if $f_1(x_1)$ and $f_2(x_2)$ are positive on, and only on, the respective spaces $\mathscr{A}_1$ and $\mathscr{A}_2$, then the product of $f_1(x_1)$ and $f_2(x_2)$ is positive on, and only on, the product space $\mathscr{A} = \{(x_1, x_2) : x_1 \in \mathscr{A}_1, x_2 \in \mathscr{A}_2\}$. For instance, if $\mathscr{A}_1 = \{x_1 : 0 < x_1 < 1\}$ and $\mathscr{A}_2 = \{x_2 : 0 < x_2 < 3\}$, then $\mathscr{A} = \{(x_1, x_2) : 0 < x_1 < 1, 0 < x_2 < 3\}$. The second remark pertains to the identity. The identity in Definition 2 should be interpreted as follows. There may be certain points $(x_1, x_2) \in \mathscr{A}$ at which $f(x_1, x_2) \neq f_1(x_1)f_2(x_2)$. However, if A is the set of points (x_1, x_2) at which the equality does not hold, then $P(A) = 0$. In the subsequent theorems and the subsequent generalizations, a product of nonnegative functions and an identity should be interpreted in an analogous manner.

Example 1. Let the joint p.d.f. of X_1 and X_2 be

$$f(x_1, x_2) = x_1 + x_2, \qquad 0 < x_1 < 1, \quad 0 < x_2 < 1,$$
$$= 0 \qquad \text{elsewhere.}$$

It will be shown that X_1 and X_2 are dependent. Here the marginal probability density functions are

$$f_1(x_1) = \int_{-\infty}^{\infty} f(x_1, x_2)\, dx_2 = \int_0^1 (x_1 + x_2)\, dx_2 = x_1 + \tfrac{1}{2}, \qquad 0 < x_1 < 1,$$

$$= 0 \qquad \text{elsewhere,}$$

and

$$f_2(x_2) = \int_{-\infty}^{\infty} f(x_1, x_2)\, dx_1 = \int_0^1 (x_1 + x_2)\, dx_1 = \tfrac{1}{2} + x_2, \qquad 0 < x_2 < 1,$$

$$= 0 \qquad \text{elsewhere.}$$

Since $f(x_1, x_2) \not\equiv f_1(x_1)f_2(x_2)$, the random variables X_1 and X_2 are dependent.

The following theorem makes it possible to assert, without computing the marginal probability density functions, that the random variables X_1 and X_2 of Example 1 are dependent.

Theorem 1. *Let the random variables X_1 and X_2 have the joint p.d.f. $f(x_1, x_2)$. Then X_1 and X_2 are independent if and only if $f(x_1, x_2)$ can be written as a product of a nonnegative function of x_1 alone and a nonnegative function of x_2 alone. That is,*

$$f(x_1, x_2) \equiv g(x_1)h(x_2),$$

where $g(x_1) > 0$, $x_1 \in \mathscr{A}_1$, zero elsewhere, and $h(x_2) > 0$, $x_2 \in \mathscr{A}_2$, zero elsewhere.

Proof. If X_1 and X_2 are independent, then $f(x_1, x_2) \equiv f_1(x_1)f_2(x_2)$, where $f_1(x_1)$ and $f_2(x_2)$ are the marginal probability density functions of X_1 and X_2, respectively. Thus the condition $f(x_1, x_2) \equiv g(x_1)h(x_2)$ is fulfilled.

Conversely, if $f(x_1, x_2) \equiv g(x_1)h(x_2)$, then, for random variables of the continuous type, we have

$$f_1(x_1) = \int_{-\infty}^{\infty} g(x_1)h(x_2)\, dx_2 = g(x_1) \int_{-\infty}^{\infty} h(x_2)\, dx_2 = c_1 g(x_1)$$

and

$$f_2(x_2) = \int_{-\infty}^{\infty} g(x_1)h(x_2)\, dx_1 = h(x_2) \int_{-\infty}^{\infty} g(x_1)\, dx_1 = c_2 h(x_2),$$

where c_1 and c_2 are constants, not functions of x_1 or x_2. Moreover, $c_1 c_2 = 1$ because

$$1 = \int_{-\infty}^{\infty} \int_{-\infty}^{\infty} g(x_1)h(x_2)\,dx_1\,dx_2 = \left[\int_{-\infty}^{\infty} g(x_1)\,dx_1\right]\left[\int_{-\infty}^{\infty} h(x_2)\,dx_2\right]$$

$$= c_2 c_1.$$

These results imply that

$$f(x_1, x_2) \equiv g(x_1)h(x_2) \equiv c_1 g(x_1) c_2 h(x_2) \equiv f_1(x_1) f_2(x_2).$$

Accordingly, X_1 and X_2 are independent.

If we now refer to Example 1, we see that the joint p.d.f.

$$f(x_1, x_2) = x_1 + x_2, \qquad 0 < x_1 < 1, \quad 0 < x_2 < 1,$$
$$= 0 \qquad \text{elsewhere},$$

cannot be written as the product of a nonnegative function of x_1 alone and a nonnegative function of x_2 alone. Accordingly, X_1 and X_2 are dependent.

Example 2. Let the p.d.f. of the random variables X_1 and X_2 be $f(x_1, x_2) = 8x_1x_2$, $0 < x_1 < x_2 < 1$, zero elsewhere. The formula $8x_1x_2$ might suggest to some that X_1 and X_2 are independent. However, if we consider the space $\mathscr{A} = \{(x_1, x_2) : 0 < x_1 < x_2 < 1\}$, we see that it is not a product space. This should make it clear that, in general, X_1 and X_2 must be dependent if the space of positive probability density of X_1 and X_2 is bounded by a curve that is neither a horizontal nor a vertical line.

We now give a theorem that frequently simplifies the calculations of probabilities of events which involve independent variables.

Theorem 2. *If X_1 and X_2 are independent random variables with marginal probability density functions $f_1(x_1)$ and $f_2(x_2)$, respectively, then*

$$\Pr(a < X_1 < b, c < X_2 < d) = \Pr(a < X_1 < b) \Pr(c < X_2 < d)$$

for every $a < b$ and $c < d$, where a, b, c, and d are constants.

Proof. From the independence of X_1 and X_2, the joint p.d.f. of X_1 and X_2 is $f_1(x_1)f_2(x_2)$. Accordingly, in the continuous case,

$$\Pr(a < X_1 < b, c < X_2 < d) = \int_a^b \int_c^d f_1(x_1)f_2(x_2)\,dx_2\,dx_1$$
$$= \left[\int_a^b f_1(x_1)\,dx_1\right]\left[\int_c^d f_2(x_2)\,dx_2\right]$$
$$= \Pr(a < X_1 < b)\Pr(c < X_2 < d);$$

or, in the discrete case,

$$\Pr(a < X_1 < b, c < X_2 < d) = \sum_{a < x_1 < b}\ \sum_{c < x_2 < d} f_1(x_1)f_2(x_2)$$
$$= \left[\sum_{a < x_1 < b} f_1(x_1)\right]\left[\sum_{c < x_2 < d} f_2(x_2)\right]$$
$$= \Pr(a < X_1 < b)\Pr(c < X_2 < d),$$

as was to be shown.

Example 3. In Example 1, X_1 and X_2 were found to be dependent. There, in general,

$$\Pr(a < X_1 < b, c < X_2 < d) \neq \Pr(a < X_1 < b)\Pr(c < X_2 < d).$$

For instance,

$$\Pr(0 < X_1 < \tfrac{1}{2}, 0 < X_2 < \tfrac{1}{2}) = \int_0^{1/2}\int_0^{1/2} (x_1 + x_2)\,dx_1\,dx_2 = \tfrac{1}{8},$$

whereas

$$\Pr(0 < X_1 < \tfrac{1}{2}) = \int_0^{1/2} (x_1 + \tfrac{1}{2})\,dx_1 = \tfrac{3}{8}$$

and

$$\Pr(0 < X_2 < \tfrac{1}{2}) = \int_0^{1/2} (\tfrac{1}{2} + x_2)\,dx_2 = \tfrac{3}{8}.$$

Not merely are calculations of some probabilities usually simpler when we have independent random variables, but many expectations, including certain moment-generating functions, have comparably simpler computations. The following result will prove so useful that we state it in the form of a theorem.

Theorem 3. *Let the independent random variables X_1 and X_2 have the marginal probability density functions $f_1(x_1)$ and $f_2(x_2)$, respectively. The expected value of the product of a function $u(X_1)$ of X_1 alone and a function $v(X_2)$ of X_2 alone is, subject to their existence, equal to* *the product of the expected value of $u(X_1)$ and the expected value of $v(X_2)$; that is,*

$$E[u(X_1)v(X_2)] = E[u(X_1)]E[v(X_2)].$$

Proof. The independence of X_1 and X_2 implies that the joint p.d.f. of X_1 and X_2 is $f_1(x_1)f_2(x_2)$. Thus we have, by definition of expectation, in the continuous case,

$$\begin{aligned} E[u(X_1)v(X_2)] &= \int_{-\infty}^{\infty}\int_{-\infty}^{\infty} u(x_1)v(x_2)f_1(x_1)f_2(x_2)\,dx_1\,dx_2 \\ &= \left[\int_{-\infty}^{\infty} u(x_1)f_1(x_1)\,dx_1\right]\left[\int_{-\infty}^{\infty} v(x_2)f_2(x_2)\,dx_2\right] \\ &= E[u(X_1)]E[v(X_2)]; \end{aligned}$$

or, in the discrete case,

$$\begin{aligned} E[u(X_1)v(X_2)] &= \sum_{x_2}\sum_{x_1} u(x_1)v(x_2)f_1(x_1)f_2(x_2) \\ &= \left[\sum_{x_1} u(x_1)f_1(x_1)\right]\left[\sum_{x_2} v(x_2)f_2(x_2)\right] \\ &= E[u(X_1)]E[v(X_2)], \end{aligned}$$

as stated in the theorem.

Example 4. Let X and Y be two independent random variables with means μ_1 and μ_2 and positive variances σ_1^2 and σ_2^2, respectively. We shall show that the independence of X and Y implies that the correlation coefficient of X and Y is zero. This is true because the covariance of X and Y is equal to

$$E[(X - \mu_1)(Y - \mu_2)] = E(X - \mu_1)E(Y - \mu_2) = 0.$$

We shall now prove a very useful theorem about independent random variables. The proof of the theorem relies heavily upon our assertion that an m.g.f., when it exists, is unique and that it uniquely determines the distribution of probability.

Theorem 4. *Let X_1 and X_2 denote random variables that have the joint p.d.f. $f(x_1, x_2)$ and the marginal probability density functions $f_1(x_1)$ and $f_2(x_2)$, respectively. Furthermore, let $M(t_1, t_2)$ denote the m.g.f. of the distribution. Then X_1 and X_2 are independent if and only if*

$$M(t_1, t_2) = M(t_1, 0)M(0, t_2).$$

Proof. If X_1 and X_2 are independent, then

$$\begin{aligned} M(t_1, t_2) &= E(e^{t_1X_1 + t_2X_2}) \\ &= E(e^{t_1X_1}e^{t_2X_2}) \\ &= E(e^{t_1X_1})E(e^{t_2X_2}) \\ &= M(t_1, 0)M(0, t_2). \end{aligned}$$

Thus the independence of X_1 and X_2 implies that the m.g.f. of the joint distribution factors into the product of the moment-generating functions of the two marginal distributions.

Suppose next that the m.g.f. of the joint distribution of X_1 and X_2 is given by $M(t_1, t_2) = M(t_1, 0)M(0, t_2)$. Now X_1 has the unique m.g.f. which, in the continuous case, is given by

$$M(t_1, 0) = \int_{-\infty}^{\infty} e^{t_1x_1}f_1(x_1)\,dx_1.$$

Similarly, the unique m.g.f. of X_2, in the continuous case, is given by

$$M(0, t_2) = \int_{-\infty}^{\infty} e^{t_2x_2}f_2(x_2)\,dx_2.$$

Thus we have

$$\begin{aligned} M(t_1, 0)M(0, t_2) &= \left[\int_{-\infty}^{\infty} e^{t_1x_1}f_1(x_1)\,dx_1\right]\left[\int_{-\infty}^{\infty} e^{t_2x_2}f_2(x_2)\,dx_2\right] \\ &= \int_{-\infty}^{\infty}\int_{-\infty}^{\infty} e^{t_1x_1 + t_2x_2}f_1(x_1)f_2(x_2)\,dx_1\,dx_2. \end{aligned}$$

We are given that $M(t_1, t_2) = M(t_1, 0)M(0, t_2)$; so

$$M(t_1, t_2) = \int_{-\infty}^{\infty}\int_{-\infty}^{\infty} e^{t_1x_1 + t_2x_2}f_1(x_1)f_2(x_2)\,dx_1\,dx_2.$$

But $M(t_1, t_2)$ is the m.g.f. of X_1 and X_2. Thus also

$$M(t_1, t_2) = \int_{-\infty}^{\infty}\int_{-\infty}^{\infty} e^{t_1x_1 + t_2x_2}f(x_1, x_2)\,dx_1\,dx_2.$$

The uniqueness of the m.g.f. implies that the two distributions of probability that are described by $f_1(x_1)f_2(x_2)$ and $f(x_1, x_2)$ are the same. Thus

$$f(x_1, x_2) \equiv f_1(x_1)f_2(x_2).$$

That is, if $M(t_1, t_2) = M(t_1, 0)M(0, t_2)$, then X_1 and X_2 are independent. This completes the proof when the random variables are of the continuous type. With random variables of the discrete type, the proof is made by using summation instead of integration.

EXERCISES

2.28. Show that the random variables X_1 and X_2 with joint p.d.f. $f(x_1, x_2) = 12x_1x_2(1 - x_2)$, $0 < x_1 < 1$, $0 < x_2 < 1$, zero elsewhere, are independent.

2.29. If the random variables X_1 and X_2 have the joint p.d.f. $f(x_1, x_2) = 2e^{-x_1 - x_2}$, $0 < x_1 < x_2$, $0 < x_2 < \infty$, zero elsewhere, show that X_1 and X_2 are dependent.

2.30. Let $f(x_1, x_2) = \frac{1}{16}$, $x_1 = 1, 2, 3, 4$, and $x_2 = 1, 2, 3, 4$, zero elsewhere, be the joint p.d.f. of X_1 and X_2. Show that X_1 and X_2 are independent.

2.31. Find $\Pr(0 < X_1 < \frac{1}{3}, 0 < X_2 < \frac{1}{3})$ if the random variables X_1 and X_2 have the joint p.d.f. $f(x_1, x_2) = 4x_1(1 - x_2)$, $0 < x_1 < 1$, $0 < x_2 < 1$, zero elsewhere.

2.32. Find the probability of the union of the events $a < X_1 < b$, $-\infty < X_2 < \infty$ and $-\infty < X_1 < \infty$, $c < X_2 < d$ if X_1 and X_2 are two independent variables with $\Pr(a < X_1 < b) = \frac{2}{3}$ and $\Pr(c < X_2 < d) = \frac{5}{8}$.

2.33. If $f(x_1, x_2) = e^{-x_1 - x_2}$, $0 < x_1 < \infty$, $0 < x_2 < \infty$, zero elsewhere, is the joint p.d.f. of the random variables X_1 and X_2, show that X_1 and X_2 are independent and that $M(t_1, t_2) = (1 - t_1)^{-1}(1 - t_2)^{-1}$, $t_2 < 1$, $t_1 < 1$. Also show that

$$E(e^{t(X_1 + X_2)}) = (1 - t)^{-2}, \quad t < 1.$$

Accordingly, find the mean and the variance of $Y = X_1 + X_2$.

2.34. Let the random variables X_1 and X_2 have the joint p.d.f. $f(x_1, x_2) = 1/\pi$, $(x_1 - 1)^2 + (x_2 + 2)^2 < 1$, zero elsewhere. Find $f_1(x_1)$ and $f_2(x_2)$. Are X_1 and X_2 independent?

2.35. Let X and Y have the joint p.d.f. $f(x, y) = 3x$, $0 < y < x < 1$, zero elsewhere. Are X and Y independent? If not, find $E(X|y)$.

2.36. Suppose that a man leaves for work between 8:00 A.M. and 8:30 A.M. and takes between 40 and 50 minutes to get to the office. Let X denote the time of departure and let Y denote the time of travel. If we assume that these random variables are independent and uniformly distributed, find the probability that he arrives at the office before 9:00 A.M.

2.5 Extension to Several Random Variables

The notions about two random variables can be extended immediately to n random variables. We make the following definition of the space of n random variables.

Definition 3. Consider a random experiment with the sample space $\mathscr{C}$. Let the random variable X_i assign to each element $c \in \mathscr{C}$ one and only one real number $X_i(c) = x_i$, $i = 1, 2, \ldots, n$. The *space* of these random variables is the set of ordered n-tuples $\mathscr{A} = \{(x_1, x_2, \ldots, x_n) : x_1 = X_1(c), \ldots, x_n = X_n(c), c \in \mathscr{C}\}$. Furthermore, let A be a subset of $\mathscr{A}$. Then $\Pr[(X_1, \ldots, X_n) \in A] = P(C)$, where $C = \{c : c \in \mathscr{C} \text{ and } [X_1(c), X_2(c), \ldots, X_n(c)] \in A\}$.

Again we should make the comment that $\Pr[(X_1, \ldots, X_n) \in A]$ could be denoted by the probability set function $P_{X_1, \ldots, X_n}(A)$. But, if there is no chance of misunderstanding, it will be written simply as $P(A)$. We say that the n random variables $X_1, X_2, \ldots, X_n$ are of the discrete type or of the continuous type, and have a distribution of that type, according as the probability set function $P(A)$, $A \subset \mathscr{A}$, can be expressed as

$$P(A) = \Pr[(X_1, \ldots, X_n) \in A] = \sum_A \cdots \sum f(x_1, \ldots, x_n),$$

or as

$$P(A) = \Pr[(X_1, \ldots, X_n) \in A] = \int \cdots \int_A f(x_1, \ldots, x_n)\, dx_1 \cdots dx_n.$$

In accordance with the convention of extending the definition of a p.d.f., it is seen that a point function f essentially satisfies the conditions of being a p.d.f. if (a) f is defined and is nonnegative for all real values of its argument(s) and if (b) its integral [for the continuous type of random variable(s)], or its sum [for the discrete type of random variable(s)] over all real values of its argument(s) is 1.

The distribution function of the n random variables $X_1, X_2, \ldots, X_n$ is the point function

$$F(x_1, x_2, \ldots, x_n) = \Pr(X_1 \le x_1, X_2 \le x_2, \ldots, X_n \le x_n).$$

An illustrative example follows.

Example 1. Let $f(x, y, z) = e^{-(x+y+z)}$, $0 < x, y, z < \infty$, zero elsewhere, be the p.d.f. of the random variables X, Y, and Z. Then the distribution function of X, Y, and Z is given by

$$\begin{aligned} F(x, y, z) &= \Pr(X \le x, Y \le y, Z \le z) \\ &= \int_0^z \int_0^y \int_0^x e^{-u-v-w}\, du\, dv\, dw \\ &= (1 - e^{-x})(1 - e^{-y})(1 - e^{-z}), \qquad 0 \le x, y, z < \infty, \end{aligned}$$

and is equal to zero elsewhere. Incidentally, except for a set of probability measure zero, we have

$$\frac{\partial^3 F(x, y, z)}{\partial x\, \partial y\, \partial z} = f(x, y, z).$$

Let $X_1, X_2, \ldots, X_n$ be random variables having joint p.d.f. $f(x_1, x_2, \ldots, x_n)$ and let $u(X_1, X_2, \ldots, X_n)$ be a function of these variables such that the n-fold integral

$$\int_{-\infty}^{\infty} \cdots \int_{-\infty}^{\infty} u(x_1, x_2, \ldots, x_n) f(x_1, x_2, \ldots, x_n)\, dx_1\, dx_2 \cdots dx_n \quad (1)$$

exists, if the random variables are of the continuous type, or such that the n-fold sum

$$\sum_{x_n} \cdots \sum_{x_1} u(x_1, x_2, \ldots, x_n) f(x_1, x_2, \ldots, x_n) \quad (2)$$

exists if the random variables are of the discrete type. The n-fold integral (or the n-fold sum, as the case may be) is called the *expectation*, denoted by $E[u(X_1, X_2, \ldots, X_n)]$, of the function $u(X_1, X_2, \ldots, X_n)$. In Section 4.7 we show this expectation to be equal to $E(Y)$, where $Y = u(X_1, X_2, \ldots, X_n)$. Of course, E is a linear operator.

We shall now discuss the notions of marginal and conditional probability density functions from the point of view of n random variables. All of the preceding definitions can be directly generalized to the case of n variables in the following manner. Let the random variables $X_1, X_2, \ldots, X_n$ have the joint p.d.f. $f(x_1, x_2, \ldots, x_n)$. If the random variables are of the continuous type, then by an argument similar to the two-variable case, we have for every $a < b$,

$$\Pr(a < X_1 < b) = \int_a^b f_1(x_1)\, dx_1,$$

where $f_1(x_1)$ is defined by the $(n - 1)$-fold integral

$$f_1(x_1) = \int_{-\infty}^{\infty} \cdots \int_{-\infty}^{\infty} f(x_1, x_2, \ldots, x_n)\, dx_2 \cdots dx_n.$$

Therefore, $f_1(x_1)$ is the p.d.f. of the one random variable X_1 and $f_1(x_1)$ is called the marginal p.d.f. of X_1. The marginal probability density functions $f_2(x_2), \ldots, f_n(x_n)$ of $X_2, \ldots, X_n$, respectively, are similar $(n - 1)$-fold integrals.

Up to this point, each marginal p.d.f. has been a p.d.f. of one random variable. It is convenient to extend this terminology to joint probability density functions, which we shall do now. Here let $f(x_1, x_2, \ldots, x_n)$ be the joint p.d.f. of the n random variables $X_1, X_2, \ldots, X_n$, just as before. Now, however, let us take any group of $k < n$ of these random variables and let us find the joint p.d.f. of them. This joint p.d.f. is called the marginal p.d.f. of this particular group of k variables. To fix the ideas, take $n = 6$, $k = 3$, and let us select the group X_2, X_4, X_5. Then the marginal p.d.f. of X_2, X_4, X_5 is the joint p.d.f. of this particular group of three variables, namely,

$$\int_{-\infty}^{\infty}\int_{-\infty}^{\infty}\int_{-\infty}^{\infty} f(x_1, x_2, x_3, x_4, x_5, x_6)\, dx_1\, dx_3\, dx_6,$$

if the random variables are of the continuous type.

Next we extend the definition of a conditional p.d.f. If $f_1(x_1) > 0$, the symbol $f_{2,\ldots,n|1}(x_2, \ldots, x_n|x_1)$ is defined by the relation

$$f_{2,\ldots,n|1}(x_2, \ldots, x_n|x_1) = \frac{f(x_1, x_2, \ldots, x_n)}{f_1(x_1)},$$

and $f_{2,\ldots,n|1}(x_2, \ldots, x_n|x_1)$ is called the *joint conditional p.d.f.* of $X_2, \ldots, X_n$, given $X_1 = x_1$. The joint conditional p.d.f. of any $n - 1$ random variables, say $X_1, \ldots, X_{i-1}, X_{i+1}, \ldots, X_n$, given $X_i = x_i$, is defined as the joint p.d.f. of $X_1, X_2, \ldots, X_n$ divided by the marginal p.d.f. $f_i(x_i)$, provided that $f_i(x_i) > 0$. More generally, the joint conditional p.d.f. of $n - k$ of the random variables, for given values of the remaining k variables, is defined as the joint p.d.f. of the n variables divided by the marginal p.d.f. of the particular group of k variables, provided that the latter p.d.f. is positive. We remark that there are many other conditional probability density functions; for instance, see Exercise 2.18.

Because a conditional p.d.f. is a p.d.f. of a certain number of random variables, the expectation of a function of these random variables has been defined. To emphasize the fact that a conditional p.d.f. is under consideration, such expectations are called conditional expectations. For instance, the conditional expectation of $u(X_2, \ldots, X_n)$ given $X_1 = x_1$, is, for random variables of the continuous type, given by

$$E[u(X_2, \ldots, X_n)|x_1] = \int_{-\infty}^{\infty} \cdots \int_{-\infty}^{\infty} u(x_2, \ldots, x_n)$$

$$\times f_{2,\ldots,n|1}(x_2, \ldots, x_n|x_1)\, dx_2 \cdots dx_n,$$

provided $f_1(x_1) > 0$ and the integral converges (absolutely). If the random variables are of the discrete type, conditional expectations are, of course, computed by using sums instead of integrals.

Let the random variables $X_1, X_2, \ldots, X_n$ have the joint p.d.f. $f(x_1, x_2, \ldots, x_n)$ and the marginal probability density functions $f_1(x_1), f_2(x_2), \ldots, f_n(x_n)$, respectively. The definition of the independence of X_1 and X_2 is generalized to the mutual independence of $X_1, X_2, \ldots, X_n$ as follows: The random variables $X_1, X_2, \ldots, X_n$ are said to be *mutually independent* if and only if

$$f(x_1, x_2, \ldots, x_n) \equiv f_1(x_1)f_2(x_2) \cdots f_n(x_n).$$

It follows immediately from this definition of the mutual independence of $X_1, X_2, \ldots, X_n$ that

$$\begin{aligned} &\Pr(a_1 < X_1 < b_1, a_2 < X_2 < b_2, \ldots, a_n < X_n < b_n) \\ &\quad = \Pr(a_1 < X_1 < b_1) \Pr(a_2 < X_2 < b_2) \cdots \Pr(a_n < X_n < b_n) \\ &\quad = \prod_{i=1}^{n} \Pr(a_i < X_i < b_i), \end{aligned}$$

where the symbol $\prod_{i=1}^{n} \varphi(i)$ is defined to be

$$\prod_{i=1}^{n} \varphi(i) = \varphi(1)\varphi(2) \cdots \varphi(n).$$

The theorem that

$$E[u(X_1)v(X_2)] = E[u(X_1)]E[v(X_2)]$$

for independent random variables X_1 and X_2 becomes, for mutually independent random variables $X_1, X_2, \ldots, X_n$,

$$E[u_1(X_1)u_2(X_2) \cdots u_n(X_n)] = E[u_1(X_1)]E[u_2(X_2)] \cdots E[u_n(X_n)],$$

or

$$E\left[\prod_{i=1}^{n} u_i(X_i)\right] = \prod_{i=1}^{n} E[u_i(X_i)].$$

The moment-generating function of the joint distribution of n random variables $X_1, X_2, \ldots, X_n$ is defined as follows. Let

$$E[\exp(t_1X_1 + t_2X_2 + \cdots + t_nX_n)]$$

exist for $-h_i < t_i < h_i$, $i = 1, 2, \ldots, n$, where each h_i is positive. This expectation is denoted by $M(t_1, t_2, \ldots, t_n)$ and it is called the m.g.f. of the joint distribution of $X_1, \ldots, X_n$ (or simply the m.g.f. of $X_1, \ldots, X_n$). As in the cases of one and two variables, this m.g.f. is unique and uniquely determines the joint distribution of the n

variables (and hence all marginal distributions). For example, the m.g.f. of the marginal distribution of X_i is $M(0, \ldots, 0, t_i, 0, \ldots, 0)$, $i = 1, 2, \ldots, n$; that of the marginal distribution of X_i and X_j is $M(0, \ldots, 0, t_i, 0, \ldots, 0, t_j, 0, \ldots, 0)$; and so on. Theorem 4 of this chapter can be generalized, and the factorization

$$M(t_1, t_2, \ldots, t_n) = \prod_{i=1}^{n} M(0, \ldots, 0, t_i, 0, \ldots, 0)$$

is a necessary and sufficient condition for the mutual independence of $X_1, X_2, \ldots, X_n$.

Remark. If X_1, X_2, and X_3 are mutually independent, they are *pairwise independent* (that is, X_i and X_j, $i \neq j$, where $i, j = 1, 2, 3$, are independent). However, the following example, due to S. Bernstein, shows that pairwise independence does not necessarily imply mutual independence. Let X_1, X_2, and X_3 have the joint p.d.f.

$$\begin{aligned} f(x_1, x_2, x_3) &= \tfrac{1}{4}, \qquad (x_1, x_2, x_3) \in \{(1, 0, 0), (0, 1, 0), (0, 0, 1), (1, 1, 1)\}, \\ &= 0 \qquad \text{elsewhere.} \end{aligned}$$

The joint p.d.f. of X_i and X_j, $i \neq j$, is

$$\begin{aligned} f_{ij}(x_i, x_j) &= \tfrac{1}{4}, \qquad (x_i, x_j) \in \{(0, 0), (1, 0), (0, 1), (1, 1)\}, \\ &= 0 \qquad \text{elsewhere,} \end{aligned}$$

whereas the marginal p.d.f. of X_i is

$$\begin{aligned} f_i(x_i) &= \tfrac{1}{2}, \qquad x_i = 0, 1, \\ &= 0 \qquad \text{elsewhere.} \end{aligned}$$

Obviously, if $i \neq j$, we have

$$f_{ij}(x_i, x_j) \equiv f_i(x_i) f_j(x_j),$$

and thus X_i and X_j are independent. However,

$$f(x_1, x_2, x_3) \not\equiv f_1(x_1) f_2(x_2) f_3(x_3).$$

Thus X_1, X_2, and X_3 are not mutually independent.

Example 2. Let X_1, X_2, and X_3 be three mutually independent random variables and let each have the p.d.f. $f(x) = 2x$, $0 < x < 1$, zero elsewhere. The joint p.d.f. of X_1, X_2, X_3 is $f(x_1)f(x_2)f(x_3) = 8x_1x_2x_3$, $0 < x_i < 1$, $i = 1, 2, 3$, zero elsewhere. Then, for illustration, the expected value of $5X_1X_2^3 + 3X_2X_3^4$ is

$$\int_0^1 \int_0^1 \int_0^1 (5x_1x_2^3 + 3x_2x_3^4) 8x_1x_2x_3 \, dx_1 \, dx_2 \, dx_3 = 2.$$

Let Y be the maximum of X_1, X_2, and X_3. Then, for instance, we have

$$\begin{aligned}\Pr\left(Y \le \tfrac{1}{2}\right) &= \Pr\left(X_1 \le \tfrac{1}{2}, X_2 \le \tfrac{1}{2}, X_3 \le \tfrac{1}{2}\right) \\ &= \int_0^{1/2}\int_0^{1/2}\int_0^{1/2} 8x_1x_2x_3\,dx_1\,dx_2\,dx_3 \\ &= \left(\tfrac{1}{2}\right)^6 = \tfrac{1}{64}.\end{aligned}$$

In a similar manner, we find that the distribution function of Y is

$$\begin{aligned}G(y) = \Pr(Y \le y) &= 0, && y < 0 \\ &= y^6, && 0 \le y < 1, \\ &= 1, && 1 \le y.\end{aligned}$$

Accordingly, the p.d.f. of Y is

$$\begin{aligned}g(y) &= 6y^5, && 0 < y < 1, \\ &= 0 && \text{elsewhere.}\end{aligned}$$

Remark. Unless there is a possible misunderstanding between *mutual* and *pairwise* independence, we usually drop the modifier *mutual*. Accordingly, using this practice in Example 2, we say that X_1, X_2, X_3 are independent random variables, meaning that they are mutually independent. Occasionally, for emphasis, we use *mutually independent* so that the reader is reminded that this is different from *pairwise independence*.

EXERCISES

2.37. Let X, Y, Z have joint p.d.f. $f(x, y, z) = 2(x + y + z)/3$, $0 < x < 1$, $0 < y < 1$, $0 < z < 1$, zero elsewhere.

(a) Find the marginal probability density functions.

(b) Compute $\Pr(0 < X < \frac{1}{2}, 0 < Y < \frac{1}{2}, 0 < Z < \frac{1}{2})$ and $\Pr(0 < X < \frac{1}{2}) = \Pr(0 < Y < \frac{1}{2}) = \Pr(0 < Z < \frac{1}{2})$.

(c) Are X, Y, and Z independent?

(d) Calculate $E(X^2YZ + 3XY^4Z^2)$.

(e) Determine the distribution function of X, Y, and Z.

(f) Find the conditional distribution of X and Y, given $Z = z$, and evaluate $E(X + Y|z)$.

(g) Determine the conditional distribution of X, given $Y = y$ and $Z = z$, and compute $E(X|y, z)$.

2.38. Let $f(x_1, x_2, x_3) = \exp[-(x_1 + x_2 + x_3)]$, $0 < x_1 < \infty$, $0 < x_2 < \infty$, $0 < x_3 < \infty$, zero elsewhere, be the joint p.d.f. of X_1, X_2, X_3.

(a) Compute $\Pr(X_1 < X_2 < X_3)$ and $\Pr(X_1 = X_2 < X_3)$.

(b) Determine the m.g.f. of X_1, X_2, and X_3. Are these random variables independent?

2.39. Let X_1, X_2, X_3, and X_4 be four independent random variables, each with p.d.f. $f(x) = 3(1 - x)^2$, $0 < x < 1$, zero elsewhere. If Y is the minimum of these four variables, find the distribution function and the p.d.f. of Y.

2.40. A fair die is cast at random three independent times. Let the random variable X_i be equal to the number of spots that appear on the ith trial, $i = 1, 2, 3$. Let the random variable Y be equal to $\max(X_i)$. Find the distribution function and the p.d.f. of Y.

Hint: $\Pr(Y \le y) = \Pr(X_i \le y, i = 1, 2, 3)$.

2.41. Let $M(t_1, t_2, t_3)$ be the m.g.f. of the random variables X_1, X_2, and X_3 of Bernstein's example, described in the remark preceding Example 2 of this section. Show that $M(t_1, t_2, 0) = M(t_1, 0, 0)M(0, t_2, 0)$, $M(t_1, 0, t_3) = M(t_1, 0, 0)M(0, 0, t_3)$, $M(0, t_2, t_3) = M(0, t_2, 0)M(0, 0, t_3)$, but $M(t_1, t_2, t_3) \neq M(t_1, 0, 0)M(0, t_2, 0)\,M(0, 0, t_3)$. Thus X_1, X_2, X_3 are pairwise independent but not mutually independent.

2.42. Let X_1, X_2, and X_3 be three random variables with means, variances, and correlation coefficients, denoted by μ_1, μ_2, μ_3; $\sigma_1^2, \sigma_2^2, \sigma_3^2$; and $\rho_{12}, \rho_{13}, \rho_{23}$, respectively. If $E(X_1 - \mu_1|x_2, x_3) = b_2(x_2 - \mu_2) + b_3(x_3 - \mu_3)$, where b_2 and b_3 are constants, determine b_2 and b_3 in terms of the variances and the correlation coefficients.

ADDITIONAL EXERCISES

2.43. Find $\Pr[X_1 X_2 \le 2]$, where X_1 and X_2 are independent and each has the distribution with p.d.f. $f(x) = 1$, $1 < x < 2$, zero elsewhere.

2.44. Let the joint p.d.f. of X and Y be given by $f(x, y) = \dfrac{2}{(1 + x + y)^3}$, $0 < x < \infty$, $0 < y < \infty$, zero elsewhere.

(a) Compute the marginal p.d.f. of X and the conditional p.d.f. of Y, given $X = x$.

(b) For a fixed $X = x$, compute $E(1 + x + Y|x)$ and use the result to compute $E(Y|x)$.

2.45. Let X_1, X_2, X_3 be independent and each have a distribution with p.d.f. $f(x) = \exp(-x)$, $0 < x < \infty$, zero elsewhere. Evaluate:

(a) $\Pr(X_1 < X_2|X_1 < 2X_2)$.

(b) $\Pr(X_1 < X_2 < X_3|X_3 < 1)$.

2.46. Let X and Y be random variables with space consisting of the four points: $(0, 0)$, $(1, 1)$, $(1, 0)$, $(1, -1)$. Assign positive probabilities to these four points so that the correlation coefficient is equal to zero. Are X and Y independent?

2.47. Two line segments, each of length 2 units, are placed along the x-axis. The midpoint of the first is between $x = 0$ and $x = 14$ and that of the second is between $x = 6$ and $x = 20$. Assuming independence and uniform distributions for these midpoints, find the probability that the line segments overlap.

2.48. Let X and Y have the joint p.d.f. $f(x, y) = \frac{1}{7}$, $(x, y) = (0, 0), (1, 0), (0, 1), (1, 1), (2, 1), (1, 2), (2, 2)$, and zero elsewhere. Find the correlation coefficient ρ.

2.49. Let X_1 and X_2 have the joint p.d.f. described by the following table:

(x_1, x_2)	$(0, 0)$	$(0, 1)$	$(0, 2)$	$(1, 1)$	$(1, 2)$	$(2, 2)$
$f(x_1, x_2)$	$\frac{1}{12}$	$\frac{2}{12}$	$\frac{1}{12}$	$\frac{3}{12}$	$\frac{4}{12}$	$\frac{1}{12}$

Find $f_1(x_1)$, $f_2(x_2)$, μ_1, μ_2, σ_1^2, σ_2^2, and ρ.

2.50. If the discrete random variables X_1 and X_2 have joint p.d.f. $f(x_1, x_2) = (3x_1 + x_2)/24$, $(x_1, x_2) = (1, 1), (1, 2), (2, 1), (2, 2)$, zero elsewhere, find the conditional mean $E(X_2|x_1)$, when $x_1 = 1$.

2.51. Let X and Y have the joint p.d.f. $f(x, y) = 21x^2y^3$, $0 < x < y < 1$, zero elsewhere. Find the conditional mean $E(Y|x)$ of Y, given $X = x$.

2.52. Let X_1 and X_2 have the p.d.f. $f(x_1, x_2) = x_1 + x_2$, $0 < x_1 < 1$, $0 < x_2 < 1$, zero elsewhere. Evaluate $\Pr(X_1/X_2 \le 2)$.

2.53. Cast a fair die and let $X = 0$ if 1, 2, or 3 spots appear, let $X = 1$ if 4 or 5 spots appear, and let $X = 2$ if 6 spots appear. Do this two independent times, obtaining X_1 and X_2. Calculate $\Pr(|X_1 - X_2| = 1)$.

2.54. Let $\sigma_1^2 = \sigma_2^2 = \sigma^2$ be the common variance of X_1 and X_2 and let ρ be the correlation coefficient of X_1 and X_2. Show that

$$\Pr[|(X_1 - \mu_1) + (X_2 - \mu_2)| \ge k\sigma] \le \frac{2(1 + \rho)}{k^2}.$$

CHAPTER 3

Some Special Distributions

3.1 The Binomial and Related Distributions

In Chapter 1 we introduced the *uniform distribution* and the *hypergeometric distribution*. In this chapter we discuss some other important distributions of random variables frequently used in statistics. We begin with the binomial and related distributions.

A *Bernoulli experiment* is a random experiment, the outcome of which can be classified in but one of two mutually exclusive and exhaustive ways, say, success or failure (e.g., female or male, life or death, nondefective or defective). A sequence of *Bernoulli trials* occurs when a Bernoulli experiment is performed several independent times so that the probability of success, say p, remains the same from trial to trial. That is, in such a sequence, we let p denote the probability of success on each trial.

Let X be a random variable associated with a Bernoulli trial by defining it as follows:

$$X(\text{success}) = 1 \qquad \text{and} \qquad X(\text{failure}) = 0.$$

That is, the two outcomes, success and failure, are denoted by one and zero, respectively. The p.d.f. of X can be written as

$$f(x) = p^x(1 - p)^{1-x}, \qquad x = 0, 1,$$

and we say that X has a *Bernoulli distribution*. The expected value of X is

$$\mu = E(X) = \sum_{x=0}^{1} xp^x(1 - p)^{1-x} = (0)(1 - p) + (1)(p) = p,$$

and the variance of X is

$$\sigma^2 = \text{var}(X) = \sum_{x=0}^{1} (x - p)^2 p^x(1 - p)^{1-x}$$

$$= p^2(1 - p) + (1 - p)^2 p = p(1 - p).$$

It follows that the standard deviation of X is $\sigma = \sqrt{p(1 - p)}$.

In a sequence of n Bernoulli trials, we shall let X_i denote the Bernoulli random variable associated with the ith trial. An observed sequence of n Bernoulli trials will then be an n-tuple of zeros and ones. In such a sequence of Bernoulli trials, we are often interested in the total number of successes and not in the order of their occurrence. If we let the random variable X equal the number of observed successes in n Bernoulli trials, the possible values of X are $0, 1, 2, \ldots, n$. If x successes occur, where $x = 0, 1, 2, \ldots, n$, then $n - x$ failures occur. The number of ways of selecting x positions for the x successes in the n trials is

$$\binom{n}{x} = \frac{n!}{x!\,(n - x)!}.$$

Since the trials are independent and since the probabilities of success and failure on each trial are, respectively, p and $1 - p$, the probability of each of these ways is $p^x(1 - p)^{n-x}$. Thus the p.d.f. of X, say $f(x)$, is the sum of the probabilities of these $\binom{n}{x}$ mutually exclusive events; that is,

$$f(x) = \binom{n}{x} p^x(1 - p)^{n-x}, \qquad x = 0, 1, 2, \ldots, n,$$

$$= 0 \qquad \text{elsewhere.}$$

Recall, if n is a positive integer, that

$$(a+b)^n = \sum_{x=0}^{n} \binom{n}{x} b^x a^{n-x}.$$

Thus it is clear that $f(x) \geq 0$ and that

$$\sum_x f(x) = \sum_{x=0}^{n} \binom{n}{x} p^x (1-p)^{n-x}$$

$$= [(1-p)+p]^n = 1.$$

That is, $f(x)$ satisfies the conditions of being a p.d.f. of a random variable X of the discrete type. A random variable X that has a p.d.f. of the form of $f(x)$ is said to have a *binomial distribution*, and any such $f(x)$ is called a *binomial p.d.f.* A binomial distribution will be denoted by the symbol $b(n, p)$. The constants n and p are called the *parameters* of the binomial distribution. Thus, if we say that X is $b(5, \frac{1}{3})$, we mean that X has the binomial p.d.f.

$$f(x) = \binom{5}{x}\left(\frac{1}{3}\right)^x\left(\frac{2}{3}\right)^{5-x}, \qquad x = 0, 1, \ldots, 5,$$

$$= 0 \qquad \text{elsewhere.}$$

The m.g.f. of a binomial distribution is easily found. It is

$$M(t) = \sum_x e^{tx} f(x) = \sum_{x=0}^{n} e^{tx} \binom{n}{x} p^x (1-p)^{n-x}$$

$$= \sum_{x=0}^{n} \binom{n}{x} (pe^t)^x (1-p)^{n-x}$$

$$= [(1-p) + pe^t]^n$$

for all real values of t. The mean μ and the variance σ^2 of X may be computed from $M(t)$. Since

$$M'(t) = n[(1-p)+pe^t]^{n-1}(pe^t)$$

and

$$M''(t) = n[(1-p)+pe^t]^{n-1}(pe^t) + n(n-1)[(1-p)+pe^t]^{n-2}(pe^t)^2,$$

it follows that

$$\mu = M'(0) = np$$

and

$$\sigma^2 = M''(0) - \mu^2 = np + n(n-1)p^2 - (np)^2 = np(1-p).$$

Example 1. Let X be the number of heads (successes) in $n = 7$ independent tosses of an unbiased coin. The p.d.f. of X is

$$f(x) = \binom{7}{x}\left(\frac{1}{2}\right)^x\left(1 - \frac{1}{2}\right)^{7-x}, \qquad x = 0, 1, 2, \ldots, 7,$$

$$= 0 \qquad \text{elsewhere.}$$

Then X has the m.g.f.

$$M(t) = (\tfrac{1}{2} + \tfrac{1}{2}e^t)^7,$$

has mean $\mu = np = \frac{7}{2}$, and has variance $\sigma^2 = np(1-p) = \frac{7}{4}$. Furthermore, we have

$$\Pr(0 \le X \le 1) = \sum_{x=0}^{1} f(x) = \frac{1}{128} + \frac{7}{128} = \frac{8}{128}$$

and

$$\Pr(X = 5) = f(5)$$

$$= \frac{7!}{5!\,2!}\left(\frac{1}{2}\right)^5\left(\frac{1}{2}\right)^2 = \frac{21}{128}.$$

Example 2. If the m.g.f. of a random variable X is

$$M(t) = (\tfrac{2}{3} + \tfrac{1}{3}e^t)^5,$$

then X has a binomial distribution with $n = 5$ and $p = \frac{1}{3}$; that is, the p.d.f. of X is

$$f(x) = \binom{5}{x}\left(\frac{1}{3}\right)^x\left(\frac{2}{3}\right)^{5-x}, \qquad x = 0, 1, 2, \ldots, 5,$$

$$= 0 \qquad \text{elsewhere.}$$

Here $\mu = np = \frac{5}{3}$ and $\sigma^2 = np(1-p) = \frac{10}{9}$.

Example 3. If Y is $b(n, \frac{1}{3})$, then $\Pr(Y \ge 1) = 1 - \Pr(Y = 0) = 1 - (\frac{2}{3})^n$. Suppose that we wish to find the smallest value of n that yields $\Pr(Y \ge 1) > 0.80$. We have $1 - (\frac{2}{3})^n > 0.80$ and $0.20 > (\frac{2}{3})^n$. Either by inspection or by use of logarithms, we see that $n = 4$ is the solution. That is, the probability of at least one success throughout $n = 4$ independent repetitions of a random experiment with probability of success $p = \frac{1}{3}$ is greater than 0.80.

Example 4. Let the random variable Y be equal to the number of successes throughout n independent repetitions of a random experiment with probability p of success. That is, Y is $b(n, p)$. The ratio Y/n is called the relative frequency of success. For every $\epsilon > 0$, we have

$$\Pr\left(\left|\frac{Y}{n} - p\right| \ge \epsilon\right) = \Pr(|Y - np| \ge \epsilon n)$$

$$= \Pr\left(|Y - \mu| \ge \epsilon\sqrt{\frac{n}{p(1-p)}}\,\sigma\right),$$

where $\mu = np$ and $\sigma^2 = np(1-p)$. In accordance with Chebyshev's inequality with $k = \epsilon\sqrt{n/p(1-p)}$, we have

$$\Pr\left(|Y-\mu| \geq \epsilon\sqrt{\frac{n}{p(1-p)}}\,\sigma\right) \leq \frac{p(1-p)}{n\epsilon^2}$$

and hence

$$\Pr\left(\left|\frac{Y}{n} - p\right| \geq \epsilon\right) \leq \frac{p(1-p)}{n\epsilon^2}.$$

Now, for every fixed $\epsilon > 0$, the right-hand member of the preceding inequality is close to zero for sufficiently large n. That is,

$$\lim_{n\to\infty} \Pr\left(\left|\frac{Y}{n} - p\right| \geq \epsilon\right) = 0$$

and

$$\lim_{n\to\infty} \Pr\left(\left|\frac{Y}{n} - p\right| < \epsilon\right) = 1.$$

Since this is true for every fixed $\epsilon > 0$, we see, in a certain sense, that the relative frequency of success is for large values of n, close to the probability p of success. This result is one form of the *law of large numbers*. It was alluded to in the initial discussion of probability in Chapter 1 and will be considered again, along with related concepts, in Chapter 5.

Example 5. Let the independent random variables X_1, X_2, X_3 have the same distribution function $F(x)$. Let Y be the middle value of X_1, X_2, X_3. To determine the distribution function of Y, say $G(y) = \Pr(Y \leq y)$, we note that $Y \leq y$ if and only if at least two of the random variables X_1, X_2, X_3 are less than or equal to y. Let us say that the ith "trial" is a success if $X_i \leq y$, $i = 1, 2, 3$; here each "trial" has the probability of success $F(y)$. In this terminology, $G(y) = \Pr(Y \leq y)$ is then the probability of at least two successes in three independent trials. Thus

$$G(y) = \binom{3}{2}[F(y)]^2[1 - F(y)] + [F(y)]^3.$$

If $F(x)$ is a continuous type of distribution function so that the p.d.f. of X is $F'(x) = f(x)$, then the p.d.f. of Y is

$$g(y) = G'(y) = 6[F(y)][1 - F(y)]\,f(y).$$

Example 6. Consider a sequence of independent repetitions of a random experiment with constant probability p of success. Let the random variable Y denote the total number of failures in this sequence before the rth success;

that is, $Y + r$ is equal to the number of trials necessary to produce exactly r successes. Here r is a fixed positive integer. To determine the p.d.f. of Y, let y be an element of $\{y : y = 0, 1, 2, \ldots\}$. Then, by the multiplication rule of probabilities, $\Pr(Y = y) = g(y)$ is equal to the product of the probability

$$\binom{y+r-1}{r-1} p^{r-1}(1-p)^y$$

of obtaining exactly $r - 1$ successes in the first $y + r - 1$ trials and the probability p of a success on the $(y + r)$th trial. Thus the p.d.f. $g(y)$ of Y is given by

$$\begin{aligned} g(y) &= \binom{y+r-1}{r-1} p^r(1-p)^y, \qquad y = 0, 1, 2, \ldots, \\ &= 0 \qquad \text{elsewhere.} \end{aligned}$$

A distribution with a p.d.f. of the form $g(y)$ is called a *negative binomial distribution*; and any such $g(y)$ is called a negative binomial p.d.f. The distribution derives its name from the fact that $g(y)$ is a general term in the expansion of $p^r[1 - (1 - p)]^{-r}$. It is left as an exercise to show that the m.g.f. of this distribution is $M(t) = p^r[1 - (1 - p)e^t]^{-r}$, for $t < -\ln(1 - p)$. If $r = 1$, then Y has the p.d.f.

$$g(y) = p(1-p)^y, \qquad y = 0, 1, 2, \ldots,$$

zero elsewhere, and the m.g.f. $M(t) = p[1 - (1 - p)e^t]^{-1}$. In this special case, $r = 1$, we say that Y has a *geometric distribution.*

The binomial distribution is generalized to the multinomial distribution as follows. Let a random experiment be repeated n independent times. On each repetition, the experiment terminates in but one of k mutually exclusive and exhaustive ways, say $C_1, C_2, \ldots, C_k$. Let p_i be the probability that the outcome is an element of C_i and let p_i remain constant throughout the n independent repetitions, $i = 1, 2, \ldots, k$. Define the random variable X_i to be equal to the number of outcomes that are elements of C_i, $i = 1, 2, \ldots, k - 1$. Furthermore, let $x_1, x_2, \ldots, x_{k-1}$ be nonnegative integers so that $x_1 + x_2 + \cdots + x_{k-1} \le n$. Then the probability that exactly x_1 terminations of the experiment are in $C_1, \ldots,$ exactly x_{k-1} terminations are in C_{k-1}, and hence exactly $n - (x_1 + \cdots + x_{k-1})$ terminations are in C_k is

$$\frac{n!}{x_1! \cdots x_{k-1}!\, x_k!} p_1^{x_1} \cdots p_{k-1}^{x_{k-1}} p_k^{x_k},$$

where x_k is merely an abbreviation for $n - (x_1 + \cdots + x_{k-1})$. This is the *multinomial p.d.f.* of $k - 1$ random variables $X_1, X_2, \ldots, X_{k-1}$ of the discrete type. To see that this is correct, note that the number of distinguishable arrangements of $x_1 C_1$'s, $x_2 C_2$'s, ..., $x_k C_k$'s is

$$\binom{n}{x_1}\binom{n - x_1}{x_2} \cdots \binom{n - x_1 - \cdots - x_{k-2}}{x_{k-1}} = \frac{n!}{x_1!\, x_2! \cdots x_k!}$$

and that the probability of each of these distinguishable arrangements is

$$p_1^{x_1} p_2^{x_2} \cdots p_k^{x_k}.$$

Hence the product of these two latter expressions gives the correct probability, which is in agreement with the formula for the multinomial p.d.f.

When $k = 3$, we often let $X = X_1$ and $Y = X_2$; then $n - X - Y = X_3$. We say that X and Y have a *trinomial distribution*. The joint p.d.f. of X and Y is

$$f(x, y) = \frac{n!}{x!\, y!\, (n - x - y)!} p_1^x p_2^y p_3^{n-x-y},$$

where x and y are nonnegative integers with $x + y \le n$, and p_1, p_2, and p_3 are positive proper fractions with $p_1 + p_2 + p_3 = 1$; and let $f(x, y) = 0$ elsewhere. Accordingly, $f(x, y)$ satisfies the conditions of being a joint p.d.f. of two random variables X and Y of the discrete type; that is, $f(x, y)$ is nonnegative and its sum over all points (x, y) at which $f(x, y)$ is positive is equal to $(p_1 + p_2 + p_3)^n = 1$.

If n is a positive integer and a_1, a_2, a_3 are fixed constants, we have

$$\begin{aligned}
&\sum_{x=0}^{n} \sum_{y=0}^{n-x} \frac{n!}{x!\, y!\, (n - x - y)!} a_1^x a_2^y a_3^{n-x-y} \\
&= \sum_{x=0}^{n} \frac{n!\, a_1^x}{x!\, (n - x)!} \sum_{y=0}^{n-x} \frac{(n - x)!}{y!\, (n - x - y)!} a_2^y a_3^{n-x-y} \\
&= \sum_{x=0}^{n} \frac{n!}{x!\, (n - x)!} a_1^x (a_2 + a_3)^{n-x} \\
&= (a_1 + a_2 + a_3)^n. \qquad (1)
\end{aligned}$$

Consequently, the m.g.f. of a trinomial distribution, in accordance with Equation (1), is given by

$$\begin{aligned}
M(t_1, t_2) &= \sum_{x=0}^{n} \sum_{y=0}^{n-x} \frac{n!}{x!\, y!\, (n - x - y)!} (p_1 e^{t_1})^x (p_2 e^{t_2})^y p_3^{n-x-y} \\
&= (p_1 e^{t_1} + p_2 e^{t_2} + p_3)^n,
\end{aligned}$$

for all real values of t_1 and t_2. The moment-generating functions of the marginal distributions of X and Y are, respectively,

$$M(t_1, 0) = (p_1e^{t_1} + p_2 + p_3)^n = [(1 - p_1) + p_1e^{t_1}]^n$$

and

$$M(0, t_2) = (p_1 + p_2e^{t_2} + p_3)^n = [(1 - p_2) + p_2e^{t_2}]^n.$$

We see immediately, from Theorem 4, Section 2.4, that X and Y are dependent random variables. In addition, X is $b(n, p_1)$ and Y is $b(n, p_2)$. Accordingly, the means and the variances of X and Y are, respectively, $\mu_1 = np_1$, $\mu_2 = np_2$, $\sigma_1^2 = np_1(1 - p_1)$, and $\sigma_2^2 = np_2(1 - p_2)$.

Consider next the conditional p.d.f. of Y, given $X = x$. We have

$$\begin{aligned} f_{2|1}(y|x) &= \frac{(n-x)!}{y!\,(n-x-y)!}\left(\frac{p_2}{1-p_1}\right)^y\left(\frac{p_3}{1-p_1}\right)^{n-x-y}, \quad y=0, 1, \ldots, n-x, \\ &= 0 \qquad \text{elsewhere.} \end{aligned}$$

Thus the conditional distribution of Y, given $X = x$, is $b[n - x, p_2/(1 - p_1)]$. Hence the conditional mean of Y, given $X = x$, is the linear function

$$E(Y|x) = (n - x)\left(\frac{p_2}{1 - p_1}\right).$$

We also find that the conditional distribution of X, given $Y = y$, is $b[n - y, p_1/(1 - p_2)]$ and thus

$$E(X|y) = (n - y)\left(\frac{p_1}{1 - p_2}\right).$$

Now recall (Example 2, Section 2.3) that the square of the correlation coefficient, say ρ^2, is equal to the product of $-p_2/(1 - p_1)$ and $-p_1/(1 - p_2)$, the coefficients of x and y in the respective conditional means. Since both of these coefficients are negative (and thus ρ is negative), we have

$$\rho = -\sqrt{\frac{p_1p_2}{(1 - p_1)(1 - p_2)}}.$$

In general, the m.g.f. of a multinomial distribution is given by

$$M(t_1, \ldots, t_{k-1}) = (p_1e^{t_1} + \cdots + p_{k-1}e^{t_{k-1}} + p_k)^n$$

for all real values of $t_1, t_2, \ldots, t_{k-1}$. Thus each one-variable marginal p.d.f. is binomial, each two-variable marginal p.d.f. is trinomial, and so on.

EXERCISES

3.1. If the m.g.f. of a random variable X is $(\frac{1}{3} + \frac{2}{3}e^t)^5$, find $\Pr(X = 2 \text{ or } 3)$.

3.2. The m.g.f. of a random variable X is $(\frac{2}{3} + \frac{1}{3}e^t)^9$. Show that

$$\Pr(\mu - 2\sigma < X < \mu + 2\sigma) = \sum_{x=1}^{5} \binom{9}{x} \left(\frac{1}{3}\right)^x \left(\frac{2}{3}\right)^{9-x}.$$

3.3. If X is $b(n, p)$, show that

$$E\left(\frac{X}{n}\right) = p \quad \text{and} \quad E\left[\left(\frac{X}{n} - p\right)^2\right] = \frac{p(1-p)}{n}.$$

3.4. Let the independent random variables X_1, X_2, X_3 have the same p.d.f. $f(x) = 3x^2, 0 < x < 1$, zero elsewhere. Find the probability that exactly two of these three variables exceed $\frac{1}{2}$.

3.5. Let Y be the number of successes in n independent repetitions of a random experiment having the probability of success $p = \frac{2}{3}$. If $n = 3$, compute $\Pr(2 \le Y)$; if $n = 5$, compute $\Pr(3 \le Y)$.

3.6. Let Y be the number of successes throughout n independent repetitions of a random experiment having probability of success $p = \frac{1}{4}$. Determine the smallest value of n so that $\Pr(1 \le Y) \ge 0.70$.

3.7. Let the independent random variables X_1 and X_2 have binomial distributions with parameters $n_1 = 3, p_1 = \frac{2}{3}$ and $n_2 = 4, p_2 = \frac{1}{2}$, respectively. Compute $\Pr(X_1 = X_2)$.

Hint: List the four mutually exclusive ways that $X_1 = X_2$ and compute the probability of each.

3.8. Toss two nickels and three dimes at random. Make appropriate assumptions and compute the probability that there are more heads showing on the nickels than on the dimes.

3.9. Let $X_1, X_2, \ldots, X_{k-1}$ have a multinomial distribution.

(a) Find the m.g.f. of $X_2, X_3, \ldots, X_{k-1}$.

(b) What is the p.d.f. of $X_2, X_3, \ldots, X_{k-1}$?

(c) Determine the conditional p.d.f. of X_1, given that

$$X_2 = x_2, \ldots, X_{k-1} = x_{k-1}.$$

(d) What is the conditional expectation $E(X_1|x_2, \ldots, x_{k-1})$?

3.10. Let X be $b(2, p)$ and let Y be $b(4, p)$. If $\Pr(X \ge 1) = \frac{5}{9}$, find $\Pr(Y \ge 1)$.

3.11. If $x = r$ is the unique mode of a distribution that is $b(n, p)$, show that

$$(n + 1)p - 1 < r < (n + 1)p.$$

Hint: Determine the values of x for which the ratio $f(x + 1)/f(x) > 1$.

3.12. Let X have a binomial distribution with parameters n and $p = \frac{1}{3}$. Determine the smallest integer n can be such that $\Pr(X \geq 1) \geq 0.85$.

3.13. Let X have the p.d.f. $f(x) = (\frac{1}{3})(\frac{2}{3})^x$, $x = 0, 1, 2, 3, \ldots$, zero elsewhere. Find the conditional p.d.f. of X, given that $X \geq 3$.

3.14. One of the numbers $1, 2, \ldots, 6$ is to be chosen by casting an unbiased die. Let this random experiment be repeated five independent times. Let the random variable X_1 be the number of terminations in the set $\{x : x = 1, 2, 3\}$ and let the random variable X_2 be the number of terminations in the set $\{x : x = 4, 5\}$. Compute $\Pr(X_1 = 2, X_2 = 1)$.

3.15. Show that the m.g.f. of the negative binomial distribution is $M(t) = p^r[1 - (1 - p)e^t]^{-r}$. Find the mean and the variance of this distribution.

Hint: In the summation representing $M(t)$, make use of the MacLaurin's series for $(1 - w)^{-r}$.

3.16. Let X_1 and X_2 have a trinomial distribution. Differentiate the moment-generating function to show that their covariance is $-np_1p_2$.

3.17. If a fair coin is tossed at random five independent times, find the conditional probability of five heads relative to the hypothesis that there are at least four heads.

3.18. Let an unbiased die be cast at random seven independent times. Compute the conditional probability that each side appears at least once relative to the hypothesis that side 1 appears exactly twice.

3.19. Compute the measures of skewness and kurtosis of the binomial distribution $b(n, p)$.

3.20. Let

$$f(x_1, x_2) = \binom{x_1}{x_2}\left(\frac{1}{2}\right)^{x_1}\left(\frac{x_1}{15}\right), \qquad \begin{array}{l} x_2 = 0, 1, \ldots, x_1, \\ x_1 = 1, 2, 3, 4, 5, \end{array}$$

zero elsewhere, be the joint p.d.f. of X_1 and X_2. Determine:

(a) $E(X_2)$.

(b) $u(x_1) = E(X_2|x_1)$.

(c) $E[u(X_1)]$.

Compare the answers of parts (a) and (c).

Hint: Note that $E(X_2) = \sum_{x_1=1}^{5} \sum_{x_2=0}^{x_1} x_2 f(x_1, x_2)$ and use the fact that $\sum_{y=0}^{n} y \binom{n}{y} (\frac{1}{2})^n = n/2$. Why?

3.21. Three fair dice are cast. In 10 independent casts, let X be the number of times all three faces are alike and let Y be the number of times only two faces are alike. Find the joint p.d.f. of X and Y and compute $E(6XY)$.

3.2 The Poisson Distribution

Recall that the series

$$1 + m + \frac{m^2}{2!} + \frac{m^3}{3!} + \cdots = \sum_{x=0}^{\infty} \frac{m^x}{x!}$$

converges, for all values of m, to e^m. Consider the function $f(x)$ defined by

$$f(x) = \frac{m^x e^{-m}}{x!}, \qquad x = 0, 1, 2, \ldots,$$

$$= 0 \qquad \text{elsewhere},$$

where $m > 0$. Since $m > 0$, then $f(x) \geq 0$ and

$$\sum_x f(x) = \sum_{x=0}^{\infty} \frac{m^x e^{-m}}{x!} = e^{-m} \sum_{x=0}^{\infty} \frac{m^x}{x!} = e^{-m} e^m = 1;$$

that is, $f(x)$ satisfies the conditions of being a p.d.f. of a discrete type of random variable. A random variable that has a p.d.f. of the form $f(x)$ is said to have a *Poisson distribution*, and any such $f(x)$ is called a *Poisson p.d.f.*

Remarks. Experience indicates that the Poisson p.d.f. may be used in a number of applications with quite satisfactory results. For example, let the random variable X denote the number of alpha particles emitted by a radioactive substance that enter a prescribed region during a prescribed interval of time. With a suitable value of m, it is found that X may be assumed to have a Poisson distribution. Again let the random variable X denote the number of defects on a manufactured article, such as a refrigerator door. Upon examining many of these doors, it is found, with an appropriate value of m, that X may be said to have a Poisson distribution. The number of automobile accidents in some unit of time (or the number of insurance claims in some unit of time) is often assumed to be a random variable which has a Poisson distribution. Each of these instances can be thought of as a process that generates a number of changes (accidents, claims, etc.) in a fixed interval (of time or space, etc.). If a process leads to a Poisson distribution, that process is called a *Poisson process*. Some assumptions that ensure a Poisson process will now be enumerated.

Let $g(x, w)$ denote the probability of x changes in each interval of length w. Furthermore, let the symbol $o(h)$ represent any function such that $\lim_{h\to 0} [o(h)/h] = 0$; for example, $h^2 = o(h)$ and $o(h) + o(h) = o(h)$. The Poisson postulates are the following:

1. $g(1, h) = \lambda h + o(h)$, where λ is a positive constant and $h > 0$.
2. $\sum_{x=2}^{\infty} g(x, h) = o(h)$.
3. The numbers of changes in nonoverlapping intervals are independent.

Postulates 1 and 3 state, in effect, that the probability of one change in a short interval h is independent of changes in other nonoverlapping intervals and is approximately proportional to the length of the interval. The substance of postulate 2 is that the probability of two or more changes in the same short interval h is essentially equal to zero. If $x = 0$, we take $g(0, 0) = 1$. In accordance with postulates 1 and 2, the probability of at least one change in an interval of length h is $\lambda h + o(h) + o(h) = \lambda h + o(h)$. Hence the probability of zero changes in this interval of length h is $1 - \lambda h - o(h)$. Thus the probability $g(0, w + h)$ of zero changes in an interval of length $w + h$ is, in accordance with postulate 3, equal to the product of the probability $g(0, w)$ of zero changes in an interval of length w and the probability $[1 - \lambda h - o(h)]$ of zero changes in a nonoverlapping interval of length h. That is,

$$g(0, w + h) = g(0, w)[1 - \lambda h - o(h)].$$

Then

$$\frac{g(0, w + h) - g(0, w)}{h} = -\lambda g(0, w) - \frac{o(h)g(0, w)}{h}.$$

If we take the limit as $h \to 0$, we have

$$D_w[g(0, w)] = -\lambda g(0, w).$$

The solution of this differential equation is

$$g(0, w) = ce^{-\lambda w}.$$

The condition $g(0, 0) = 1$ implies that $c = 1$; so

$$g(0, w) = e^{-\lambda w}.$$

If x is a positive integer, we take $g(x, 0) = 0$. The postulates imply that

$$g(x, w + h) = [g(x, w)][1 - \lambda h - o(h)] + [g(x - 1, w)][\lambda h + o(h)] + o(h).$$

Accordingly, we have

$$\frac{g(x, w+h) - g(x, w)}{h} = -\lambda g(x, w) + \lambda g(x-1, w) + \frac{o(h)}{h}$$

and

$$D_w[g(x, w)] = -\lambda g(x, w) + \lambda g(x-1, w),$$

for $x = 1, 2, 3, \ldots$. It can be shown, by mathematical induction, that the solutions to these differential equations, with boundary conditions $g(x, 0) = 0$ for $x = 1, 2, 3, \ldots$, are, respectively,

$$g(x, w) = \frac{(\lambda w)^x e^{-\lambda w}}{x!}, \qquad x = 1, 2, 3, \ldots.$$

Hence the number of changes X in an interval of length w has a Poisson distribution with parameter $m = \lambda w$.

The m.g.f. of a Poisson distribution is given by

$$M(t) = \sum_x e^{tx} f(x) = \sum_{x=0}^{\infty} e^{tx} \frac{m^x e^{-m}}{x!}$$

$$= e^{-m} \sum_{x=0}^{\infty} \frac{(me^t)^x}{x!}$$

$$= e^{-m} e^{me^t} = e^{m(e^t - 1)}$$

for all real values of t. Since

$$M'(t) = e^{m(e^t - 1)}(me^t)$$

and

$$M''(t) = e^{m(e^t - 1)}(me^t) + e^{m(e^t - 1)}(me^t)^2,$$

then

$$\mu = M'(0) = m$$

and

$$\sigma^2 = M''(0) - \mu^2 = m + m^2 - m^2 = m.$$

That is, a Poisson distribution has $\mu = \sigma^2 = m > 0$. On this account, a Poisson p.d.f. is frequently written

$$f(x) = \frac{\mu^x e^{-\mu}}{x!}, \qquad x = 0, 1, 2, \ldots,$$

$$= 0 \qquad \text{elsewhere.}$$

Thus the parameter m in a Poisson p.d.f. is the mean μ. Table I in Appendix B gives approximately the distribution for various values of the parameter $m = \mu$.

Example 1. Suppose that X has a Poisson distribution with $\mu = 2$. Then the p.d.f. of X is

$$f(x) = \frac{2^x e^{-2}}{x!}, \qquad x = 0, 1, 2, \ldots,$$
$$= 0 \qquad \text{elsewhere.}$$

The variance of this distribution is $\sigma^2 = \mu = 2$. If we wish to compute $\Pr(1 \le X)$, we have

$$\Pr(1 \le X) = 1 - \Pr(X = 0)$$
$$= 1 - f(0) = 1 - e^{-2} = 0.865,$$

approximately, by Table I of Appendix B.

Example 2. If the m.g.f. of a random variable X is

$$M(t) = e^{4(e^t - 1)},$$

then X has a Poisson distribution with $\mu = 4$. Accordingly, by way of example,

$$\Pr(X = 3) = \frac{4^3 e^{-4}}{3!} = \frac{32}{3} e^{-4};$$

or, by Table I,

$$\Pr(X = 3) = \Pr(X \le 3) - \Pr(X \le 2) = 0.433 - 0.238 = 0.195.$$

Example 3. Let the probability of exactly one blemish in 1 foot of wire be about $\frac{1}{1000}$ and let the probability of two or more blemishes in that length be, for all practical purposes, zero. Let the random variable X be the number of blemishes in 3000 feet of wire. If we assume the independence of the numbers of blemishes in nonoverlapping intervals, then the postulates of the Poisson process are approximated, with $\lambda = \frac{1}{1000}$ and $w = 3000$. Thus X has an approximate Poisson distribution with mean $3000(\frac{1}{1000}) = 3$. For example, the probability that there are exactly five blemishes in 3000 feet of wire is

$$\Pr(X = 5) = \frac{3^5 e^{-3}}{5!}$$

and by Table I,

$$\Pr(X = 5) = \Pr(X \le 5) - \Pr(X \le 4) = 0.101,$$

approximately.

EXERCISES

3.22. If the random variable X has a Poisson distribution such that $\Pr(X = 1) = \Pr(X = 2)$, find $\Pr(X = 4)$.

3.23. The m.g.f. of a random variable X is $e^{4(e^t - 1)}$. Show that $\Pr(\mu - 2\sigma < X < \mu + 2\sigma) = 0.931$.

3.24. In a lengthy manuscript, it is discovered that only 13.5 percent of the pages contain no typing errors. If we assume that the number of errors per page is a random variable with a Poisson distribution, find the percentage of pages that have exactly one error.

3.25. Let the p.d.f. $f(x)$ be positive on and only on the nonnegative integers. Given that $f(x) = (4/x)f(x - 1)$, $x = 1, 2, 3, \ldots$. Find $f(x)$.

Hint: Note that $f(1) = 4f(0)$, $f(2) = (4^2/2!)f(0)$, and so on. That is, find each $f(x)$ in terms of $f(0)$ and then determine $f(0)$ from

$$1 = f(0) + f(1) + f(2) + \cdots.$$

3.26. Let X have a Poisson distribution with $\mu = 100$. Use Chebyshev's inequality to determine a lower bound for $\Pr(75 < X < 125)$.

3.27. Given that $g(x, 0) = 0$ and that

$$D_w[g(x, w)] = -\lambda g(x, w) + \lambda g(x - 1, w)$$

for $x = 1, 2, 3, \ldots$. If $g(0, w) = e^{-\lambda w}$, show, by mathematical induction, that

$$g(x, w) = \frac{(\lambda w)^x e^{-\lambda w}}{x!}, \qquad x = 1, 2, 3, \ldots.$$

3.28. Let the number of chocolate drops in a certain type of cookie have a Poisson distribution. We want the probability that a cookie of this type contains at least two chocolate drops to be greater than 0.99. Find the smallest value that the mean of the distribution can take.

3.29. Compute the measures of skewness and kurtosis of the Poisson distribution with mean μ.

3.30. On the average a grocer sells 3 of a certain article per week. How many of these should he have in stock so that the chance of his running out within a week will be less than 0.01? Assume a Poisson distribution.

3.31. Let X have a Poisson distribution. If $\Pr(X = 1) = \Pr(X = 3)$, find the mode of the distribution.

3.32. Let X have a Poisson distribution with mean 1. Compute, if it exists, the expected value $E(X!)$.

3.33. Let X and Y have the joint p.d.f. $f(x, y) = e^{-2}/[x!(y-x)!]$, $y = 0, 1, 2, \ldots$; $x = 0, 1, \ldots, y$, zero elsewhere.
(a) Find the m.g.f. $M(t_1, t_2)$ of this joint distribution.
(b) Compute the means, the variances, and the correlation coefficient of X and Y.
(c) Determine the conditional mean $E(X|y)$.
Hint: Note that

$$\sum_{x=0}^{y} [\exp(t_1 x)]y!/[x!\,(y-x)!] = [1 + \exp(t_1)]^y.$$

Why?

3.3 The Gamma and Chi-Square Distributions

In this section we introduce the gamma and chi-square distributions. It is proved in books on advanced calculus that the integral

$$\int_0^\infty y^{\alpha-1}e^{-y}\,dy$$

exists for $\alpha > 0$ and that the value of the integral is a positive number. The integral is called the gamma function of α, and we write

$$\Gamma(\alpha) = \int_0^\infty y^{\alpha-1}e^{-y}\,dy.$$

If $\alpha = 1$, clearly

$$\Gamma(1) = \int_0^\infty e^{-y}\,dy = 1.$$

If $\alpha > 1$, an integration by parts shows that

$$\Gamma(\alpha) = (\alpha-1)\int_0^\infty y^{\alpha-2}e^{-y}\,dy = (\alpha-1)\Gamma(\alpha-1).$$

Accordingly, if α is a positive integer greater than 1,

$$\Gamma(\alpha) = (\alpha-1)(\alpha-2)\cdots(3)(2)(1)\Gamma(1) = (\alpha-1)!.$$

Since $\Gamma(1) = 1$, this suggests that we take $0! = 1$, as we have done.

In the integral that defines $\Gamma(\alpha)$, let us introduce a new variable x by writing $y = x/\beta$, where $\beta > 0$. Then

$$\Gamma(\alpha) = \int_0^\infty \left(\frac{x}{\beta}\right)^{\alpha-1} e^{-x/\beta}\left(\frac{1}{\beta}\right) dx,$$

or, equivalently,

$$1 = \int_0^\infty \frac{1}{\Gamma(\alpha)\beta^\alpha} x^{\alpha-1} e^{-x/\beta}\, dx.$$

Since $\alpha > 0$, $\beta > 0$, and $\Gamma(\alpha) > 0$, we see that

$$f(x) = \frac{1}{\Gamma(\alpha)\beta^\alpha} x^{\alpha-1} e^{-x/\beta}, \qquad 0 < x < \infty,$$

$$= 0 \qquad \text{elsewhere,}$$

is a p.d.f. of a random variable of the continuous type. A random variable X that has a p.d.f. of this form is said to have a *gamma distribution* with parameters α and β; and any such $f(x)$ is called a *gamma-type p.d.f.*

Remark. The gamma distribution is frequently the probability model for waiting times; for instance, in life testing, the waiting time until "death" is the random variable which frequently has a gamma distribution. To see this, let us assume the postulates of a Poisson process and let the interval of length w be a time interval. Specifically, let the random variable W be the time that is needed to obtain exactly k changes (possibly deaths), where k is a fixed positive integer. Then the distribution function of W is

$$G(w) = \Pr(W \le w) = 1 - \Pr(W > w).$$

However, the event $W > w$, for $w > 0$, is equivalent to the event in which there are less than k changes in a time interval of length w. That is, if the random variable X is the number of changes in an interval of length w, then

$$\Pr(W > w) = \sum_{x=0}^{k-1} \Pr(X = x) = \sum_{x=0}^{k-1} \frac{(\lambda w)^x e^{-\lambda w}}{x!}.$$

It is left as an exercise to verify that

$$\int_{\lambda w}^\infty \frac{z^{k-1} e^{-z}}{(k-1)!}\, dz = \sum_{x=0}^{k-1} \frac{(\lambda w)^x e^{-\lambda w}}{x!}.$$

If, momentarily, we accept this result, we have, for $w > 0$,

$$G(w) = 1 - \int_{\lambda w}^\infty \frac{z^{k-1} e^{-z}}{\Gamma(k)}\, dz = \int_0^{\lambda w} \frac{z^{k-1} e^{-z}}{\Gamma(k)}\, dz,$$

and for $w \le 0$, $G(w) = 0$. If we change the variable of integration in the integral that defines $G(w)$ by writing $z = \lambda y$, then

$$G(w) = \int_0^w \frac{\lambda^k y^{k-1} e^{-\lambda y}}{\Gamma(k)}\, dy, \qquad w > 0,$$

and $G(w) = 0$, $w \le 0$. Accordingly, the p.d.f. of W is

$$g(w) = G'(w) = \frac{\lambda^k w^{k-1} e^{-\lambda w}}{\Gamma(k)}, \qquad 0 < w < \infty,$$
$$= 0 \qquad \text{elsewhere.}$$

That is, W has a gamma distribution with $\alpha = k$ and $\beta = 1/\lambda$. If W is the waiting time until the first change, that is, if $k = 1$, the p.d.f. of W is

$$g(w) = \lambda e^{-\lambda w}, \qquad 0 < w < \infty,$$
$$= 0 \qquad \text{elsewhere,}$$

and W is said to have an *exponential distribution* with mean $\beta = 1/\lambda$.

We now find the m.g.f. of a gamma distribution. Since

$$M(t) = \int_0^\infty e^{tx} \frac{1}{\Gamma(\alpha)\beta^\alpha} x^{\alpha-1} e^{-x/\beta}\, dx$$
$$= \int_0^\infty \frac{1}{\Gamma(\alpha)\beta^\alpha} x^{\alpha-1} e^{-x(1-\beta t)/\beta}\, dx,$$

we may set $y = x(1 - \beta t)/\beta$, $t < 1/\beta$, or $x = \beta y/(1 - \beta t)$, to obtain

$$M(t) = \int_0^\infty \frac{\beta/(1-\beta t)}{\Gamma(\alpha)\beta^\alpha} \left(\frac{\beta y}{1-\beta t}\right)^{\alpha-1} e^{-y}\, dy.$$

That is,

$$M(t) = \left(\frac{1}{1-\beta t}\right)^\alpha \int_0^\infty \frac{1}{\Gamma(\alpha)} y^{\alpha-1} e^{-y}\, dy$$
$$= \frac{1}{(1-\beta t)^\alpha}, \qquad t < \frac{1}{\beta}.$$

Now

$$M'(t) = (-\alpha)(1 - \beta t)^{-\alpha-1}(-\beta)$$

and

$$M''(t) = (-\alpha)(-\alpha - 1)(1 - \beta t)^{-\alpha-2}(-\beta)^2.$$

Hence, for a gamma distribution, we have

$$\mu = M'(0) = \alpha\beta$$

and

$$\sigma^2 = M''(0) - \mu^2 = \alpha(\alpha + 1)\beta^2 - \alpha^2\beta^2 = \alpha\beta^2.$$

Example 1. Let the waiting time W have a gamma p.d.f. with $\alpha = k$ and $\beta = 1/\lambda$. Accordingly, $E(W) = k/\lambda$. If $k = 1$, then $E(W) = 1/\lambda$; that is, the expected waiting time for $k = 1$ changes is equal to the reciprocal of λ.

Example 2. Let X be a random variable such that

$$E(X^m) = \frac{(m+3)!}{3!} 3^m, \qquad m = 1, 2, 3, \ldots .$$

Then the m.g.f. of X is given by the series

$$M(t) = 1 + \frac{4!\ 3}{3!\ 1!} t + \frac{5!\ 3^2}{3!\ 2!} t^2 + \frac{6!\ 3^3}{3!\ 3!} t^3 + \cdots .$$

This, however, is the Maclaurin's series for $(1 - 3t)^{-4}$, provided that $-1 < 3t < 1$. Accordingly, X has a gamma distribution with $\alpha = 4$ and $\beta = 3$.

Remark. The gamma distribution is not only a good model for waiting times, but one for many nonnegative random variables of the continuous type. For illustration, the distribution of certain incomes could be modeled satisfactorily by the gamma distribution, since the two parameters α and β provide a great deal of flexibility. Several gamma probability density functions are depicted in Figure 3.1.

Let us now consider the special case of the gamma distribution in which $\alpha = r/2$, where r is a positive integer, and $\beta = 2$. A random variable X of the continuous type that has the p.d.f.

$$\begin{aligned} f(x) &= \frac{1}{\Gamma(r/2)2^{r/2}} x^{r/2-1} e^{-x/2}, \qquad 0 < x < \infty, \\ &= 0 \qquad \text{elsewhere,} \end{aligned}$$

and the m.g.f.

$$M(t) = (1 - 2t)^{-r/2}, \qquad t < \tfrac{1}{2},$$

is said to have a *chi-square distribution*, and any $f(x)$ of this form is called a *chi-square p.d.f.* The mean and the variance of a chi-square distribution are $\mu = \alpha\beta = (r/2)2 = r$ and $\sigma^2 = \alpha\beta^2 = (r/2)2^2 = 2r$, respectively. For no obvious reason, we call the parameter r the number of degrees of freedom of the chi-square distribution (or of the chi-square p.d.f.). Because the chi-square distribution has an important role in statistics and occurs so frequently, we write, for brevity, that X is $\chi^2(r)$ to mean that the random variable X has a chi-square distribution with r degrees of freedom.

Example 3. If X has the p.d.f.

$$f(x) = \tfrac{1}{4}xe^{-x/2}, \qquad 0 < x < \infty,$$
$$= 0 \quad \text{elsewhere},$$

then X is $\chi^2(4)$. Hence $\mu = 4$, $\sigma^2 = 8$, and $M(t) = (1 - 2t)^{-2}$, $t < \frac{1}{2}$.

Example 4. If X has the m.g.f. $M(t) = (1 - 2t)^{-8}$, $t < \frac{1}{2}$, then X is $\chi^2(16)$.

If the random variable X is $\chi^2(r)$, then, with $c_1 < c_2$, we have

$$\Pr(c_1 \le X \le c_2) = \Pr(X \le c_2) - \Pr(X \le c_1),$$

since $\Pr(X = c_1) = 0$. To compute such a probability, we need the value of an integral like

$$\Pr(X \le x) = \int_0^x \frac{1}{\Gamma(r/2)2^{r/2}} w^{r/2-1}e^{-w/2}\,dw.$$

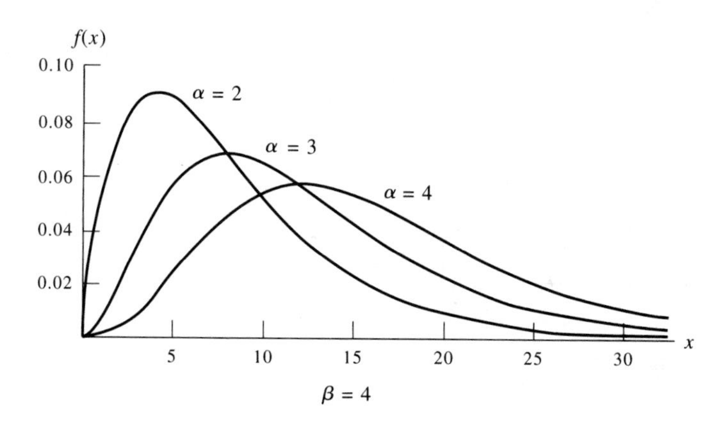

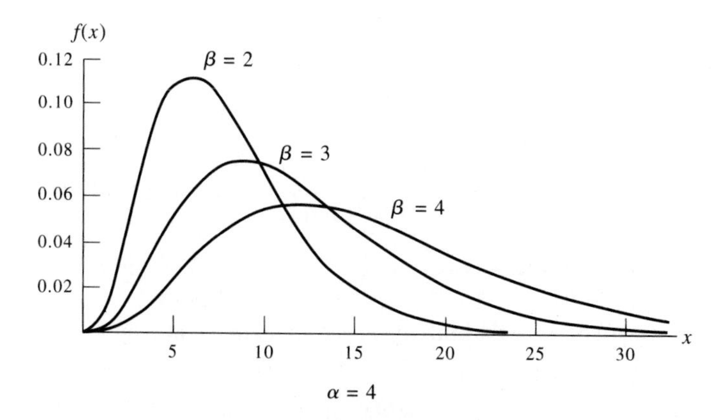

FIGURE 3.1

Tables of this integral for selected values of r and x have been prepared and are partially reproduced in Table II in Appendix B.

Example 5. Let X be $\chi^2(10)$. Then, by Table II of Appendix B, with $r = 10$,

$$\begin{aligned}\Pr(3.25 \le X \le 20.5) &= \Pr(X \le 20.5) - \Pr(X \le 3.25)\\ &= 0.975 - 0.025 = 0.95.\end{aligned}$$

Again, by way of example, if $\Pr(a < X) = 0.05$, then $\Pr(X \le a) = 0.95$, and thus $a = 18.3$ from Table II with $r = 10$.

Example 6. Let X have a gamma distribution with $\alpha = r/2$, where r is a positive integer, and $\beta > 0$. Define the random variable $Y = 2X/\beta$. We seek the p.d.f. of Y. Now the distribution function of Y is

$$G(y) = \Pr(Y \le y) = \Pr\left(X \le \frac{\beta y}{2}\right).$$

If $y \le 0$, then $G(y) = 0$; but if $y > 0$, then

$$G(y) = \int_0^{\beta y/2} \frac{1}{\Gamma(r/2)\beta^{r/2}} x^{r/2-1} e^{-x/\beta}\, dx.$$

Accordingly, the p.d.f. of Y is

$$\begin{aligned}g(y) = G'(y) &= \frac{\beta/2}{\Gamma(r/2)\beta^{r/2}} (\beta y/2)^{r/2-1} e^{-y/2}\\ &= \frac{1}{\Gamma(r/2)2^{r/2}} y^{r/2-1} e^{-y/2}\end{aligned}$$

if $y > 0$. That is, Y is $\chi^2(r)$.

EXERCISES

3.34. If $(1 - 2t)^{-6}$, $t < \frac{1}{2}$, is the m.g.f. of the random variable X, find $\Pr(X < 5.23)$.

3.35. If X is $\chi^2(5)$, determine the constants c and d so that $\Pr(c < X < d) = 0.95$ and $\Pr(X < c) = 0.025$.

3.36. If X has a gamma distribution with $\alpha = 3$ and $\beta = 4$, find $\Pr(3.28 < X < 25.2)$.

Hint: Consider the probability of the equivalent event $1.64 < Y < 12.6$, where $Y = 2X/4 = X/2$.

3.37. Let X be a random variable such that $E(X^m) = (m + 1)!\, 2^m$, $m = 1, 2, 3, \ldots$. Determine the m.g.f. and the distribution of X.

3.38. Show that

$$\int_{\mu}^{\infty} \frac{1}{\Gamma(k)} z^{k-1}e^{-z}\,dz = \sum_{x=0}^{k-1} \frac{\mu^x e^{-\mu}}{x!}, \qquad k = 1, 2, 3, \ldots .$$

This demonstrates the relationship between the distribution functions of the gamma and Poisson distributions.

Hint: Either integrate by parts $k - 1$ times or simply note that the "antiderivative" of $z^{k-1}e^{-z}$ is

$$-z^{k-1}e^{-z} - (k-1)z^{k-2}e^{-z} - \cdots - (k-1)!\,e^{-z}$$

by differentiating the latter expression.

3.39. Let X_1, X_2, and X_3 be independent random variables, each with p.d.f. $f(x) = e^{-x}$, $0 < x < \infty$, zero elsewhere. Find the distribution of $Y = \text{minimum}\,(X_1, X_2, X_3)$.

Hint: $\Pr(Y \le y) = 1 - \Pr(Y > y) = 1 - \Pr(X_i > y, i = 1, 2, 3)$.

3.40. Let X have a gamma distribution with p.d.f.

$$f(x) = \frac{1}{\beta^2} x e^{-x/\beta}, \qquad 0 < x < \infty,$$

zero elsewhere. If $x = 2$ is the unique mode of the distribution, find the parameter β and $\Pr(X < 9.49)$.

3.41. Compute the measures of skewness and kurtosis of a gamma distribution with parameters α and β.

3.42. Let X have a gamma distribution with parameters α and β. Show that $\Pr(X \ge 2\alpha\beta) \le (2/e)^\alpha$.

Hint: Use the result of Exercise 1.115.

3.43. Give a reasonable definition of a chi-square distribution with zero degrees of freedom.

Hint: Work with the m.g.f. of a distribution that is $\chi^2(r)$ and let $r = 0$.

3.44. In the Poisson postulates on page 127, let λ be a nonnegative function of w, say $\lambda(w)$, such that $D_w[g(0, w)] = -\lambda(w)g(0, w)$. Suppose that $\lambda(w) = krw^{r-1}$, $r \ge 1$.

(a) Find $g(0, w)$ noting that $g(0, 0) = 1$.

(b) Let W be the time that is needed to obtain exactly one change. Then find the distribution function of W, namely $G(w) = \Pr(W \le w) = 1 - \Pr(W > w) = 1 - g(0, w)$, $0 \le w$, and then find the p.d.f. of W. This p.d.f. is that of the *Weibull distribution*, which is used in the study of breaking strengths of materials.

3.45. Let X have a Poisson distribution with parameter m. If m is an experimental value of a random variable having a gamma distribution with $\alpha = 2$ and $\beta = 1$, compute $\Pr(X = 0, 1, 2)$.

3.46. Let X have the uniform distribution with p.d.f. $f(x) = 1, 0 < x < 1$, zero elsewhere. Find the distribution function of $Y = -2 \ln X$. What is the p.d.f. of Y?

3.47. Find the uniform distribution of the continuous type that has the same mean and the same variance as those of a chi-square distribution with 8 degrees of freedom.

3.4 The Normal Distribution

Consider the integral

$$I = \int_{-\infty}^{\infty} \exp\left(\frac{-y^2}{2}\right) dy.$$

This integral exists because the integrand is a positive continuous function which is bounded by an integrable function; that is,

$$0 < \exp\left(-\frac{y^2}{2}\right) < \exp(-|y| + 1), \qquad -\infty < y < \infty,$$

and

$$\int_{-\infty}^{\infty} \exp(-|y| + 1)\, dy = 2e.$$

To evaluate the integral I, we note that $I > 0$ and that I^2 may be written

$$I^2 = \int_{-\infty}^{\infty} \int_{-\infty}^{\infty} \exp\left(-\frac{y^2 + z^2}{2}\right) dy\, dz.$$

This iterated integral can be evaluated by changing to polar coordinates. If we set $y = r \cos \theta$ and $z = r \sin \theta$, we have

$$I^2 = \int_0^{2\pi} \int_0^{\infty} e^{-r^2/2} r\, dr\, d\theta$$

$$= \int_0^{2\pi} d\theta = 2\pi.$$

Accordingly, $I = \sqrt{2\pi}$ and

$$\int_{-\infty}^{\infty} \frac{1}{\sqrt{2\pi}} e^{-y^2/2}\, dy = 1.$$

If we introduce a new variable of integration, say x, by writing

$$y = \frac{x - a}{b}, \qquad b > 0,$$

the preceding integral becomes

$$\int_{-\infty}^{\infty} \frac{1}{b\sqrt{2\pi}} \exp\left[-\frac{(x-a)^2}{2b^2}\right] dx = 1.$$

Since $b > 0$, this implies that

$$f(x) = \frac{1}{b\sqrt{2\pi}} \exp\left[-\frac{(x-a)^2}{2b^2}\right], \qquad -\infty < x < \infty$$

satisfies the conditions of being a p.d.f. of a continuous type of random variable. A random variable of the continuous type that has a p.d.f. of the form of $f(x)$ is said to have a *normal distribution*, and any $f(x)$ of this form is called a normal p.d.f.

We can find the m.g.f. of a normal distribution as follows. In

$$\begin{aligned} M(t) &= \int_{-\infty}^{\infty} e^{tx} \frac{1}{b\sqrt{2\pi}} \exp\left[-\frac{(x-a)^2}{2b^2}\right] dx \\ &= \int_{-\infty}^{\infty} \frac{1}{b\sqrt{2\pi}} \exp\left(-\frac{-2b^2tx + x^2 - 2ax + a^2}{2b^2}\right) dx \end{aligned}$$

we complete the square in the exponent. Thus $M(t)$ becomes

$$\begin{aligned} M(t) &= \exp\left[-\frac{a^2 - (a + b^2t)^2}{2b^2}\right] \int_{-\infty}^{\infty} \frac{1}{b\sqrt{2\pi}} \\ &\quad \times \exp\left[-\frac{(x - a - b^2t)^2}{2b^2}\right] dx \\ &= \exp\left(at + \frac{b^2t^2}{2}\right) \end{aligned}$$

because the integrand of the last integral can be thought of as a normal p.d.f. with a replaced by $a + b^2t$, and hence it is equal to 1.

The mean μ and variance σ^2 of a normal distribution will be calculated from $M(t)$. Now

$$M'(t) = M(t)(a + b^2t)$$

and

$$M''(t) = M(t)(b^2) + M(t)(a + b^2t)^2.$$

Thus

$$\mu = M'(0) = a$$

and

$$\sigma^2 = M''(0) - \mu^2 = b^2 + a^2 - a^2 = b^2.$$

This permits us to write a normal p.d.f. in the form of

$$f(x) = \frac{1}{\sigma\sqrt{2\pi}} \exp\left[-\frac{(x-\mu)^2}{2\sigma^2}\right], \qquad -\infty < x < \infty,$$

a form that shows explicitly the values of μ and σ^2. The m.g.f. $M(t)$ can be written

$$M(t) = \exp\left(\mu t + \frac{\sigma^2 t^2}{2}\right).$$

Example 1. If X has the m.g.f.

$$M(t) = e^{2t + 32t^2},$$

then X has a normal distribution with $\mu = 2$, $\sigma^2 = 64$.

The normal p.d.f. occurs so frequently in certain parts of statistics that we denote it, for brevity, by $N(\mu, \sigma^2)$. Thus, if we say that the random variable X is $N(0, 1)$, we mean that X has a normal distribution with mean $\mu = 0$ and variance $\sigma^2 = 1$, so that the p.d.f. of X is

$$f(x) = \frac{1}{\sqrt{2\pi}} e^{-x^2/2}, \qquad -\infty < x < \infty.$$

If we say that X is $N(5, 4)$, we mean that X has a normal distribution with mean $\mu = 5$ and variance $\sigma^2 = 4$, so that the p.d.f. of X is

$$f(x) = \frac{1}{2\sqrt{2\pi}} \exp\left[-\frac{(x-5)^2}{2(4)}\right], \qquad -\infty < x < \infty.$$

Moreover, if

$$M(t) = e^{t^2/2},$$

then X is $N(0, 1)$.

The graph of

$$f(x) = \frac{1}{\sigma\sqrt{2\pi}} \exp\left[-\frac{(x-\mu)^2}{2\sigma^2}\right], \qquad -\infty < x < \infty,$$

is seen (1) to be symmetric about a vertical axis through $x = \mu$, (2) to have its maximum of $1/(\sigma\sqrt{2\pi})$ at $x = \mu$, and (3) to have the x-axis as a horizontal asymptote. It should also be verified that (4) there are points of inflection at $x = \mu \pm \sigma$.

Remark. Each of the special distributions considered thus far has been "justified" by some derivation that is based upon certain concepts found in elementary probability theory. Such a motivation for the normal distribution is not given at this time; a motivation is presented in Chapter 5. However, the normal distribution is one of the more widely used distributions in applications of statistical methods. Variables that are often assumed to be random variables having normal distributions (with appropriate values of μ and σ) are the diameter of a hole made by a drill press, the score on a test, the yield of a grain on a plot of ground, and the length of a newborn child.

We now prove a very useful theorem.

Theorem 1. *If the random variable X is $N(\mu, \sigma^2)$, $\sigma^2 > 0$, then the random variable $W = (X - \mu)/\sigma$ is $N(0, 1)$.*

Proof. The distribution function $G(w)$ of W is, since $\sigma > 0$,

$$G(w) = \Pr\left(\frac{X - \mu}{\sigma} \le w\right) = \Pr\,(X \le w\sigma + \mu).$$

That is,

$$G(w) = \int_{-\infty}^{w\sigma + \mu} \frac{1}{\sigma\sqrt{2\pi}} \exp\left[-\frac{(x - \mu)^2}{2\sigma^2}\right] dx.$$

If we change the variable of integration by writing $y = (x - \mu)/\sigma$, then

$$G(w) = \int_{-\infty}^{w} \frac{1}{\sqrt{2\pi}} e^{-y^2/2}\, dy.$$

Accordingly, the p.d.f. $g(w) = G'(w)$ of the continuous-type random variable W is

$$g(w) = \frac{1}{\sqrt{2\pi}} e^{-w^2/2}, \qquad -\infty < w < \infty.$$

Thus W is $N(0, 1)$, which is the desired result (see also Exercise 3.100).

This fact considerably simplifies the calculations of probabilities concerning normally distributed variables, as will be seen presently.

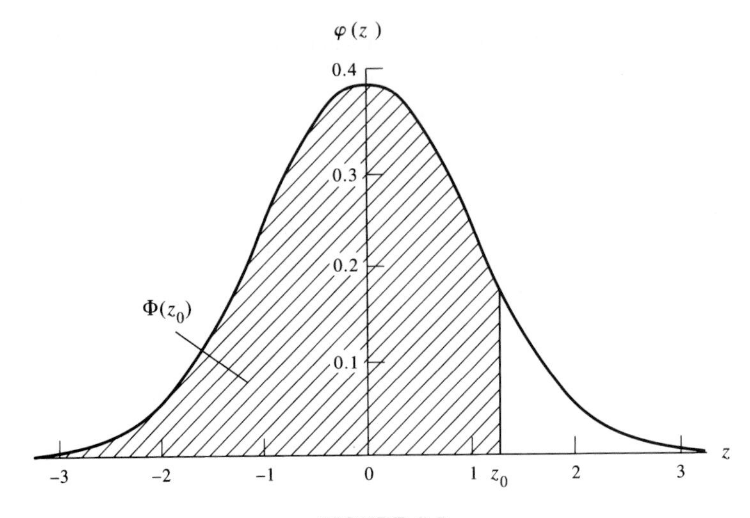

FIGURE 3.2

Suppose that X is $N(\mu, \sigma^2)$. Then, with $c_1 < c_2$ we have, since $\Pr(X = c_1) = 0$,

$$\begin{aligned}\Pr(c_1 < X < c_2) &= \Pr(X < c_2) - \Pr(X < c_1)\\ &= \Pr\left(\frac{X-\mu}{\sigma} < \frac{c_2-\mu}{\sigma}\right) - \Pr\left(\frac{X-\mu}{\sigma} < \frac{c_1-\mu}{\sigma}\right)\\ &= \int_{-\infty}^{(c_2-\mu)/\sigma} \frac{1}{\sqrt{2\pi}} e^{-w^2/2}\,dw - \int_{-\infty}^{(c_1-\mu)/\sigma} \frac{1}{\sqrt{2\pi}} e^{-w^2/2}\,dw\end{aligned}$$

because $W = (X - \mu)/\sigma$ is $N(0, 1)$. That is, probabilities concerning X, which is $N(\mu, \sigma^2)$, can be expressed in terms of probabilities concerning W, which is $N(0, 1)$.

An integral such as

$$\int_{-\infty}^{k} \frac{1}{\sqrt{2\pi}} e^{-w^2/2}\,dw$$

cannot be evaluated by the fundamental theorem of calculus because an "antiderivative" of $e^{-w^2/2}$ is not expressible as an elementary function. Instead, tables of the approximate value of this integral for various values of k have been prepared and are partially reproduced in Table III in Appendix B. We use the notation

$$\Phi(z) = \int_{-\infty}^{z} \frac{1}{\sqrt{2\pi}} e^{-w^2/2}\,dw.$$

Moreover, we say that $\Phi(z)$ and its derivative $\Phi'(z) = \varphi(z)$ are, respectively, the distribution function and p.d.f. of a *standard normal distribution* $N(0, 1)$. These are depicted in Figure 3.2.

To summarize, we have shown that if X is $N(\mu, \sigma^2)$, then

$$\Pr(c_1 < X < c_2) = \Pr\left(\frac{X-\mu}{\sigma} < \frac{c_2-\mu}{\sigma}\right) - \Pr\left(\frac{X-\mu}{\sigma} < \frac{c_1-\mu}{\sigma}\right)$$

$$= \Phi\left(\frac{c_2-\mu}{\sigma}\right) - \Phi\left(\frac{c_1-\mu}{\sigma}\right).$$

It is left as an exercise to show that $\Phi(-x) = 1 - \Phi(x)$.

Example 2. Let X be $N(2, 25)$. Then, by Table III,

$$\Pr(0 < X < 10) = \Phi\left(\frac{10-2}{5}\right) - \Phi\left(\frac{0-2}{5}\right)$$

$$= \Phi(1.6) - \Phi(-0.4)$$

$$= 0.945 - (1 - 0.655) = 0.600$$

and

$$\Pr(-8 < X < 1) = \Phi\left(\frac{1-2}{5}\right) - \Phi\left(\frac{-8-2}{5}\right)$$

$$= \Phi(-0.2) - \Phi(-2)$$

$$= (1 - 0.579) - (1 - 0.977) = 0.398.$$

Example 3. Let X be $N(\mu, \sigma^2)$. Then, by Table III,

$$\Pr(\mu - 2\sigma < X < \mu + 2\sigma) = \Phi\left(\frac{\mu + 2\sigma - \mu}{\sigma}\right) - \Phi\left(\frac{\mu - 2\sigma - \mu}{\sigma}\right)$$

$$= \Phi(2) - \Phi(-2)$$

$$= 0.977 - (1 - 0.977) = 0.954.$$

Example 4. Suppose that 10 percent of the probability for a certain distribution that is $N(\mu, \sigma^2)$ is below 60 and that 5 percent is above 90. What are the values of μ and σ? We are given that the random variable X is $N(\mu, \sigma^2)$ and that $\Pr(X \le 60) = 0.10$ and $\Pr(X \le 90) = 0.95$. Thus $\Phi[(60 - \mu)/\sigma] = 0.10$ and $\Phi[(90 - \mu)/\sigma] = 0.95$. From Table III we have

$$\frac{60-\mu}{\sigma} = -1.282, \qquad \frac{90-\mu}{\sigma} = 1.645.$$

These conditions require that $\mu = 73.1$ and $\sigma = 10.2$ approximately.

Remark. In this chapter we have illustrated three types of *parameters* associated with distributions. The mean μ of $N(\mu, \sigma^2)$ is called a *location* *parameter* because changing its value simply changes the location of the middle of the normal p.d.f.; that is, the graph of the p.d.f. looks exactly the same except for a shift in location. The *standard deviation* σ of $N(\mu, \sigma^2)$ is called a *scale parameter* because changing its value changes the spread of the distribution. That is, a small value of σ requires the graph of the normal p.d.f. to be tall and narrow, while a large value of σ requires it to spread out and not be so tall. No matter what the values of μ and σ, however, the graph of the normal p.d.f. will be that familiar "bell shape." Incidentally, the β of the gamma distribution is also a scale parameter. On the other hand, the α of the gamma distribution is called a *shape parameter*, as changing its value modifies the shape of the graph of the p.d.f. as can be seen by referring to Figure 3.1. The parameters p and μ of the binomial and Poisson distributions, respectively, are also shape parameters.

We close this section with an important theorem.

Theorem 2. *If the random variable* X is $N(\mu, \sigma^2)$, $\sigma^2 > 0$, *then the random variable* $V = (X - \mu)^2/\sigma^2$ *is* $\chi^2(1)$.

Proof. Because $V = W^2$, where $W = (X - \mu)/\sigma$ is $N(0, 1)$, the distribution function $G(v)$ of V is, for $v \geq 0$,

$$G(v) = \Pr(W^2 \leq v) = \Pr(-\sqrt{v} \leq W \leq \sqrt{v}).$$

That is,

$$G(v) = 2\int_0^{\sqrt{v}} \frac{1}{\sqrt{2\pi}} e^{-w^2/2}\, dw, \qquad 0 \leq v,$$

and

$$G(v) = 0, \qquad v < 0.$$

If we change the variable of integration by writing $w = \sqrt{y}$, then

$$G(v) = \int_0^{v} \frac{1}{\sqrt{2\pi}\sqrt{y}} e^{-y/2}\, dy, \qquad 0 \leq v.$$

Hence the p.d.f. $g(v) = G'(v)$ of the continuous-type random variable V is

$$g(v) = \frac{1}{\sqrt{\pi}\sqrt{2}} v^{1/2 - 1} e^{-v/2}, \qquad 0 < v < \infty,$$
$$= 0 \qquad \text{elsewhere.}$$

Since $g(v)$ is a p.d.f. and hence

$$\int_0^{\infty} g(v)\, dv = 1,$$

it must be that $\Gamma(\frac{1}{2}) = \sqrt{\pi}$ and thus V is $\chi^2(1)$.

EXERCISES

3.48. If

$$\Phi(z) = \int_{-\infty}^{z} \frac{1}{\sqrt{2\pi}} e^{-w^2/2}\, dw,$$

show that $\Phi(-z) = 1 - \Phi(z)$.

3.49. If X is $N(75, 100)$, find $\Pr(X < 60)$ and $\Pr(70 < X < 100)$.

3.50. If X is $N(\mu, \sigma^2)$, find b so that $\Pr[-b < (X - \mu)/\sigma < b] = 0.90$.

3.51. Let X be $N(\mu, \sigma^2)$ so that $\Pr(X < 89) = 0.90$ and $\Pr(X < 94) = 0.95$. Find μ and σ^2.

3.52. Show that the constant c can be selected so that $f(x) = c2^{-x^2}$, $-\infty < x < \infty$, satisfies the conditions of a normal p.d.f.

Hint: Write $2 = e^{\ln 2}$.

3.53. If X is $N(\mu, \sigma^2)$, show that $E(|X - \mu|) = \sigma\sqrt{2/\pi}$.

3.54. Show that the graph of a p.d.f. $N(\mu, \sigma^2)$ has points of inflection at $x = \mu - \sigma$ and $x = \mu + \sigma$.

3.55. Evaluate $\int_2^3 \exp[-2(x - 3)^2]\, dx$.

3.56. Determine the ninetieth percentile of the distribution, which is $N(65, 25)$.

3.57. If $e^{3t + 8t^2}$ is the m.g.f. of the random variable X, find $\Pr(-1 < X < 9)$.

3.58. Let the random variable X have the p.d.f.

$$f(x) = \frac{2}{\sqrt{2\pi}} e^{-x^2/2}, \qquad 0 < x < \infty, \quad \text{zero elsewhere.}$$

Find the mean and variance of X.

Hint: Compute $E(X)$ directly and $E(X^2)$ by comparing that integral with the integral representing the variance of a variable that is $N(0, 1)$.

3.59. Let X be $N(5, 10)$. Find $\Pr[0.04 < (X - 5)^2 < 38.4]$.

3.60. If X is $N(1, 4)$, compute the probability $\Pr(1 < X^2 < 9)$.

3.61. If X is $N(75, 25)$, find the conditional probability that X is greater than 80 relative to the hypothesis that X is greater than 77. See Exercise 2.18.

3.62. Let X be a random variable such that $E(X^{2m}) = (2m)!/(2^m m!)$, $m = 1, 2, 3, \ldots$ and $E(X^{2m-1}) = 0$, $m = 1, 2, 3, \ldots$. Find the m.g.f. and the p.d.f. of X.

3.63. Let the mutually independent random variables X_1, X_2, and X_3 be $N(0, 1)$, $N(2, 4)$, and $N(-1, 1)$, respectively. Compute the probability that exactly two of these three variables are less than zero.

3.64. Compute the measures of skewness and kurtosis of a distribution which is $N(\mu, \sigma^2)$.

3.65. Let the random variable X have a distribution that is $N(\mu, \sigma^2)$.
(a) Does the random variable $Y = X^2$ also have a normal distribution?
(b) Would the random variable $Y = aX + b$, a and b nonzero constants, have a normal distribution?
Hint: In each case, first determine $\Pr(Y \le y)$.

3.66. Let the random variable X be $N(\mu, \sigma^2)$. What would this distribution be if $\sigma^2 = 0$?
Hint: Look at the m.g.f. of X for $\sigma^2 > 0$ and investigate its limit as $\sigma^2 \to 0$.

3.67. Let $\varphi(x)$ and $\Phi(x)$ be the p.d.f. and distribution function of a standard normal distribution. Let Y have a *truncated* distribution with p.d.f. $g(y) = \varphi(y)/[\Phi(b) - \Phi(a)]$, $a < y < b$, zero elsewhere. Show that $E(Y)$ is equal to $[\varphi(a) - \varphi(b)]/[\Phi(b) - \Phi(a)]$.

3.68. Let $f(x)$ and $F(x)$ be the p.d.f. and the distribution function of a distribution of the continuous type such that $f'(x)$ exists for all x. Let the mean of the truncated distribution that has p.d.f. $g(y) = f(y)/F(b)$, $-\infty < y < b$, zero elsewhere, be equal to $-f(b)/F(b)$ for all real b. Prove that $f(x)$ is a p.d.f. of a standard normal distribution.

3.69. Let X and Y be independent random variables, each with a distribution that is $N(0, 1)$. Let $Z = X + Y$. Find the integral that represents the distribution function $G(z) = \Pr(X + Y \le z)$ of Z. Determine the p.d.f. of Z.
Hint: We have that $G(z) = \int_{-\infty}^{\infty} H(x, z)\, dx$, where

$$H(x, z) = \int_{-\infty}^{z-x} \frac{1}{2\pi} \exp[-(x^2 + y^2)/2]\, dy.$$

Find $G'(z)$ by evaluating $\int_{-\infty}^{\infty} [\partial H(x, z)/\partial z]\, dx$.

3.5 The Bivariate Normal Distribution

Remark. If the reader with an adequate background in matrix algebra so chooses, this section can be omitted at this point and Section 4.10 can be considered later. If this decision is made, only an example in Section 4.7 and a few exercises need be skipped because the bivariate normal distribution would not be known. Many statisticians, however, find it easier to remember the multivariate (including the bivariate) normal p.d.f. and m.g.f. using matrix notation that is used in Section 4.10. Moreover, that section provides an excellent example of a transformation (in particular, an orthogonal one) and a good illustration of the moment-generating function technique; these are two of the major concepts introduced in Chapter 4.

Let us investigate the function

$$f(x, y) = \frac{1}{2\pi\sigma_1\sigma_2\sqrt{1-\rho^2}} e^{-q/2}, \qquad -\infty < x < \infty, \qquad -\infty < y < \infty,$$

where, with $\sigma_1 > 0$, $\sigma_2 > 0$, and $-1 < \rho < 1$,

$$q = \frac{1}{1-\rho^2}\left[\left(\frac{x-\mu_1}{\sigma_1}\right)^2 - 2\rho\left(\frac{x-\mu_1}{\sigma_1}\right)\left(\frac{y-\mu_2}{\sigma_2}\right) + \left(\frac{y-\mu_2}{\sigma_2}\right)^2\right].$$

At this point we do not know that the constants μ_1, μ_2, σ_1^2, σ_2^2, and ρ are those respective parameters of a distribution. As a matter of fact, we do not know that $f(x, y)$ has the properties of a joint p.d.f. It will be shown that:

1. $f(x, y)$ is a joint p.d.f.
2. X is $N(\mu_1, \sigma_1^2)$ and Y is $N(\mu_2, \sigma_2^2)$.
3. ρ is the correlation coefficient of X and Y.

A joint p.d.f. of this form is called a *bivariate normal p.d.f.*, and the random variables X and Y are said to have a *bivariate normal distribution*.

That the nonnegative function $f(x, y)$ is actually a joint p.d.f. can be seen as follows. Define $f_1(x)$ by

$$f_1(x) = \int_{-\infty}^{\infty} f(x, y)\, dy.$$

Now

$$(1-\rho^2)q = \left[\left(\frac{y-\mu_2}{\sigma_2}\right) - \rho\left(\frac{x-\mu_1}{\sigma_1}\right)\right]^2 + (1-\rho^2)\left(\frac{x-\mu_1}{\sigma_1}\right)^2$$

$$= \left(\frac{y-b}{\sigma_2}\right)^2 + (1-\rho^2)\left(\frac{x-\mu_1}{\sigma_1}\right)^2,$$

where $b = \mu_2 + \rho(\sigma_2/\sigma_1)(x - \mu_1)$. Thus

$$f_1(x) = \frac{\exp[-(x-\mu_1)^2/2\sigma_1^2]}{\sigma_1\sqrt{2\pi}} \int_{-\infty}^{\infty} \frac{\exp\{-(y-b)^2/[2\sigma_2^2(1-\rho^2)]\}}{\sigma_2\sqrt{1-\rho^2}\sqrt{2\pi}}\, dy.$$

For the purpose of integration, the integrand of the integral in this expression for $f_1(x)$ may be considered a normal p.d.f. with mean b and variance $\sigma_2^2(1-\rho^2)$. Thus this integral is equal to 1 and

$$f_1(x) = \frac{1}{\sigma_1\sqrt{2\pi}} \exp\left[-\frac{(x-\mu_1)^2}{2\sigma_1^2}\right], \qquad -\infty < x < \infty.$$

Since

$$\int_{-\infty}^{\infty}\int_{-\infty}^{\infty} f(x, y)\,dy\,dx = \int_{-\infty}^{\infty} f_1(x)\,dx = 1,$$

the nonnegative function $f(x, y)$ is a joint p.d.f. of two continuous-type random variables X and Y. Accordingly, the function $f_1(x)$ is the marginal p.d.f. of X, and X is seen to be $N(\mu_1, \sigma_1^2)$. In like manner, we see that Y is $N(\mu_2, \sigma_2^2)$.

Moreover, from the development above, we note that

$$f(x, y) = f_1(x)\left(\frac{1}{\sigma_2\sqrt{1-\rho^2}\sqrt{2\pi}} \exp\left[-\frac{(y-b)^2}{2\sigma_2^2(1-\rho^2)}\right]\right),$$

where $b = \mu_2 + \rho(\sigma_2/\sigma_1)(x-\mu_1)$. Accordingly, the second factor in the right-hand member of the equation above is the conditional p.d.f. of Y, given that $X = x$. That is, the conditional p.d.f. of Y, given $X = x$, is itself normal with mean $\mu_2 + \rho(\sigma_2/\sigma_1)(x-\mu_1)$ and variance $\sigma_2^2(1-\rho^2)$. Thus, with a bivariate normal distribution, the conditional mean of Y, given that $X = x$, is linear in x and is given by

$$E(Y|x) = \mu_2 + \rho\frac{\sigma_2}{\sigma_1}(x-\mu_1).$$

Since the coefficient of x in this linear conditional mean $E(Y|x)$ is $\rho\sigma_2/\sigma_1$, and since σ_1 and σ_2 represent the respective standard deviations, the number ρ is, in fact, the correlation coefficient of X and Y. This follows from the result, established in Section 2.3, that the coefficient of x in a general linear conditional mean $E(Y|x)$ is the product of the correlation coefficient and the ratio σ_2/σ_1.

Although the mean of the conditional distribution of Y, given $X = x$, depends upon x (unless $\rho = 0$), the variance $\sigma_2^2(1-\rho^2)$ is the same for all real values of x. Thus, by way of example, given that $X = x$, the conditional probability that Y is within $(2.576)\sigma_2\sqrt{1-\rho^2}$ units of the conditional mean is 0.99, whatever the value of x may be. In this sense, most of the probability for the distribution of X and Y lies in the band

$$\mu_2 + \rho \frac{\sigma_2}{\sigma_1}(x - \mu_1) \pm (2.576)\sigma_2\sqrt{1 - \rho^2}$$

about the graph of the linear conditional mean. For every fixed positive σ_2, the width of this band depends upon ρ. Because the band is narrow when ρ^2 is nearly 1, we see that ρ does measure the intensity of the concentration of the probability for X and Y about the linear conditional mean. This is the fact to which we alluded in the remark of Section 2.3.

In a similar manner we can show that the conditional distribution of X, given $Y = y$, is the normal distribution

$$N\left[\mu_1 + \rho \frac{\sigma_1}{\sigma_2}(y - \mu_2), \sigma_1^2(1 - \rho^2)\right].$$

Example 1. Let us assume that in a certain population of married couples the height X_1 of the husband and the height X_2 of the wife have a bivariate normal distribution with parameters $\mu_1 = 5.8$ feet, $\mu_2 = 5.3$ feet, $\sigma_1 = \sigma_2 = 0.2$ foot, and $\rho = 0.6$. The conditional p.d.f. of X_2, given $X_1 = 6.3$, is normal with mean $5.3 + (0.6)(6.3 - 5.8) = 5.6$ and standard deviation $(0.2)\sqrt{(1 - 0.36)} = 0.16$. Accordingly, given that the height of the husband is 6.3 feet, the probability that his wife has a height between 5.28 and 5.92 feet is

$$\Pr(5.28 < X_2 < 5.92 | X_1 = 6.3) = \Phi(2) - \Phi(-2) = 0.954.$$

The interval (5.28, 5.92) could be thought of as a 95.4 percent *prediction interval* for the wife's height, given $X_1 = 6.3$.

The m.g.f. of a bivariate normal distribution can be determined as follows. We have

$$M(t_1, t_2) = \int_{-\infty}^{\infty}\int_{-\infty}^{\infty} e^{t_1x + t_2y}f(x, y)\, dx\, dy$$

$$= \int_{-\infty}^{\infty} e^{t_1x}f_1(x)\left[\int_{-\infty}^{\infty} e^{t_2y}f_{2|1}(y|x)\, dy\right] dx$$

for all real values of t_1 and t_2. The integral within the brackets is the m.g.f. of the conditional p.d.f. $f_{2|1}(y|x)$. Since $f_{2|1}(y|x)$ is a normal p.d.f. with mean $\mu_2 + \rho(\sigma_2/\sigma_1)(x - \mu_1)$ and variance $\sigma_2^2(1 - \rho^2)$, then

$$\int_{-\infty}^{\infty} e^{t_2y}f_{2|1}(y|x)\, dy = \exp\left\{t_2\left[\mu_2 + \rho\frac{\sigma_2}{\sigma_1}(x - \mu_1)\right] + \frac{t_2^2\sigma_2^2(1 - \rho^2)}{2}\right\}.$$

Accordingly, $M(t_1, t_2)$ can be written in the form

$$\exp\left\{t_2\mu_2 - t_2\rho\frac{\sigma_2}{\sigma_1}\mu_1 + \frac{t_2^2\sigma_2^2(1-\rho^2)}{2}\right\}\int_{-\infty}^{\infty}\exp\left[\left(t_1 + t_2\rho\frac{\sigma_2}{\sigma_1}\right)x\right]f_1(x)\,dx.$$

But $E(e^{tX}) = \exp[\mu_1 t + (\sigma_1^2 t^2)/2]$ for all real values of t. Accordingly, if we set $t = t_1 + t_2\rho(\sigma_2/\sigma_1)$, we see that $M(t_1, t_2)$ is given by

$$\exp\left\{t_2\mu_2 - t_2\rho\frac{\sigma_2}{\sigma_1}\mu_1 + \frac{t_2^2\sigma_2^2(1-\rho^2)}{2} + \mu_1\left(t_1 + t_2\rho\frac{\sigma_2}{\sigma_1}\right) + \sigma_1^2\frac{\left(t_1 + t_2\rho\frac{\sigma_2}{\sigma_1}\right)^2}{2}\right\}$$

or, equivalently,

$$M(t_1, t_2) = \exp\left(\mu_1 t_1 + \mu_2 t_2 + \frac{\sigma_1^2 t_1^2 + 2\rho\sigma_1\sigma_2 t_1 t_2 + \sigma_2^2 t_2^2}{2}\right).$$

It is interesting to note that if, in this m.g.f. $M(t_1, t_2)$, the correlation coefficient ρ is set equal to zero, then

$$M(t_1, t_2) = M(t_1, 0)M(0, t_2).$$

Thus X and Y are independent when $\rho = 0$. If, conversely,

$$M(t_1, t_2) \equiv M(t_1, 0)M(0, t_2),$$

we have $e^{\rho\sigma_1\sigma_2 t_1 t_2} = 1$. Since each of σ_1 and σ_2 is positive, then $\rho = 0$. Accordingly, we have the following theorem.

Theorem 3. *Let X and Y have a bivariate normal distribution with means μ_1 and μ_2, positive variances σ_1^2 and σ_2^2, and correlation coefficient ρ. Then X and Y are independent if and only if $\rho = 0$.*

As a matter of fact, if any two random variables are independent and have positive standard deviations, we have noted in Example 4 of Section 2.4 that $\rho = 0$. However, $\rho = 0$ does not in general imply that two variables are independent; this can be seen in Exercises 2.20 (c) and 2.25. The importance of Theorem 3 lies in the fact that we now know when and only when two random variables that have a bivariate normal distribution are independent.

EXERCISES

3.70. Let X and Y have a bivariate normal distribution with respective parameters $\mu_X = 2.8$, $\mu_Y = 110$, $\sigma_X^2 = 0.16$, $\sigma_Y^2 = 100$, and $\rho = 0.6$. Compute:
(a) $\Pr(106 < Y < 124)$.
(b) $\Pr(106 < Y < 124 | X = 3.2)$.

3.71. Let X and Y have a bivariate normal distribution with parameters $\mu_1 = 3$, $\mu_2 = 1$, $\sigma_1^2 = 16$, $\sigma_2^2 = 25$, and $\rho = \frac{3}{5}$. Determine the following probabilities:
(a) $\Pr(3 < Y < 8)$.
(b) $\Pr(3 < Y < 8 | X = 7)$.
(c) $\Pr(-3 < X < 3)$.
(d) $\Pr(-3 < X < 3 | Y = -4)$.

3.72. If $M(t_1, t_2)$ is the m.g.f. of a bivariate normal distribution, compute the covariance by using the formula

$$\frac{\partial^2 M(0, 0)}{\partial t_1 \, \partial t_2} - \frac{\partial M(0, 0)}{\partial t_1} \frac{\partial M(0, 0)}{\partial t_2}.$$

Now let $\psi(t_1, t_2) = \ln M(t_1, t_2)$. Show that $\partial^2 \psi(0, 0)/\partial t_1 \, \partial t_2$ gives this covariance directly.

3.73. Let X and Y have a bivariate normal distribution with parameters $\mu_1 = 5$, $\mu_2 = 10$, $\sigma_1^2 = 1$, $\sigma_2^2 = 25$, and $\rho > 0$. If $\Pr(4 < Y < 16 | X = 5) = 0.954$, determine ρ.

3.74. Let X and Y have a bivariate normal distribution with parameters $\mu_1 = 20$, $\mu_2 = 40$, $\sigma_1^2 = 9$, $\sigma_2^2 = 4$, and $\rho = 0.6$. Find the shortest interval for which 0.90 is the conditional probability that Y is in this interval, given that $X = 22$.

3.75. Say the correlation coefficient between the heights of husbands and wives is 0.70 and the mean male height is 5 feet 10 inches with standard deviation 2 inches, and the mean female height is 5 feet 4 inches with standard deviation $1\frac{1}{2}$ inches. Assuming a bivariate normal distribution, what is the best guess of the height of a woman whose husband's height is 6 feet? Find a 95 percent prediction interval for her height.

3.76. Let

$$f(x, y) = (1/2\pi) \exp\left[-\tfrac{1}{2}(x^2 + y^2)\right]\{1 + xy \exp\left[-\tfrac{1}{2}(x^2 + y^2 - 2)\right]\},$$

where $-\infty < x < \infty$, $-\infty < y < \infty$. If $f(x, y)$ is a joint p.d.f., it is not a normal bivariate p.d.f. Show that $f(x, y)$ actually is a joint p.d.f. and that each marginal p.d.f. is normal. Thus the fact that each marginal p.d.f. is normal does not imply that the joint p.d.f. is bivariate normal.

3.77. Let X, Y, and Z have the joint p.d.f.

$$\left(\frac{1}{2\pi}\right)^{3/2} \exp\left(-\frac{x^2+y^2+z^2}{2}\right)\left[1 + xyz\exp\left(-\frac{x^2+y^2+z^2}{2}\right)\right],$$

where $-\infty < x < \infty$, $-\infty < y < \infty$, and $-\infty < z < \infty$. While X, Y, and Z are obviously dependent, show that X, Y, and Z are pairwise independent and that each pair has a bivariate normal distribution.

3.78. Let X and Y have a bivariate normal distribution with parameters $\mu_1 = \mu_2 = 0, \sigma_1^2 = \sigma_2^2 = 1$, and correlation coefficient ρ. Find the distribution of the random variable $Z = aX + bY$ in which a and b are nonzero constants.

Hint: Write $G(z) = \Pr(Z \le z)$ as an iterated integral and compute $G'(z) = g(z)$ by differentiating under the first integral sign and then evaluating the resulting integral by completing the square in the exponent.

ADDITIONAL EXERCISES

3.79. Let X have a binomial distribution with parameters $n = 288$ and $p = \frac{1}{3}$. Use Chebyshev's inequality to determine a lower bound for $\Pr(76 < X < 116)$.

3.80. Let $f(x) = \dfrac{e^{-\mu}\mu^x}{x!}$, $x = 0, 1, 2, \ldots$, zero elsewhere. Find the values of μ so that $x = 1$ is the unique mode; that is, $f(0) < f(1)$ and $f(1) > f(2) > f(3) > \cdots$.

3.81. Let X and Y be two independent binomial variables with parameters $n = 4$, $p = \frac{1}{2}$ and $n = 3$, $p = \frac{2}{3}$, respectively. Determine $\Pr(X - Y = 3)$.

3.82. Let X and Y be two independent binomial variables, both with parameters n and $p = \frac{1}{2}$. Show that

$$\Pr(X - Y = 0) = \frac{(2n)!}{n!\, n!\, (2^{2n})}.$$

3.83. Two people toss a coin five independent times each. Find the probability that they will obtain the same number of heads.

3.84. Color blindness appears in 1 percent of the people in a certain population. How large must a sample with replacement be if the probability of its containing at least one color-blind person is to be at least 0.95? Assume a binomial distribution $b(n, p = 0.01)$ and find n.

3.85. Assume that the number X of hours of sunshine per day in a certain place has a chi-square distribution with 10 degrees of freedom. The profit of a certain outdoor activity depends upon the number of hours of sunshine through the formula

$$\text{profit} = 1000(1 - e^{-X/10}).$$

Find the expected level of the profit.

3.86. Place five similar balls (each either red or blue) in a bowl at random as follows: A coin is flipped 5 independent times and a red ball is placed in the bowl for each head and a blue ball for each tail. The bowl is then taken and two balls are selected at random without replacement. Given that each of those two balls is red, compute the conditional probability that 5 red balls were placed in the bowl at random.

3.87. If a die is rolled four independent times, what is the probability of one four, two fives, and one six, given that at least one six is produced?

3.88. Let the p.d.f. $f(x)$ be positive on, and only on, the integers 0, 1, 2, 3, 4, 5, 6, 7, 8, 9, 10, so that $f(x) = [(11 - x)/x]\, f(x - 1)$, $x = 1, 2, 3, \ldots, 10$. Find $f(x)$.

3.89. Let X and Y have a bivariate normal distribution with $\mu_1 = 5$, $\mu_2 = 10$, $\sigma_1^2 = 1$, $\sigma_2^2 = 25$, and $\rho = \frac{4}{5}$. Compute $\Pr(7 < Y < 19|x = 5)$.

3.90. Say that Jim has three cents and that Bill has seven cents. A coin is tossed ten independent times. For each head that appears, Bill pays Jim two cents, and for each tail that appears, Jim pays Bill one cent. What is the probability that neither person is in debt after the ten trials?

3.91. If $E(X^r) = [(r + 1)!](2^r)$, $r = 1, 2, 3, \ldots,$ find the m.g.f. and p.d.f. of X.

3.92. For a biased coin, say that the probability of exactly two heads in three independent tosses is $\frac{4}{9}$. What is the probability of exactly six heads in nine independent tosses of this coin?

3.93. It is discovered that 75 percent of the pages of a certain book contain no errors. If we assume that the number of errors per page follows a Poisson distribution, find the percentage of pages that have exactly one error.

3.94. Let X have a Poisson distribution with double mode at $x = 1$ and $x = 2$. Find $\Pr[X = 0]$.

3.95. Let X and Y be jointly normally distributed with $\mu_X = 20$, $\mu_Y = 40$, $\sigma_X = 3$, $\sigma_Y = 2$, $\rho = 0.6$. Find a symmetric interval about the conditional mean, so that the probability is 0.90 that Y lies in that interval given that X equals 25.

3.96. Let $f(x) = \binom{10}{x} p^x(1-p)^{10-x}$, $x = 0, 1, \ldots, 10$, zero elsewhere. Find the values of p, so that $f(0) \geq f(1) \geq \cdots \geq f(10)$.

3.97. Let $f(x, y)$ be a bivariate normal p.d.f. and let c be a positive constant so that $c < (2\pi\sigma_1\sigma_2\sqrt{1-\rho^2})^{-1}$. Show that $c = f(x, y)$ defines an ellipse in the xy-plane.

3.98. Let $f_1(x, y)$ and $f_2(x, y)$ be two bivariate normal probability density functions, each having means equal to zero and variances equal to 1. The respective correlation coefficients are ρ and $-\rho$. Consider the joint distribution of X and Y defined by the joint p.d.f. $[f_1(x, y) + f_2(x, y)]/2$. Show that the two marginal distributions are both $N(0, 1)$, X and Y are dependent, and $E(XY) = 0$ and hence the correlation coefficient of X and Y is zero.

3.99. Let X be $N(\mu, \sigma^2)$. Define the random variable $Y = e^X$ and find its p.d.f. by differentiating $G(y) = \Pr(e^X \leq y) = \Pr(X \leq \ln y)$. This is the p.d.f. of a *lognormal distribution*.

3.100. In the proof of Theorem 1 of Section 3.4, we could let

$$G(w) = \Pr(X \leq w\sigma + \mu) = F(w\sigma + \mu),$$

where F and $F' = f$ are the distribution function and p.d.f. of X, respectively. Then, by the chain rule,

$$g(w) = G'(w) = [F'(w\sigma + \mu)]\sigma.$$

Show that the right-hand member is the p.d.f. of a standard normal distribution; thus this provides another proof of Theorem 1.

CHAPTER 4

Distributions of Functions of Random Variables

4.1 Sampling Theory

Let $X_1, X_2, \ldots, X_n$ denote n random variables that have the joint p.d.f. $f(x_1, x_2, \ldots, x_n)$. These variables may or may not be independent. Problems such as the following are very interesting in themselves; but more important, their solutions often provide the basis for making statistical inferences. Let Y be a random variable that is defined by a function of $X_1, X_2, \ldots, X_n$, say $Y = u(X_1, X_2, \ldots, X_n)$. Once the p.d.f. $f(x_1, x_2, \ldots, x_n)$ is given, can we find the p.d.f. of Y? In some of the preceding chapters, we have solved a few of these problems. Among them are the following two. If $n = 1$ and if X_1 is $N(\mu, \sigma^2)$, then $Y = (X_1 - \mu)/\sigma$ is $N(0, 1)$. Let n be a positive integer and let the random variables X_i, $i = 1, 2, \ldots, n$, be independent, each having the same p.d.f. $f(x) = p^x(1-p)^{1-x}$, $x = 0, 1$, and zero elsewhere. If $Y = \sum_1^n X_i$, then Y is $b(n, p)$. It should be observed that $Y = u(X_1) = (X_1 - \mu)/\sigma$ is a function of X_1 that depends upon the two parameters of the normal distribution; whereas $Y = u(X_1, X_2, \ldots, X_n) = \sum_1^n X_i$ does not depend upon p, the parameter of the common p.d.f. of the X_i, $i = 1, 2, \ldots, n$. The distinction that we make between these functions is brought out in the following definition.

Definition 1. A function of one or more random variables that does not depend upon any *unknown* parameter is called a *statistic.*

In accordance with this definition, the random variable $Y = \sum_1^n X_i$ discussed above is a statistic. But the random variable $Y = (X_1 - \mu)/\sigma$ is not a statistic unless μ and σ are known numbers. It should be noted that, although a statistic does not depend upon any unknown parameter, the *distribution* of the statistic may very well depend upon unknown parameters.

Remark. We remark, for the benefit of the more advanced reader, that a statistic is usually defined to be a measurable function of the random variables. In this book, however, we wish to minimize the use of measure theoretic terminology, so we have suppressed the modifier "measurable." It is quite clear that a statistic is a random variable. In fact, some probabilists avoid the use of the word "statistic" altogether, and they refer to a measurable function of random variables as a random variable. We decided to use the word "statistic" because the reader will encounter it so frequently in books and journals.

We can motivate the study of the distribution of a statistic in the following way. Let a random variable X be defined on a sample space $\mathscr{C}$ and let the space of X be denoted by $\mathscr{A}$. In many situations confronting us, the distribution of X is not completely known. For instance, we may know the distribution except for the value of an unknown parameter. To obtain more information about this distribution (or the unknown parameter), we shall repeat under identical conditions the random experiment n independent times. Let the random variable X_i be a function of the ith outcome, $i = 1, 2, \ldots, n$. Then we call $X_1, X_2, \ldots, X_n$ the *observations* of a random sample from the distribution under consideration. Suppose that we can define a statistic $Y=u(X_1, X_2, \ldots, X_n)$ whose p.d.f. is found to be $g(y)$. Perhaps this p.d.f. shows that there is a great probability that Y has a value close to the unknown parameter. Once the experiment has been repeated in the manner indicated and we have $X_1=x_1, \ldots, X_n=x_n$, then $y=u(x_1, x_2, \ldots, x_n)$ is a known number. It is to be hoped that this known number can in some manner be used to elicit information about the unknown parameter. Thus a statistic may prove to be useful.

Remarks. Let the random variable X be defined as the diameter of a hole to be drilled by a certain drill press and let it be assumed that X has a normal distribution. Past experience with many drill presses makes this assumption plausible; but the assumption does not specify the mean μ nor the variance σ^2 of this normal distribution. The only way to obtain information about μ and σ^2 is to have recourse to experimentation. Thus we shall drill a number, say $n=20$, of these holes whose diameters will be $X_1, X_2, \ldots, X_{20}$. Then $X_1, X_2, \ldots, X_{20}$ is a random sample from the normal distribution under consideration. Once the holes have been drilled and the diameters measured, the 20 numbers may be used, as will be seen later, to elicit information about μ and σ^2.

The term "random sample" is now defined in a more formal manner.

Definition 2. Let $X_1, X_2, \ldots, X_n$ denote n independent random variables, each of which has the same but possibly unknown p.d.f. $f(x)$; that is, the probability density functions of $X_1, X_2, \ldots, X_n$ are, respectively, $f_1(x_1)=f(x_1)$, $f_2(x_2)=f(x_2), \ldots, f_n(x_n)=f(x_n)$, so that the joint p.d.f. is $f(x_1)f(x_2)\cdots f(x_n)$. The random variables $X_1, X_2, \ldots, X_n$ are then said to constitute a *random sample* from a distribution that has p.d.f. $f(x)$. That is, the observations of a random sample are *independent and identically distributed* (often abbreviated i.i.d.).

Later we shall define what we mean by a random sample from a distribution of more than one random variable.

Sometimes it is convenient to refer to a random sample of size n from a given distribution and, as has been remarked, to refer to $X_1, X_2, \ldots, X_n$ as the observations of the random sample. A reexamination of Example 2 of Section 2.5 reveals that we found the p.d.f. of the statistic, which is the maximum of the observations of a random sample of size $n=3$, from a distribution with p.d.f.

$f(x) = 2x$, $0 < x < 1$, zero elsewhere. In Section 3.1 we found the p.d.f. of the statistic, which is the sum of the observations of a random sample of size n from a distribution that has p.d.f. $f(x) = p^x(1-p)^{1-x}$, $x = 0, 1$, zero elsewhere. This fact was also referred to at the beginning of this section.

In this book, most of the statistics that we shall encounter will be functions of the observations of a random sample from a given distribution. Next, we define two important statistics of this type.

Definition 3. Let $X_1, X_2, \ldots, X_n$ denote a random sample of size n from a given distribution. The statistic

$$\bar{X} = \frac{X_1 + X_2 + \cdots + X_n}{n} = \sum_{i=1}^{n} \frac{X_i}{n}$$

is called the *mean* of the random sample, and the statistic

$$S^2 = \sum_{i=1}^{n} \frac{(X_i - \bar{X})^2}{n} = \sum_{i=1}^{n} \frac{X_i^2}{n} - \bar{X}^2$$

is called the *variance* of the random sample.

Remarks. Many writers do not define the variance of a random sample as we have done but, instead, they take $S^2 = \sum_1^n (X_i - \bar{X})^2/(n-1)$. There are good reasons for doing this. But a certain price has to be paid, as we shall indicate. Let $x_1, x_2, \ldots, x_n$ denote experimental values of the random variable X that has the p.d.f. $f(x)$ and the distribution function $F(x)$. Thus we may look upon $x_1, x_2, \ldots, x_n$ as the experimental values of a random sample of size n from the given distribution. The *distribution of the sample* is then defined to be the distribution obtained by assigning a probability of $1/n$ to each of the points $x_1, x_2, \ldots, x_n$. This is a distribution of the discrete type. The corresponding distribution function will be denoted by $F_n(x)$ and it is a step function. If we let f_x denote the number of sample values that are less than or equal to x, then $F_n(x) = f_x/n$, so that $F_n(x)$ gives the relative frequency of the event $X \le x$ in the set of n observations. The function $F_n(x)$ is often called the "empirical distribution function" and it has a number of uses.

Because the distribution of the sample is a discrete distribution, the mean and the variance have been defined and are, respectively, $\sum_1^n x_i/n = \bar{x}$ and $\sum_1^n (x_i - \bar{x})^2/n = s^2$. Thus, if one finds the distribution of the sample and the associated empirical distribution function to be useful concepts, it would seem logically inconsistent to define the variance of a random sample in any way other than we have.

We have also defined $\bar{X}$ and S^2 only for observations that are i.i.d., that is, when $X_1, X_2, \ldots, X_n$ denote a random sample. However, statisticians often use these symbols, $\bar{X}$ and S^2, even if the assumption of independence is dropped. For example, suppose that $X_1, X_2, \ldots, X_n$ were the observations taken at random from a finite collection of numbers *without replacement*. These observations could be thought of as a sample and its mean $\bar{X}$ and variance S^2 computed; yet $X_1, X_2, \ldots, X_n$ are dependent. Moreover, the n observations could simply be some values, not necessarily taken from a distribution, and we could compute the mean $\bar{X}$ and the variance S^2 associated with these n values. If we do these things, however, we must recognize the conditions under which the observations were obtained, and we cannot make the same statements that are associated with the mean and the variance of what we call a random sample.

Random sampling distribution theory means the general problem of finding distributions of functions of the observations of a random sample. Up to this point, the only method, other than direct probabilistic arguments, of finding the distribution of a function of one or more random variables is the *distribution function technique*. That is, if $X_1, X_2, \ldots, X_n$ are random variables, the distribution of $Y = u(X_1, X_2, \ldots, X_n)$ is determined by computing the distribution function of Y,

$$G(y) = \Pr [u(X_1, X_2, \ldots, X_n) \le y].$$

Even in what superficially appears to be a very simple problem, this can be quite tedious. This fact is illustrated in the next paragraph.

Let X_1, X_2, X_3 denote a random sample of size 3 from a standard normal distribution. Let Y denote the statistic that is the sum of the squares of the sample observations. The distribution function of Y is

$$G(y) = \Pr (X_1^2 + X_2^2 + X_3^2 \le y).$$

If $y < 0$, then $G(y) = 0$. However, if $y \ge 0$, then

$$G(y) = \iiint\limits_A \frac{1}{(2\pi)^{3/2}} \exp\left[-\frac{1}{2}(x_1^2 + x_2^2 + x_3^2)\right] dx_1\, dx_2\, dx_3,$$

where A is the set of points (x_1, x_2, x_3) interior to, or on the surface of, a sphere with center at $(0, 0, 0)$ and radius equal to $\sqrt{y}$. This is not a simple integral. We might hope to make progress by changing to spherical coordinates:

$$x_1 = \rho \cos\theta \sin\varphi, \qquad x_2 = \rho \sin\theta \sin\varphi, \qquad x_3 = \rho \cos\varphi,$$

where $\rho \geq 0$, $0 \leq \theta < 2\pi$, $0 \leq \varphi \leq \pi$. Then, for $y \geq 0$,

$$G(y) = \int_0^{\sqrt{y}} \int_0^{2\pi} \int_0^{\pi} \frac{1}{(2\pi)^{3/2}} e^{-\rho^2/2} \rho^2 \sin\varphi \, d\varphi \, d\theta \, d\rho$$

$$= \sqrt{\frac{2}{\pi}} \int_0^{\sqrt{y}} \rho^2 e^{-\rho^2/2} \, d\rho.$$

If we change the variable of integration by setting $\rho = \sqrt{w}$, we have

$$G(y) = \sqrt{\frac{2}{\pi}} \int_0^{y} \frac{\sqrt{w}}{2} e^{-w/2} \, dw,$$

for $y \geq 0$. Since Y is a random variable of the continuous type, the p.d.f. of Y is $g(y) = G'(y)$. Thus

$$g(y) = \frac{1}{\sqrt{2\pi}} y^{3/2 - 1} e^{-y/2}, \qquad 0 < y < \infty,$$

$$= 0 \qquad \text{elsewhere.}$$

Because $\Gamma(\frac{3}{2}) = (\frac{1}{2})\Gamma(\frac{1}{2}) = (\frac{1}{2})\sqrt{\pi}$, and thus $\sqrt{2\pi} = \Gamma(\frac{3}{2})2^{3/2}$, we see that Y is $\chi^2(3)$.

The problem that we have just solved highlights the desirability of having, if possible, various methods of determining the distribution of a function of random variables. We shall find that other techniques are available and that often a particular technique is vastly superior to the others in a given situation. These techniques will be discussed in subsequent sections.

Example 1. Let the random variable Y be distributed uniformly over the unit interval $0 < y < 1$; that is, the distribution function of Y is

$$G(y) = 0, \qquad y \leq 0,$$

$$= y, \qquad 0 < y < 1,$$

$$= 1, \qquad 1 \leq y.$$

Suppose that $F(x)$ is a distribution function of the continuous type which is strictly increasing when $0 < F(x) < 1$. If we define the random variable X by the relationship $Y = F(X)$, we now show that X has a distribution which corresponds to $F(x)$. If $0 < F(x) < 1$, the inequalities $X \le x$ and $F(X) \le F(x)$ are equivalent. Thus, with $0 < F(x) < 1$, the distribution of X is

$$\Pr (X \le x) = \Pr [F(X) \le F(x)] = \Pr [Y \le F(x)]$$

because $Y = F(X)$. However, $\Pr (Y \le y) = G(y)$, so we have

$$\Pr (X \le x) = G[F(x)] = F(x), \qquad 0 < F(x) < 1.$$

That is, the distribution function of X is $F(x)$.

This result permits us to *simulate* random variables of different types. This is done by simply determining values of the uniform variable Y, usually with a computer. Then, after determining the observed value $Y = y$, solve the equation $y = F(x)$, either explicitly or by numerical methods. This yields the inverse function $x = F^{-1}(y)$. By the preceding result, this number x will be an observed value of X that has distribution function $F(x)$.

It is also interesting to note that the converse of this result is true. If X has distribution function $F(x)$ of the continuous type, then $Y = F(X)$ is uniformly distributed over $0 < y < 1$. The reason for this is, for $0 < y < 1$, that

$$\Pr (Y \le y) = \Pr [F(X) \le y] = \Pr [X \le F^{-1}(y)].$$

However, it is given that $\Pr (X \le x) = F(x)$, so

$$\Pr (Y \le y) = F[F^{-1}(y)] = y, \qquad 0 < y < 1.$$

This is the distribution function of a random variable that is distributed uniformly on the interval (0, 1).

EXERCISES

4.1. Show that

$$S^2 = \frac{1}{n}\sum_1^n (X_i - \bar{X})^2 = \frac{1}{n}\sum_1^n X_i^2 - \bar{X}^2,$$

where $\bar{X} = \sum_1^n X_i/n$.

4.2. Find the probability that exactly four observations of a random sample of size 5 from the distribution having p.d.f. $f(x) = (x + 1)/2$, $-1 < x < 1$, zero elsewhere, exceed zero.

4.3. Let X_1, X_2, X_3 be a random sample of size 3 from a distribution that is $N(6, 4)$. Determine the probability that the largest sample observation exceeds 8.

4.4. What is the probability that at least one observation of a random sample of size $n = 5$ from a continuous-type distribution exceeds the 90th percentile?

4.5. Let X have the p.d.f. $f(x) = 4x^3$, $0 < x < 1$, zero elsewhere. Show that $Y = -2 \ln X^4$ is $\chi^2(2)$.

4.6. Let X_1, X_2 be a random sample of size $n = 2$ from a distribution with p.d.f. $f(x) = 4x^3, 0 < x < 1$, zero elsewhere. Find the mean and the variance of the ratio $Y = X_1/X_2$.

Hint: First find the distribution function $\Pr(Y \le y)$ when $0 < y < 1$ and then when $1 \le y$.

4.7. Let X_1, X_2 be a random sample from the distribution having p.d.f. $f(x) = 2x$, $0 < x < 1$, zero elsewhere. Find $\Pr(X_1/X_2 \le \frac{1}{2})$ and $\Pr(X_1 X_2 \ge \frac{1}{4})$.

4.8. If the sample size is $n = 2$, find the constant c so that $S^2 = c(X_1 - X_2)^2$.

4.9. If $x_i = i$, $i = 1, 2, \ldots, n$, compute the values of $\bar{x} = \Sigma\, x_i/n$ and $s^2 = \Sigma\,(x_i - \bar{x})^2/n$.

4.10. Let $y_i = a + bx_i$, $i = 1, 2, \ldots, n$, where a and b are constants. Find $\bar{y} = \Sigma\, y_i/n$ and $s_y^2 = \Sigma\,(y_i - \bar{y})^2/n$ in terms of a, b, $\bar{x} = \Sigma\, x_i/n$, and $s_x^2 = \Sigma\,(x_i - \bar{x})^2/n$.

4.11. Let X_1 and X_2 denote two i.i.d. random variables, each from a distribution that is $N(0, 1)$. Find the p.d.f. of $Y = X_1^2 + X_2^2$.

Hint: In the double integral representing $\Pr(Y \le y)$, use polar coordinates.

4.12. The four values $y_1 = 0.42$, $y_2 = 0.31$, $y_3 = 0.87$, and $y_4 = 0.65$ represent the observed values of a random sample of size $n = 4$ from the uniform distribution over $0 < y < 1$. Using these four values, find a corresponding observed random sample from a distribution that has p.d.f. $f(x) = e^{-x}$, $0 < x < \infty$, zero elsewhere.

4.13. Let X_1, X_2 denote a random sample of size 2 from a distribution with p.d.f. $f(x) = \frac{1}{2}, 0 < x < 2$, zero elsewhere. Find the joint p.d.f. of X_1 and X_2. Let $Y = X_1 + X_2$. Find the distribution function and the p.d.f. of Y.

4.14. Let X_1, X_2 denote a random sample of size 2 from a distribution with p.d.f. $f(x) = 1$, $0 < x < 1$, zero elsewhere. Find the distribution function and the p.d.f. of $Y = X_1/X_2$.

4.15. Let X_1, X_2, X_3 be three i.i.d. random variables, each from a distribution having p.d.f. $f(x) = 5x^4$, $0 < x < 1$, zero elsewhere. Let Y be the largest observation in the sample. Find the distribution function and p.d.f. of Y.

4.16. Let X_1 and X_2 be observations of a random sample from a distribution with p.d.f. $f(x) = 2x$, $0 < x < 1$, zero elsewhere. Evaluate the conditional probability $\Pr(X_1 < X_2 | X_1 < 2X_2)$.

4.2 Transformations of Variables of the Discrete Type

An alternative method of finding the distribution of a function of one or more random variables is called the *change-of-variable technique*. There are some delicate questions (with particular reference to random variables of the continuous type) involved in this technique, and these make it desirable for us first to consider special cases.

Let X have the Poisson p.d.f.

$$f(x) = \frac{\mu^x e^{-\mu}}{x!}, \qquad x = 0, 1, 2, \ldots,$$

$$= 0 \qquad \text{elsewhere.}$$

As we have done before, let $\mathscr{A}$ denote the space $\mathscr{A} = \{x : x = 0, 1, 2, \ldots\}$, so that $\mathscr{A}$ is the set where $f(x) > 0$. Define a new random variable Y by $Y = 4X$. We wish to find the p.d.f. of Y by the change-of-variable technique. Let $y = 4x$. We call $y = 4x$ a transformation from x to y, and we say that the transformation maps the space $\mathscr{A}$ onto the space $\mathscr{B} = \{y : y = 0, 4, 8, 12, \ldots\}$. The space $\mathscr{B}$ is obtained by transforming each point in $\mathscr{A}$ in accordance with $y = 4x$. We note two things about this transformation. It is such that to each point in $\mathscr{A}$ there corresponds one, and only one, point in $\mathscr{B}$; and conversely, to each point in $\mathscr{B}$ there corresponds one, and only one, point in $\mathscr{A}$. That is, the transformation $y = 4x$ sets up a one-to-one correspondence between the points of $\mathscr{A}$ and those of $\mathscr{B}$. Any function $y = u(x)$ (not merely $y = 4x$) that maps a space $\mathscr{A}$ (not merely our $\mathscr{A}$) onto a space $\mathscr{B}$ (not merely our $\mathscr{B}$) such that there is a one-to-one correspondence between the points of $\mathscr{A}$ and those of $\mathscr{B}$ is called a *one-to-one transformation*. It is important to note that a one-to-one transformation, $y = u(x)$, implies that x is a single-valued function of y. In our case this is obviously true, since $y = 4x$ requires that $x = (\frac{1}{4})y$.

Our problem is that of finding the p.d.f. $g(y)$ of the discrete type of random variable $Y = 4X$. Now $g(y) = \Pr(Y = y)$. Because there is a one-to-one correspondence between the points of $\mathscr{A}$ and those of

$\mathscr{B}$, the event $Y = y$ or $4X = y$ can occur when, and only when, the event $X = (\frac{1}{4})y$ occurs. That is, the two events are equivalent and have the same probability. Hence

$$g(y) = \Pr(Y = y) = \Pr\left(X = \frac{y}{4}\right) = \frac{\mu^{y/4}e^{-\mu}}{(y/4)!}, \qquad y = 0, 4, 8, \ldots,$$

$$= 0 \qquad \text{elsewhere.}$$

The foregoing detailed discussion should make the subsequent text easier to read. Let X be a random variable of the discrete type, having p.d.f. $f(x)$. Let $\mathscr{A}$ denote the set of discrete points, at each of which $f(x) > 0$, and let $y = u(x)$ define a one-to-one transformation that maps $\mathscr{A}$ onto $\mathscr{B}$. If we solve $y = u(x)$ for x in terms of y, say, $x = w(y)$, then for each $y \in \mathscr{B}$, we have $x = w(y) \in \mathscr{A}$. Consider the random variable $Y = u(X)$. If $y \in \mathscr{B}$, then $x = w(y) \in \mathscr{A}$, and the events $Y = y$ [or $u(X) = y$] and $X = w(y)$ are equivalent. Accordingly, the p.d.f. of Y is

$$g(y) = \Pr(Y = y) = \Pr[X = w(y)] = f[w(y)], \qquad y \in \mathscr{B},$$

$$= 0 \qquad \text{elsewhere.}$$

Example 1. Let X have the binomial p.d.f.

$$f(x) = \frac{3!}{x!(3-x)!}\left(\frac{2}{3}\right)^x\left(\frac{1}{3}\right)^{3-x}, \qquad x = 0, 1, 2, 3,$$

$$= 0 \qquad \text{elsewhere.}$$

We seek the p.d.f. $g(y)$ of the random variable $Y = X^2$. The transformation $y = u(x) = x^2$ maps $\mathscr{A} = \{x : x = 0, 1, 2, 3\}$ onto $\mathscr{B} = \{y : y = 0, 1, 4, 9\}$. In general, $y = x^2$ does not define a one-to-one transformation; here, however, it does, for there are no negative values of x in $\mathscr{A} = \{x : x = 0, 1, 2, 3\}$. That is, we have the single-valued inverse function $x = w(y) = \sqrt{y}$ (not $-\sqrt{y}$), and so

$$g(y) = f(\sqrt{y}) = \frac{3!}{(\sqrt{y})!\,(3-\sqrt{y})!}\left(\frac{2}{3}\right)^{\sqrt{y}}\left(\frac{1}{3}\right)^{3-\sqrt{y}}, \qquad y = 0, 1, 4, 9,$$

$$= 0 \qquad \text{elsewhere.}$$

There are no essential difficulties involved in a problem like the following. Let $f(x_1, x_2)$ be the joint p.d.f. of two discrete-type random variables X_1 and X_2 with $\mathscr{A}$ the (two-dimensional) set of points at which $f(x_1, x_2) > 0$. Let $y_1 = u_1(x_1, x_2)$ and $y_2 = u_2(x_1, x_2)$ define a one-to-one transformation that maps $\mathscr{A}$ onto $\mathscr{B}$. The joint p.d.f. of the two new random variables $Y_1 = u_1(X_1, X_2)$ and $Y_2 = u_2(X_1, X_2)$ is given by

$$g(y_1, y_2) = f[w_1(y_1, y_2), w_2(y_1, y_2)], \qquad (y_1, y_2) \in \mathscr{B},$$

$$= 0 \qquad \text{elsewhere,}$$

where $x_1 = w_1(y_1, y_2)$, $x_2 = w_2(y_1, y_2)$ is the single-valued inverse of $y_1 = u_1(x_1, x_2)$, $y_2 = u_2(x_1, x_2)$. From this joint p.d.f. $g(y_1, y_2)$ we may obtain the marginal p.d.f. of Y_1 by summing on y_2 or the marginal p.d.f. of Y_2 by summing on y_1.

Perhaps it should be emphasized that the technique of change of variables involves the introduction of as many "new" variables as there were "old" variables. That is, suppose that $f(x_1, x_2, x_3)$ is the joint p.d.f. of X_1, X_2, and X_3, with $\mathscr{A}$ the set where $f(x_1, x_2, x_3) > 0$. Let us say we seek the p.d.f. of $Y_1 = u_1(X_1, X_2, X_3)$. We would then define (if possible) $Y_2 = u_2(X_1, X_2, X_3)$ and $Y_3 = u_3(X_1, X_2, X_3)$, so that $y_1 = u_1(x_1, x_2, x_3)$, $y_2 = u_2(x_1, x_2, x_3)$, $y_3 = u_3(x_1, x_2, x_3)$ define a one-to-one transformation of $\mathscr{A}$ onto $\mathscr{B}$. This would enable us to find the joint p.d.f. of Y_1, Y_2, and Y_3 from which we would get the marginal p.d.f. of Y_1 by summing on y_2 and y_3.

Example 2. Let X_1 and X_2 be two independent random variables that have Poisson distributions with means μ_1 and μ_2, respectively. The joint p.d.f. of X_1 and X_2 is

$$\frac{\mu_1^{x_1}\mu_2^{x_2}e^{-\mu_1-\mu_2}}{x_1!\,x_2!}, \qquad x_1 = 0, 1, 2, 3, \ldots, \qquad x_2 = 0, 1, 2, 3, \ldots,$$

and is zero elsewhere. Thus the space $\mathscr{A}$ is the set of points (x_1, x_2), where each of x_1 and x_2 is a nonnegative integer. We wish to find the p.d.f. of $Y_1 = X_1 + X_2$. If we use the change of variable technique, we need to define a second random variable Y_2. Because Y_2 is of no interest to us, let us choose it in such a way that we have a simple one-to-one transformation. For example, take $Y_2 = X_2$. Then $y_1 = x_1 + x_2$ and $y_2 = x_2$ represent a one-to-one transformation that maps $\mathscr{A}$ onto

$$\mathscr{B} = \{(y_1, y_2) : y_2 = 0, 1, \ldots, y_1 \quad \text{and} \quad y_1 = 0, 1, 2, \ldots\}.$$

Note that, if $(y_1, y_2) \in \mathscr{B}$, then $0 \le y_2 \le y_1$. The inverse functions are given by $x_1 = y_1 - y_2$ and $x_2 = y_2$. Thus the joint p.d.f. of Y_1 and Y_2 is

$$g(y_1, y_2) = \frac{\mu_1^{y_1-y_2}\mu_2^{y_2}e^{-\mu_1-\mu_2}}{(y_1-y_2)!\,y_2!}, \qquad (y_1, y_2) \in \mathscr{B},$$

and is zero elsewhere. Consequently, the marginal p.d.f. of Y_1 is given by

$$g_1(y_1) = \sum_{y_2=0}^{y_1} g(y_1, y_2)$$

$$= \frac{e^{-\mu_1-\mu_2}}{y_1!} \sum_{y_2=0}^{y_1} \frac{y_1!}{(y_1-y_2)!\,y_2!} \mu_1^{y_1-y_2}\mu_2^{y_2}$$

$$= \frac{(\mu_1+\mu_2)^{y_1}e^{-\mu_1-\mu_2}}{y_1!}, \qquad y_1 = 0, 1, 2, \ldots,$$

and is zero elsewhere. That is, $Y_1 = X_1 + X_2$ has a Poisson distribution with parameter $\mu_1 + \mu_2$.

Remark. It should be noted that Example 2 essentially illustrates the distribution function technique too. That is, without defining $Y_2 = X_2$, we have that the distribution function of $Y_1 = X_1 + X_2$ is

$$G_1(y_1) = \Pr(X_1 + X_2 \le y_1).$$

In this discrete case, with $y_1 = 0, 1, 2, \ldots$, the p.d.f. of Y_1 is equal to

$$g_1(y_1) = G_1(y_1) - G_1(y_1 - 1) = \Pr(X_1 + X_2 = y_1).$$

That is,

$$g_1(y_1) = \sum\sum_{x_1+x_2=y_1} \frac{\mu_1^{x_1}\mu_2^{x_2}e^{-\mu_1-\mu_2}}{x_1!\,x_2!}.$$

This summation is over all points of $\mathscr{A}$ such that $x_1 + x_2 = y_1$ and thus can be written as

$$g_1(y_1) = \sum_{x_2=0}^{y_1} \frac{\mu_1^{y_1-x_2}\mu_2^{x_2}e^{-\mu_1-\mu_2}}{(y_1-x_2)!\,x_2!},$$

which is exactly the summation given in Example 2.

Example 3. In Section 4.1, we found that we could simulate a continuous-type random variable X with distribution function $F(x)$ through $X = F^{-1}(Y)$, where Y has a uniform distribution on $0 < y < 1$. In a sense, we can simulate a discrete-type random variable X in much the same way, but we must understand what $X = F^{-1}(Y)$ means in this case. Here $F(x)$ is a step function with the height of the step at $x = x_0$ equal to $\Pr(X = x_0)$. For illustration, in Example 3 of Section 1.5, $\Pr(X = 3) = \frac{3}{6}$ is the height of the step at $x = 3$ in Figure 1.3 that depicts the distribution function. If we now think of selecting a random point Y, having the uniform distribution on $0 < y \le 1$, on the vertical axis of Figure 1.3, the probability of falling between $\frac{3}{6}$ and $\frac{6}{6}$ is $\frac{3}{6}$. However, if it falls between those two values, the horizontal line drawn from it would "hit" the step at $x = 3$. That is, for $\frac{3}{6} < y \le \frac{6}{6}$, then $F^{-1}(y) = 3$. Of course, if $\frac{1}{6} < y \le \frac{3}{6}$, then $F^{-1}(y) = 2$; and if $0 < y \le \frac{1}{6}$, we have $F^{-1}(y) = 1$. Thus, with this procedure, we can generate the numbers $x = 1$,

$x = 2$, and $x = 3$ with respective probabilities $\frac{1}{6}$, $\frac{2}{6}$, and $\frac{3}{6}$, as we desired. Clearly, this procedure can be generalized to simulate any random variable X of the discrete type.

EXERCISES

4.17. Let X have a p.d.f. $f(x) = \frac{1}{3}$, $x = 1, 2, 3$, zero elsewhere. Find the p.d.f. of $Y = 2X + 1$.

4.18. If $f(x_1, x_2) = (\frac{2}{3})^{x_1 + x_2}(\frac{1}{3})^{2 - x_1 - x_2}$, $(x_1, x_2) = (0, 0), (0, 1), (1, 0), (1, 1)$, zero elsewhere, is the joint p.d.f. of X_1 and X_2, find the joint p.d.f. of $Y_1 = X_1 - X_2$ and $Y_2 = X_1 + X_2$.

4.19. Let X have the p.d.f. $f(x) = (\frac{1}{2})^x$, $x = 1, 2, 3, \ldots$, zero elsewhere. Find the p.d.f. of $Y = X^3$.

4.20. Let X_1 and X_2 have the joint p.d.f. $f(x_1, x_2) = x_1 x_2/36$, $x_1 = 1, 2, 3$ and $x_2 = 1, 2, 3$, zero elsewhere. Find first the joint p.d.f. of $Y_1 = X_1 X_2$ and $Y_2 = X_2$, and then find the marginal p.d.f. of Y_1.

4.21. Let the independent random variables X_1 and X_2 be $b(n_1, p)$ and $b(n_2, p)$, respectively. Find the joint p.d.f. of $Y_1 = X_1 + X_2$ and $Y_2 = X_2$, and then find the marginal p.d.f. of Y_1.
Hint: Use the fact that

$$\sum_{w=0}^{k} \binom{n_1}{w}\binom{n_2}{k-w} = \binom{n_1 + n_2}{k}.$$

This can be proved by comparing the coefficients of x^k in each member of the identity $(1 + x)^{n_1}(1 + x)^{n_2} \equiv (1 + x)^{n_1 + n_2}$.

4.22. Let X_1 and X_2 be independent random variables of the discrete type with joint p.d.f. $f_1(x_1)f_2(x_2)$, $(x_1, x_2) \in \mathcal{A}$. Let $y_1 = u_1(x_1)$ and $y_2 = u_2(x_2)$ denote a one-to-one transformation that maps $\mathcal{A}$ onto $\mathcal{B}$. Show that $Y_1 = u_1(X_1)$ and $Y_2 = u_2(X_2)$ are independent.

4.23. Consider the random variable X with p.d.f. $f(x) = x/15$, $x = 1, 2, 3, 4, 5$, and zero elsewhere.

(a) Graph the distribution function $F(x)$ of X.

(b) Using a computer or a table of random numbers, determine 30 values of Y, which has the (approximate) uniform distribution on $0 < y < 1$.

(c) From these 30 values of Y, find the corresponding 30 values of X and determine the relative frequencies of $x = 1$, $x = 2$, $x = 3$, $x = 4$, and $x = 5$. How do these compare to the respective probabilities of $\frac{1}{15}$, $\frac{2}{15}$, $\frac{3}{15}$, $\frac{4}{15}$, $\frac{5}{15}$?

4.24. Using the technique given in Example 3 and Exercise 4.23, generate 50 values having a Poisson distribution with $\mu = 1$.
Hint: Use Table I in Appendix B.

4.3 Transformations of Variables of the Continuous Type

In the preceding section we introduced the notion of a one-to-one transformation and the mapping of a set $\mathcal{A}$ onto a set $\mathcal{B}$ under that transformation. Those ideas were sufficient to enable us to find the distribution of a function of several random variables of the discrete type. In this section we examine the same problem when the random variables are of the continuous type. It is again helpful to begin with a special problem.

Example 1. Let X be a random variable of the continuous type, having p.d.f.

$$\begin{aligned} f(x) &= 2x, \qquad 0 < x < 1, \\ &= 0 \qquad \text{elsewhere.} \end{aligned}$$

Here $\mathcal{A}$ is the space $\{x : 0 < x < 1\}$, where $f(x) > 0$. Define the random variable Y by $Y = 8X^3$ and consider the transformation $y = 8x^3$. Under the transformation $y = 8x^3$, the set $\mathcal{A}$ is mapped onto the set $\mathcal{B} = \{y : 0 < y < 8\}$, and, moreover, the transformation is one-to-one. For every $0 < a < b < 8$, the event $a < Y < b$ will occur when, and only when, the event $\frac{1}{2}\sqrt[3]{a} < X < \frac{1}{2}\sqrt[3]{b}$ occurs because there is a one-to-one correspondence between the points of $\mathcal{A}$ and $\mathcal{B}$. Thus

$$\begin{aligned} \Pr(a < Y < b) &= \Pr\left(\tfrac{1}{2}\sqrt[3]{a} < X < \tfrac{1}{2}\sqrt[3]{b}\right) \\ &= \int_{\sqrt[3]{a}/2}^{\sqrt[3]{b}/2} 2x\,dx. \end{aligned}$$

Let us rewrite this integral by changing the variable of integration from x to y by writing $y = 8x^3$ or $x = \frac{1}{2}\sqrt[3]{y}$. Now

$$\frac{dx}{dy} = \frac{1}{6y^{2/3}},$$

and, accordingly, we have

$$\begin{aligned} \Pr(a < Y < b) &= \int_a^b 2\left(\frac{\sqrt[3]{y}}{2}\right)\left(\frac{1}{6y^{2/3}}\right) dy \\ &= \int_a^b \frac{1}{6y^{1/3}}\,dy. \end{aligned}$$

Since this is true for every $0 < a < b < 8$, the p.d.f. $g(y)$ of Y is the integrand; that is,

$$\begin{aligned} g(y) &= \frac{1}{6y^{1/3}}, \qquad 0 < y < 8, \\ &= 0 \qquad \text{elsewhere.} \end{aligned}$$

It is worth noting that we found the p.d.f. of the random variable $Y = 8X^3$ by using a theorem on the change of variable in a definite integral. However, to obtain $g(y)$ we actually need only two things: (1) the set $\mathscr{B}$ of points y where $g(y) > 0$ and (2) the integrand of the integral on y to which $\Pr(a < Y < b)$ is equal. These can be found by two simple rules:

1. Verify that the transformation $y = 8x^3$ maps $\mathscr{A} = \{x : 0 < x < 1\}$ onto $\mathscr{B} = \{y : 0 < y < 8\}$ and that the transformation is one-to-one.
2. Determine $g(y)$ on this set $\mathscr{B}$ by substituting $\frac{1}{2}\sqrt[3]{y}$ for x in $f(x)$ and then multiplying this result by the derivative of $\frac{1}{2}\sqrt[3]{y}$. That is,

$$g(y) = f\left(\frac{\sqrt[3]{y}}{2}\right)\frac{d[(\frac{1}{2})\sqrt[3]{y}]}{dy} = \frac{1}{6y^{1/3}}, \qquad 0 < y < 8,$$

$$= 0 \qquad \text{elsewhere.}$$

We shall accept a theorem in analysis on the change of variable in a definite integral to enable us to state a more general result. Let X be a random variable of the continuous type having p.d.f. $f(x)$. Let $\mathscr{A}$ be the one-dimensional space where $f(x) > 0$. Consider the random variable $Y = u(X)$, where $y = u(x)$ defines a one-to-one transformation that maps the set $\mathscr{A}$ onto the set $\mathscr{B}$. Let the inverse of $y = u(x)$ be denoted by $x = w(y)$, and let the derivative $dx/dy = w'(y)$ be continuous and not equal zero for all points y in $\mathscr{B}$. Then the p.d.f. of the random variable $Y = u(X)$ is given by

$$g(y) = f[w(y)]|w'(y)|, \qquad y \in \mathscr{B},$$

$$= 0 \qquad \text{elsewhere,}$$

where $|w'(y)|$ represents the absolute value of $w'(y)$. This is precisely what we did in Example 1 of this section, except there we deliberately chose $y = 8x^3$ to be an increasing function so that

$$\frac{dx}{dy} = w'(y) = \frac{1}{6y^{2/3}}, \qquad 0 < y < 8,$$

is positive, and hence

$$\left|\frac{1}{6y^{2/3}}\right| = \frac{1}{6y^{2/3}}, \qquad 0 < y < 8.$$

Henceforth, we shall refer to $dx/dy = w'(y)$ as the Jacobian (denoted by J) of the transformation. In most mathematical areas, $J = w'(y)$ is referred to as the Jacobian of the inverse transformation $x = w(y)$, but in this book it will be called the Jacobian of the transformation, simply for convenience.

Example 2. Let X have the p.d.f.

$$\begin{aligned} f(x) &= 1, \quad && 0 < x < 1, \\ &= 0 && \text{elsewhere.} \end{aligned}$$

We are to show that the random variable $Y = -2 \ln X$ has a chi-square distribution with 2 degrees of freedom. Here the transformation is $y = u(x) = -2 \ln x$, so that $x = w(y) = e^{-y/2}$. The space $\mathscr{A}$ is $\mathscr{A} = \{x : 0 < x < 1\}$, which the one-to-one transformation $y = -2 \ln x$ maps onto $\mathscr{B} = \{y : 0 < y < \infty\}$. The Jacobian of the transformation is

$$J = \frac{dx}{dy} = w'(y) = -\frac{1}{2} e^{-y/2}.$$

Accordingly, the p.d.f. $g(y)$ of $Y = -2 \ln X$ is

$$\begin{aligned} g(y) &= f(e^{-y/2})|J| = \tfrac{1}{2} e^{-y/2}, \quad && 0 < y < \infty, \\ &= 0 && \text{elsewhere,} \end{aligned}$$

a p.d.f. that is chi-square with 2 degrees of freedom. Note that this problem was first proposed in Exercise 3.46.

This method of finding the p.d.f. of a function of one random variable of the continuous type will now be extended to functions of two random variables of this type. Again, only functions that define a one-to-one transformation will be considered at this time. Let $y_1 = u_1(x_1, x_2)$ and $y_2 = u_2(x_1, x_2)$ define a one-to-one transformation that maps a (two-dimensional) set $\mathscr{A}$ in the x_1x_2-plane onto a (two-dimensional) set $\mathscr{B}$ in the y_1y_2-plane. If we express each of x_1 and x_2 in terms of y_1 and y_2, we can write $x_1 = w_1(y_1, y_2)$, $x_2 = w_2(y_1, y_2)$. The determinant of order 2,

$$\begin{vmatrix} \dfrac{\partial x_1}{\partial y_1} & \dfrac{\partial x_1}{\partial y_2} \\ \dfrac{\partial x_2}{\partial y_1} & \dfrac{\partial x_2}{\partial y_2} \end{vmatrix},$$

is called the *Jacobian* of the transformation and will be denoted by the symbol J. It will be assumed that these first-order partial derivatives are continuous and that the Jacobian J is not identically equal to zero in $\mathscr{B}$. An illustrative example may be desirable before we proceed with the extension of the change of variable technique to two random variables of the continuous type.

Example 3. Let $\mathscr{A}$ be the set $\mathscr{A} = \{(x_1, x_2) : 0 < x_1 < 1, 0 < x_2 < 1\}$ depicted in Figure 4.1. We wish to determine the set $\mathscr{B}$ in the y_1y_2-plane that is the mapping of $\mathscr{A}$ under the one-to-one transformation

$$y_1 = u_1(x_1, x_2) = x_1 + x_2,$$

$$y_2 = u_2(x_1, x_2) = x_1 - x_2,$$

and we wish to compute the Jacobian of the transformation. Now

$$x_1 = w_1(y_1, y_2) = \tfrac{1}{2}(y_1 + y_2),$$

$$x_2 = w_2(y_1, y_2) = \tfrac{1}{2}(y_1 - y_2).$$

To determine the set $\mathscr{B}$ in the y_1y_2-plane onto which $\mathscr{A}$ is mapped under the transformation, note that the boundaries of $\mathscr{A}$ are transformed as follows into the boundaries of $\mathscr{B}$;

$$x_1 = 0 \quad \text{into} \quad 0 = \tfrac{1}{2}(y_1 + y_2),$$

$$x_1 = 1 \quad \text{into} \quad 1 = \tfrac{1}{2}(y_1 + y_2),$$

$$x_2 = 0 \quad \text{into} \quad 0 = \tfrac{1}{2}(y_1 - y_2),$$

$$x_2 = 1 \quad \text{into} \quad 1 = \tfrac{1}{2}(y_1 - y_2).$$

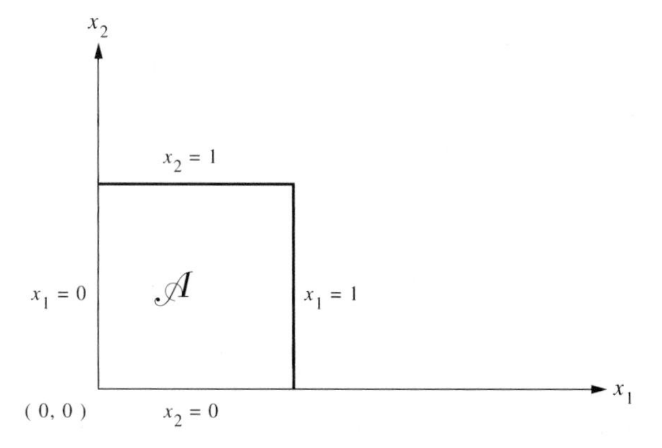

FIGURE 4.1

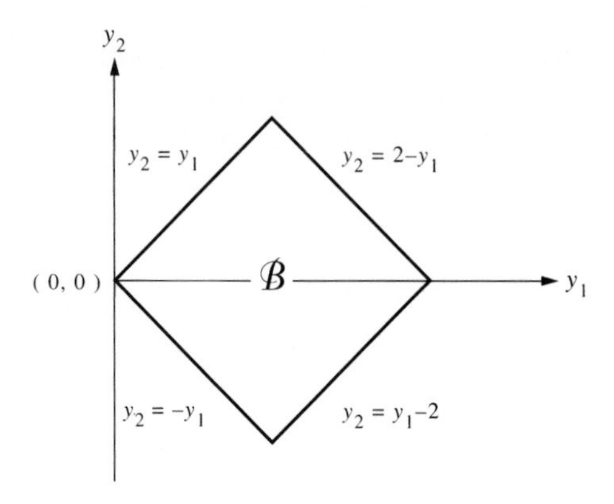

FIGURE 4.2

Accordingly, $\mathscr{B}$ is shown in Figure 4.2. Finally,

$$J = \begin{vmatrix} \dfrac{\partial x_1}{\partial y_1} & \dfrac{\partial x_1}{\partial y_2} \\ \dfrac{\partial x_2}{\partial y_1} & \dfrac{\partial x_2}{\partial y_2} \end{vmatrix} = \begin{vmatrix} \dfrac{1}{2} & \dfrac{1}{2} \\ \dfrac{1}{2} & -\dfrac{1}{2} \end{vmatrix} = -\frac{1}{2}.$$

Remark. Although, in Example 3, we suggest transforming the boundaries of $\mathscr{A}$, others might want to use the inequalities

$$0 < x_1 < 1 \quad \text{and} \quad 0 < x_2 < 1$$

directly. These four inequalities become

$$0 < \tfrac{1}{2}(y_1 + y_2) < 1 \quad \text{and} \quad 0 < \tfrac{1}{2}(y_1 - y_2) < 1.$$

It is easy to see that these are equivalent to

$$-y_1 < y_2, \quad y_2 < 2 - y_1, \quad y_2 < y_1, \quad y_1 - 2 < y_2;$$

and they define the set $\mathscr{B}$. In this example, these methods were rather simple and essentially the same. Other examples could present more complicated transformations, and only experience can help one decide which is the best method in each case.

We now proceed with the problem of finding the joint p.d.f. of two functions of two continuous-type random variables. Let X_1 and X_2 be random variables of the continuous type, having joint p.d.f. $h(x_1, x_2)$. Let $\mathscr{A}$ be the two-dimensional set in the x_1x_2-plane where $h(x_1, x_2) > 0$. Let $Y_1 = u_1(X_1, X_2)$ be a random variable whose p.d.f. is to be found. If $y_1 = u_1(x_1, x_2)$ and $y_2 = u_2(x_1, x_2)$ define a one-to-one transformation of $\mathscr{A}$ onto a set $\mathscr{B}$ in the y_1y_2-plane (with nonidentically zero Jacobian), we can find, by use of a theorem in analysis, the joint p.d.f. of $Y_1 = u_1(X_1, X_2)$ and $Y_2 = u_2(X_1, X_2)$. Let A be a subset of $\mathscr{A}$, and let B denote the mapping of A under the one-to-one transformation (see Figure 4.3). The events $(X_1, X_2) \in A$ and $(Y_1, Y_2) \in B$ are equivalent. Hence

$$\begin{aligned}\Pr[(Y_1, Y_2) \in B] &= \Pr[(X_1, X_2) \in A] \\ &= \iint_A h(x_1, x_2)\, dx_1\, dx_2.\end{aligned}$$

We wish now to change variables of integration by writing $y_1 = u_1(x_1, x_2), y_2 = u_2(x_1, x_2)$, or $x_1 = w_1(y_1, y_2), x_2 = w_2(y_1, y_2)$. It has been proved in analysis that this change of variables requires

$$\iint_A h(x_1, x_2)\, dx_1\, dx_2 = \iint_B h[w_1(y_1, y_2), w_2(y_1, y_2)]|J|\, dy_1\, dy_2.$$

Thus, for every set B in $\mathscr{B}$,

$$\Pr[(Y_1, Y_2) \in B] = \iint_B h[w_1(y_1, y_2), w_2(y_1, y_2)]|J|\, dy_1\, dy_2,$$

which implies that the joint p.d.f. $g(y_1, y_2)$ of Y_1 and Y_2 is

$$\begin{aligned}g(y_1, y_2) &= h[w_1(y_1, y_2), w_2(y_1, y_2)]|J|, \qquad (y_1, y_2) \in \mathscr{B}, \\ &= 0 \qquad \text{elsewhere.}\end{aligned}$$

Accordingly, the marginal p.d.f. $g_1(y_1)$ of Y_1 can be obtained from the joint p.d.f. $g(y_1, y_2)$ in the usual manner by integrating on y_2. Several examples of this result will be given.

Example 4. Let the random variable X have the p.d.f.

$$\begin{aligned}f(x) &= 1, \qquad 0 < x < 1, \\ &= 0 \qquad \text{elsewhere,}\end{aligned}$$

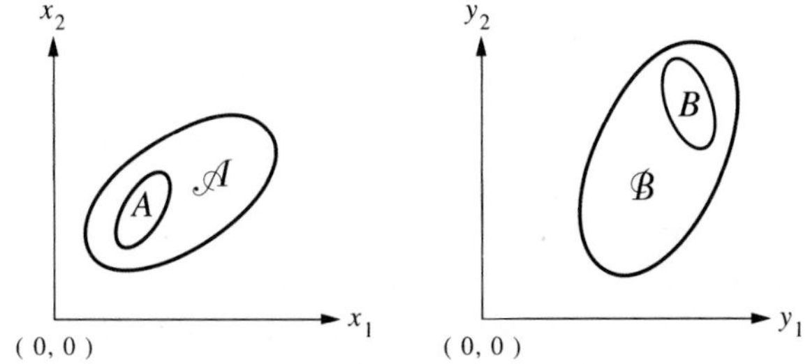

FIGURE 4.3

and let X_1, X_2 denote a random sample from this distribution. The joint p.d.f. of X_1 and X_2 is then

$$h(x_1, x_2) = f(x_1)f(x_2) = 1, \qquad 0 < x_1 < 1, \quad 0 < x_2 < 1,$$
$$= 0 \qquad \text{elsewhere.}$$

Consider the two random variables $Y_1 = X_1 + X_2$ and $Y_2 = X_1 - X_2$. We wish to find the joint p.d.f. of Y_1 and Y_2. Here the two-dimensional space $\mathscr{A}$ in the x_1x_2-plane is that of Example 3 of this section. The one-to-one transformation $y_1 = x_1 + x_2$, $y_2 = x_1 - x_2$ maps $\mathscr{A}$ onto the space $\mathscr{B}$ of that example. Moreover, the Jacobian of that transformation has been shown to be $J = -\frac{1}{2}$. Thus

$$g(y_1, y_2) = h[\tfrac{1}{2}(y_1 + y_2), \tfrac{1}{2}(y_1 - y_2)]|J|$$
$$= f[\tfrac{1}{2}(y_1 + y_2)]f[\tfrac{1}{2}(y_1 - y_2)]|J| = \tfrac{1}{2}, \qquad (y_1, y_2) \in \mathscr{B},$$
$$= 0 \qquad \text{elsewhere.}$$

Because $\mathscr{B}$ is not a product space, the random variables Y_1 and Y_2 are dependent. The marginal p.d.f. of Y_1 is given by

$$g_1(y_1) = \int_{-\infty}^{\infty} g(y_1, y_2)\, dy_2.$$

If we refer to Figure 4.2, it is seen that

$$g_1(y_1) = \int_{-y_1}^{y_1} \tfrac{1}{2}\, dy_2 = y_1, \qquad 0 < y_1 \le 1,$$
$$= \int_{y_1 - 2}^{2 - y_1} \tfrac{1}{2}\, dy_2 = 2 - y_1, \qquad 1 < y_1 < 2,$$
$$= 0 \qquad \text{elsewhere.}$$

In a similar manner, the marginal p.d.f. $g_2(y_2)$ is given by

$$g_2(y_2) = \int_{-y_2}^{y_2 + 2} \tfrac{1}{2}\, dy_1 = y_2 + 1, \qquad -1 < y_2 \le 0,$$
$$= \int_{y_2}^{2 - y_2} \tfrac{1}{2}\, dy_1 = 1 - y_2, \qquad 0 < y_2 < 1,$$
$$= 0 \quad \text{elsewhere.}$$

Example 5. Let X_1, X_2 be a random sample of size $n = 2$ from a standard normal distribution. Say that we are interested in the distribution of $Y_1 = X_1/X_2$. Often in selecting the second random variable, we use the denominator of the ratio or a function of that denominator. So let $Y_2 = X_2$. With the set $\{(x_1, x_2) : -\infty < x_1 < \infty, -\infty < x_2 < \infty\}$, we note that the ratio is not defined at $x_2 = 0$. However, $\Pr(X_2 = 0) = 0$; so we take the p.d.f. of X_2 to be zero at $x_2 = 0$. This results in the set

$$\mathscr{A} = \{(x_1, x_2) : -\infty < x_1 < \infty, \quad -\infty < x_2 < 0 \quad \text{or} \quad 0 < x_2 < \infty\}.$$

With $y_1 = x_1/x_2$, $y_2 = x_2$ or, equivalently, $x_1 = y_1 y_2$, $x_2 = y_2$, $\mathscr{A}$ maps onto

$$\mathscr{B} = \{(y_1, y_2) : -\infty < y_1 < \infty, \quad -\infty < y_2 < 0 \quad \text{or} \quad 0 < y_2 < \infty\}.$$

Also,

$$J = \begin{vmatrix} y_2 & y_1 \\ 0 & 1 \end{vmatrix} = y_2 \not\equiv 0.$$

Since

$$h(x_1, x_2) = \frac{1}{2\pi} \exp\left[-\frac{1}{2}(x_1^2 + x_2^2)\right], \quad (x_1, x_2) \in \mathscr{A},$$

we have that the joint p.d.f. of Y_1 and Y_2 is

$$g(y_1, y_2) = \frac{1}{2\pi} \exp\left[-\frac{1}{2} y_2^2(1 + y_1^2)\right] |y_2|, \quad (y_1, y_2) \in \mathscr{B}.$$

Thus

$$g_1(y_1) = \int_{-\infty}^{0} g(y_1, y_2)\, dy_2 + \int_{0}^{\infty} g(y_1, y_2)\, dy_2.$$

Since $g(y_1, y_2)$ is an even function of y_2, we can write

$$g_1(y_1) = 2 \int_0^\infty \frac{1}{2\pi} \exp\left[-\frac{1}{2} y_2^2(1 + y_1^2)\right] (y_2)\, dy_2$$

$$= \frac{1}{\pi} \left\{ \frac{-\exp\left[-\frac{1}{2} y_2^2(1 + y_1^2)\right]}{1 + y_1^2} \right\}_0^\infty = \frac{1}{\pi(1 + y_1^2)}, \quad -\infty < y_1 < \infty.$$

This marginal p.d.f. of $Y_1 = X_1/X_2$ is that of a *Cauchy distribution*. Although the Cauchy p.d.f. is symmetric about $y_1 = 0$, the mean does not exist because the integral

$$\int_{-\infty}^{\infty} |y_1| g_1(y_1)\, dy_1$$

does not exist. The median and the mode, however, are both equal to zero.

Example 6. Let $Y_1 = \frac{1}{2}(X_1 - X_2)$, where X_1 and X_2 are i.i.d. random variables, each being $\chi^2(2)$. The joint p.d.f. of X_1 and X_2 is

$$f(x_1)f(x_2) = \frac{1}{4} \exp\left(-\frac{x_1 + x_2}{2}\right), \quad 0 < x_1 < \infty, \quad 0 < x_2 < \infty,$$

$$= 0 \quad \text{elsewhere}.$$

Let $Y_2 = X_2$ so that $y_1 = \frac{1}{2}(x_1 - x_2)$, $y_2 = x_2$ or $x_1 = 2y_1 + y_2$, $x_2 = y_2$ define a one-to-one transformation from $\mathscr{A} = \{(x_1, x_2) : 0 < x_1 < \infty, 0 < x_2 < \infty\}$ onto $\mathscr{B} = \{(y_1, y_2) : -2y_1 < y_2 \text{ and } 0 < y_2, -\infty < y_1 < \infty\}$. The Jacobian of the transformation is

$$J = \begin{vmatrix} 2 & 1 \\ 0 & 1 \end{vmatrix} = 2;$$

hence the joint p.d.f. of Y_1 and Y_2 is

$$g(y_1, y_2) = \frac{|2|}{4} e^{-y_1 - y_2}, \qquad (y_1, y_2) \in \mathscr{B},$$

$$= 0 \qquad \text{elsewhere.}$$

Thus the p.d.f. of Y_1 is given by

$$g_1(y_1) = \int_{-2y_1}^{\infty} \tfrac{1}{2} e^{-y_1 - y_2}\, dy_2 = \tfrac{1}{2} e^{y_1}, \qquad -\infty < y_1 < 0,$$

$$= \int_{0}^{\infty} \tfrac{1}{2} e^{-y_1 - y_2}\, dy_2 = \tfrac{1}{2} e^{-y_1}, \qquad 0 \le y_1 < \infty,$$

or

$$g_1(y_1) = \tfrac{1}{2} e^{-|y_1|}, \qquad -\infty < y_1 < \infty.$$

This p.d.f. is now frequently called the *double exponential p.d.f.*

***Example* 7.** In this example a rather important result is established. Let X_1 and X_2 be independent random variables of the continuous type with joint p.d.f. $f_1(x_1)f_2(x_2)$ that is positive on the two-dimensional space $\mathscr{A}$. Let $Y_1 = u_1(X_1)$, a function of X_1 alone, and $Y_2 = u_2(X_2)$, a function of X_2 alone. We assume for the present that $y_1 = u_1(x_1)$, $y_2 = u_2(x_2)$ define a one-to-one transformation from $\mathscr{A}$ onto a two-dimensional set $\mathscr{B}$ in the y_1y_2-plane. Solving for x_1 and x_2 in terms of y_1 and y_2, we have $x_1 = w_1(y_1)$ and $x_2 = w_2(y_2)$, so

$$J = \begin{vmatrix} w_1'(y_1) & 0 \\ 0 & w_2'(y_2) \end{vmatrix} = w_1'(y_1)w_2'(y_2) \not\equiv 0.$$

Hence the joint p.d.f. of Y_1 and Y_2 is

$$g(y_1, y_2) = f_1[w_1(y_1)]f_2[w_2(y_2)]|w_1'(y_1)w_2'(y_2)|, \qquad (y_1, y_2) \in \mathscr{B},$$

$$= 0 \qquad \text{elsewhere.}$$

However, from the procedure for changing variables in the case of one random variable, we see that the marginal probability density functions of Y_1 and Y_2 are, respectively, $g_1(y_1) = f_1[w_1(y_1)]|w_1'(y_1)|$ and

$g_2(y_2) = f_2[w_2(y_2)]|w_2'(y_2)|$ for y_1 and y_2 in some appropriate sets. Consequently,

$$g(y_1, y_2) \equiv g_1(y_1)g_2(y_2).$$

Thus, summarizing, we note that if X_1 and X_2 are independent random variables, then $Y_1 = u_1(X_1)$ and $Y_2 = u_2(X_2)$ are also independent random variables. It has been seen that the result holds if X_1 and X_2 are of the discrete type; see Exercise 4.22.

In the *simulation* of random variables using uniform random variables, it is frequently difficult to solve $y = F(x)$ for x. Thus other methods are necessary. For instance, consider the important normal case in which we desire to determine X so that it is $N(0, 1)$. Of course, once X is determined, other normal variables can then be obtained through X by the transformation $Z = \sigma X + \mu$.

To simulate normal variables, Box and Muller suggested the following procedure. Let Y_1, Y_2 be a random sample from the uniform distribution over $0 < y < 1$. Define X_1 and X_2 by

$$X_1 = (-2 \ln Y_1)^{1/2} \cos (2\pi Y_2),$$

$$X_2 = (-2 \ln Y_1)^{1/2} \sin (2\pi Y_2).$$

The corresponding transformation is one-to-one and maps $\{(y_1, y_2): 0 < y_1 < 1,\ 0 < y_2 < 1\}$ onto $\{(x_1, x_2): -\infty < x_1 < \infty, -\infty < x_2 < \infty\}$ except for sets involving $x_1 = 0$ and $x_2 = 0$, which have probability zero. The inverse transformation is given by

$$y_1 = \exp\left(-\frac{x_1^2 + x_2^2}{2}\right),$$

$$y_2 = \frac{1}{2\pi} \arctan \frac{x_2}{x_1}.$$

This has the Jacobian

$$J = \begin{vmatrix} (-x_1)\exp\left(-\frac{x_1^2 + x_2^2}{2}\right) & (-x_2)\exp\left(-\frac{x_1^2 + x_2^2}{2}\right) \\ \frac{-x_2/x_1^2}{(2\pi)(1 + x_2^2/x_1^2)} & \frac{1/x_1}{(2\pi)(1 + x_2^2/x_1^2)} \end{vmatrix}$$

$$= \frac{-(1 + x_2^2/x_1^2)\exp\left(-\frac{x_1^2 + x_2^2}{2}\right)}{(2\pi)(1 + x_2^2/x_1^2)} = \frac{-\exp\left(-\frac{x_1^2 + x_2^2}{2}\right)}{2\pi}.$$

Since the joint p.d.f. of Y_1 and Y_2 is 1 on $0 < y_1 < 1$, $0 < y_2 < 1$, and zero elsewhere, the joint p.d.f. of X_1 and X_2 is

$$\frac{\exp\left(-\dfrac{x_1^2 + x_2^2}{2}\right)}{2\pi}, \qquad -\infty < x_1 < \infty, \quad -\infty < x_2 < \infty.$$

That is, X_1 and X_2 are independent standard normal random variables.

We close this section by observing a way of finding the p.d.f. of a sum of two independent random variables. Let X_1 and X_2 be independent with respective probability density functions $f_1(x_1)$ and $f_2(x_2)$. Let $Y_1 = X_1 + X_2$ and $Y_2 = X_2$. Thus we have the one-to-one transformation $x_1 = y_1 - y_2$ and $x_2 = y_2$ with Jacobian $J = 1$. Here we say that $\mathscr{A} = \{(x_1, x_2): -\infty < x_1 < \infty, -\infty < x_2 < \infty\}$ maps onto $\mathscr{B} = \{(y_1, y_2): -\infty < y_1 < \infty, \ -\infty < y_2 < \infty\}$, but we recognize that in a particular problem the joint p.d.f. might equal zero on some part of these sets. Thus the joint p.d.f. of Y_1 and Y_2 is

$$g(y_1, y_2) = f_1(y_1 - y_2)f_2(y_2), \qquad (y_1, y_2) \in \mathscr{B},$$

and the marginal p.d.f. of $Y_1 = X_1 + X_2$ is given by

$$g_1(y_1) = \int_{-\infty}^{\infty} f_1(y_1 - y_2)f_2(y_2)\, dy_2,$$

which is the well-known *convolution formula.*

EXERCISES

4.25. Let X have the p.d.f. $f(x) = x^2/9$, $0 < x < 3$, zero elsewhere. Find the p.d.f. of $Y = X^3$.

4.26. If the p.d.f. of X is $f(x) = 2xe^{-x^2}$, $0 < x < \infty$, zero elsewhere, determine the p.d.f. of $Y = X^2$.

4.27. Let X have the *logistic p.d.f.* $f(x) = e^{-x}/(1 + e^{-x})^2$, $-\infty < x < \infty$.

(a) Show that the graph of $f(x)$ is symmetric about the vertical axis through $x = 0$.

(b) Find the distribution function of X.

(c) Find the p.d.f. of $Y = e^{-X}$.

(d) Show that the m.g.f. $M(t)$ of X is $\Gamma(1 - t)\Gamma(1 + t)$, $-1 < t < 1$. *Hint:* In the integral representing $M(t)$, let $y = (1 + e^{-x})^{-1}$.

4.28. Let X have the uniform distribution over the interval $(-\pi/2, \pi/2)$. Show that $Y = \tan X$ has a Cauchy distribution.

4.29. Let X_1 and X_2 be two independent normal random variables, each with mean zero and variance one (possibly resulting from a Box–Muller transformation). Show that

$$Z_1 = \mu_1 + \sigma_1 X_1,$$

$$Z_2 = \mu_2 + \rho\sigma_2 X_1 + \sigma_2\sqrt{1-\rho^2}X_2,$$

where $0 < \sigma_1$, $0 < \sigma_2$, and $0 < \rho < 1$, have a bivariate normal distribution with respective parameters μ_1, μ_2, σ_1^2, σ_2^2, and ρ.

4.30. Let X_1 and X_2 denote a random sample of size 2 from a distribution that is $N(\mu, \sigma^2)$. Let $Y_1 = X_1 + X_2$ and $Y_2 = X_1 - X_2$. Find the joint p.d.f. of Y_1 and Y_2 and show that these random variables are independent.

4.31. Let X_1 and X_2 denote a random sample of size 2 from a distribution that is $N(\mu, \sigma^2)$. Let $Y_1 = X_1 + X_2$ and $Y_2 = X_1 + 2X_2$. Show that the joint p.d.f. of Y_1 and Y_2 is bivariate normal with correlation coefficient $3/\sqrt{10}$.

4.32. Use the convolution formula to determine the p.d.f. of $Y_1 = X_1 + X_2$, where X_1 and X_2 are i.i.d. random variables, each with p.d.f. $f(x) = e^{-x}$, $0 < x < \infty$, zero elsewhere.

Hint: Note that the integral on y_2 has limits of 0 and y_1, where $0 < y_1 < \infty$. Why?

4.33. Let X_1 and X_2 have the joint p.d.f. $h(x_1, x_2) = 2e^{-x_1-x_2}$, $0 < x_1 < x_2 < \infty$, zero elsewhere. Find the joint p.d.f. of $Y_1 = 2X_1$ and $Y_2 = X_2 - X_1$ and argue that Y_1 and Y_2 are independent.

4.34. Let X_1 and X_2 have the joint p.d.f. $h(x_1, x_2) = 8x_1x_2$, $0 < x_1 < x_2 < 1$, zero elsewhere. Find the joint p.d.f. of $Y_1 = X_1/X_2$ and $Y_2 = X_2$ and argue that Y_1 and Y_2 are independent.

Hint: Use the inequalities $0 < y_1y_2 < y_2 < 1$ in considering the mapping from $\mathscr{A}$ onto $\mathscr{B}$.

4.4 The Beta, *t*, and *F* Distributions

It is the purpose of this section to define three additional distributions quite useful in certain problems of statistical inference. These are called, respectively, the beta distribution, the (Student's) *t*-distribution, and the *F*-distribution.

The beta distribution. Let X_1 and X_2 be two independent random variables that have gamma distributions and joint p.d.f.

$$h(x_1, x_2) = \frac{1}{\Gamma(\alpha)\Gamma(\beta)} x_1^{\alpha-1} x_2^{\beta-1} e^{-x_1-x_2}, \qquad 0 < x_1 < \infty, \quad 0 < x_2 < \infty,$$

zero elsewhere, where $\alpha > 0$, $\beta > 0$. Let $Y_1 = X_1 + X_2$ and $Y_2 = X_1/(X_1 + X_2)$. We shall show that Y_1 and Y_2 are independent.

The space $\mathscr{A}$ is, exclusive of the points on the coordinate axes, the first quadrant of the x_1x_2-plane. Now

$$y_1 = u_1(x_1, x_2) = x_1 + x_2,$$

$$y_2 = u_2(x_1, x_2) = \frac{x_1}{x_1 + x_2}$$

may be written $x_1 = y_1y_2$, $x_2 = y_1(1 - y_2)$, so

$$J = \begin{vmatrix} y_2 & y_1 \\ 1 - y_2 & -y_1 \end{vmatrix} = -y_1 \not\equiv 0.$$

The transformation is one-to-one, and it maps $\mathscr{A}$ onto $\mathscr{B} = \{(y_1, y_2) : 0 < y_1 < \infty, 0 < y_2 < 1\}$ in the y_1y_2-plane. The joint p.d.f. of Y_1 and Y_2 is then

$$\begin{aligned} g(y_1, y_2) &= (y_1) \frac{1}{\Gamma(\alpha)\Gamma(\beta)} (y_1y_2)^{\alpha-1}[y_1(1 - y_2)]^{\beta-1} e^{-y_1} \\ &= \frac{y_2^{\alpha-1}(1 - y_2)^{\beta-1}}{\Gamma(\alpha)\Gamma(\beta)} y_1^{\alpha+\beta-1} e^{-y_1}, \qquad 0 < y_1 < \infty, \quad 0 < y_2 < 1, \\ &= 0 \qquad \text{elsewhere.} \end{aligned}$$

In accordance with Theorem 1, Section 2.4, the random variables are independent. The marginal p.d.f. of Y_2 is

$$\begin{aligned} g_2(y_2) &= \frac{y_2^{\alpha-1}(1 - y_2)^{\beta-1}}{\Gamma(\alpha)\Gamma(\beta)} \int_0^\infty y_1^{\alpha+\beta-1} e^{-y_1}\, dy_1, \\ &= \frac{\Gamma(\alpha + \beta)}{\Gamma(\alpha)\Gamma(\beta)} y_2^{\alpha-1}(1 - y_2)^{\beta-1}, \qquad 0 < y_2 < 1, \\ &= 0 \qquad \text{elsewhere.} \end{aligned}$$

This p.d.f. is that of the *beta distribution* with parameters α and β. Since $g(y_1, y_2) \equiv g_1(y_1)g_2(y_2)$, it must be that the p.d.f. of Y_1 is

$$\begin{aligned} g_1(y_1) &= \frac{1}{\Gamma(\alpha + \beta)} y_1^{\alpha+\beta-1} e^{-y_1}, \qquad 0 < y_1 < \infty, \\ &= 0 \qquad \text{elsewhere,} \end{aligned}$$

which is that of a gamma distribution with parameter values of $\alpha + \beta$ and 1.

It is an easy exercise to show that the mean and the variance of Y_2, which has a beta distribution with parameters α and β, are, respectively,

$$\mu = \frac{\alpha}{\alpha + \beta}, \qquad \sigma^2 = \frac{\alpha\beta}{(\alpha + \beta + 1)(\alpha + \beta)^2}.$$

The *t*-distribution. Let W denote a random variable that is $N(0, 1)$; let V denote a random variable that is $\chi^2(r)$; and let W and V be independent. Then the joint p.d.f. of W and V, say $h(w, v)$, is the product of the p.d.f. of W and that of V or

$$h(w,v) = \frac{1}{\sqrt{2\pi}} e^{-w^2/2} \frac{1}{\Gamma(r/2)2^{r/2}} v^{r/2-1} e^{-v/2}, \qquad -\infty < w < \infty, \quad 0 < v < \infty,$$

$$= 0 \qquad \text{elsewhere.}$$

Define a new random variable T by writing

$$T = \frac{W}{\sqrt{V/r}}.$$

The change-of-variable technique will be used to obtain the p.d.f. $g_1(t)$ of T. The equations

$$t = \frac{w}{\sqrt{v/r}} \qquad \text{and} \qquad u = v$$

define a one-to-one transformation that maps $\mathscr{A} = \{(w, v) : -\infty < w < \infty,\ 0 < v < \infty\}$ onto $\mathscr{B} = \{(t, u) : -\infty < t < \infty,\ 0 < u < \infty\}$. Since $w = t\sqrt{u}/\sqrt{r}$, $v = u$, the absolute value of the Jacobian of the transformation is $|J| = \sqrt{u}/\sqrt{r}$. Accordingly, the joint p.d.f. of T and $U = V$ is given by

$$g(t, u) = h\left(\frac{t\sqrt{u}}{\sqrt{r}}, u\right)|J|$$

$$= \frac{1}{\sqrt{2\pi}\,\Gamma(r/2)2^{r/2}} u^{r/2-1} \exp\left[-\frac{u}{2}\left(1 + \frac{t^2}{r}\right)\right]\frac{\sqrt{u}}{\sqrt{r}}, \qquad -\infty < t < \infty, \quad 0 < u < \infty,$$

$$= 0 \qquad \text{elsewhere.}$$

The marginal p.d.f. of T is then

$$g_1(t) = \int_{-\infty}^{\infty} g(t, u)\, du$$

$$= \int_0^{\infty} \frac{1}{\sqrt{2\pi r}\, \Gamma(r/2)2^{r/2}} u^{(r+1)/2 - 1} \exp\left[-\frac{u}{2}\left(1 + \frac{t^2}{r}\right)\right] du.$$

In this integral let $z = u[1 + (t^2/r)]/2$, and it is seen that

$$g_1(t) = \int_0^{\infty} \frac{1}{\sqrt{2\pi r}\, \Gamma(r/2)2^{r/2}} \left(\frac{2z}{1 + t^2/r}\right)^{(r+1)/2 - 1} e^{-z}\left(\frac{2}{1 + t^2/r}\right) dz$$

$$= \frac{\Gamma[(r+1)/2]}{\sqrt{\pi r}\, \Gamma(r/2)} \frac{1}{(1 + t^2/r)^{(r+1)/2}}, \qquad -\infty < t < \infty.$$

Thus, if W is $N(0, 1)$, if V is $\chi^2(r)$, and if W and V are independent, then

$$T = \frac{W}{\sqrt{V/r}}$$

has the immediately preceding p.d.f. $g_1(t)$. The distribution of the random variable T is usually called a *t-distribution*. It should be observed that a t-distribution is completely determined by the parameter r, the number of degrees of freedom of the random variable that has the chi-square distribution. Some approximate values of

$$\Pr(T \le t) = \int_{-\infty}^{t} g_1(w)\, dw$$

for selected values of r and t can be found in Table IV in Appendix B.

Remark. This distribution was first discovered by W. S. Gosset when he was working for an Irish brewery. Because that brewery did not want other breweries to know that statistical methods were being used, Gosset published under the pseudonym Student. Thus this distribution is often known as Student's t-distribution.

The *F*-distribution. Next consider two independent chi-square random variables U and V having r_1 and r_2 degrees of freedom, respectively. The joint p.d.f. $h(u, v)$ of U and V is then

$$h(u, v) = \frac{1}{\Gamma(r_1/2)\Gamma(r_2/2)2^{(r_1+r_2)/2}} u^{r_1/2-1}v^{r_2/2-1}e^{-(u+v)/2},$$

$$0 < u < \infty, \quad 0 < v < \infty,$$

$$= 0 \qquad \text{elsewhere.}$$

We define the new random variable

$$W = \frac{U/r_1}{V/r_2}$$

and we propose finding the p.d.f. $g_1(w)$ of W. The equations

$$w = \frac{u/r_1}{v/r_2}, \qquad z = v,$$

define a one-to-one transformation that maps the set $\mathscr{A} = \{(u, v) : 0 < u < \infty, 0 < v < \infty\}$ onto the set $\mathscr{B} = \{(w, z) : 0 < w < \infty, 0 < z < \infty\}$, Since $u = (r_1/r_2)zw$, $v = z$, the absolute value of the Jacobian of the transformation is $|J| = (r_1/r_2)z$. The joint p.d.f. $g(w, z)$ of the random variables W and $Z = V$ is then

$$g(w, z) = \frac{1}{\Gamma(r_1/2)\Gamma(r_2/2)2^{(r_1+r_2)/2}}\left(\frac{r_1 zw}{r_2}\right)^{r_1/2-1} z^{r_2/2-1}$$

$$\times \exp\left[-\frac{z}{2}\left(\frac{r_1 w}{r_2}+1\right)\right]\frac{r_1 z}{r_2},$$

provided that $(w, z) \in \mathscr{B}$, and zero elsewhere. The marginal p.d.f. $g_1(w)$ of W is then

$$g_1(w) = \int_{-\infty}^{\infty} g(w, z)\, dz$$

$$= \int_0^{\infty} \frac{(r_1/r_2)^{r_1/2}(w)^{r_1/2-1}}{\Gamma(r_1/2)\Gamma(r_2/2)2^{(r_1+r_2)/2}} z^{(r_1+r_2)/2-1}$$

$$\times \exp\left[-\frac{z}{2}\left(\frac{r_1 w}{r_2}+1\right)\right] dz.$$

If we change the variable of integration by writing

$$y = \frac{z}{2}\left(\frac{r_1 w}{r_2}+1\right),$$

it can be seen that

$$g_1(w) = \int_0^\infty \frac{(r_1/r_2)^{r_1/2}(w)^{r_1/2-1}}{\Gamma(r_1/2)\Gamma(r_2/2)2^{(r_1+r_2)/2}} \left(\frac{2y}{r_1 w/r_2 + 1}\right)^{(r_1+r_2)/2-1} e^{-y}$$

$$\times \left(\frac{2}{r_1 w/r_2 + 1}\right) dy$$

$$= \frac{\Gamma[(r_1+r_2)/2](r_1/r_2)^{r_1/2}}{\Gamma(r_1/2)\Gamma(r_2/2)} \frac{(w)^{r_1/2-1}}{(1 + r_1 w/r_2)^{(r_1+r_2)/2}}, \qquad 0 < w < \infty,$$

$$= 0 \qquad \text{elsewhere.}$$

Accordingly, if U and V are independent chi-square variables with r_1 and r_2 degrees of freedom, respectively, then

$$W = \frac{U/r_1}{V/r_2}$$

has the immediately preceding p.d.f. $g_1(w)$. The distribution of this random variable is usually called an *F-distribution*; and we often call the ratio, which we have denoted by W, F. That is,

$$F = \frac{U/r_1}{V/r_2}.$$

It should be observed that an F-distribution is completely determined by the two parameters r_1 and r_2. Table V in Appendix B gives some approximate values of

$$\Pr(F \le b) = \int_0^b g_1(w)\, dw$$

for selected values of r_1, r_2, and b.

EXERCISES

4.35. Find the mean and variance of the beta distribution.

Hint: From that p.d.f., we know that

$$\int_0^1 y^{\alpha-1}(1-y)^{\beta-1}\, dy = \frac{\Gamma(\alpha)\Gamma(\beta)}{\Gamma(\alpha+\beta)}$$

for all $\alpha > 0$, $\beta > 0$.

4.36. Determine the constant c in each of the following so that each $f(x)$ is a *beta p.d.f.*

(a) $f(x) = cx(1-x)^3$, $0 < x < 1$, zero elsewhere.

(b) $f(x) = cx^4(1 - x)^5$, $0 < x < 1$, zero elsewhere.
(c) $f(x) = cx^2(1 - x)^8$, $0 < x < 1$, zero elsewhere.

4.37. Determine the constant c so that $f(x) = cx(3 - x)^4$, $0 < x < 3$, zero elsewhere, is a p.d.f.

4.38. Show that the graph of the beta p.d.f. is symmetric about the vertical line through $x = \frac{1}{2}$ if $\alpha = \beta$.

4.39. Show, for $k = 1, 2, \ldots, n$, that

$$\int_p^1 \frac{n!}{(k-1)!\,(n-k)!} z^{k-1}(1-z)^{n-k}\,dz = \sum_{x=0}^{k-1} \binom{n}{x} p^x(1-p)^{n-x}.$$

This demonstrates the relationship between the distribution functions of the beta and binomial distributions.

4.40. Let T have a t-distribution with 10 degrees of freedom. Find $\Pr(|T| > 2.228)$ from Table IV.

4.41. Let T have a t-distribution with 14 degrees of freedom. Determine b so that $\Pr(-b < T < b) = 0.90$.

4.42. Let F have an F-distribution with parameters r_1 and r_2. Prove that $1/F$ has an F-distribution with parameters r_2 and r_1.

4.43. If F has an F-distribution with parameters $r_1 = 5$ and $r_2 = 10$, find a and b so that $\Pr(F \le a) = 0.05$ and $\Pr(F \le b) = 0.95$, and, accordingly, $\Pr(a < F < b) = 0.90$.

Hint: Write $\Pr(F \le a) = \Pr(1/F \ge 1/a) = 1 - \Pr(1/F \le 1/a)$, and use the result of Exercise 4.42 and Table V.

4.44. Let $T = W/\sqrt{V/r}$, where the independent variables W and V are, respectively, normal with mean zero and variance 1 and chi-square with r degrees of freedom. Show that T^2 has an F-distribution with parameters $r_1 = 1$ and $r_2 = r$.

Hint: What is the distribution of the numerator of T^2?

4.45. Show that the t-distribution with $r = 1$ degree of freedom and the Cauchy distribution are the same.

4.46. Show that

$$Y = \frac{1}{1 + (r_1/r_2)W},$$

where W has an F-distribution with parameters r_1 and r_2, has a beta distribution.

4.47. Let X_1, X_2 be a random sample from a distribution having the p.d.f. $f(x) = e^{-x}$, $0 < x < \infty$, zero elsewhere. Show that $Z = X_1/X_2$ has an F-distribution.

4.5 Extensions of the Change-of-Variable Technique

In Section 4.3 it was seen that the determination of the joint p.d.f. of two functions of two random variables of the continuous type was essentially a corollary to a theorem in analysis having to do with the change of variables in a twofold integral. This theorem has a natural extension to n-fold integrals. This extension is as follows. Consider an integral of the form

$$\int \cdots \int_A h(x_1, x_2, \ldots, x_n)\, dx_1\, dx_2 \cdots dx_n$$

taken over a subset A of an n-dimensional space $\mathscr{A}$. Let

$$y_1 = u_1(x_1, x_2, \ldots, x_n), \qquad y_2 = u_2(x_1, x_2, \ldots, x_n), \ldots,$$
$$y_n = u_n(x_1, \ldots, x_n),$$

together with the inverse functions

$$x_1 = w_1(y_1, y_2, \ldots, y_n), \qquad x_2 = w_2(y_1, y_2, \ldots, y_n), \ldots,$$
$$x_n = w_n(y_1, y_2, \ldots, y_n)$$

define a one-to-one transformation that maps $\mathscr{A}$ onto $\mathscr{B}$ in the $y_1, y_2, \ldots, y_n$ space (and hence maps the subset A of $\mathscr{A}$ onto a subset B of $\mathscr{B}$). Let the first partial derivatives of the inverse functions be continuous and let the n by n determinant (called the Jacobian)

$$J = \begin{vmatrix} \dfrac{\partial x_1}{\partial y_1} & \dfrac{\partial x_1}{\partial y_2} & \cdots & \dfrac{\partial x_1}{\partial y_n} \\ \dfrac{\partial x_2}{\partial y_1} & \dfrac{\partial x_2}{\partial y_2} & \cdots & \dfrac{\partial x_2}{\partial y_n} \\ \vdots & \vdots & & \vdots \\ \dfrac{\partial x_n}{\partial y_1} & \dfrac{\partial x_n}{\partial y_2} & \cdots & \dfrac{\partial x_n}{\partial y_n} \end{vmatrix}$$

not be identically zero in $\mathscr{B}$. Then

$$\int \cdots \int_A h(x_1, x_2, \ldots, x_n)\, dx_1\, dx_2 \cdots dx_n$$
$$= \int \cdots \int_B h[w_1(y_1, \ldots, y_n), w_2(y_1, \ldots, y_n), \ldots, w_n(y_1, \ldots, y_n)]$$
$$\times |J|\, dy_1\, dy_2 \cdots dy_n.$$

Whenever the conditions of this theorem are satisfied, we can determine the joint p.d.f. of n functions of n random variables. Appropriate changes of notation in Section 4.3 (to indicate n-space as opposed to 2-space) are all that is needed to show that the joint p.d.f. of the random variables $Y_1 = u_1(X_1, X_2, \ldots, X_n)$, $Y_2 = u_2(X_1, X_2, \ldots, X_n)$, $\ldots, Y_n = u_n(X_1, X_2, \ldots, X_n)$—where the joint p.d.f. of $X_1, X_2, \ldots, X_n$ is $h(x_1, \ldots, x_n)$—is given by

$$g(y_1, y_2, \ldots, y_n) = |J| h[w_1(y_1, \ldots, y_n), \ldots, w_n(y_1, \ldots, y_n)],$$

when $(y_1, y_2, \ldots, y_n) \in \mathscr{B}$, and is zero elsewhere.

Example 1. Let $X_1, X_2, \ldots, X_{k+1}$ be independent random variables, each having a gamma distribution with $\beta = 1$. The joint p.d.f. of these variables may be written as

$$h(x_1, x_2, \ldots, x_{k+1}) = \prod_{i=1}^{k+1} \frac{1}{\Gamma(\alpha_i)} x_i^{\alpha_i - 1} e^{-x_i}, \qquad 0 < x_i < \infty,$$

$$= 0 \qquad \text{elsewhere.}$$

Let

$$Y_i = \frac{X_i}{X_1 + X_2 + \cdots + X_{k+1}}, \qquad i = 1, 2, \ldots, k,$$

and $Y_{k+1} = X_1 + X_2 + \cdots + X_{k+1}$ denote $k + 1$ new random variables. The associated transformation maps $\mathscr{A} = \{(x_1, \ldots, x_{k+1}) : 0 < x_i < \infty,\ i = 1, \ldots, k + 1\}$ onto the space

$$\mathscr{B} = \{(y_1, \ldots, y_k, y_{k+1}) : 0 < y_i,\ i = 1, \ldots, k,$$
$$y_1 + \cdots + y_k < 1,\ 0 < y_{k+1} < \infty\}.$$

The single-valued inverse functions are $x_1 = y_1 y_{k+1}, \ldots,\ x_k = y_k y_{k+1}$, $x_{k+1} = y_{k+1}(1 - y_1 - \cdots - y_k)$, so that the Jacobian is

$$J = \begin{vmatrix} y_{k+1} & 0 & \cdots & 0 & y_1 \\ 0 & y_{k+1} & \cdots & 0 & y_2 \\ \vdots & \vdots & & \vdots & \vdots \\ 0 & 0 & \cdots & y_{k+1} & y_k \\ -y_{k+1} & -y_{k+1} & \cdots & -y_{k+1} & (1 - y_1 - \cdots - y_k) \end{vmatrix} = y_{k+1}^k.$$

Hence the joint p.d.f. of $Y_1, \ldots, Y_k, Y_{k+1}$ is given by

$$\frac{y_{k+1}^{\alpha_1 + \cdots + \alpha_{k+1} - 1} y_1^{\alpha_1 - 1} \cdots y_k^{\alpha_k - 1}(1 - y_1 - \cdots - y_k)^{\alpha_{k+1} - 1} e^{-y_{k+1}}}{\Gamma(\alpha_1) \cdots \Gamma(\alpha_k)\Gamma(\alpha_{k+1})},$$

provided that $(y_1, \ldots, y_k, y_{k+1}) \in \mathscr{B}$ and is equal to zero elsewhere. The joint p.d.f. of $Y_1, \ldots, Y_k$ is seen by inspection to be given by

$$g(y_1, \ldots, y_k) = \frac{\Gamma(\alpha_1 + \cdots + \alpha_{k+1})}{\Gamma(\alpha_1) \cdots \Gamma(\alpha_{k+1})} y_1^{\alpha_1 - 1} \cdots y_k^{\alpha_k - 1}(1 - y_1 - \cdots - y_k)^{\alpha_{k+1} - 1},$$

when $0 < y_i$, $i = 1, \ldots, k$, $y_1 + \cdots + y_k < 1$, while the function g is equal to zero elsewhere. Random variables $Y_1, \ldots, Y_k$ that have a joint p.d.f. of this form are said to have a *Dirichlet distribution* with parameters $\alpha_1, \ldots, \alpha_k, \alpha_{k+1}$, and any such $g(y_1, \ldots, y_k)$ is called a *Dirichlet p.d.f.* It is seen, in the special case of $k = 1$, that the Dirichlet p.d.f. becomes a beta p.d.f. Moreover, it is also clear from the joint p.d.f. of $Y_1, \ldots, Y_k, Y_{k+1}$ that Y_{k+1} has a gamma distribution with parameters $\alpha_1 + \cdots + \alpha_k + \alpha_{k+1}$ and $\beta = 1$ and that Y_{k+1} is independent of $Y_1, Y_2, \ldots, Y_k$.

We now consider some other problems that are encountered when transforming variables. Let X have the Cauchy p.d.f.

$$f(x) = \frac{1}{\pi(1 + x^2)}, \qquad -\infty < x < \infty,$$

and let $Y = X^2$. We seek the p.d.f. $g(y)$ of Y. Consider the transformation $y = x^2$. This transformation maps the space of X, $\mathscr{A} = \{x : -\infty < x < \infty\}$, onto $\mathscr{B} = \{y : 0 \le y < \infty\}$. However, the transformation is not one-to-one. To each $y \in \mathscr{B}$, with the exception of $y = 0$, there correspond two points $x \in \mathscr{A}$. For example, if $y = 4$, we may have either $x = 2$ or $x = -2$. In such an instance, we represent $\mathscr{A}$ as the union of two disjoint sets A_1 and A_2 such that $y = x^2$ defines a one-to-one transformation that maps each of A_1 and A_2 onto $\mathscr{B}$. If we take A_1 to be $\{x : -\infty < x < 0\}$ and A_2 to be $\{x : 0 \le x < \infty\}$, we see that A_1 is mapped onto $\{y : 0 < y < \infty\}$, whereas A_2 is mapped onto $\{y : 0 \le y < \infty\}$, and these sets are not the same. Our difficulty is caused by the fact that $x = 0$ is an element of $\mathscr{A}$. Why, then, do we not return to the Cauchy p.d.f. and take $f(0) = 0$? Then our new $\mathscr{A}$ is $\mathscr{A} = \{-\infty < x < \infty \text{ but } x \ne 0\}$. We then take $A_1 = \{x : -\infty < x < 0\}$ and $A_2 = \{x : 0 < x < \infty\}$. Thus $y = x^2$, with the inverse $x = -\sqrt{y}$, maps A_1 onto $\mathscr{B} = \{y : 0 < y < \infty\}$ and the transformation is one-to-one. Moreover, the transformation $y = x^2$, with inverse $x = \sqrt{y}$, maps A_2 onto $\mathscr{B} = \{y : 0 < y < \infty\}$ and the transformation is one-to-one. Consider the probability $\Pr(Y \in B)$, where $B \subset \mathscr{B}$. Let $A_3 = \{x : x = -\sqrt{y}, y \in B\} \subset A_1$ and let $A_4 = \{x : x = \sqrt{y}, y \in B\} \subset A_2$. Then $Y \in B$ when and only when

$X \in A_3$ or $X \in A_4$. Thus we have

$$\Pr(Y \in B) = \Pr(X \in A_3) + \Pr(X \in A_4)$$
$$= \int_{A_3} f(x)\, dx + \int_{A_4} f(x)\, dx.$$

In the first of these integrals, let $x = -\sqrt{y}$. Thus the Jacobian, say J_1, is $-1/2\sqrt{y}$; moreover, the set A_3 is mapped onto B. In the second integral let $x = \sqrt{y}$. Thus the Jacobian, say J_2, is $1/2\sqrt{y}$; moreover, the set A_4 is also mapped onto B. Finally,

$$\Pr(Y \in B) = \int_B f(-\sqrt{y}) \left| -\frac{1}{2\sqrt{y}} \right| dy + \int_B f(\sqrt{y}) \frac{1}{2\sqrt{y}}\, dy$$
$$= \int_B [f(-\sqrt{y}) + f(\sqrt{y})] \frac{1}{2\sqrt{y}}\, dy.$$

Hence the p.d.f. of Y is given by

$$g(y) = \frac{1}{2\sqrt{y}} [f(-\sqrt{y}) + f(\sqrt{y})], \qquad y \in B.$$

With $f(x)$ the Cauchy p.d.f. we have

$$g(y) = \frac{1}{\pi(1 + y)\sqrt{y}}, \qquad 0 < y < \infty,$$
$$= 0 \qquad \text{elsewhere.}$$

In the preceding discussion of a random variable of the continuous type, we had two inverse functions, $x = -\sqrt{y}$ and $x = \sqrt{y}$. That is why we sought to partition $\mathscr{A}$ (or a modification of $\mathscr{A}$) into two disjoint subsets such that the transformation $y = x^2$ maps each onto the same $\mathscr{B}$. Had there been three inverse functions, we would have sought to partition $\mathscr{A}$ (or a modified form of $\mathscr{A}$) into three disjoint subsets, and so on. It is hoped that this detailed discussion will make the following paragraph easier to read.

Let $h(x_1, x_2, \ldots, x_n)$ be the joint p.d.f. of $X_1, X_2, \ldots, X_n$, which are random variables of the continuous type. Let $\mathscr{A}$ be the n-dimensional space where $h(x_1, x_2, \ldots, x_n) > 0$, and consider the transformation $y_1 = u_1(x_1, x_2, \ldots, x_n)$, $y_2 = u_2(x_1, x_2, \ldots, x_n)$, $\ldots$, $y_n = u_n(x_1, x_2, \ldots, x_n)$, which maps $\mathscr{A}$ onto $\mathscr{B}$ in the $y_1, y_2, \ldots, y_n$ space. To each point of $\mathscr{A}$ there will correspond, of course, but one point in $\mathscr{B}$; but to a point in $\mathscr{B}$ there may correspond more than one point in $\mathscr{A}$. That is, the transformation may not be one-to-one.

Suppose, however, that we can represent $\mathscr{A}$ as the union of a finite number, say k, of mutually disjoint sets $A_1, A_2, \ldots, A_k$ so that

$$y_1 = u_1(x_1, x_2, \ldots, x_n), \ldots, \qquad y_n = u_n(x_1, x_2, \ldots, x_n)$$

define a one-to-one transformation of each A_i onto $\mathscr{B}$. Thus, to each point in $\mathscr{B}$ there will correspond exactly one point in each of $A_1, A_2, \ldots, A_k$. Let

$$\begin{aligned} x_1 &= w_{1i}(y_1, y_2, \ldots, y_n), \\ x_2 &= w_{2i}(y_1, y_2, \ldots, y_n), \\ &\vdots \\ x_n &= w_{ni}(y_1, y_2, \ldots, y_n), \end{aligned} \qquad i = 1, 2, \ldots, k,$$

denote the k groups of n inverse functions, one group for each of these k transformations. Let the first partial derivatives be continuous and let each

$$J_i = \begin{vmatrix} \frac{\partial w_{1i}}{\partial y_1} & \frac{\partial w_{1i}}{\partial y_2} & \cdots & \frac{\partial w_{1i}}{\partial y_n} \\ \frac{\partial w_{2i}}{\partial y_1} & \frac{\partial w_{2i}}{\partial y_2} & \cdots & \frac{\partial w_{2i}}{\partial y_n} \\ \vdots & \vdots & & \vdots \\ \frac{\partial w_{ni}}{\partial y_1} & \frac{\partial w_{ni}}{\partial y_2} & \cdots & \frac{\partial w_{ni}}{\partial y_n} \end{vmatrix}, \quad i = 1, 2, \ldots, k,$$

be not identically equal to zero in $\mathscr{B}$. From a consideration of the probability of the union of k mutually exclusive events and by applying the change of variable technique to the probability of each of these events, it can be seen that the joint p.d.f. of $Y_1 = u_1(X_1, X_2, \ldots, X_n)$, $Y_2 = u_2(X_1, X_2, \ldots, X_n), \ldots, Y_n = u_n(X_1, X_2, \ldots, X_n)$, is given by

$$g(y_1, y_2, \ldots, y_n) = \sum_{i=1}^{k} |J_i| h[w_{1i}(y_1, \ldots, y_n), \ldots, w_{ni}(y_1, \ldots, y_n)],$$

provided that $(y_1, y_2, \ldots, y_n) \in \mathscr{B}$, and equals zero elsewhere. The p.d.f. of any Y_i, say Y_1, is then

$$g_1(y_1) = \int_{-\infty}^{\infty} \cdots \int_{-\infty}^{\infty} g(y_1, y_2, \ldots, y_n)\, dy_2 \cdots dy_n.$$

An illustrative example follows.

Example 2. To illustrate the result just obtained, take $n = 2$ and let X_1, X_2 denote a random sample of size 2 from a standard normal distribution. The joint p.d.f. of X_1 and X_2 is

$$f(x_1, x_2) = \frac{1}{2\pi} \exp\left(-\frac{x_1^2 + x_2^2}{2}\right), \qquad -\infty < x_1 < \infty, \quad -\infty < x_2 < \infty.$$

Let Y_1 denote the mean and let Y_2 denote twice the variance of the random sample. The associated transformation is

$$y_1 = \frac{x_1 + x_2}{2},$$

$$y_2 = \frac{(x_1 - x_2)^2}{2}.$$

This transformation maps $\mathscr{A} = \{(x_1, x_2) : -\infty < x_1 < \infty, -\infty < x_2 < \infty\}$ onto $\mathscr{B} = \{(y_1, y_2) : -\infty < y_1 < \infty, 0 \le y_2 < \infty\}$. But the transformation is not one-to-one because, to each point in $\mathscr{B}$, exclusive of points where $y_2 = 0$, there correspond two points in $\mathscr{A}$. In fact, the two groups of inverse functions are

$$x_1 = y_1 - \sqrt{\frac{y_2}{2}}, \qquad x_2 = y_1 + \sqrt{\frac{y_2}{2}}$$

and

$$x_1 = y_1 + \sqrt{\frac{y_2}{2}}, \qquad x_2 = y_1 - \sqrt{\frac{y_2}{2}}.$$

Moreover, the set $\mathscr{A}$ cannot be represented as the union of two disjoint sets, each of which under our transformation maps onto $\mathscr{B}$. Our difficulty is caused by those points of $\mathscr{A}$ that lie on the line whose equation is $x_2 = x_1$. At each of these points, we have $y_2 = 0$. However, we can define $f(x_1, x_2)$ to be zero at each point where $x_1 = x_2$. We can do this without altering the distribution of probability, because the probability measure of this set is zero. Thus we have a new $\mathscr{A} = \{(x_1, x_2) : -\infty < x_1 < \infty, -\infty < x_2 < \infty$, but $x_1 \neq x_2\}$. This space is the union of the two disjoint sets $A_1 = \{(x_1, x_2) : x_2 > x_1\}$ and $A_2 = \{(x_1, x_2) : x_2 < x_1\}$. Moreover, our transformation now defines a one-to-one transformation of each A_i, $i = 1, 2$, onto the new $\mathscr{B} = \{(y_1, y_2) : -\infty < y_1 < \infty, 0 < y_2 < \infty\}$. We can now find the joint p.d.f., say $g(y_1, y_2)$, of the mean Y_1 and twice the variance Y_2 of our random sample. An easy computation shows that $|J_1| = |J_2| = 1/\sqrt{2y_2}$. Thus

$$\begin{aligned} g(y_1, y_2) &= \frac{1}{2\pi} \exp\left[-\frac{(y_1 - \sqrt{y_2/2})^2}{2} - \frac{(y_1 + \sqrt{y_2/2})^2}{2}\right] \frac{1}{\sqrt{2y_2}} \\ &\quad + \frac{1}{2\pi} \exp\left[-\frac{(y_1 + \sqrt{y_2/2})^2}{2} - \frac{(y_1 - \sqrt{y_2/2})^2}{2}\right] \frac{1}{\sqrt{2y_2}} \\ &= \sqrt{\frac{2}{2\pi}}\, e^{-y_1^2} \frac{1}{\sqrt{2}\Gamma(\frac{1}{2})} y_2^{1/2-1} e^{-y_2/2}, \qquad -\infty < y_1 < \infty, \quad 0 < y_2 < \infty. \end{aligned}$$

We can make three interesting observations. The mean Y_1 of our random sample is $N(0, \frac{1}{2})$; Y_2, which is twice the variance of our sample, is $\chi^2(1)$; and the two are independent. Thus the mean and the variance of our sample are independent.

EXERCISES

4.48. Let X_1, X_2, X_3 denote a random sample from a standard normal distribution. Let the random variables Y_1, Y_2, Y_3 be defined by

$$X_1 = Y_1 \cos Y_2 \sin Y_3, \qquad X_2 = Y_1 \sin Y_2 \sin Y_3, \qquad X_3 = Y_1 \cos Y_3,$$

where $0 \le Y_1 < \infty$, $0 \le Y_2 < 2\pi$, $0 \le Y_3 \le \pi$. Show that Y_1, Y_2, Y_3 are mutually independent.

4.49. Let X_1, X_2, X_3 be i.i.d., each with the distribution having p.d.f. $f(x) = e^{-x}$, $0 < x < \infty$, zero elsewhere. Show that

$$Y_1 = \frac{X_1}{X_1 + X_2}, \qquad Y_2 = \frac{X_1 + X_2}{X_1 + X_2 + X_3}, \qquad Y_3 = X_1 + X_2 + X_3$$

are mutually independent.

4.50. Let $X_1, X_2, \ldots, X_r$ be r independent gamma variables with parameters $\alpha = \alpha_i$ and $\beta = 1$, $i = 1, 2, \ldots, r$, respectively. Show that $Y_1 = X_1 + X_2 + \cdots + X_r$ has a gamma distribution with parameters $\alpha = \alpha_1 + \cdots + \alpha_r$ and $\beta = 1$.

Hint: Let $Y_2 = X_2 + \cdots + X_r$, $Y_3 = X_3 + \cdots + X_r, \ldots, Y_r = X_r$.

4.51. Let $Y_1, \ldots, Y_k$ have a Dirichlet distribution with parameters $\alpha_1, \ldots, \alpha_k, \alpha_{k+1}$.

(a) Show that Y_1 has a beta distribution with parameters $\alpha = \alpha_1$ and $\beta = \alpha_2 + \cdots + \alpha_{k+1}$.

(b) Show that $Y_1 + \cdots + Y_r$, $r \le k$, has a beta distribution with parameters $\alpha = \alpha_1 + \cdots + \alpha_r$ and $\beta = \alpha_{r+1} + \cdots + \alpha_{k+1}$.

(c) Show that $Y_1 + Y_2, Y_3 + Y_4, Y_5, \ldots, Y_k$, $k \ge 5$, have a Dirichlet distribution with parameters $\alpha_1 + \alpha_2, \alpha_3 + \alpha_4, \alpha_5, \ldots, \alpha_k, \alpha_{k+1}$.

Hint: Recall the definition of Y_i in Example 1 and use the fact that the sum of several independent gamma variables with $\beta = 1$ is a gamma variable (Exercise 4.50).

4.52. Let X_1, X_2, and X_3 be three independent chi-square variables with r_1, r_2, and r_3 degrees of freedom, respectively.

(a) Show that $Y_1 = X_1/X_2$ and $Y_2 = X_1 + X_2$ are independent and that Y_2 is $\chi^2(r_1 + r_2)$.

(b) Deduce that

$$\frac{X_1/r_1}{X_2/r_2} \quad \text{and} \quad \frac{X_3/r_3}{(X_1+X_2)/(r_1+r_2)}$$

are independent F-variables.

4.53. If $f(x) = \frac{1}{2}$, $-1 < x < 1$, zero elsewhere, is the p.d.f. of the random variable X, find the p.d.f. of $Y = X^2$.

4.54. If X_1, X_2 is a random sample from a standard normal distribution, find the joint p.d.f. of $Y_1 = X_1^2 + X_2^2$ and $Y_2 = X_2$ and the marginal p.d.f. of Y_1.

Hint: Note that the space of Y_1 and Y_2 is given by $-\sqrt{y_1} < y_2 < \sqrt{y_1}$, $0 < y_1 < \infty$.

4.55. If X has the p.d.f. $f(x) = \frac{1}{4}$, $-1 < x < 3$, zero elsewhere, find the p.d.f. of $Y = X^2$.

Hint: Here $\mathscr{B} = \{y : 0 \le y < 9\}$ and the event $Y \in B$ is the union of two mutually exclusive events if $B = \{y : 0 < y < 1\}$.

4.6 Distributions of Order Statistics

In this section the notion of an order statistic will be defined and we shall investigate some of the simpler properties of such a statistic. These statistics have in recent times come to play an important role in statistical inference partly because some of their properties do not depend upon the distribution from which the random sample is obtained.

Let $X_1, X_2, \ldots, X_n$ denote a random sample from a distribution of the *continuous type* having a p.d.f. $f(x)$ that is positive, provided that $a < x < b$. Let Y_1 be the smallest of these X_i, Y_2 the next X_i in order of magnitude, . . . , and Y_n the largest X_i. That is, $Y_1 < Y_2 < \cdots < Y_n$ represent $X_1, X_2, \ldots, X_n$ when the latter are arranged in ascending order of magnitude. Then Y_i, $i = 1, 2, \ldots, n$, is called the ith order statistic of the random sample $X_1, X_2, \ldots, X_n$. It will be shown that the joint p.d.f. of $Y_1, Y_2, \ldots, Y_n$ is given by

$$\begin{aligned} g(y_1, y_2, \ldots, y_n) &= (n!)f(y_1)f(y_2)\cdots f(y_n), \\ &\qquad\qquad a < y_1 < y_2 < \cdots < y_n < b, \\ &= 0 \qquad \text{elsewhere.} \end{aligned} \tag{1}$$

We shall prove this only for the case $n = 3$, but the argument is seen to be entirely general. With $n = 3$, the joint p.d.f. of X_1, X_2, X_3 is

$f(x_1)f(x_2)f(x_3)$. Consider a probability such as $\Pr(a < X_1 = X_2 < b, a < X_3 < b)$. This probability is given by

$$\int_a^b \int_a^b \int_{x_2}^{x_2} f(x_1)f(x_2)f(x_3)\, dx_1\, dx_2\, dx_3 = 0,$$

since

$$\int_{x_2}^{x_2} f(x_1)\, dx_1$$

is defined in calculus to be zero. As has been pointed out, we may, without altering the distribution of X_1, X_2, X_3, define the joint p.d.f. $f(x_1)f(x_2)f(x_3)$ to be zero at all points (x_1, x_2, x_3) that have at least two of their coordinates equal. Then the set $\mathscr{A}$, where $f(x_1)f(x_2)f(x_3) > 0$, is the union of the six mutually disjoint sets:

$$A_1 = \{(x_1, x_2, x_3) : a < x_1 < x_2 < x_3 < b\},$$

$$A_2 = \{(x_1, x_2, x_3) : a < x_2 < x_1 < x_3 < b\},$$

$$A_3 = \{(x_1, x_2, x_3) : a < x_1 < x_3 < x_2 < b\},$$

$$A_4 = \{(x_1, x_2, x_3) : a < x_2 < x_3 < x_1 < b\},$$

$$A_5 = \{(x_1, x_2, x_3) : a < x_3 < x_1 < x_2 < b\},$$

$$A_6 = \{(x_1, x_2, x_3) : a < x_3 < x_2 < x_1 < b\}.$$

There are six of these sets because we can arrange x_1, x_2, x_3 in precisely $3! = 6$ ways. Consider the functions y_1 = minimum of x_1, x_2, x_3; y_2 = middle in magnitude of x_1, x_2, x_3; and y_3 = maximum of x_1, x_2, x_3. These functions define one-to-one transformations that map each of $A_1, A_2, \ldots, A_6$ onto the same set $\mathscr{B} = \{(y_1, y_2, y_3) : a < y_1 < y_2 < y_3 < b\}$. The inverse functions are, for points in A_1, $x_1 = y_1$, $x_2 = y_2$, $x_3 = y_3$; for points in A_2, they are $x_1 = y_2$, $x_2 = y_1$, $x_3 = y_3$; and so on, for each of the remaining four sets. Then we have that

$$J_1 = \begin{vmatrix} 1 & 0 & 0 \\ 0 & 1 & 0 \\ 0 & 0 & 1 \end{vmatrix} = 1$$

and

$$J_2 = \begin{vmatrix} 0 & 1 & 0 \\ 1 & 0 & 0 \\ 0 & 0 & 1 \end{vmatrix} = -1.$$

It is easily verified that the absolute value of each of the $3! = 6$ Jacobians is $+1$. Thus the joint p.d.f. of the three order statistics $Y_1 =$ minimum of X_1, X_2, X_3; $Y_2 =$ middle in magnitude of X_1, X_2, X_3; $Y_3 =$ maximum of X_1, X_2, X_3 is

$$\begin{aligned} g(y_1, y_2, y_3) &= |J_1|\, f(y_1)f(y_2)f(y_3) + |J_2|\, f(y_2)f(y_1)f(y_3) + \cdots \\ &\quad + |J_6|\, f(y_3)f(y_2)f(y_1), \qquad a < y_1 < y_2 < y_3 < b, \\ &= (3!)f(y_1)f(y_2)f(y_3), \qquad a < y_1 < y_2 < y_3 < b, \\ &= 0 \qquad \text{elsewhere.} \end{aligned}$$

This is Equation (1) with $n = 3$.

In accordance with the natural extension of Theorem 1, Section 2.4, to distributions of more than two random variables, it is seen that the order statistics, unlike the items of the random sample, are dependent.

Example 1. Let X denote a random variable of the continuous type with a p.d.f. $f(x)$ that is positive and continuous, provided that $a < x < b$ and is zero elsewhere. The distribution function $F(x)$ of X may be written

$$F(x) = \int_a^x f(w)\, dw, \qquad a < x < b.$$

If $x \le a$, $F(x) = 0$; and if $b \le x$, $F(x) = 1$. Thus there is a unique median m of the distribution with $F(m) = \frac{1}{2}$. Let X_1, X_2, X_3 denote a random sample from this distribution and let $Y_1 < Y_2 < Y_3$ denote the order statistics of the sample. We shall compute the probability that $Y_2 \le m$. The joint p.d.f. of the three order statistics is

$$\begin{aligned} g(y_1, y_2, y_3) &= 6f(y_1)f(y_2)f(y_3), \qquad a < y_1 < y_2 < y_3 < b, \\ &= 0 \qquad \text{elsewhere.} \end{aligned}$$

The p.d.f. of Y_2 is then

$$\begin{aligned} h(y_2) &= 6f(y_2) \int_{y_2}^{b} \int_a^{y_2} f(y_1)f(y_3)\, dy_1\, dy_3, \\ &= 6f(y_2)F(y_2)[1 - F(y_2)], \qquad a < y_2 < b, \\ &= 0 \qquad \text{elsewhere.} \end{aligned}$$

Accordingly,

$$\begin{aligned} \Pr(Y_2 \le m) &= 6 \int_a^m \{F(y_2)f(y_2) - [F(y_2)]^2 f(y_2)\}\, dy_2 \\ &= 6\left\{\frac{[F(y_2)]^2}{2} - \frac{[F(y_2)]^3}{3}\right\}_a^m = \frac{1}{2}. \end{aligned}$$

The procedure used in Example 1 can be used to obtain general formulas for the marginal probability density functions of the order statistics. We shall do this now. Let X denote a random variable of the continuous type having a p.d.f. $f(x)$ that is positive and continuous, provided that $a < x < b$, and is zero elsewhere. Then the distribution function $F(x)$ may be written

$$\begin{aligned} F(x) &= 0, \qquad x \le a, \\ &= \int_a^x f(w)\, dw, \qquad a < x < b, \\ &= 1, \qquad b \le x. \end{aligned}$$

Accordingly, $F'(x) = f(x)$, $a < x < b$. Moreover, if $a < x < b$,

$$\begin{aligned} 1 - F(x) &= F(b) - F(x) \\ &= \int_a^b f(w)\, dw - \int_a^x f(w)\, dw \\ &= \int_x^b f(w)\, dw. \end{aligned}$$

Let $X_1, X_2, \ldots, X_n$ denote a random sample of size n from this distribution, and let $Y_1, Y_2, \ldots, Y_n$ denote the order statistics of this random sample. Then the joint p.d.f. of $Y_1, Y_2, \ldots, Y_n$ is

$$\begin{aligned} g(y_1, y_2, \ldots, y_n) &= n!\, f(y_1)f(y_2)\cdots f(y_n), \quad a < y_1 < y_2 < \cdots < y_n < b, \\ &= 0 \qquad \text{elsewhere.} \end{aligned}$$

It will first be shown how the marginal p.d.f. of Y_n may be expressed in terms of the distribution function $F(x)$ and the p.d.f. $f(x)$ of the random variable X. If $a < y_n < b$, the marginal p.d.f. of y_n is given by

$$\begin{aligned} &g_n(y_n) \\ &= \int_a^{y_n} \cdots \int_a^{y_4} \int_a^{y_3} \int_a^{y_2} n!\, f(y_1)f(y_2)\cdots f(y_n)\, dy_1\, dy_2\, dy_3 \cdots dy_{n-1} \\ &= \int_a^{y_n} \cdots \int_a^{y_4} \int_a^{y_3} n! \left(\int_a^{y_2} f(y_1)\, dy_1 \right) f(y_2)\cdots f(y_n)\, dy_2 \cdots dy_{n-1} \\ &= \int_a^{y_n} \cdots \int_a^{y_4} \int_a^{y_3} n!\, F(y_2)f(y_2)\cdots f(y_n)\, dy_2 \cdots dy_{n-1}, \end{aligned}$$

since $F(x) = \int_a^x f(w)\,dw$. Now

$$\int_a^{y_3} F(y_2)f(y_2)\,dy_2 = \frac{[F(y_2)]^2}{2}\bigg|_a^{y_3}$$

$$= \frac{[F(y_3)]^2}{2},$$

since $F(a) = 0$. Thus

$$g_n(y_n) = \int_a^{y_n} \cdots \int_a^{y_4} n!\,\frac{[F(y_3)]^2}{2} f(y_3)\cdots f(y_n)\,dy_3 \cdots dy_{n-1}.$$

But

$$\int_a^{y_4} \frac{[F(y_3)]^2}{2} f(y_3)\,dy_3 = \frac{[F(y_3)]^3}{2\cdot 3}\bigg|_a^{y_4} = \frac{[F(y_4)]^3}{2\cdot 3},$$

so

$$g_n(y_n) = \int_a^{y_n} \cdots \int_a^{y_5} n!\,\frac{[F(y_4)]^3}{3!} f(y_4)\cdots f(y_n)\,dy_4 \cdots dy_{n-1}.$$

If the successive integrations on $y_4, \ldots, y_{n-1}$ are carried out, it is seen that

$$g_n(y_n) = n!\,\frac{[F(y_n)]^{n-1}}{(n-1)!} f(y_n)$$

$$= n[F(y_n)]^{n-1}f(y_n), \qquad a < y_n < b,$$

$$= 0 \qquad \text{elsewhere.}$$

It will next be shown how to express the marginal p.d.f. of Y_1 in terms of $F(x)$ and $f(x)$. We have, for $a < y_1 < b$,

$$g_1(y_1) = \int_{y_1}^{b} \cdots \int_{y_{n-3}}^{b} \int_{y_{n-2}}^{b} \int_{y_{n-1}}^{b} n!\,f(y_1)f(y_2)\cdots f(y_n)\,dy_n\,dy_{n-1}\cdots dy_2$$

$$= \int_{y_1}^{b} \cdots \int_{y_{n-3}}^{b} \int_{y_{n-2}}^{b} n!\,f(y_1)f(y_2)\cdots$$

$$f(y_{n-1})[1 - F(y_{n-1})]\,dy_{n-1}\cdots dy_2.$$

But

$$\int_{y_{n-2}}^{b} [1 - F(y_{n-1})]f(y_{n-1})\, dy_{n-1} = -\frac{[1 - F(y_{n-1})]^2}{2}\Bigg|_{y_{n-2}}^{b}$$

$$= \frac{[1 - F(y_{n-2})]^2}{2},$$

so that

$$g_1(y_1) = \int_{y_1}^{b} \cdots \int_{y_{n-3}}^{b} n!\, f(y_1) \cdots f(y_{n-2}) \frac{[1 - F(y_{n-2})]^2}{2}\, dy_{n-2} \cdots dy_2.$$

Upon completing the integrations, it is found that

$$g_1(y_1) = n[1 - F(y_1)]^{n-1} f(y_1), \qquad a < y_1 < b,$$
$$= 0 \qquad \text{elsewhere.}$$

Once it is observed that

$$\int_a^x [F(w)]^{\alpha - 1} f(w)\, dw = \frac{[F(x)]^\alpha}{\alpha}, \qquad \alpha > 0$$

and that

$$\int_y^b [1 - F(w)]^{\beta - 1} f(w)\, dw = \frac{[1 - F(y)]^\beta}{\beta}, \qquad \beta > 0,$$

it is easy to express the marginal p.d.f. of any order statistic, say Y_k, in terms of $F(x)$ and $f(x)$. This is done by evaluating the integral

$$g_k(y_k) = \int_a^{y_k} \cdots \int_a^{y_2} \int_{y_k}^{b} \cdots \int_{y_{n-1}}^{b} n!\, f(y_1) f(y_2) \cdots f(y_n)\, dy_n \cdots$$
$$dy_{k+1}\, dy_1 \cdots dy_{k-1}.$$

The result is

$$g_k(y_k) = \frac{n!}{(k-1)!\,(n-k)!} [F(y_k)]^{k-1} [1 - F(y_k)]^{n-k} f(y_k),$$
$$a < y_k < b,$$
$$= 0 \qquad \text{elsewhere.} \tag{2}$$

Example 2. Let $Y_1 < Y_2 < Y_3 < Y_4$ denote the order statistics of a random sample of size 4 from a distribution having p.d.f.

$$f(x) = 2x, \qquad 0 < x < 1,$$
$$= 0 \qquad \text{elsewhere.}$$

We shall express the p.d.f. of Y_3 in terms of $f(x)$ and $F(x)$ and then compute $\Pr(\frac{1}{2} < Y_3)$. Here $F(x) = x^2$, provided that $0 < x < 1$, so that

$$g_3(y_3) = \frac{4!}{2!\,1!}(y_3^2)^2(1 - y_3^2)(2y_3), \qquad 0 < y_3 < 1,$$

$$= 0 \qquad \text{elsewhere.}$$

Thus

$$\Pr(\tfrac{1}{2} < Y_3) = \int_{1/2}^{\infty} g_3(y_3)\,dy_3$$

$$= \int_{1/2}^{1} 24(y_3^5 - y_3^7)\,dy_3 = \tfrac{243}{256}.$$

Finally, the joint p.d.f. of any two order statistics, say $Y_i < Y_j$, is as easily expressed in terms of $F(x)$ and $f(x)$. We have

$$g_{ij}(y_i, y_j) = \int_a^{y_i} \cdots \int_a^{y_2} \int_{y_i}^{y_j} \cdots \int_{y_{j-2}}^{y_j} \int_{y_j}^{b} \cdots \int_{y_{n-1}}^{b} n!\, f(y_1) \cdots$$

$$f(y_n)\,dy_n \cdots dy_{j+1}\,dy_{j-1} \cdots dy_{i+1}\,dy_1 \cdots dy_{i-1}.$$

Since, for $\gamma > 0$,

$$\int_x^y [F(y) - F(w)]^{\gamma - 1} f(w)\,dw = -\frac{[F(y) - F(w)]^{\gamma}}{\gamma}\Bigg|_x^y$$

$$= \frac{[F(y) - F(x)]^{\gamma}}{\gamma},$$

it is found that

$$g_{ij}(y_i, y_j) = \frac{n!}{(i-1)!\,(j-i-1)!\,(n-j)!}$$

$$\times [F(y_i)]^{i-1}[F(y_j) - F(y_i)]^{j-i-1}[1 - F(y_j)]^{n-j} f(y_i) f(y_j) \qquad (3)$$

for $a < y_i < y_j < b$, and zero elsewhere.

Remark. There is an easy method of remembering a p.d.f. like that given in Formula (3). The probability $\Pr(y_i < Y_i < y_i + \Delta_i,\ y_j < Y_j < y_j + \Delta_j)$, where Δ_i and Δ_j are small, can be approximated by the following multinomial probability. In n independent trials, $i - 1$ outcomes must be less than y_i (an event that has probability $p_1 = F(y_i)$ on each trial); $j - i - 1$ outcomes must be between $y_i + \Delta_i$ and y_j [an event with approximate probability $p_2 = F(y_j) - F(y_i)$ on each trial]; $n - j$ outcomes must be greater than $y_j + \Delta_j$ (an event with approximate probability $p_3 = 1 - F(y_j)$ on each trial); one outcome must be between y_i and $y_i + \Delta_i$ (an event with approximate probability $p_4 = f(y_i)\,\Delta_i$ on each trial); and finally one outcome must be between y_j and $y_j + \Delta_j$ [an event with approximate probability $p_5 = f(y_j)\Delta_j$ on each trial]. This multinomial probability is

$$\frac{n!}{(i-1)!\,(j-i-1)!\,(n-j)!\,1!\,1!}\,p_1^{i-1}p_2^{j-i-1}p_3^{n-j}p_4p_5,$$

which is $g_{i,j}(y_i, y_j)\Delta_i\Delta_j$.

Certain functions of the order statistics $Y_1, Y_2, \ldots, Y_n$ are important statistics themselves. A few of these are: (a) $Y_n - Y_1$, which is called the range of the random sample; (b) $(Y_1 + Y_n)/2$, which is called the midrange of the random sample; and (c) if n odd, $Y_{(n+1)/2}$, which is called the median of the random sample.

Example 3. Let Y_1, Y_2, Y_3 be the order statistics of a random sample of size 3 from a distribution having p.d.f.

$$\begin{aligned} f(x) &= 1, \qquad 0 < x < 1, \\ &= 0 \qquad \text{elsewhere.} \end{aligned}$$

We seek the p.d.f. of the sample range $Z_1 = Y_3 - Y_1$. Since $F(x) = x$, $0 < x < 1$, the joint p.d.f. of Y_1 and Y_3 is

$$\begin{aligned} g_{13}(y_1, y_3) &= 6(y_3 - y_1), \qquad 0 < y_1 < y_3 < 1, \\ &= 0 \qquad \text{elsewhere.} \end{aligned}$$

In addition to $Z_1 = Y_3 - Y_1$, let $Z_2 = Y_3$. Consider the functions $z_1 = y_3 - y_1$, $z_2 = y_3$, and their inverses $y_1 = z_2 - z_1$, $y_3 = z_2$, so that the corresponding Jacobian of the one-to-one transformation is

$$J = \begin{vmatrix} \dfrac{\partial y_1}{\partial z_1} & \dfrac{\partial y_1}{\partial z_2} \\ \dfrac{\partial y_3}{\partial z_1} & \dfrac{\partial y_3}{\partial z_2} \end{vmatrix} = \begin{vmatrix} -1 & 1 \\ 0 & 1 \end{vmatrix} = -1.$$

Thus the joint p.d.f. of Z_1 and Z_2 is

$$\begin{aligned} h(z_1, z_2) &= |-1|6z_1 = 6z_1, \qquad 0 < z_1 < z_2 < 1. \\ &= 0 \qquad \text{elsewhere.} \end{aligned}$$

Accordingly, the p.d.f. of the range $Z_1 = Y_3 - Y_1$ of the random sample of size 3 is

$$\begin{aligned} h_1(z_1) &= \int_{z_1}^{1} 6z_1\,dz_2 = 6z_1(1 - z_1), \qquad 0 < z_1 < 1, \\ &= 0 \qquad \text{elsewhere.} \end{aligned}$$

EXERCISES

4.56. Let $Y_1 < Y_2 < Y_3 < Y_4$ be the order statistics of a random sample of size 4 from the distribution having p.d.f. $f(x) = e^{-x}$, $0 < x < \infty$, zero elsewhere. Find $\Pr(3 \le Y_4)$.

4.57. Let X_1, X_2, X_3 be a random sample from a distribution of the continuous type having p.d.f. $f(x) = 2x$, $0 < x < 1$, zero elsewhere.

(a) Compute the probability that the smallest of these X_i exceeds the median of the distribution.

(b) If $Y_1 < Y_2 < Y_3$ are the order statistics, find the correlation between Y_2 and Y_3.

4.58. Let $f(x) = \frac{1}{6}$, $x = 1, 2, 3, 4, 5, 6$, zero elsewhere, be the p.d.f. of a distribution of the discrete type. Show that the p.d.f. of the smallest observation of a random sample of size 5 from this distribution is

$$g_1(y_1) = \left(\frac{7 - y_1}{6}\right)^5 - \left(\frac{6 - y_1}{6}\right)^5, \qquad y_1 = 1, 2, \ldots, 6,$$

zero elsewhere. Note that in this exercise the random sample is from a distribution of the discrete type. All formulas in the text were derived under the assumption that the random sample is from a distribution of the continuous type and are not applicable. Why?

4.59. Let $Y_1 < Y_2 < Y_3 < Y_4 < Y_5$ denote the order statistics of a random sample of size 5 from a distribution having p.d.f. $f(x) = e^{-x}$, $0 < x < \infty$, zero elsewhere. Show that $Z_1 = Y_2$ and $Z_2 = Y_4 - Y_2$ are independent.

Hint: First find the joint p.d.f. of Y_2 and Y_4.

4.60. Let $Y_1 < Y_2 < \cdots < Y_n$ be the order statistics of a random sample of size n from a distribution with p.d.f. $f(x) = 1$, $0 < x < 1$, zero elsewhere. Show that the kth order statistic Y_k has a beta p.d.f. with parameters $\alpha = k$ and $\beta = n - k + 1$.

4.61. Let $Y_1 < Y_2 < \cdots < Y_n$ be the order statistics from a Weibull distribution, Exercise 3.44, Section 3.3. Find the distribution function and p.d.f. of Y_1.

4.62. Find the probability that the range of a random sample of size 4 from the uniform distribution having the p.d.f. $f(x) = 1$, $0 < x < 1$, zero elsewhere, is less than $\frac{1}{2}$.

4.63. Let $Y_1 < Y_2 < Y_3$ be the order statistics of a random sample of size 3 from a distribution having the p.d.f. $f(x) = 2x$, $0 < x < 1$, zero elsewhere. Show that $Z_1 = Y_1/Y_2$, $Z_2 = Y_2/Y_3$, and $Z_3 = Y_3$ are mutually independent.

4.64. If a random sample of size 2 is taken from a distribution having p.d.f. $f(x) = 2(1 - x)$, $0 < x < 1$, zero elsewhere, compute the probability that one sample observation is at least twice as large as the other.

4.65. Let $Y_1 < Y_2 < Y_3$ denote the order statistics of a random sample of size 3 from a distribution with p.d.f. $f(x) = 1$, $0 < x < 1$, zero elsewhere. Let $Z = (Y_1 + Y_3)/2$ be the midrange of the sample. Find the p.d.f. of Z.

4.66. Let $Y_1 < Y_2$ denote the order statistics of a random sample of size 2 from $N(0, \sigma^2)$.

(a) Show that $E(Y_1) = -\sigma/\sqrt{\pi}$.
Hint: Evaluate $E(Y_1)$ by using the joint p.d.f. of Y_1 and Y_2, and first integrating on y_1.

(b) Find the covariance of Y_1 and Y_2.

4.67. Let $Y_1 < Y_2$ be the order statistics of a random sample of size 2 from a distribution of the continuous type which has p.d.f. $f(x)$ such that $f(x) > 0$, provided that $x \geq 0$, and $f(x) = 0$ elsewhere. Show that the independence of $Z_1 = Y_1$ and $Z_2 = Y_2 - Y_1$ characterizes the gamma p.d.f. $f(x)$, which has parameters $\alpha = 1$ and $\beta > 0$.

Hint: Use the change-of-variable technique to find the joint p.d.f. of Z_1 and Z_2 from that of Y_1 and Y_2. Accept the fact that the functional equation $h(0)h(x + y) \equiv h(x)h(y)$ has the solution $h(x) = c_1 e^{c_2 x}$, where c_1 and c_2 are constants.

4.68. Let $Y_1 < Y_2 < Y_3 < Y_4$ be the order statistics of a random sample of size $n = 4$ from a distribution with p.d.f. $f(x) = 2x$, $0 < x < 1$.

(a) Find the joint p.d.f. of Y_3 and Y_4.

(b) Find the conditional p.d.f. of Y_3, given $Y_4 = y_4$.

(c) Evaluate $E(Y_3|y_4)$.

4.69. Two numbers are selected at random from the interval (0, 1). If these values are uniformly and independently distributed, compute the probability that the three resulting line segments, by cutting the interval at the numbers, can form a triangle.

4.70. Let X and Y denote independent random variables with respective probability density functions $f(x) = 2x$, $0 < x < 1$, zero elsewhere, and $g(y) = 3y^2$, $0 < y < 1$, zero elsewhere. Let $U = \min(X, Y)$ and $V = \max(X, Y)$. Find the joint p.d.f. of U and V.

Hint: Here the two inverse transformations are given by $x = u$, $y = v$ and $x = v$, $y = u$.

4.71. Let the joint p.d.f. of X and Y be $f(x, y) = \frac{12}{7}x(x + y)$, $0 < x < 1$, $0 < y < 1$, zero elsewhere. Let $U = \min(X, Y)$ and $V = \max(X, Y)$. Find the joint p.d.f. of U and V.

4.72. Let $X_1, X_2, \ldots, X_n$ be a random sample from a distribution of either type. A measure of spread is *Gini's mean difference*

$$G = \sum_{j=2}^{n} \sum_{i=1}^{j-1} |X_i - X_j| \bigg/ \binom{n}{2}.$$

(a) If $n = 10$, find $a_1, a_2, \ldots, a_{10}$ so that $G = \sum_{i=1}^{10} a_i Y_i$, where $Y_1, Y_2, \ldots, Y_{10}$ are the order statistics of the sample.

(b) Show that $E(G) = 2\sigma/\sqrt{\pi}$ if the sample arises from the normal distribution $N(\mu, \sigma^2)$.

4.73. Let $Y_1 < Y_2 < \cdots < Y_n$ be the order statistics of a random sample of size n from the exponential distribution with p.d.f. $f(x) = e^{-x}$, $0 < x < \infty$, zero elsewhere.

(a) Show that $Z_1 = nY_1$, $Z_2 = (n-1)(Y_2 - Y_1)$, $Z_3 = (n-2)(Y_3 - Y_2)$, $\ldots, Z_n = Y_n - Y_{n-1}$ are independent and that each Z_i has the exponential distribution.

(b) Demonstrate that all linear functions of $Y_1, Y_2, \ldots, Y_n$, such as $\sum_1^n a_i Y_i$, can be expressed as linear functions of independent random variables.

4.74. In the Program Evaluation and Review Technique (PERT), we are interested in the total time to complete a project that is comprised of a large number of subprojects. For illustration, let X_1, X_2, X_3 be three independent random times for three subprojects. If these subprojects are in series (the first one must be completed before the second starts, etc.), then we are interested in the sum $Y = X_1 + X_2 + X_3$. If these are in parallel (can be worked on simultaneously), then we are interested in $Z = \max(X_1, X_2, X_3)$. In the case each of these random variables has the uniform distribution with p.d.f. $f(x) = 1$, $0 < x < 1$, zero elsewhere, find (a) the p.d.f. of Y and (b) the p.d.f. of Z.

4.7 The Moment-Generating-Function Technique

The change-of-variable procedure has been seen, in certain cases, to be an effective method of finding the distribution of a function of several random variables. An alternative procedure, built around the concept of the m.g.f. of a distribution, will be presented in this section. This procedure is particularly effective in certain instances. We should recall that an m.g.f., when it exists, is unique and that it uniquely determines the distribution of probability.

Let $h(x_1, x_2, \ldots, x_n)$ denote the joint p.d.f. of the n random variables $X_1, X_2, \ldots, X_n$. These random variables may or may not be the observations of a random sample from some distribution that has a given p.d.f. $f(x)$. Let $Y_1 = u_1(X_1, X_2, \ldots, X_n)$. We seek $g(y_1)$, the p.d.f. of the random variable Y_1. Consider the m.g.f. of Y_1. If it exists, it is given by

$$M(t) = E(e^{tY_1}) = \int_{-\infty}^{\infty} e^{ty_1} g(y_1)\, dy_1$$

in the continuous case. It would seem that we need to know $g(y_1)$ before we can compute $M(t)$. That this is not the case is a fundamental fact. To see this consider

$$\int_{-\infty}^{\infty} \cdots \int_{-\infty}^{\infty} \exp\,[tu_1(x_1, \ldots, x_n)]h(x_1, \ldots, x_n)\, dx_1 \cdots dx_n, \tag{1}$$

which we assume to exist for $-h < t < h$. We shall introduce n new variables of integration. They are $y_1 = u_1(x_1, x_2, \ldots, x_n), \ldots, y_n = u_n(x_1, x_2, \ldots, x_n)$. Momentarily, we assume that these functions define a one-to-one transformation. Let $x_i = w_i(y_1, y_2, \ldots, y_n)$, $i = 1, 2, \ldots, n$, denote the inverse functions and let J denote the Jacobian. Under this transformation, display (1) becomes

$$\int_{-\infty}^{\infty} \cdots \int_{-\infty}^{\infty} e^{ty_1}|J|h(w_1, \ldots, w_n)\, dy_2 \cdots dy_n\, dy_1. \tag{2}$$

In accordance with Section 4.5,

$$|J|h[w_1(y_1, y_2, \ldots, y_n), \ldots, w_n(y_1, y_2, \ldots, y_n)]$$

is the joint p.d.f. of $Y_1, Y_2, \ldots, Y_n$. The marginal p.d.f. $g(y_1)$ of Y_1 is obtained by integrating this joint p.d.f. on $y_2, \ldots, y_n$. Since the factor e^{ty_1} does not involve the variables $y_2, \ldots, y_n$, display (2) may be written as

$$\int_{-\infty}^{\infty} e^{ty_1} g(y_1)\, dy_1. \tag{3}$$

But this is by definition the m.g.f. $M(t)$ of the distribution of Y_1. That is, we can compute $E\{\exp\,[tu_1(X_1, \ldots, X_n)]\}$ and have the value of $E(e^{tY_1})$, where $Y_1 = u_1(X_1, \ldots, X_n)$. This fact provides another technique to help us find the p.d.f. of a function of several random variables. For if the m.g.f. of Y_1 is seen to be that of a certain kind of distribution, the uniqueness property makes it certain that Y_1 has that kind of distribution. When the p.d.f. of Y_1 is obtained in this manner, we say that we use the *moment-generating-function technique*.

The reader will observe that we have assumed the transformation to be one-to-one. We did this for simplicity of presentation. If the transformation is not one-to-one, let

$$x_j = w_{ji}(y_1, \ldots, y_n), \qquad j = 1, 2, \ldots, n, \quad i = 1, 2, \ldots, k,$$

denote the k groups of n inverse functions each. Let J_i, $i = 1, 2, \ldots, k$, denote the k Jacobians. Then

$$\sum_{i=1}^{k} |J_i| h[w_{1i}(y_1, \ldots, y_n), \ldots, w_{ni}(y_1, \ldots, y_n)] \tag{4}$$

is the joint p.d.f. of $Y_1, \ldots, Y_n$. Then display (1) becomes display (2) with $|J|h(w_1, \ldots, w_n)$ replaced by display (4). Hence our result is valid if the transformation is not one-to-one. It seems evident that we can treat the discrete case in an analogous manner with the same result.

It should be noted that the expectation of Y_1 can be computed in like manner. That is,

$$\begin{aligned} E(Y_1) &= \int_{-\infty}^{\infty} y_1 g(y_1)\, dy_1 \\ &= \int_{-\infty}^{\infty} \cdots \int_{-\infty}^{\infty} u_1(x_1, \ldots, x_n) h(x_1, \ldots, x_n)\, dx_1 \cdots dx_n, \end{aligned}$$

and this fact has been mentioned earlier in the book. Moreover, this holds for the expectation of any function of Y_1, say $w(Y_1)$; that is,

$$\begin{aligned} E[w(Y_1)] &= \int_{-\infty}^{\infty} w(y_1) g(y_1)\, dy_1 \\ &= \int_{-\infty}^{\infty} \cdots \int_{-\infty}^{\infty} w[u_1(x_1, \ldots, x_n)] h(x_1, \ldots, x_n)\, dx_1 \cdots dx_n. \end{aligned}$$

We shall now give some examples and prove some theorems where we use the moment-generating-function technique. In the first example, to emphasize the nature of the problem, we find the distribution of a rather simple statistic both by a direct probabilistic argument and by the moment-generating-function technique.

Example 1. Let the independent random variables X_1 and X_2 have the same p.d.f.

$$\begin{aligned} f(x) &= \frac{x}{6}, \qquad x = 1, 2, 3, \\ &= 0 \qquad \text{elsewhere;} \end{aligned}$$

so the joint p.d.f. of X_1 and X_2 is

$$f(x_1)f(x_2) = \frac{x_1x_2}{36}, \qquad x_1 = 1, 2, 3, \quad x_2 = 1, 2, 3,$$
$$= 0 \qquad \text{elsewhere.}$$

A probability, such as $\Pr(X_1 = 2, X_2 = 3)$, can be seen immediately to be $(2)(3)/36 = \frac{1}{6}$. However, consider a probability such as $\Pr(X_1 + X_2 = 3)$. The computation can be made by first observing that the event $X_1 + X_2 = 3$ is the union, exclusive of the events with probability zero, of the two mutually exclusive events $(X_1 = 1, X_2 = 2)$ and $(X_1 = 2, X_2 = 1)$. Thus

$$\Pr(X_1 + X_2 = 3) = \Pr(X_1 = 1, X_2 = 2) + \Pr(X_1 = 2, X_2 = 1)$$
$$= \frac{(1)(2)}{36} + \frac{(2)(1)}{36} = \frac{4}{36}.$$

More generally, let y represent any of the numbers 2, 3, 4, 5, 6. The probability of each of the events $X_1 + X_2 = y$, $y = 2, 3, 4, 5, 6$, can be computed as in the case $y = 3$. Let $g(y) = \Pr(X_1 + X_2 = y)$. Then the table

y	2	3	4	5	6
$g(y)$	$\frac{1}{36}$	$\frac{4}{36}$	$\frac{10}{36}$	$\frac{12}{36}$	$\frac{9}{36}$

gives the values of $g(y)$ for $y = 2, 3, 4, 5, 6$. For all other values of y, $g(y) = 0$. What we have actually done is to define a new random variable Y by $Y = X_1 + X_2$, and we have found the p.d.f. $g(y)$ of this random variable Y. We shall now solve the same problem, and by the moment-generating-function technique.

Now the m.g.f. of Y is

$$M(t) = E(e^{t(X_1 + X_2)})$$
$$= E(e^{tX_1}e^{tX_2})$$
$$= E(e^{tX_1})E(e^{tX_2}),$$

since X_1 and X_2 are independent. In this example X_1 and X_2 have the same distribution, so they have the same m.g.f.; that is,

$$E(e^{tX_1}) = E(e^{tX_2}) = \tfrac{1}{6}e^t + \tfrac{2}{6}e^{2t} + \tfrac{3}{6}e^{3t}.$$

Thus

$$M(t) = (\tfrac{1}{6}e^t + \tfrac{2}{6}e^{2t} + \tfrac{3}{6}e^{3t})^2$$
$$= \tfrac{1}{36}e^{2t} + \tfrac{4}{36}e^{3t} + \tfrac{10}{36}e^{4t} + \tfrac{12}{36}e^{5t} + \tfrac{9}{36}e^{6t}.$$

This form of $M(t)$ tells us immediately that the p.d.f. $g(y)$ of Y is zero except at $y = 2, 3, 4, 5, 6$, and that $g(y)$ assumes the values $\frac{1}{36}$, $\frac{4}{36}$, $\frac{10}{36}$, $\frac{12}{36}$, $\frac{9}{36}$, respectively, at these points where $g(y) > 0$. This is, of course, the same result that was obtained in the first solution. There appears here to be little, if any, preference for one solution over the other. But in more complicated situations, and particularly with random variables of the continuous type, the moment-generating-function technique can prove very powerful.

Example 2. Let X_1 and X_2 be independent with normal distributions $N(\mu_1, \sigma_1^2)$ and $N(\mu_2, \sigma_2^2)$, respectively. Define the random variable Y by $Y = X_1 - X_2$. The problem is to find $g(y)$, the p.d.f. of Y. This will be done by first finding the m.g.f. of Y. It is

$$\begin{aligned} M(t) &= E(e^{t(X_1 - X_2)}) \\ &= E(e^{tX_1}e^{-tX_2}) \\ &= E(e^{tX_1})E(e^{-tX_2}), \end{aligned}$$

since X_1 and X_2 are independent. It is known that

$$E(e^{tX_1}) = \exp\left(\mu_1 t + \frac{\sigma_1^2 t^2}{2}\right)$$

and that

$$E(e^{tX_2}) = \exp\left(\mu_2 t + \frac{\sigma_2^2 t^2}{2}\right)$$

for all real t. Then $E(e^{-tX_2})$ can be obtained from $E(e^{tX_2})$ by replacing t by $-t$. That is,

$$E(e^{-tX_2}) = \exp\left(-\mu_2 t + \frac{\sigma_2^2 t^2}{2}\right).$$

Finally, then,

$$\begin{aligned} M(t) &= \exp\left(\mu_1 t + \frac{\sigma_1^2 t^2}{2}\right)\exp\left(-\mu_2 t + \frac{\sigma_2^2 t^2}{2}\right) \\ &= \exp\left((\mu_1 - \mu_2)t + \frac{(\sigma_1^2 + \sigma_2^2)t^2}{2}\right). \end{aligned}$$

The distribution of Y is completely determined by its m.g.f. $M(t)$, and it is seen that Y has the p.d.f. $g(y)$, which is $N(\mu_1 - \mu_2, \sigma_1^2 + \sigma_2^2)$. That is, the difference between two independent, normally distributed, random variables is itself a random variable which is normally distributed with mean equal to the difference of the means (in the order indicated) and the variance equal to the sum of the variances.

The following theorem, which is a generalization of Example 2, is very important in distribution theory.

Theorem 1. *Let $X_1, X_2, \ldots, X_n$ be independent random variables having, respectively, the normal distributions $N(\mu_1, \sigma_1^2)$, $N(\mu_2, \sigma_2^2), \ldots,$ and $N(\mu_n, \sigma_n^2)$. The random variable $Y = k_1X_1 + k_2X_2 + \cdots + k_nX_n$, where $k_1, k_2, \ldots, k_n$ are real constants, is normally distributed with mean $k_1\mu_1 + \cdots + k_n\mu_n$ and variance $k_1^2\sigma_1^2 + \cdots + k_n^2\sigma_n^2$. That is, Y is $N\left(\sum_1^n k_i\mu_i, \sum_1^n k_i^2\sigma_i^2\right)$.*

Proof. Because $X_1, X_2, \ldots, X_n$ are independent, the m.g.f. of Y is given by

$$\begin{aligned} M(t) &= E\{\exp[t(k_1X_1 + k_2X_2 + \cdots + k_nX_n)]\} \\ &= E(e^{tk_1X_1})E(e^{tk_2X_2}) \cdots E(e^{tk_nX_n}). \end{aligned}$$

Now

$$E(e^{tX_i}) = \exp\left(\mu_i t + \frac{\sigma_i^2 t^2}{2}\right),$$

for all real t, $i = 1, 2, \ldots, n$. Hence we have

$$E(e^{tk_iX_i}) = \exp\left[\mu_i(k_it) + \frac{\sigma_i^2(k_it)^2}{2}\right].$$

That is, the m.g.f. of Y is

$$\begin{aligned} M(t) &= \prod_{i=1}^n \exp\left[(k_i\mu_i)t + \frac{(k_i^2\sigma_i^2)t^2}{2}\right] \\ &= \exp\left[\left(\sum_1^n k_i\mu_i\right)t + \frac{\left(\sum_1^n k_i^2\sigma_i^2\right)t^2}{2}\right]. \end{aligned}$$

But this is the m.g.f. of a distribution that is $N\left(\sum_1^n k_i\mu_i, \sum_1^n k_i^2\sigma_i^2\right)$.

This is the desired result.

The next theorem is a generalization of Theorem 1.

Theorem 2. If *$X_1, X_2, \ldots, X_n$ are independent random variables with respective moment-generating functions $M_i(t)$, $i = 1, 2, 3, \ldots, n$,* *then the moment-generating function of*

$$Y = \sum_{i=1}^{n} a_i X_i,$$

where $a_1, a_2, \ldots, a_k$ are real constants, is

$$M_Y(t) = \prod_{i=1}^{n} M_i(a_i t).$$

Proof. The m.g.f. of Y is given by

$$\begin{aligned} M_Y(t) &= E[e^{tY}] = E[e^{t(a_1X_1 + a_2X_2 + \cdots + a_nX_n)}] \\ &= E[e^{a_1tX_1}e^{a_2tX_2} \cdots e^{a_ntX_n}] \\ &= E[e^{a_1tX_1}]E[e^{a_2tX_2}] \cdots E[e^{a_ntX_n}] \end{aligned}$$

because $X_1, X_2, \ldots, X_n$ are independent. However, since

$$E(^{tX_i}) = M_i(t),$$

then

$$E(e^{a_itX_i}) = M_i(a_it).$$

Thus we have that

$$\begin{aligned} M_Y(t) &= M_1(a_1t)M_2(a_2t) \cdots M_n(a_nt) \\ &= \prod_{i=1}^{n} M_i(a_it). \end{aligned}$$

A corollary follows immediately, and it will be used in some important examples.

Corollary. *If $X_1, X_2, \ldots, X_n$ are observations of a random sample from a distribution with moment-generating function $M(t)$, then*

(a) *The moment-generating function of $Y = \sum_{i=1}^{n} X_i$ is*

$$M_Y(t) = \prod_{i=1}^{n} M(t) = [M(t)]^n;$$

(b) *The moment-generating function of $\bar{X} = \sum_{i=1}^{n} (1/n)X_i$ is*

$$M_{\bar{X}}(t) = \prod_{i=1}^{n} M\left(\frac{t}{n}\right) = \left[M\left(\frac{t}{n}\right)\right]^n.$$

Proof. For (a), let $a_i = 1$, $i = 1, 2, \ldots, n$, in Theorem 2. For (b), take $a_i = 1/n$, $i = 1, 2, \ldots, n$.

The following examples and the exercises give some important applications of Theorem 2 and its corollary.

Example 3. Let $X_1, X_2, \ldots, X_n$ denote the outcomes on n Bernoulli trials. The m.g.f. of X_i, $i = 1, 2, \ldots, n$, is

$$M(t) = 1 - p + pe^t.$$

If $Y = \sum_{i=1}^{n} X_i$, then

$$M_Y(t) = \prod_{i=1}^{n} (1 - p + pe^t) = (1 - p + pe^t)^n.$$

Thus we again see that Y is $b(n, p)$.

Example 4. Let X_1, X_2, X_3 be the observations of a random sample of size $n = 3$ from the exponential distribution having mean β and, of course, m.g.f. $M(t) = 1/(1 - \beta t)$, $t < 1/\beta$. The m.g.f. of $Y = X_1 + X_2 + X_3$ is

$$M_Y(t) = [(1 - \beta t)^{-1}]^3 = (1 - \beta t)^{-3}, \qquad t < 1/\beta,$$

which is that of a gamma distribution with parameters $\alpha = 3$ and β. Thus Y has this distribution. On the other hand, the m.g.f. of $\bar{X}$ is

$$M_{\bar{X}}(t) = \left[\left(1 - \frac{\beta t}{3}\right)^{-1}\right]^3 = \left(1 - \frac{\beta t}{3}\right)^{-3}, \qquad t < 3/\beta;$$

and hence the distribution of $\bar{X}$ is gamma with parameters $\alpha = 3$ and $\beta/3$, respectively.

The next example is so important that we state it as a theorem.

Theorem 3. *Let $X_1, X_2, \ldots, X_n$ be independent variables that have, respectively, the chi-square distributions $\chi^2(r_1), \chi^2(r_2), \ldots,$ and $\chi^2(r_n)$. Then the random variable $Y = X_1 + X_2 + \cdots + X_n$ has a chi-square distribution with $r_1 + \cdots + r_n$ degrees of freedom; that is, Y is*

$$\chi^2(r_1 + \cdots + r_n).$$

Proof. Since

$$M_i(t) = E(e^{tX_i}) = (1 - 2t)^{-r_i/2}, \qquad t < \tfrac{1}{2}, \quad i = 1, 2, \ldots, n,$$

we have, using Theorem 2 with $a_1 = \cdots = a_n = 1$,

$$M(t) = (1 - 2t)^{-(r_1 + r_2 + \cdots + r_n)/2}, \qquad t < \tfrac{1}{2}.$$

But this is the m.g.f. of a distribution that is $\chi^2(r_1 + r_2 + \cdots + r_n)$. Accordingly, Y has this chi-square distribution.

Next, let $X_1, X_2, \ldots, X_n$ be a random sample of size n from a distribution that is $N(\mu, \sigma^2)$. In accordance with Theorem 2 of

Section 3.4, each of the random variables $(X_i - \mu)^2/\sigma^2$, $i = 1, 2, \ldots, n$, is $\chi^2(1)$. Moreover, these n random variables are independent. Accordingly, by Theorem 3, the random variable $Y = \sum_1^n [(X_i - \mu)/\sigma]^2$ is $\chi^2(n)$. This proves the following theorem.

Theorem 4. *Let $X_1, X_2, \ldots, X_n$ denote a random sample of size n from a distribution that is $N(\mu, \sigma^2)$. The random variable*

$$Y = \sum_1^n \left(\frac{X_i - \mu}{\sigma}\right)^2$$

has a chi-square distribution with n degrees of freedom.

Not always do we sample from a distribution of one random variable. Let the random variables X and Y have the joint p.d.f. $f(x, y)$ and let the $2n$ random variables $(X_1, Y_1), (X_2, Y_2), \ldots, (X_n, Y_n)$ have the joint p.d.f.

$$f(x_1, y_1)f(x_2, y_2)\cdots f(x_n, y_n).$$

The n random pairs $(X_1, Y_1), (X_2, Y_2), \ldots, (X_n, Y_n)$ are then independent and are said to constitute a *random sample* of size n from the distribution of X and Y. In the next paragraph we shall take $f(x, y)$ to be the normal bivariate p.d.f., and we shall solve a problem in sampling theory when we are sampling from this two-variable distribution.

Let $(X_1, Y_1), (X_2, Y_2), \ldots, (X_n, Y_n)$ denote a random sample of size n from a bivariate normal distribution with p.d.f. $f(x, y)$ and parameters μ_1, μ_2, σ_1^2, σ_2^2, and ρ. We wish to find the joint p.d.f. of the two statistics $\bar{X} = \sum_1^n X_i/n$ and $\bar{Y} = \sum_1^n Y_i/n$. We call $\bar{X}$ the mean of $X_1, \ldots, X_n$ and $\bar{Y}$ the mean of $Y_1, \ldots, Y_n$. Since the joint p.d.f. of the $2n$ random variables (X_i, Y_i), $i = 1, 2, \ldots, n$, is given by

$$h = f(x_1, y_1)f(x_2, y_2)\cdots f(x_n, y_n),$$

the m.g.f. of the two means $\bar{X}$ and $\bar{Y}$ is given by

$$M(t_1, t_2) = \int_{-\infty}^{\infty}\cdots\int_{-\infty}^{\infty} \exp\left(\frac{t_1 \sum_1^n x_i}{n} + \frac{t_2 \sum_1^n y_i}{n}\right) h \, dx_1 \cdots dy_n$$

$$= \prod_{i=1}^n \left[\int_{-\infty}^{\infty}\int_{-\infty}^{\infty} \exp\left(\frac{t_1 x_i}{n} + \frac{t_2 y_i}{n}\right) f(x_i, y_i)\, dx_i\, dy_i\right].$$

The justification of the form of the right-hand member of the second equality is that each pair (X_i, Y_i) has the same p.d.f. and that these n pairs are independent. The twofold integral in the brackets in the last equality is the joint m.g.f. of X_i and Y_i (see Section 3.5) with t_1 replaced by t_1/n and t_2 replaced by t_2/n. Accordingly,

$$M(t_1, t_2) = \prod_{i=1}^{n} \exp\left[\frac{t_1\mu_1}{n} + \frac{t_2\mu_2}{n} + \frac{\sigma_1^2(t_1/n)^2 + 2\rho\sigma_1\sigma_2(t_1/n)(t_2/n) + \sigma_2^2(t_2/n)^2}{2}\right.$$

$$= \exp\left[t_1\mu_1 + t_2\mu_2 + \frac{(\sigma_1^2/n)t_1^2 + 2\rho(\sigma_1\sigma_2/n)t_1t_2 + (\sigma_2^2/n)t_2^2}{2}\right].$$

But this is the m.g.f. of a bivariate normal distribution with means μ_1 and μ_2, variances σ_1^2/n and σ_2^2/n, and correlation coefficient ρ; therefore, $\bar{X}$ and $\bar{Y}$ have this joint distribution.

EXERCISES

4.75. Let the i.i.d. random variables X_1 and X_2 have the same p.d.f. $f(x) = \frac{1}{6}$, $x = 1, 2, 3, 4, 5, 6$, zero elsewhere. Find the p.d.f. of $Y = X_1 + X_2$. Note, under appropriate assumptions, that Y may be interpreted as the sum of the spots that appear when two dice are cast.

4.76. Let X_1 and X_2 be independent with normal distributions $N(6, 1)$ and $N(7, 1)$, respectively. Find $\Pr(X_1 > X_2)$.

Hint: Write $\Pr(X_1 > X_2) = \Pr(X_1 - X_2 > 0)$ and determine the distribution of $X_1 - X_2$.

4.77. Let X_1 and X_2 be independent random variables. Let X_1 and $Y = X_1 + X_2$ have chi-square distributions with r_1 and r degrees of freedom, respectively. Here $r_1 < r$. Show that X_2 has a chi-square distribution with $r - r_1$ degrees of freedom.

Hint: Write $M(t) = E(e^{t(X_1 + X_2)})$ and make use of the independence of X_1 and X_2.

4.78. Let the independent random variables X_1 and X_2 have binomial distributions with parameters n_1, $p_1 = \frac{1}{2}$ and n_2, $p_2 = \frac{1}{2}$, respectively. Show that $Y = X_1 - X_2 + n_2$ has a binomial distribution with parameters $n = n_1 + n_2$, $p = \frac{1}{2}$.

4.79. Let X_1, X_2, X_3 be a random sample of size $n = 3$ from $N(1, 4)$. Compute $P(X_1 + 2X_2 - 2X_3 > 7)$.

4.80. Let X_1 and X_2 be two independent random variables. Let X_1 and $Y = X_1 + X_2$ have Poisson distributions with means μ_1 and $\mu > \mu_1$, respectively. Find the distribution of X_2.

4.81. Let X_1, X_2 be two independent gamma random variables with parameters $\alpha_1 = 3$, $\beta_1 = 3$ and $\alpha_2 = 5$, $\beta_2 = 1$, respectively.
(a) Find the m.g.f. of $Y = 2X_1 + 6X_2$.
(b) What is the distribution of Y?

4.82. A certain job is completed in three steps in series. The means and standard deviations for the steps are (in minutes):

Step	Mean	Standard Deviation
1	17	2
2	13	1
3	13	2

Assuming independent steps and normal distributions, compute the probability that the job will take less than 40 minutes to complete.

4.83. Let X be $N(0, 1)$. Use the moment-generating-function technique to show that $Y = X^2$ is $\chi^2(1)$.

Hint: Evaluate the integral that represents $E(e^{tX^2})$ by writing $w = x\sqrt{1 - 2t}$, $t < \frac{1}{2}$.

4.84. Let $X_1, X_2, \ldots, X_n$ denote n mutually independent random variables with the moment-generating functions $M_1(t), M_2(t), \ldots, M_n(t)$, respectively.
(a) Show that $Y = k_1X_1 + k_2X_2 + \cdots + k_nX_n$, where $k_1, k_2, \ldots, k_n$ are real constants, has the m.g.f. $M(t) = \prod_1^n M_i(k_it)$.
(b) If each $k_i = 1$ and if X_i is Poisson with mean μ_i, $i = 1, 2, \ldots, n$, prove that Y is Poisson with mean $\mu_1 + \cdots + \mu_n$.

4.85. If $X_1, X_2, \ldots, X_n$ is a random sample from a distribution with m.g.f. $M(t)$, show that the moment-generating functions of $\sum_1^n X_i$ and $\sum_1^n X_i/n$ are, respectively, $[M(t)]^n$ and $[M(t/n)]^n$.

4.86. In Exercise 4.74 concerning PERT, assume that each of the three independent variables has the p.d.f. $f(x) = e^{-x}$, $0 < x < \infty$, zero elsewhere. Find:
(a) The p.d.f. of Y.
(b) The p.d.f. of Z.

4.87. If X and Y have a bivariate normal distribution with parameters $\mu_1, \mu_2, \sigma_1^2, \sigma_2^2$, and ρ, show that $Z = aX + bY + c$ is

$$N(a\mu_1 + b\mu_2 + c,\ a^2\sigma_1^2 + 2ab\rho\sigma_1\sigma_2 + b^2\sigma_2^2),$$

where a, b, and c are constants.

Hint: Use the m.g.f. $M(t_1, t_2)$ of X and Y to find the m.g.f. of Z.

4.88. Let X and Y have a bivariate normal distribution with parameters $\mu_1 = 25$, $\mu_2 = 35$, $\sigma_1^2 = 4$, $\sigma_2^2 = 16$, and $\rho = \frac{17}{32}$. If $Z = 3X - 2Y$, find $\Pr(-2 < Z < 19)$.

4.89. Let U and V be independent random variables, each having a standard normal distribution. Show that the m.g.f. $E(e^{t(UV)})$ of the product UV is $(1 - t^2)^{-1/2}$, $-1 < t < 1$.

Hint: Compare $E(e^{tUV})$ with the integral of a bivariate normal p.d.f. that has means equal to zero.

4.90. Let X and Y have a bivariate normal distribution with the parameters $\mu_1, \mu_2, \sigma_1^2, \sigma_2^2$, and ρ. Show that

$$W = X - \mu_1 \quad \text{and} \quad Z = (Y - \mu_2) - \rho(\sigma_2/\sigma_1)(X - \mu_1)$$

are independent normal variables.

4.91. Let X_1, X_2, X_3 be a random sample of size $n = 3$ from the standard normal distribution.

(a) Show that $Y_1 = X_1 + \delta X_3$, $Y_2 = X_2 + \delta X_3$ has a bivariate normal distribution.

(b) Find the value of δ so that the correlation coefficient $\rho = \frac{1}{2}$.

(c) What additional transformation involving Y_1 and Y_2 would produce a bivariate normal distribution with means μ_1 and μ_2, variances σ_1^2 and σ_2^2, and the same correlation coefficient ρ?

4.92. Let $X_1, X_2, \ldots, X_n$ be a random sample of size n from the normal distribution $N(\mu, \sigma^2)$. Find the joint distribution of $Y = \sum_1^n a_i X_i$ and $Z = \sum_1^n b_i X_i$, where the a_i and b_i are real constants. When, and only when, are Y and Z independent?

Hint: Note that the joint m.g.f. $E\left[\exp\left(t_1 \sum_1^n a_i X_i + t_2 \sum_1^n b_i X_i\right)\right]$ is that of a bivariate normal distribution.

4.93. Let X_1, X_2 be a random sample of size 2 from a distribution with positive variance and m.g.f. $M(t)$. If $Y = X_1 + X_2$ and $Z = X_1 - X_2$ are independent, prove that the distribution from which the sample is taken is a normal distribution.

Hint: Show that

$$m(t_1, t_2) = E\{\exp[t_1(X_1 + X_2) + t_2(X_1 - X_2)]\} = M(t_1 + t_2)M(t_1 - t_2).$$

Express each member of $m(t_1, t_2) = m(t_1, 0)m(0, t_2)$ in terms of M; differentiate twice with respect to t_2; set $t_2 = 0$; and solve the resulting differential equation in M.

4.8 The Distributions of $\bar{X}$ and nS^2/σ^2

Let $X_1, X_2, \ldots, X_n$ denote a random sample of size $n \geq 2$ from a distribution that is $N(\mu, \sigma^2)$. In this section we shall investigate the distributions of the mean and the variance of this random sample, that is, the distributions of the two statistics $\bar{X} = \sum_1^n X_i/n$ and $S^2 = \sum_1^n (X_i - \bar{X})^2/n$.

The problem of the distribution of $\bar{X}$, the mean of the sample, is solved by the use of Theorem 1 of Section 4.7. We have here, in the notation of the statement of that theorem, $\mu_1 = \mu_2 = \cdots = \mu_n = \mu$, $\sigma_1^2 = \sigma_2^2 = \cdots = \sigma_n^2 = \sigma^2$, and $k_1 = k_2 = \cdots = k_n = 1/n$. Accordingly, $Y = \bar{X}$ has a normal distribution with mean and variance given by

$$\sum_1^n \left(\frac{1}{n}\mu\right) = \mu, \qquad \sum_1^n \left[\left(\frac{1}{n}\right)^2 \sigma^2\right] = \frac{\sigma^2}{n},$$

respectively. That is, $\bar{X}$ is $N(\mu, \sigma^2/n)$.

Example 1. Let $\bar{X}$ be the mean of a random sample of size 25 from a distribution that is $N(75, 100)$. Thus $\bar{X}$ is $N(75, 4)$. Then, for instance,

$$\begin{aligned} \Pr(71 < \bar{X} < 79) &= \Phi\left(\frac{79 - 75}{2}\right) - \Phi\left(\frac{71 - 75}{2}\right) \\ &= \Phi(2) - \Phi(-2) = 0.954. \end{aligned}$$

We now take up the problem of the distribution of S^2, the variance of a random sample $X_1, \ldots, X_n$ from a distribution that is $N(\mu, \sigma^2)$. To do this, let us first consider the joint distribution of $Y_1 = \bar{X}$, $Y_2 = X_2 - \bar{X}$, $Y_3 = X_3 - \bar{X}, \ldots, Y_n = X_n - \bar{X}$. The corresponding inverse transformation

$$\begin{aligned} x_1 &= y_1 - y_2 - y_3 - \cdots - y_n \\ x_2 &= y_1 + y_2 \\ x_3 &= y_1 + y_3 \\ &\vdots \qquad \vdots \\ x_n &= y_1 + y_n \end{aligned}$$

has Jacobian n. Since

$$\begin{aligned} \sum_1^n (x_i - \mu)^2 &= \sum_1^n (x_i - \bar{x} + \bar{x} - \mu)^2 \\ &= \sum_1^n (x_i - \bar{x})^2 + n(\bar{x} - \mu)^2 \end{aligned}$$

because $2(\bar{x} - \mu) \sum_1^n (x_i - \bar{x}) = 0$, the joint p.d.f. of $X_1, X_2, \ldots, X_n$ can be written

$$\left(\frac{1}{\sqrt{2\pi}\,\sigma}\right)^n \exp\left[-\frac{\sum (x_i - \bar{x})^2}{2\sigma^2} - \frac{n(\bar{x} - \mu)^2}{2\sigma^2}\right],$$

where $\bar{x}$ represents $(x_1 + x_2 + \cdots + x_n)/n$ and $-\infty < x_i < \infty$, $i = 1, 2, \ldots, n$. Accordingly, with $y_1 = \bar{x}$ and $x_1 - \bar{x} = -y_2 - y_3 - \cdots - y_n$, we find that the joint p.d.f. of $Y_1, Y_2, \ldots, Y_n$ is

$$(n)\left(\frac{1}{\sqrt{2\pi}\,\sigma}\right)^n \exp\left[-\frac{(-y_2 - \cdots - y_n)^2}{2\sigma^2} - \frac{\sum_2^n y_i^2}{2\sigma^2} - \frac{n(y_1 - \mu)^2}{2\sigma^2}\right],$$

$-\infty < y_i < \infty$, $i = 1, 2, \ldots, n$. Note that this is the product of the p.d.f. of Y_1, namely,

$$\frac{1}{\sqrt{2\pi\sigma^2/n}} \exp\left[-\frac{(y_1 - \mu)^2}{2\sigma^2/n}\right], \qquad -\infty < y_1 < \infty,$$

and a function of $y_2, \ldots, y_n$. Thus Y_1 must be independent of the $n - 1$ random variables $Y_2, Y_3, \ldots, Y_n$ and that function of $y_2, \ldots, y_n$ is the joint p.d.f. of $Y_2, Y_3, \ldots, Y_n$. Moreover, this means that $Y_1 = \bar{X}$ and thus

$$\frac{n(Y_1 - \mu)^2}{\sigma^2} = \frac{n(\bar{X} - \mu)^2}{\sigma^2} = W_1$$

are independent of

$$\frac{(-Y_2 - \cdots - Y_n)^2 + \sum_2^n Y_i^2}{\sigma^2} = \frac{\sum_1^n (X_i - \bar{X})^2}{\sigma^2} = W_2.$$

Since W_1 is the square of a standard normal variable, it is distributed as $\chi^2(1)$. Also, we know that

$$W = \sum_1^n \left(\frac{X_i - \mu}{\sigma}\right)^2 = W_1 + W_2$$

is $\chi^2(n)$. From the independence of W_1 and W_2, we have

$$E(e^{tW}) = E(e^{tW_1})E(e^{tW_2})$$

or, equivalently,

$$(1 - 2t)^{-n/2} = (1 - 2t)^{-1/2}E(e^{tW_2}), \qquad t < \tfrac{1}{2}.$$

Thus

$$E(e^{tW_2}) = (1 - 2t)^{-(n-1)/2}, \qquad t < \tfrac{1}{2},$$

and hence $W_2 = nS^2/\sigma^2$ is $\chi^2(n-1)$. The determination of the p.d.f. of S^2 is an easy exercise from this result (see Exercise 4.99).

To summarize, we have established, in this section, three important properties of $\bar{X}$ and S^2 when the sample arises from a distribution which is $N(\mu, \sigma^2)$:

1. $\bar{X}$ is $N(\mu, \sigma^2/n)$.
2. nS^2/σ^2 is $\chi^2(n-1)$.
3. $\bar{X}$ and S^2 are independent.

For illustration, as the result of properties (1), (2), and (3), we have that $\sqrt{n}(\bar{X} - \mu)/\sigma$ is $N(0, 1)$. Thus, from the definition of Student's t,

$$T = \frac{(\bar{X} - \mu)/(\sigma/\sqrt{n})}{\sqrt{nS^2/\sigma^2(n-1)}} = \frac{\bar{X} - \mu}{S/\sqrt{n-1}}$$

has a t-distribution with $n - 1$ degrees of freedom. It was a random variable like this one that motivated Gosset's search for the distribution of T. This t-statistic will play an important role in statistical applications.

EXERCISES

4.94. Let $\bar{X}$ be the mean of a random sample of size 5 from a normal distribution with $\mu = 0$ and $\sigma^2 = 125$. Determine c so that $\Pr(\bar{X} < c) = 0.90$.

4.95. If $\bar{X}$ is the mean of a random sample of size n from a normal distribution with mean μ and variance 100, find n so that $\Pr(\mu - 5 < \bar{X} < \mu + 5) = 0.954$.

4.96. Let $X_1, X_2, \ldots, X_{25}$ and $Y_1, Y_2, \ldots, Y_{25}$ be two independent random samples from two normal distributions $N(0, 16)$ and $N(1, 9)$, respectively. Let $\bar{X}$ and $\bar{Y}$ denote the corresponding sample means. Compute $\Pr(\bar{X} > \bar{Y})$.

4.97. Find the mean and variance of $S^2 = \sum_1^n (X_i - \bar{X})^2/n$, where $X_1, X_2, \ldots, X_n$ is a random sample from $N(\mu, \sigma^2)$.

Hint: Find the mean and variance of nS^2/σ^2.

4.98. Let S^2 be the variance of a random sample of size 6 from the normal distribution $N(\mu, 12)$. Find $\Pr(2.30 < S^2 < 22.2)$.

4.99. Find the p.d.f. of the sample variance $V = S^2$, provided that the distribution from which the sample arises is $N(\mu, \sigma^2)$.

4.100. Let $\bar{X}$ and $\bar{Y}$ be the respective means of two independent random samples, each of size 4, from the two respective normal distributions $N(10, 9)$ and $N(3, 4)$. Compute $\Pr(\bar{X} > 2\bar{Y})$.

4.101. Let $X_1, X_2, \ldots, X_5$ be a random sample of size $n = 5$ from $N(0, \sigma^2)$. (a) Find the constant c so that $c(X_1 - X_2)/\sqrt{X_3^2 + X_4^2 + X_5^2}$ has a t-distribution. (b) How many degrees of freedom are associated with this T?

4.102. If a random sample of size 2 is taken from a normal distribution with mean 7 and variance 8, find the probability that the absolute value of the difference of these two observations exceeds 2.

4.103. Let $\bar{X}$ and S^2 be the mean and the variance of a random sample of size 25 from a distribution that is $N(3, 100)$. Then evaluate $\Pr(0 < \bar{X} < 6, 55.2 < S^2 < 145.6)$.

4.9 Expectations of Functions of Random Variables

Let $X_1, X_2, \ldots, X_n$ denote random variables that have the joint p.d.f. $f(x_1, x_2, \ldots, x_n)$. Let the random variable Y be defined by $Y = u(X_1, X_2, \ldots, X_n)$. In Section 4.7, we found that we could compute expectations of functions of Y without first finding the p.d.f. of Y. Indeed, this fact was the basis of the moment-generating-function procedure for finding the p.d.f. of Y. We can take advantage of this fact in a number of other instances. Some illustrative examples will be given.

Example 1. Say that W is $N(0, 1)$, that V is $\chi^2(r)$ with $r \geq 2$, and that W and V are independent. The mean of the random variable $T = W\sqrt{r/V}$ exists and is zero because the graph of the p.d.f. of T (see Section 4.4) is symmetric about the vertical axis through $t = 0$. The variance of T, when it exists, could be computed by integrating the product of t^2 and the p.d.f. of T. But it seems much simpler to compute

$$\sigma_T^2 = E(T^2) = E\left(W^2 \frac{r}{V}\right) = E(W^2)E\left(\frac{r}{V}\right).$$

Now W^2 is $\chi^2(1)$, so $E(W^2) = 1$. Furthermore,

$$E\left(\frac{r}{V}\right) = \int_0^\infty \frac{r}{v} \frac{1}{2^{r/2}\Gamma(r/2)} v^{r/2-1} e^{-v/2}\, dv$$

exists if $r > 2$ and is given by

$$\frac{r\Gamma[(r-2)/2]}{2\Gamma(r/2)} = \frac{r\Gamma[(r-2)/2]}{2[(r-2)/2]\Gamma[(r-2)/2]} = \frac{r}{r-2}.$$

Thus $\sigma_T^2 = r/(r-2)$, $r > 2$.

Example 2. Let X_i denote a random variable with mean μ_i and variance σ_i^2, $i = 1, 2, \ldots, n$. Let $X_1, X_2, \ldots, X_n$ be independent and let $k_1, k_2, \ldots, k_n$ denote real constants. We shall compute the mean and variance of a linear function $Y = k_1X_1 + k_2X_2 + \cdots + k_nX_n$. Because E is a linear operator, the mean of Y is given by

$$\begin{aligned}\mu_Y &= E(k_1X_1 + k_2X_2 + \cdots + k_nX_n)\\ &= k_1E(X_1) + k_2E(X_2) + \cdots + k_nE(X_n)\\ &= k_1\mu_1 + k_2\mu_2 + \cdots + k_n\mu_n = \sum_1^n k_i\mu_i.\end{aligned}$$

The variance of Y is given by

$$\begin{aligned}\sigma_Y^2 &= E\{[(k_1X_1 + \cdots + k_nX_n) - (k_1\mu_1 + \cdots + k_n\mu_n)]^2\}\\ &= E\{[k_1(X_1 - \mu_1) + \cdots + k_n(X_n - \mu_n)]^2\}\\ &= E\left\{\sum_{i=1}^n k_i^2(X_i - \mu_i)^2 + 2\sum_{i<j}\sum k_ik_j(X_i - \mu_i)(X_j - \mu_j)\right\}\\ &= \sum_{i=1}^n k_i^2 E[(X_i - \mu_i)^2] + 2\sum_{i<j}\sum k_ik_jE[(X_i - \mu_i)(X_j - \mu_j)].\end{aligned}$$

Consider $E[(X_i - \mu_i)(X_j - \mu_j)]$, $i < j$. Because X_i and X_j are independent, we have

$$E[(X_i - \mu_i)(X_j - \mu_j)] = E(X_i - \mu_i)E(X_j - \mu_j) = 0.$$

Finally, then,

$$\sigma_Y^2 = \sum_{i=1}^n k_i^2 E[(X_i - \mu_i)^2] = \sum_{i=1}^n k_i^2\sigma_i^2.$$

We can obtain a more general result if, in Example 2, we remove the hypothesis of independence of $X_1, X_2, \ldots, X_n$. We shall do this and we shall let ρ_{ij} denote the correlation coefficient of X_i and X_j. Thus for easy reference to Example 2, we write

$$E[(X_i - \mu_i)(X_j - \mu_j)] = \rho_{ij}\sigma_i\sigma_j, \qquad i < j.$$

If we refer to Example 2, we see that again $\mu_Y = \sum_1^n k_i\mu_i$. But now

$$\sigma_Y^2 = \sum_1^n k_i^2\sigma_i^2 + 2\sum_{i<j}\sum k_ik_j\rho_{ij}\sigma_i\sigma_j.$$

Thus we have the following theorem.

Theorem 5. *Let $X_1, \ldots, X_n$ denote random variables that have means $\mu_1, \ldots, \mu_n$ and variances $\sigma_1^2, \ldots, \sigma_n^2$. Let ρ_{ij}, $i \neq j$, denote the correlation coefficient of X_i and X_j and let $k_1, \ldots, k_n$ denote real constants. The mean and the variance of the linear function*

$$Y = \sum_1^n k_iX_i$$

are, respectively,

$$\mu_Y = \sum_1^n k_i\mu_i$$

and

$$\sigma_Y^2 = \sum_1^n k_i^2\sigma_i^2 + 2\sum_{i<j}\sum k_ik_j\rho_{ij}\sigma_i\sigma_j.$$

The following corollary of this theorem is quite useful.

Corollary. *Let $X_1, \ldots, X_n$ denote the observations of a random sample of size n from a distribution that has mean μ and variance σ^2. The mean and the variance of $Y = \sum_1^n k_iX_i$ are, respectively, $\mu_Y = \left(\sum_1^n k_i\right)\mu$ and $\sigma_Y^2 = \left(\sum_1^n k_i^2\right)\sigma^2$.*

Example 3. Let $\bar{X} = \sum_1^n X_i/n$ denote the mean of a random sample of size n from a distribution that has mean μ and variance σ^2. In accordance with the corollary, we have $\mu_{\bar{X}} = \mu\sum_1^n (1/n) = \mu$ and $\sigma_{\bar{X}}^2 = \sigma^2\sum_1^n (1/n)^2 = \sigma^2/n$. We have seen, in Section 4.8, that if our sample is from a distribution that is $N(\mu, \sigma^2)$, then $\bar{X}$ is $N(\mu, \sigma^2/n)$. It is interesting that $\mu_{\bar{X}} = \mu$ and $\sigma_{\bar{X}}^2 = \sigma^2/n$ whether the sample is or is not from a normal distribution.

EXERCISES

4.104. Let X_1, X_2, X_3, X_4 be four i.i.d. random variables having the same p.d.f. $f(x) = 2x$, $0 < x < 1$, zero elsewhere. Find the mean and variance of the sum Y of these four random variables.

4.105. Let X_1 and X_2 be two independent random variables so that the variances of X_1 and X_2 are $\sigma_1^2 = k$ and $\sigma_2^2 = 2$, respectively. Given that the variance of $Y = 3X_2 - X_1$ is 25, find k.

4.106. If the independent variables X_1 and X_2 have means μ_1, μ_2 and variances σ_1^2, σ_2^2, respectively, show that the mean and variance of the product $Y = X_1X_2$ are $\mu_1\mu_2$ and $\sigma_1^2\sigma_2^2 + \mu_1^2\sigma_2^2 + \mu_2^2\sigma_1^2$, respectively.

4.107. Find the mean and variance of the sum Y of the observations of a random sample of size 5 from the distribution having p.d.f. $f(x) = 6x(1-x)$, $0 < x < 1$, zero elsewhere.

4.108. Determine the mean and variance of the mean $\bar{X}$ of a random sample of size 9 from a distribution having p.d.f. $f(x) = 4x^3$, $0 < x < 1$, zero elsewhere.

4.109. Let X and Y be random variables with $\mu_1 = 1$, $\mu_2 = 4$, $\sigma_1^2 = 4$, $\sigma_2^2 = 6$, $\rho = \frac{1}{2}$. Find the mean and variance of $Z = 3X - 2Y$.

4.110. Let X and Y be independent random variables with means μ_1, μ_2 and variances σ_1^2, σ_2^2. Determine the correlation coefficient of X and $Z = X - Y$ in terms of $\mu_1, \mu_2, \sigma_1^2, \sigma_2^2$.

4.111. Let μ and σ^2 denote the mean and variance of the random variable X. Let $Y = c + bX$, where b and c are real constants. Show that the mean and the variance of Y are, respectively, $c + b\mu$ and $b^2\sigma^2$.

4.112. Find the mean and the variance of $Y = X_1 - 2X_2 + 3X_3$, where X_1, X_2, X_3 are observations of a random sample from a chi-square distribution with 6 degrees of freedom.

4.113. Let X and Y be random variables such that var $(X) = 4$, var $(Y) = 2$, and var $(X + 2Y) = 15$. Determine the correlation coefficient of X and Y.

4.114. Let X and Y be random variables with means μ_1, μ_2; variances σ_1^2, σ_2^2; and correlation coefficient ρ. Show that the correlation coefficient of $W = aX + b$, $a > 0$, and $Z = cY + d$, $c > 0$, is ρ.

4.115. A person rolls a die, tosses a coin, and draws a card from an ordinary deck. He receives \$3 for each point up on the die, \$10 for a head, \$0 for a tail, and \$1 for each spot on the card (jack = 11, queen = 12, king = 13). If we assume that the three random variables involved are independent and uniformly distributed, compute the mean and variance of the amount to be received.

4.116. Let U and V be two independent chi-square variables with r_1 and r_2 degrees of freedom, respectively. Find the mean and variance of $F = (r_2U)/(r_1V)$. What restriction is needed on the parameters r_1 and r_2 in order to ensure the existence of both the mean and the variance of F?

4.117. Let $X_1, X_2, \ldots, X_n$ be a random sample of size n from a distribution with mean μ and variance σ^2. Show that $E(S^2) = (n-1)\sigma^2/n$, where S^2 is the variance of the random sample.

Hint: Write $S^2 = (1/n)\sum_1^n (X_i - \mu)^2 - (\bar{X} - \mu)^2$.

4.118. Let X_1 and X_2 be independent random variables with nonzero variances. Find the correlation coefficient of $Y = X_1X_2$ and X_1 in terms of the means and variances of X_1 and X_2.

4.119. Let X_1 and X_2 have a joint distribution with parameters $\mu_1, \mu_2, \sigma_1^2, \sigma_2^2$, and ρ. Find the correlation coefficient of the linear functions $Y = a_1X_1 + a_2X_2$ and $Z = b_1X_1 + b_2X_2$ in terms of the real constants a_1, a_2, b_1, b_2, and the parameters of the distribution.

4.120. Let $X_1, X_2, \ldots, X_n$ be a random sample of size n from a distribution which has mean μ and variance σ^2. Use Chebyshev's inequality to show, for every $\epsilon > 0$, that $\lim_{n\to\infty} \Pr(|\bar{X} - \mu| < \epsilon) = 1$; this is another form of the law of large numbers.

4.121. Let X_1, X_2, and X_3 be random variables with equal variances but with correlation coefficients $\rho_{12} = 0.3$, $\rho_{13} = 0.5$, and $\rho_{23} = 0.2$. Find the correlation coefficient of the linear functions $Y = X_1 + X_2$ and $Z = X_2 + X_3$.

4.122. Find the variance of the sum of 10 random variables if each has variance 5 and if each pair has correlation coefficient 0.5.

4.123. Let X and Y have the parameters $\mu_1, \mu_2, \sigma_1^2, \sigma_2^2$, and ρ. Show that the correlation coefficient of X and $[Y - \rho(\sigma_2/\sigma_1)X]$ is zero.

4.124. Let X_1 and X_2 have a bivariate normal distribution with parameters μ_1, $\mu_2, \sigma_1^2, \sigma_2^2$, and ρ. Compute the means, the variances, and the correlation coefficient of $Y_1 = \exp(X_1)$ and $Y_2 = \exp(X_2)$.

Hint: Various moments of Y_1 and Y_2 can be found by assigning appropriate values to t_1 and t_2 in $E[\exp(t_1X_1 + t_2X_2)]$.

4.125. Let X be $N(\mu, \sigma^2)$ and consider the transformation $X = \ln Y$ or, equivalently, $Y = e^X$.

(a) Find the mean and the variance of Y by first determining $E(e^X)$ and $E[(e^X)^2]$.

Hint: Use the m.g.f. of X.

(b) Find the p.d.f. of Y. This is the p.d.f. of the *lognormal distribution*.

4.126. Let X_1 and X_2 have a trinomial distribution with parameters n, p_1, p_2.

(a) What is the distribution of $Y = X_1 + X_2$?

(b) From the equality $\sigma_Y^2 = \sigma_1^2 + \sigma_2^2 + 2\rho\sigma_1\sigma_2$, once again determine the correlation coefficient ρ of X_1 and X_2.

4.127. Let $Y_1 = X_1 + X_2$ and $Y_2 = X_2 + X_3$, where X_1, X_2, and X_3 are three independent random variables. Find the joint m.g.f. and the correlation coefficient of Y_1 and Y_2 provided that:
(a) X_i has a Poisson distribution with mean μ_i, $i = 1, 2, 3$.
(b) X_i is $N(\mu_i, \sigma_i^2)$, $i = 1, 2, 3$.

4.128. Let $X_1, \ldots, X_n$ be random variables that have means $\mu_1, \ldots, \mu_n$ and variances $\sigma_1^2, \ldots, \sigma_n^2$. Let ρ_{ij}, $i \neq j$, denote the correlation coefficient of X_i and X_j. Let $a_1, \ldots, a_n$ and $b_1, \ldots, b_n$ be real constants. Show that the covariance of $Y = \sum_{i=1}^{n} a_i X_i$ and $Z = \sum_{j=1}^{n} b_j X_j$ is $\sum_{j=1}^{n} \sum_{i=1}^{n} a_i b_j \sigma_i \sigma_j \rho_{ij}$, where $\rho_{ii} = 1$, $i = 1, 2, \ldots, n$.

*4.10 The Multivariate Normal Distribution

We have studied in some detail normal distributions of one random variable. In this section we investigate a joint distribution of n random variables that will be called a *multivariate normal* distribution. This investigation assumes that the student is familiar with elementary matrix algebra, with real symmetric quadratic forms, and with orthogonal transformations. Henceforth, the expression *quadratic form* means a quadratic form in a prescribed number of variables whose matrix is real and symmetric. All symbols that represent matrices will be set in boldface type.

Let $\mathbf{A}$ denote an $n \times n$ real symmetric matrix which is positive definite. Let $\boldsymbol{\mu}$ denote the $n \times 1$ matrix such that $\boldsymbol{\mu}'$, the transpose of $\boldsymbol{\mu}$, is $\boldsymbol{\mu}' = [\mu_1, \mu_2, \ldots, \mu_n]$, where each μ_i is a real constant. Finally, let $\mathbf{x}$ denote the $n \times 1$ matrix such that $\mathbf{x}' = [x_1, x_2, \ldots, x_n]$. We shall show that if C is an appropriately chosen positive constant, the nonnegative function

$$f(x_1, x_2, \ldots, x_n) = C \exp\left[-\frac{(\mathbf{x} - \boldsymbol{\mu})'\mathbf{A}(\mathbf{x} - \boldsymbol{\mu})}{2}\right],$$

$$-\infty < x_i < \infty, \quad i = 1, 2, \ldots, n,$$

is a joint p.d.f. of n random variables $X_1, X_2, \ldots, X_n$ that are of the continuous type. Thus we need to show that

$$\int_{-\infty}^{\infty} \cdots \int_{-\infty}^{\infty} f(x_1, x_2, \ldots, x_n)\, dx_1\, dx_2 \cdots dx_n = 1. \tag{1}$$

Let $\mathbf{t}$ denote the $n \times 1$ matrix such that $\mathbf{t}' = [t_1, t_2, \ldots, t_n]$, where $t_1, t_2, \ldots, t_n$ are arbitrary real numbers. We shall evaluate the integral

$$C \int_{-\infty}^{\infty} \cdots \int_{-\infty}^{\infty} \exp\left[\mathbf{t}'\mathbf{x} - \frac{(\mathbf{x} - \boldsymbol{\mu})'\mathbf{A}(\mathbf{x} - \boldsymbol{\mu})}{2}\right] dx_1 \cdots dx_n, \quad (2)$$

and then we shall subsequently set $t_1 = t_2 = \cdots = t_n = 0$, and thus establish Equation (1). First, we change the variables of integration in integral (2) from $x_1, x_2, \ldots, x_n$ to $y_1, y_2, \ldots, y_n$ by writing $\mathbf{x} - \boldsymbol{\mu} = \mathbf{y}$, where $\mathbf{y}' = [y_1, y_2, \ldots, y_n]$. The Jacobian of the transformation is one and the n-dimensional x-space is mapped onto an n-dimensional y-space, so that integral (2) may be written as

$$C \exp(\mathbf{t}'\boldsymbol{\mu}) \int_{-\infty}^{\infty} \cdots \int_{-\infty}^{\infty} \exp\left(\mathbf{t}'\mathbf{y} - \frac{\mathbf{y}'\mathbf{A}\mathbf{y}}{2}\right) dy_1 \cdots dy_n. \quad (3)$$

Because the real symmetric matrix $\mathbf{A}$ is positive definite, the n characteristic numbers (proper values, latent roots, or eigenvalues) $a_1, a_2, \ldots, a_n$ of $\mathbf{A}$ are positive. There exists an appropriately chosen $n \times n$ real orthogonal matrix $\mathbf{L}$ ($\mathbf{L}' = \mathbf{L}^{-1}$, where $\mathbf{L}^{-1}$ is the inverse of $\mathbf{L}$) such that

$$\mathbf{L}'\mathbf{A}\mathbf{L} = \begin{bmatrix} a_1 & 0 & \cdots & 0 \\ 0 & a_2 & \cdots & 0 \\ \vdots & \vdots & & \vdots \\ 0 & 0 & \cdots & a_n \end{bmatrix},$$

for a suitable ordering of $a_1, a_2, \ldots, a_n$. We shall sometimes write $\mathbf{L}'\mathbf{A}\mathbf{L} = \operatorname{diag}[a_1, a_2, \ldots, a_n]$. In integral (3), we shall change the variables of integration from $y_1, y_2, \ldots, y_n$ to $z_1, z_2, \ldots, z_n$ by writing $\mathbf{y} = \mathbf{L}\mathbf{z}$, where $\mathbf{z}' = [z_1, z_2, \ldots, z_n]$. The Jacobian of the transformation is the determinant of the orthogonal matrix $\mathbf{L}$. Since $\mathbf{L}'\mathbf{L} = \mathbf{I}_n$, where $\mathbf{I}_n$ is the unit matrix of order n, we have the determinant $|\mathbf{L}'\mathbf{L}| = 1$ and $|\mathbf{L}|^2 = 1$. Thus the absolute value of the Jacobian is one. Moreover, the n-dimensional y-space is mapped onto an n-dimensional z-space. The integral (3) becomes

$$C \exp(\mathbf{t}'\boldsymbol{\mu}) \int_{-\infty}^{\infty} \cdots \int_{-\infty}^{\infty} \exp\left[\mathbf{t}'\mathbf{L}\mathbf{z} - \frac{\mathbf{z}'(\mathbf{L}'\mathbf{A}\mathbf{L})\mathbf{z}}{2}\right] dz_1 \cdots dz_n. \quad (4)$$

It is computationally convenient to write, momentarily, $\mathbf{t}'\mathbf{L} = \mathbf{w}'$, where $\mathbf{w}' = [w_1, w_2, \ldots, w_n]$. Then

$$\exp[\mathbf{t}'\mathbf{L}\mathbf{z}] = \exp[\mathbf{w}'\mathbf{z}] = \exp\left(\sum_1^n w_i z_i\right).$$

Moreover,

$$\exp\left[-\frac{\mathbf{z}'(\mathbf{L}'\mathbf{A}\mathbf{L})\mathbf{z}}{2}\right] = \exp\left[-\frac{\sum_1^n a_i z_i^2}{2}\right].$$

Then integral (4) may be written as the product of n integrals in the following manner:

$$C \exp(\mathbf{w}'\mathbf{L}'\boldsymbol{\mu}) \prod_{i=1}^{n} \left[\int_{-\infty}^{\infty} \exp\left(w_i z_i - \frac{a_i z_i^2}{2}\right) dz_i\right]$$

$$= C \exp(\mathbf{w}'\mathbf{L}'\boldsymbol{\mu}) \prod_{i=1}^{n} \left[\sqrt{\frac{2\pi}{a_i}} \int_{-\infty}^{\infty} \frac{\exp\left(w_i z_i - \dfrac{a_i z_i^2}{2}\right)}{\sqrt{2\pi/a_i}}\, dz_i\right]. \quad (5)$$

The integral that involves z_i can be treated as the m.g.f., with the more familiar symbol t replaced by w_i, of a distribution which is $N(0, 1/a_i)$. Thus the right-hand member of Equation (5) is equal to

$$C \exp(\mathbf{w}'\mathbf{L}'\boldsymbol{\mu}) \prod_{i=1}^{n} \left[\sqrt{\frac{2\pi}{a_i}} \exp\left(\frac{w_i^2}{2a_i}\right)\right]$$

$$= C \exp(\mathbf{w}'\mathbf{L}'\boldsymbol{\mu}) \sqrt{\frac{(2\pi)^n}{a_1 a_2 \cdots a_n}} \exp\left(\sum_1^n \frac{w_i^2}{2a_i}\right). \quad (6)$$

Now, because $\mathbf{L}^{-1} = \mathbf{L}'$, we have

$$(\mathbf{L}'\mathbf{A}\mathbf{L})^{-1} = \mathbf{L}'\mathbf{A}^{-1}\mathbf{L} = \operatorname{diag}\left[\frac{1}{a_1}, \frac{1}{a_2}, \ldots, \frac{1}{a_n}\right].$$

Thus

$$\sum_1^n \frac{w_i^2}{a_i} = \mathbf{w}'(\mathbf{L}'\mathbf{A}^{-1}\mathbf{L})\mathbf{w} = (\mathbf{L}\mathbf{w})'\mathbf{A}^{-1}(\mathbf{L}\mathbf{w}) = \mathbf{t}'\mathbf{A}^{-1}\mathbf{t}.$$

Moreover, the determinant $|\mathbf{A}^{-1}|$ of $\mathbf{A}^{-1}$ is

$$|\mathbf{A}^{-1}| = |\mathbf{L}'\mathbf{A}^{-1}\mathbf{L}| = \frac{1}{a_1 a_2 \cdots a_n}.$$

Accordingly, the right-hand member of Equation (6), which is equal to integral (2), may be written as

$$C e^{\mathbf{t}'\boldsymbol{\mu}} \sqrt{(2\pi)^n |\mathbf{A}^{-1}|} \exp\left(\frac{\mathbf{t}'\mathbf{A}^{-1}\mathbf{t}}{2}\right). \quad (7)$$

If, in this function, we set $t_1 = t_2 = \cdots = t_n = 0$, we have the value of the left-hand member of Equation (1). Thus we have

$$C\sqrt{(2\pi)^n|\mathbf{A}^{-1}|} = 1.$$

Accordingly, the function

$$f(x_1, x_2, \ldots, x_n) = \frac{1}{(2\pi)^{n/2}\sqrt{|\mathbf{A}^{-1}|}} \exp\left[-\frac{(\mathbf{x}-\boldsymbol{\mu})'\mathbf{A}(\mathbf{x}-\boldsymbol{\mu})}{2}\right],$$

$-\infty < x_i < \infty$, $i = 1, 2, \ldots, n$, is a joint p.d.f. of n random variables $X_1, X_2, \ldots, X_n$ that are of the continuous type. Such a p.d.f. is called a *nonsingular multivariate normal* p.d.f.

We have now proved that $f(x_1, x_2, \ldots, x_n)$ is a p.d.f. However, we have proved more than that. Because $f(x_1, x_2, \ldots, x_n)$ is a p.d.f., integral (2) is the m.g.f. $M(t_1, t_2, \ldots, t_n)$ of this joint distribution of probability. Since integral (2) is equal to function (7), the m.g.f. of the multivariate normal distribution is given by

$$M(t_1, t_2, \ldots, t_n) = \exp\left(\mathbf{t}'\boldsymbol{\mu} + \frac{\mathbf{t}'\mathbf{A}^{-1}\mathbf{t}}{2}\right).$$

Let the elements of the real, symmetric, and positive definite matrix $\mathbf{A}^{-1}$ be denoted by σ_{ij}, $i, j = 1, 2, \ldots, n$. Then

$$M(0, \ldots, 0, t_i, 0, \ldots, 0) = \exp\left(t_i\mu_i + \frac{\sigma_{ii}t_i^2}{2}\right)$$

is the m.g.f. of X_i, $i = 1, 2, \ldots, n$. Thus X_i is $N(\mu_i, \sigma_{ii})$, $i = 1, 2, \ldots, n$. Moreover, with $i \neq j$, we see that $M(0, \ldots, 0, t_i, 0, \ldots, t_j, 0, \ldots, 0)$, the m.g.f. of X_i and X_j, is equal to

$$\exp\left(t_i\mu_i + t_j\mu_j + \frac{\sigma_{ii}t_i^2 + 2\sigma_{ij}t_it_j + \sigma_{jj}t_j^2}{2}\right),$$

which is the m.g.f. of a *bivariate normal distribution*. In Exercise 4.131 the reader is asked to show that σ_{ij} is the covariance of the random variables X_i and X_j. Thus the matrix $\boldsymbol{\mu}$, where $\boldsymbol{\mu}' = [\mu_1, \mu_2, \ldots, \mu_n]$, is the matrix of the means of the random variables $X_1, \ldots, X_n$. Moreover, the elements on the principal diagonal of $\mathbf{A}^{-1}$ are, respectively, the variances $\sigma_{ii} = \sigma_i^2$, $i = 1, 2, \ldots, n$, and the elements not on the principal diagonal of $\mathbf{A}^{-1}$ are, respectively, the covariances

$\sigma_{ij} = \rho_{ij}\sigma_i\sigma_j$, $i \neq j$, of the random variables $X_1, X_2, \ldots, X_n$. We call the matrix $\mathbf{A}^{-1}$, which is given by

$$\begin{bmatrix} \sigma_{11} & \sigma_{12} & \cdots & \sigma_{1n} \\ \sigma_{12} & \sigma_{22} & \cdots & \sigma_{2n} \\ \vdots & \vdots & & \vdots \\ \sigma_{1n} & \sigma_{2n} & \cdots & \sigma_{nn} \end{bmatrix},$$

the *covariance matrix* of the multivariate normal distribution and henceforth we shall denote this matrix by the symbol $\mathbf{V}$. In terms of the positive definite covariance matrix $\mathbf{V}$, the multivariate normal p.d.f. is written

$$\frac{1}{(2\pi)^{n/2}\sqrt{|\mathbf{V}|}} \exp\left[-\frac{(\mathbf{x}-\boldsymbol{\mu})'\mathbf{V}^{-1}(\mathbf{x}-\boldsymbol{\mu})}{2}\right], \qquad -\infty < x_i < \infty,$$

$i = 1, 2, \ldots, n$, and the m.g.f. of this distribution is given by

$$\exp\left(\mathbf{t}'\boldsymbol{\mu} + \frac{\mathbf{t}'\mathbf{V}\mathbf{t}}{2}\right)$$

for all real values of $\mathbf{t}$.

Note that this m.g.f. equals the product of n functions, where the first is a function of t_1 alone, the second is a function of t_2 alone, and so on, if and only if $\mathbf{V}$ is a diagonal matrix. This condition, $\sigma_{ij} = \rho_{ij}\sigma_i\sigma_j = 0$, means $\rho_{ij} = 0$, $i \neq j$. That is, the multivariate normal random variables are independent if and only if $\rho_{ij} = 0$ for all $i \neq j$.

Example 1. Let $X_1, X_2, \ldots, X_n$ have a multivariate normal distribution with matrix $\boldsymbol{\mu}$ of means and positive definite covariance matrix $\mathbf{V}$. If we let $\mathbf{X}' = [X_1, X_2, \ldots, X_n]$, then the m.g.f. $M(t_1, t_2, \ldots, t_n)$ of this joint distribution of probability is

$$E(e^{\mathbf{t}'\mathbf{X}}) = \exp\left(\mathbf{t}'\boldsymbol{\mu} + \frac{\mathbf{t}'\mathbf{V}\mathbf{t}}{2}\right). \tag{8}$$

Consider a linear function Y of $X_1, X_2, \ldots, X_n$ which is defined by $Y = \mathbf{c}'\mathbf{X} = \sum_1^n c_iX_i$, where $\mathbf{c}' = [c_1, c_2, \ldots, c_n]$ and the several c_i are real and not all zero. We wish to find the p.d.f. of Y. The m.g.f. $m(t)$ of the distribution of Y is given by

$$m(t) = E(e^{tY}) = E(e^{t\mathbf{c}'\mathbf{X}}).$$

Now the expectation (8) exists for all real values of $\mathbf{t}$. Thus we can replace $\mathbf{t}'$ in expectation (8) by $t\mathbf{c}'$ and obtain

$$m(t) = \exp\left(t\mathbf{c}'\boldsymbol{\mu} + \frac{\mathbf{c}'\mathbf{V}\mathbf{c}t^2}{2}\right).$$

Thus the random variable Y is $N(\mathbf{c}'\boldsymbol{\mu}, \mathbf{c}'\mathbf{V}\mathbf{c})$.

EXERCISES

4.129. Let $X_1, X_2, \ldots, X_n$ have a multivariate normal distribution with positive definite covariance matrix $\mathbf{V}$. Prove that these random variables are mutually independent if and only if $\mathbf{V}$ is a diagonal matrix.

4.130. Let $n = 2$ and take

$$\mathbf{V} = \begin{bmatrix} \sigma_1^2 & \rho\sigma_1\sigma_2 \\ \rho\sigma_1\sigma_2 & \sigma_2^2 \end{bmatrix}.$$

Determine $|\mathbf{V}|$, $\mathbf{V}^{-1}$, and $(\mathbf{x} - \boldsymbol{\mu})'\mathbf{V}^{-1}(\mathbf{x} - \boldsymbol{\mu})$. Compare the bivariate normal p.d.f. of Section 3.5 with this multivariate normal p.d.f. when $n = 2$.

4.131. Let $m(t_i, t_j)$ represent the m.g.f. of X_i and X_j as given in the text. Show that

$$\frac{\partial^2 m(0, 0)}{\partial t_i \partial t_j} - \left[\frac{\partial m(0, 0)}{\partial t_i}\right]\left[\frac{\partial m(0, 0)}{\partial t_j}\right] = \sigma_{ij};$$

that is, prove that the covariance of X_i and X_j is σ_{ij}, which appears in that formula for $m(t_i, t_j)$.

4.132. Let $X_1, X_2, \ldots, X_n$ have a multivariate normal distribution, where $\boldsymbol{\mu}$ is the matrix of the means and $\mathbf{V}$ is the positive definite covariance matrix. Let $Y = \mathbf{c}'\mathbf{X}$ and $Z = \mathbf{d}'\mathbf{X}$, where $\mathbf{X}' = [X_1, \ldots, X_n]$, $\mathbf{c}' = [c_1, \ldots, c_n]$, and $\mathbf{d}' = [d_1, \ldots, d_n]$ are real matrices.

(a) Find $m(t_1, t_2) = E(e^{t_1 Y + t_2 Z})$ to see that Y and Z have a bivariate normal distribution.

(b) Prove that Y and Z are independent if and only if $\mathbf{c}'\mathbf{V}\mathbf{d} = 0$.

(c) If $X_1, X_2, \ldots, X_n$ are independent random variables which have the same variance σ^2, show that the necessary and sufficient condition of part (b) becomes $\mathbf{c}'\mathbf{d} = 0$.

4.133. Let $\mathbf{X}' = [X_1, X_2, \ldots, X_n]$ have the multivariate normal distribution of Exercise 4.132. Consider the p linear functions of $X_1, \ldots, X_n$ defined by $\mathbf{W} = \mathbf{B}\mathbf{X}$, where $\mathbf{W}' = [W_1, \ldots, W_p]$, $p \le n$, and $\mathbf{B}$ is a $p \times n$ real matrix of rank p. Find $m(v_1, \ldots, v_p) = E(e^{\mathbf{v}'\mathbf{W}})$, where $\mathbf{v}'$ is a real matrix $[v_1, \ldots, v_p]$, to see that $W_1, \ldots, W_p$ have a p-variate normal distribution which has $\mathbf{B}\boldsymbol{\mu}$ for the matrix of the means and $\mathbf{B}\mathbf{V}\mathbf{B}'$ for the covariance matrix.

4.134. Let $\mathbf{X}' = [X_1, X_2, \ldots, X_n]$ have the n-variate normal distribution of Exercise 4.132. Show that $X_1, X_2, \ldots, X_p$, $p < n$, have a p-variate normal distribution. What submatrix of $\mathbf{V}$ is the covariance matrix of $X_1, X_2, \ldots, X_p$?

Hint: In the m.g.f. $M(t_1, t_2, \ldots, t_n)$ of $X_1, X_2, \ldots, X_n$, let $t_{p+1} = \cdots = t_n = 0$.

ADDITIONAL EXERCISES

4.135. If X has the p.d.f. $f(x) = \frac{1}{3}$, $-1 < x < 2$, zero elsewhere, find the p.d.f. of $Y = X^4$.

4.136. The continuous random variable X has a p.d.f. given by $f(x) = 1$, $0 < x < 1$, zero elsewhere. The random variable Y is such that $Y = -2 \ln X$. What is the distribution of Y? What are the mean and the variance of Y?

4.137. Let X_1, X_2 be a random sample of size $n = 2$ from a Poisson distribution with mean μ. If $\Pr(X_1 + X_2 = 3) = (\frac{32}{3})e^{-4}$, compute $\Pr(X_1 = 2, X_2 = 4)$.

4.138. Let $X_1, X_2, \ldots, X_{25}$ be a random sample of size $n = 25$ from a distribution with p.d.f. $f(x) = 3/x^4$, $1 < x < \infty$, zero elsewhere. Let Y equal the number of these X values less than or equal to 2. What is the distribution of Y?

4.139. Find the probability that the range of a random sample of size 3 from the uniform distribution over the interval $(-5, 5)$ is less than 7.

4.140. Let $Y_1 < Y_2 < Y_3$ be the order statistics of a sample of size 3 from a distribution having p.d.f. $f(x) = \frac{1}{3}$, $-1 < x < 2$, zero elsewhere. Determine $\Pr[-\frac{1}{2} < Y_2 < \frac{1}{2}]$.

4.141. Let X and Y be random variables so that $Z = X - 2Y$ has variance equal to 28. If $\sigma_X^2 = 4$ and $\rho_{XY} = \frac{1}{2}$, find the variance σ_Y^2 of Y.

4.142. Let $Y_1 < Y_2 < Y_3 < Y_4$ be the order statistics of a random sample of size $n = 4$ from a distribution with p.d.f. $f(x) = 2(1 - x)$, $0 < x < 1$, zero elsewhere. Compute $\Pr(Y_1 < 0.1)$.

4.143. A certain job is completed in three steps in series. The means and standard deviations for the steps are (in hours):

Step	Mean	Standard Deviation
1	3	0.2
2	1	0.1
3	4	0.2

Assuming normal distributions and independent steps, compute the probability that the job will take less than 7.6 hours to complete.

4.144. Let $X_1, X_2, \ldots, X_n$ be a random sample of size n from a distribution having mean μ and variance 25. Use Chebyshev's inequality to determine the smallest value of n so that 0.75 is a lower bound for $\Pr[|\bar{X} - \mu| \le 1]$.

4.145. Let X_1 and X_2 be independent random variables with joint p.d.f.

$$f(x_1, x_2) = \frac{x_1(4 - x_2)}{36}, \qquad x_1 = 1, 2, 3, \quad x_2 = 1, 2, 3,$$

and zero elsewhere. Find the p.d.f. of $Y = X_1 - X_2$.

4.146. An unbiased die is cast eight independent times. Let Y be the smallest of the eight numbers obtained. Find the p.d.f. of Y.

4.147. Let X_1, X_2, X_3 be i.i.d. $N(\mu, \sigma^2)$ and define

$$Y_1 = X_1 + \delta X_3$$

and

$$Y_2 = X_2 + \delta X_3.$$

(a) Find the means and variances of Y_1 and Y_2 and their correlation coefficient.
(b) Find the joint m.g.f. of Y_1 and Y_2.

4.148. The following were obtained from two sets of data:

$$n_1 = 20, \qquad \bar{x} = 25, \qquad s_x^2 = 5,$$

$$n_2 = 30, \qquad \bar{y} = 20, \qquad s_y^2 = 4.$$

Find the mean and variance of the combined sample.

4.149. Let $Y_1 < Y_2 < \cdots < Y_5$ be the order statistics of a random sample of size 5 from a distribution that has the p.d.f. $f(x) = 1$, $0 < x < 1$, zero elsewhere. Compute $\Pr(Y_1 < \frac{1}{5}, Y_5 > \frac{3}{5})$.

4.150. Let $M(t) = (1 - t)^{-3}$, $t < 1$, be the m.g.f. of X. Find the m.g.f. of $Y = \dfrac{X - 10}{25}$.

4.151. Let $\bar{X}$ be the mean of a random sample of size n from a normal distribution with mean μ and variance $\sigma^2 = 64$. Find n so that

$$\Pr(\mu - 6 < \bar{X} < \mu + 6) = 0.9973.$$

4.152. Find the probability of obtaining a total of 14 in one toss of four dice.

4.153. Two independent random samples, each of size 6, are taken from two normal distributions having common variance σ^2. If W_1 and W_2 are the variances of these respective samples, find the constant k such that

$$\Pr\left[\min\left(\frac{W_1}{W_2}, \frac{W_2}{W_1}\right) < k\right] = 0.10.$$

4.154. The mean and variance of 9 observations are 4 and 14, respectively. We find that a tenth observation equals 6. Find the mean and the variance of the 10 observations.

4.155. Draw 15 cards at random and without replacement from a pack of 25 cards numbered 1, 2, 3, . . . , 25. Find the probability that 10 is the median of the cards selected.

4.156. Let $Y_1 < Y_2 < Y_3 < Y_4$ be the order statistics of a random sample of size $n = 4$ from a uniform distribution over the interval (0, 1).

(a) Find the joint p.d.f. of Y_1 and Y_4.

(b) Determine the conditional p.d.f. of Y_2 and Y_3, given $Y_1 = y_1$ and $Y_4 = y_4$.

(c) Find the joint p.d.f. of $Z_1 = Y_1/Y_4$ and $Z_2 = Y_4$.

4.157. Let $X_1, X_2, \ldots, X_n$ be a random sample from a distribution with mean μ and variance σ^2. Consider the second differences

$$Z_j = X_{j+2} - 2X_{j+1} + X_j, \qquad j = 1, 2, \ldots, n-2.$$

Compute the variance of the average, $\sum_{j=1}^{n-2} Z_j/(n-2)$, of the second differences.

4.158. Let X and Y have a bivariate normal distribution. Show that $X + Y$ and $X - Y$ are independent if and only if $\sigma_1^2 = \sigma_2^2$.

4.159. Let X be a Poisson random variable with mean μ. If the conditional distribution of Y, given $X = x$, is $b(x, p)$. Show that Y has a Poisson distribution and is independent of $X - Y$.

4.160. Let $X_1, X_2, \ldots, X_n$ be a random sample from $N(\mu, \sigma^2)$. Show that the sample mean $\bar{X}$ and each $X_i - \bar{X}$, $i = 1, 2, \ldots, n$, are independent. Actually $\bar{X}$ and the vector $(X_1 - \bar{X}, X_2 - \bar{X}, \ldots, X_n - \bar{X})$ are independent and this implies that $\bar{X}$ and $\sum_{i=1}^{n} (X_i - \bar{X})^2$ are independent. Thus we could find the joint distribution of $\bar{X}$ and nS^2/σ^2 using this result.

4.161. Let $X_1, X_2, \ldots, X_n$ be a random sample from a distribution with p.d.f. $f(x) = \frac{1}{6}$, $x = 1, 2, \ldots, 6$, zero elsewhere. Let $Y = \min(X_i)$ and $Z = \max(X_i)$. Say that the joint distribution function of Y and Z is $G(y, z) = \Pr(Y \le y, Z \le z)$, where y and z are nonnegative integers such that $1 \le y \le z \le 6$.

(a) Show that

$$G(y, z) = F^n(z) - [F(z) - F(y)]^n, \qquad 1 \le y \le z \le 6,$$

where $F(x)$ is the distribution function associated with $f(x)$.
Hint: Note that the event $(Z \le z) = (Y \le y, Z \le z) \cup (y < Y, Z \le z)$

(b) Find the joint p.d.f. of Y and Z by evaluating

$$g(y, z) = G(y, z) - G(y - 1, z) - G(y, z - 1) + G(y - 1, z - 1).$$

4.162. Let $\mathbf{X} = (X_1, X_2, X_3)'$ have a multivariate normal distribution with mean vector $\boldsymbol{\mu} = (6, -2, 1)'$ and covariance matrix

$$\mathbf{V} = \begin{bmatrix} 1 & 0 & -1 \\ 0 & 2 & 1 \\ -1 & 1 & 3 \end{bmatrix}.$$

Find the joint p.d.f. of

$$Y_1 = 3X_1 + X_2 - 2X_3 \qquad \text{and} \qquad Y_2 = X_1 - 5X_2 + X_3.$$

4.163. If

$$\mathbf{V} = \begin{bmatrix} 1 & \rho & \rho \\ \rho & 1 & \rho \\ \rho & \rho & 1 \end{bmatrix}$$

is a covariance matrix, what can be said about the value of ρ?

CHAPTER 5

Limiting Distributions

5.1 Convergence in Distribution

In some of the preceding chapters it has been demonstrated by example that the distribution of a random variable (perhaps a statistic) often depends upon a positive integer n. For example, if the random variable X is $b(n, p)$, the distribution of X depends upon n. If $\bar{X}$ is the mean of a random sample of size n from a distribution that is $N(\mu, \sigma^2)$, then $\bar{X}$ is itself $N(\mu, \sigma^2/n)$ and the distribution of $\bar{X}$ depends upon n. If S^2 is the variance of this random sample from the normal distribution to which we have just referred, the random variable nS^2/σ^2 is $\chi^2(n-1)$, and so the distribution of this random variable depends upon n.

We know from experience that the determination of the probability density function of a random variable can, upon occasion, present rather formidable computational difficulties. For example, if $\bar{X}$ is the mean of a random sample $X_1, X_2, \ldots, X_n$ from a distribution that has the following p.d.f.

$$f(x) = 1, \qquad 0 < x < 1,$$
$$= 0 \qquad \text{elsewhere,}$$

then (Exercise 4.85) the m.g.f. of $\bar{X}$ is given by $[M(t/n)]^n$, where here

$$M(t) = \int_0^1 e^{tx}\,dx = \frac{e^t - 1}{t}, \qquad t \neq 0,$$
$$= 1, \qquad t = 0.$$

Hence

$$E(e^{t\bar{X}}) = \left(\frac{e^{t/n} - 1}{t/n}\right)^n, \qquad t \neq 0,$$
$$= 1, \qquad t = 0.$$

Since the m.g.f. of $\bar{X}$ depends upon n, the distribution of $\bar{X}$ depends upon n. It is true that various mathematical techniques can be used to determine the p.d.f. of $\bar{X}$ for a fixed, but arbitrarily fixed, positive integer n. But the p.d.f. is so complicated that few, if any, of us would be interested in using it to compute probabilities about $\bar{X}$. One of the purposes of this chapter is to provide ways of approximating, for large values of n, some of these complicated probability density functions.

Consider a distribution that depends upon the positive integer n. Clearly, the distribution function F of that distribution will also depend upon n. Throughout this chapter, we denote this fact by writing the distribution function as F_n and the corresponding p.d.f. as f_n. Moreover, to emphasize the fact that we are working with sequences of distribution functions and random variables, we place a subscript n on the random variables. For example, we shall write

$$F_n(\bar{x}) = \int_{-\infty}^{\bar{x}} \frac{1}{\sqrt{1/n}\sqrt{2\pi}} e^{-nw^2/2}\,dw$$

for the distribution function of the mean $\bar{X}_n$ of a random sample of size n from a normal distribution with mean zero and variance 1.

We now define convergence in distribution of a sequence of random variables.

Definition 1. Let the distribution function $F_n(y)$ of the random variable Y_n depend upon n, $n = 1, 2, 3, \ldots$. If $F(y)$ is a distribution function and if $\lim_{n\to\infty} F_n(y) = F(y)$ for every point y at which $F(y)$ is continuous, then the sequence of random variables, $Y_1, Y_2, \ldots,$ *converges in distribution* to a random variable with distribution function $F(y)$.

The following examples are illustrative of this convergence in distribution.

Example 1. Let Y_n denote the nth order statistic of a random sample $X_1, X_2, \ldots, X_n$ from a distribution having p.d.f.

$$
\begin{aligned}
f(x) &= \frac{1}{\theta}, \qquad 0 < x < \theta, \quad 0 < \theta < \infty, \\
&= 0 \qquad \text{elsewhere.}
\end{aligned}
$$

The p.d.f. of Y_n is

$$
\begin{aligned}
g_n(y) &= \frac{ny^{n-1}}{\theta^n}, \qquad 0 < y < \theta, \\
&= 0 \qquad \text{elsewhere,}
\end{aligned}
$$

and the distribution function of Y_n is

$$
\begin{aligned}
F_n(y) &= 0, \qquad y < 0, \\
&= \int_0^y \frac{nz^{n-1}}{\theta^n}\,dz = \left(\frac{y}{\theta}\right)^n, \qquad 0 \le y < \theta, \\
&= 1, \qquad \theta \le y < \infty.
\end{aligned}
$$

Then

$$
\begin{aligned}
\lim_{n\to\infty} F_n(y) &= 0, \qquad -\infty < y < \theta, \\
&= 1, \qquad \theta \le y < \infty.
\end{aligned}
$$

Now

$$
\begin{aligned}
F(y) &= 0, \qquad -\infty < y < \theta, \\
&= 1, \qquad \theta \le y < \infty,
\end{aligned}
$$

is a distribution function. Moreover, $\lim_{n\to\infty} F_n(y) = F(y)$ at each point of continuity of $F(y)$. Recall that a distribution of the discrete type which has a probability of 1 at a single point has been called a *degenerate distribution*. Thus, in this example, the sequence of the nth order statistics, Y_n, $n = 1, 2, 3, \ldots,$ converges in distribution to a random variable that has a degenerate distribution at the point $y = \theta$.

Example 2. Let $\overline{X}_n$ have the distribution function

$$F_n(\overline{x}) = \int_{-\infty}^{\overline{x}} \frac{1}{\sqrt{1/n}\sqrt{2\pi}} e^{-nw^2/2}\, dw.$$

If the change of variable $v = \sqrt{n}w$ is made, we have

$$F_n(\overline{x}) = \int_{-\infty}^{\sqrt{n}\overline{x}} \frac{1}{\sqrt{2\pi}} e^{-v^2/2}\, dv.$$

It is clear that

$$\begin{aligned} \lim_{n\to\infty} F_n(\overline{x}) &= 0, & \overline{x} &< 0, \\ &= \tfrac{1}{2}, & \overline{x} &= 0, \\ &= 1, & \overline{x} &> 0. \end{aligned}$$

Now the function

$$\begin{aligned} F(\overline{x}) &= 0, & \overline{x} &< 0, \\ &= 1, & \overline{x} &\geq 0, \end{aligned}$$

is a distribution function and $\lim_{n\to\infty} F_n(\overline{x}) = F(\overline{x})$ at every point of continuity of $F(\overline{x})$. To be sure, $\lim_{n\to\infty} F_n(0) \neq F(0)$, but $F(\overline{x})$ is not continuous at $\overline{x} = 0$. Accordingly, the sequence $\overline{X}_1, \overline{X}_2, \overline{X}_3, \ldots$ converges in distribution to a random variable that has a degenerate distribution at $\overline{x} = 0$.

Example 3. Even if a sequence $X_1, X_2, X_3, \ldots$ converges in distribution to a random variable X, we cannot in general determine the distribution of X by taking the limit of the p.d.f. of X_n. This is illustrated by letting X_n have the p.d.f.

$$\begin{aligned} f_n(x) &= 1, & x &= 2 + \frac{1}{n}, \\ &= 0 & &\text{elsewhere.} \end{aligned}$$

Clearly, $\lim_{n\to\infty} f_n(x) = 0$ for all values of x. This may suggest that X_n, $n = 1, 2, 3, \ldots$, does not converge in distribution. However, the distribution function of X_n is

$$\begin{aligned} F_n(x) &= 0, & x &< 2 + \frac{1}{n}, \\ &= 1, & x &\geq 2 + \frac{1}{n}, \end{aligned}$$

and

$$\lim_{n\to\infty} F_n(x) = 0, \qquad x \le 2,$$
$$= 1, \qquad x > 2.$$

Since

$$F(x) = 0, \qquad x < 2,$$
$$= 1, \qquad x \ge 2,$$

is a distribution function, and since $\lim_{n\to\infty} F_n(x) = F(x)$ at all points of continuity of $F(x)$, the sequence $X_1, X_2, X_3, \ldots$ converges in distribution to a random variable with distribution function $F(x)$.

It is interesting to note that although we refer to a sequence of random variables, $X_1, X_2, X_3, \ldots$, converging in distribution to a random variable X having some distribution function $F(x)$, it is actually the distribution functions $F_1, F_2, F_3, \ldots$ that converge. That is,

$$\lim_{n\to\infty} F_n(x) = F(x)$$

at all points x for which $F(x)$ is continuous. For that reason we often find it convenient to refer to $F(x)$ as the *limiting distribution*. Moreover, it is then a little easier to say that X_n, representing the sequence $X_1, X_2, X_3, \ldots$, has a limiting distribution with distribution function $F(x)$. Henceforth, we use this terminology.

Example 4. Let Y_n denote the nth order statistic of a random sample from the uniform distribution of Example 1. Let $Z_n = n(\theta - Y_n)$. The p.d.f. of Z_n is

$$h_n(z) = \frac{(\theta - z/n)^{n-1}}{\theta^n}, \qquad 0 < z < n\theta,$$
$$= 0 \qquad \text{elsewhere},$$

and the distribution function of Z_n is

$$G_n(z) = 0, \qquad z < 0,$$
$$= \int_0^z \frac{(\theta - w/n)^{n-1}}{\theta^n}\, dw = 1 - \left(1 - \frac{z}{n\theta}\right)^n, \qquad 0 \le z < n\theta,$$
$$= 1, \qquad n\theta \le z.$$

Hence

$$\lim_{n\to\infty} G_n(z) = 0, \qquad z \le 0,$$
$$= 1 - e^{-z/\theta}, \qquad 0 < z < \infty.$$

Now

$$G(z) = 0, \qquad z < 0,$$
$$= 1 - e^{-z/\theta}, \qquad 0 \le z,$$

is a distribution function that is everywhere continuous and $\lim_{n\to\infty} G_n(z) = G(z)$ at all points. Thus Z_n has a limiting distribution with distribution function $G(z)$. This affords us an example of a limiting distribution that is not degenerate.

Example 5. Let T_n have a t-distribution with n degrees of freedom, $n = 1, 2, 3, \ldots$. Thus its distribution function is

$$F_n(t) = \int_{-\infty}^{t} \frac{\Gamma[(n+1)/2]}{\sqrt{\pi n}\,\Gamma(n/2)} \frac{1}{(1 + y^2/n)^{(n+1)/2}}\, dy,$$

where the integrand is the p.d.f. $f_n(y)$ of T_n. Accordingly,

$$\lim_{n\to\infty} F_n(t) = \lim_{n\to\infty} \int_{-\infty}^{t} f_n(y)\, dy$$

$$= \int_{-\infty}^{t} \lim_{n\to\infty} f_n(y)\, dy.$$

The change of the order of the limit and integration is justified because $|f_n(y)|$ is dominated by a function, like $10 f_1(y)$, with a finite integral. That is,

$$|f_n(y)| \le 10 f_1(y)$$

and

$$\int_{-\infty}^{t} 10 f_1(y)\, dy = \frac{10}{\pi} \arctan t < \infty,$$

for all real t. Hence, here we can find the limiting distribution by finding the limit of the p.d.f. of T_n. It is

$$\lim_{n\to\infty} f_n(y) = \lim_{n\to\infty} \left[\frac{\Gamma[(n+1)/2]}{\sqrt{n/2}\,\Gamma(n/2)}\right]$$

$$\times \lim_{n\to\infty} \frac{1}{(1 + y^2/n)^{1/2}} \lim_{n\to\infty} \left\{\frac{1}{\sqrt{2\pi}} \left[\left(1 + \frac{y^2}{n}\right)\right]^{-n/2}\right\}.$$

Using the fact from elementary calculus that

$$\lim_{n\to\infty}\left(1+\frac{y^2}{n}\right)^n = e^{y^2},$$

the limit associated with the third factor is clearly the p.d.f. of the standard normal distribution. The second limit obviously equals 1. If we knew more about the gamma function, it is easy to show that the first limit also equals 1. Thus we have

$$\lim_{n\to\infty} F_n(t) = \int_{-\infty}^{t} \frac{1}{\sqrt{2\pi}} e^{-y^2/2}\, dy,$$

and hence T_n has a limiting standard normal distribution.

EXERCISES

5.1. Let $\bar{X}_n$ denote the mean of a random sample of size n from a distribution that is $N(\mu, \sigma^2)$. Find the limiting distribution of $\bar{X}_n$.

5.2. Let Y_1 denote the first order statistic of a random sample of size n from a distribution that has the p.d.f. $f(x) = e^{-(x-\theta)}$, $\theta < x < \infty$, zero elsewhere. Let $Z_n = n(Y_1 - \theta)$. Investigate the limiting distribution of Z_n.

5.3. Let Y_n denote the nth order statistic of a random sample from a distribution of the continuous type that has distribution function $F(x)$ and p.d.f. $f(x) = F'(x)$. Find the limiting distribution of $Z_n = n[1 - F(Y_n)]$.

5.4. Let Y_2 denote the second order statistic of a random sample of size n from a distribution of the continuous type that has distribution function $F(x)$ and p.d.f. $f(x) = F'(x)$. Find the limiting distribution of $W_n = nF(Y_2)$.

5.5. Let the p.d.f. of Y_n be $f_n(y) = 1$, $y = n$, zero elsewhere. Show that Y_n does not have a limiting distribution. (In this case, the probability has "escaped" to infinity.)

5.6. Let $X_1, X_2, \ldots, X_n$ be a random sample of size n from a distribution that is $N(\mu, \sigma^2)$, where $\sigma^2 > 0$. Show that the sum $Z_n = \sum_1^n X_i$ does not have a limiting distribution.

5.2 Convergence in Probability

In the discussion concerning convergence in distribution, it was noted that it was really the sequence of distribution functions that converges to what we call a limiting distribution function.

Convergence in probability is quite different, although we demonstrate that in a special case there is a relationship between the two concepts.

Definition 2. A sequence of random variables $X_1, X_2, X_3, \ldots$ *converges in probability* to a random variable X if, for every $\epsilon > 0$,

$$\lim_{n\to\infty} \Pr(|X_n - X| < \epsilon) = 1,$$

or equivalently,

$$\lim_{n\to\infty} \Pr(|X_n - X| \geq \epsilon) = 0.$$

Statisticians are usually interested in this convergence when the random variable X is a constant, that is, when the random variable X has a degenerate distribution at that constant. Hence we concentrate on that situation.

Example 1. Let $\bar{X}_n$ denote the mean of a random sample of size n from a distribution that has mean μ and positive variance σ^2. Then the mean and variance of $\bar{X}_n$ are μ and σ^2/n. Consider, for every fixed $\epsilon > 0$, the probability

$$\Pr(|\bar{X}_n - \mu| \geq \epsilon) = \Pr\left(|\bar{X}_n - \mu| \geq \frac{k\sigma}{\sqrt{n}}\right),$$

where $k = \epsilon\sqrt{n}/\sigma$. In accordance with the inequality of Chebyshev, this probability is less than or equal to $1/k^2 = \sigma^2/n\epsilon^2$. So, for every fixed $\epsilon > 0$, we have

$$\lim_{n\to\infty} \Pr(|\bar{X}_n - \mu| \geq \epsilon) \leq \lim_{n\to\infty} \frac{\sigma^2}{n\epsilon^2} = 0.$$

Hence $\bar{X}_n$, $n = 1, 2, 3, \ldots$, converges in probability to μ if σ^2 is finite. (In a more advanced course, the student will learn that μ finite is sufficient to ensure this convergence in probability.) This result is called the *weak law of large numbers.*

Remark. A stronger type of convergence is given by $\Pr(\lim_{n\to\infty} Y_n = c) = 1$; in this case we say that Y_n, $n = 1, 2, 3, \ldots$, converges to c *with probability 1.* Although we do not consider this type of convergence, it is known that the mean $\bar{X}_n$, $n = 1, 2, 3, \ldots$, of a random sample converges with probability 1 to the mean μ of the distribution, provided that the latter exists. This is one form of the *strong law of large numbers.*

We prove a theorem that relates a certain limiting distribution to convergence in probability to a constant.

Theorem 1. *Let $F_n(y)$ denote the distribution function of a random variable Y_n whose distribution depends upon the positive integer n. Let c denote a constant which does not depend upon n. The sequence Y_n, $n = 1, 2, 3, \ldots$, converges in probability to the constant c if and only if the limiting distribution of Y_n is degenerate at $y = c$.*

Proof. First, assume that the $\lim_{n\to\infty} \Pr(|Y_n - c| < \epsilon) = 1$ for every $\epsilon > 0$. We are to prove that the random variable Y_n is such that

$$\begin{aligned} \lim_{n\to\infty} F_n(y) &= 0, \qquad y < c, \\ &= 1, \qquad y > c. \end{aligned}$$

Note that we do not need to know anything about the $\lim_{n\to\infty} F_n(c)$. For if the limit of $F_n(y)$ is as indicated, then Y_n has a limiting distribution with distribution function

$$\begin{aligned} F(y) &= 0, \qquad y < c, \\ &= 1, \qquad y \geq c. \end{aligned}$$

Now

$$\Pr(|Y_n - c| < \epsilon) = F_n[(c + \epsilon) -] - F_n(c - \epsilon),$$

where $F_n[(c + \epsilon) -]$ is the left-hand limit of $F_n(y)$ at $y = c + \epsilon$. Thus we have

$$1 = \lim_{n\to\infty} \Pr(|Y_n - c| < \epsilon) = \lim_{n\to\infty} F_n[(c + \epsilon) -] - \lim_{n\to\infty} F_n(c - \epsilon).$$

Because $0 \leq F_n(y) \leq 1$ for all values of y and for every positive integer n, it must be that

$$\lim_{n\to\infty} F_n(c - \epsilon) = 0, \qquad \lim_{n\to\infty} F_n[(c + \epsilon) -] = 1.$$

Since this is true for every $\epsilon > 0$, we have

$$\begin{aligned} \lim_{n\to\infty} F_n(y) &= 0, \qquad y < c, \\ &= 1, \qquad y > c, \end{aligned}$$

as we were required to show.

To complete the proof of Theorem 1, we assume that

$$\lim_{n\to\infty} F_n(y) = 0, \qquad y < c,$$
$$= 1, \qquad y > c.$$

We are to prove that $\lim_{n\to\infty} \Pr(|Y_n - c| < \epsilon) = 1$ for every $\epsilon > 0$. Because

$$\Pr(|Y_n - c| < \epsilon) = F_n[(c + \epsilon) -] - F_n(c - \epsilon),$$

and because it is given that

$$\lim_{n\to\infty} F_n[(c + \epsilon) -] = 1,$$

$$\lim_{n\to\infty} F_n(c - \epsilon) = 0,$$

for every $\epsilon > 0$, we have the desired result. This completes the proof of the theorem.

For convenience, in the notation of Theorem 1, we sometimes say that Y_n, rather than the sequence $Y_1, Y_2, Y_3, \ldots$, converges in probability to the constant c.

EXERCISES

5.7. Let the random variable Y_n have a distribution that is $b(n, p)$.
(a) Prove that Y_n/n converges in probability to p. This result is one form of the weak law of large numbers.
(b) Prove that $1 - Y_n/n$ converges in probability to $1 - p$.

5.8. Let S_n^2 denote the variance of a random sample of size n from a distribution that is $N(\mu, \sigma^2)$. Prove that $nS_n^2/(n - 1)$ converges in probability to σ^2.

5.9. Let W_n denote a random variable with mean μ and variance b/n^p, where $p > 0$, μ, and b are constants (not functions of n). Prove that W_n converges in probability to μ.

Hint: Use Chebyshev's inequality.

5.10. Let Y_n denote the nth order statistic of a random sample of size n from a uniform distribution on the interval $(0, \theta)$, as in Example 1 of Section 5.1. Prove that $Z_n = \sqrt{Y_n}$ converges in probability to $\sqrt{\theta}$.

5.3 Limiting Moment-Generating Functions

To find the limiting distribution function of a random variable Y_n by use of the definition of limiting distribution function obviously requires that we know $F_n(y)$ for each positive integer n. But, as indicated in the introductory remarks of Section 5.1, this is precisely the problem we should like to avoid. If it exists, the moment-generating function that corresponds to the distribution function $F_n(y)$ often provides a convenient method of determining the limiting distribution function. To emphasize that the distribution of a random variable Y_n depends upon the positive integer n, in this chapter we shall write the moment-generating function of Y_n in the form $M(t; n)$.

The following theorem, which is essentially Curtiss' modification of a theorem of Lévy and Cramér, explains how the moment-generating function may be used in problems of limiting distributions. A proof of the theorem requires a knowledge of that same facet of analysis that permitted us to assert that a moment-generating function, when it exists, uniquely determines a distribution. Accordingly, no proof of the theorem will be given.

Theorem 2. *Let the random variable Y_n have the distribution function $F_n(y)$ and the moment-generating function $M(t; n)$ that exists for $-h < t < h$ for all n. If there exists a distribution function $F(y)$, with corresponding moment-generating function $M(t)$, defined for $|t| \le h_1 < h$, such that $\lim_{n\to\infty} M(t; n) = M(t)$, then Y_n has a limiting distribution with distribution function $F(y)$.*

In this and the subsequent sections are several illustrations of the use of Theorem 2. In some of these examples it is convenient to use a certain limit that is established in some courses in advanced calculus. We refer to a limit of the form

$$\lim_{n\to\infty}\left[1 + \frac{b}{n} + \frac{\psi(n)}{n}\right]^{cn},$$

where b and c do not depend upon n and where $\lim_{n\to\infty} \psi(n) = 0$. Then

$$\lim_{n\to\infty}\left[1 + \frac{b}{n} + \frac{\psi(n)}{n}\right]^{cn} = \lim_{n\to\infty}\left(1 + \frac{b}{n}\right)^{cn} = e^{bc}.$$

For example,

$$\lim_{n\to\infty}\left(1-\frac{t^2}{n}+\frac{t^3}{n^{3/2}}\right)^{-n/2}=\lim_{n\to\infty}\left(1-\frac{t^2}{n}+\frac{t^3/\sqrt{n}}{n}\right)^{-n/2}.$$

Here $b = -t^2$, $c = -\frac{1}{2}$, and $\psi(n) = t^3/\sqrt{n}$. Accordingly, for every fixed value of t, the limit is $e^{t^2/2}$.

Example 1. Let Y_n have a distribution that is $b(n, p)$. Suppose that the mean $\mu = np$ is the same for every n; that is, $p = \mu/n$, where μ is a constant. We shall find the limiting distribution of the binomial distribution, when $p = \mu/n$, by finding the limit of $M(t; n)$. Now

$$M(t; n) = E(e^{tY_n}) = [(1-p) + pe^t]^n = \left[1 + \frac{\mu(e^t - 1)}{n}\right]^n$$

for all real values of t. Hence we have

$$\lim_{n\to\infty} M(t; n) = e^{\mu(e^t - 1)}$$

for all real values of t. Since there exists a distribution, namely the Poisson distribution with mean μ, that has this m.g.f. $e^{\mu(e^t - 1)}$, then, in accordance with the theorem and under the conditions stated, it is seen that Y_n has a limiting Poisson distribution with mean μ.

Whenever a random variable has a limiting distribution, we may, if we wish, use the limiting distribution as an approximation to the exact distribution function. The result of this example enables us to use the Poisson distribution as an approximation to the binomial distribution when n is large and p is small. This is clearly an advantage, for it is easy to provide tables for the one-parameter Poisson distribution. On the other hand, the binomial distribution has two parameters, and tables for this distribution are very ungainly. To illustrate the use of the approximation, let Y have a binomial distribution with $n = 50$ and $p = \frac{1}{25}$. Then

$$\Pr(Y \le 1) = (\tfrac{24}{25})^{50} + 50(\tfrac{1}{25})(\tfrac{24}{25})^{49} = 0.400,$$

approximately. Since $\mu = np = 2$, the Poisson approximation to this probability is

$$e^{-2} + 2e^{-2} = 0.406.$$

Example 2. Let Z_n be $\chi^2(n)$. Then the m.g.f. of Z_n is $(1 - 2t)^{-n/2}$, $t < \frac{1}{2}$. The mean and the variance of Z_n are, respectively, n and $2n$. The limiting distribution of the random variable $Y_n = (Z_n - n)/\sqrt{2n}$ will be investigated. Now the m.g.f. of Y_n is

$$M(t; n) = E\left\{\exp\left[t\left(\frac{Z_n - n}{\sqrt{2n}}\right)\right]\right\}$$

$$= e^{-tn/\sqrt{2n}}E\left(e^{tZ_n/\sqrt{2n}}\right)$$

$$= \exp\left[-\left(t\sqrt{\frac{2}{n}}\right)\left(\frac{n}{2}\right)\right]\left(1 - 2\frac{t}{\sqrt{2n}}\right)^{-n/2}, \qquad t < \frac{\sqrt{2n}}{2}.$$

This may be written in the form

$$M(t; n) = \left(e^{t\sqrt{2/n}} - t\sqrt{\frac{2}{n}}\,e^{t\sqrt{2/n}}\right)^{-n/2}, \qquad t < \sqrt{\frac{n}{2}}.$$

In accordance with Taylor's formula, there exists a number $\xi(n)$, between 0 and $t\sqrt{2/n}$, such that

$$e^{t\sqrt{2/n}} = 1 + t\sqrt{\frac{2}{n}} + \frac{1}{2}\left(t\sqrt{\frac{2}{n}}\right)^2 + \frac{e^{\xi(n)}}{6}\left(t\sqrt{\frac{2}{n}}\right)^3.$$

If this sum is substituted for $e^{t\sqrt{2/n}}$ in the last expression for $M(t; n)$, it is seen that

$$M(t; n) = \left(1 - \frac{t^2}{n} + \frac{\psi(n)}{n}\right)^{-n/2},$$

where

$$\psi(n) = \frac{\sqrt{2}t^3e^{\xi(n)}}{3\sqrt{n}} - \frac{\sqrt{2}t^3}{\sqrt{n}} - \frac{2t^4e^{\xi(n)}}{3n}.$$

Since $\xi(n) \to 0$ as $n \to \infty$, then $\lim \psi(n) = 0$ for every fixed value of t. In accordance with the limit proposition cited earlier in this section, we have

$$\lim_{n\to\infty} M(t; n) = e^{t^2/2}$$

for all real values of t. That is, the random variable $Y_n = (Z_n - n)/\sqrt{2n}$ has a limiting standard normal distribution.

EXERCISES

5.11. Let X_n have a gamma distribution with parameter $\alpha = n$ and β, where β is not a function of n. Let $Y_n = X_n/n$. Find the limiting distribution of Y_n.

5.12. Let Z_n be $\chi^2(n)$ and let $W_n = Z_n/n^2$. Find the limiting distribution of W_n.

5.13. Let X be $\chi^2(50)$. Approximate Pr $(40 < X < 60)$.

5.14. Let $p = 0.95$ be the probability that a man, in a certain age group, lives at least 5 years.

(a) If we are to observe 60 such men and if we assume independence, find the probability that at least 56 of them live 5 or more years.

(b) Find an approximation to the result of part (a) by using the Poisson distribution.

Hint: Redefine p to be 0.05 and $1 - p = 0.95$.

5.15. Let the random variable Z_n have a Poisson distribution with parameter $\mu = n$. Show that the limiting distribution of the random variable $Y_n = (Z_n - n)/\sqrt{n}$ is normal with mean zero and variance 1.

5.16. Let S_n^2 denote the variance of a random sample of size n from a distribution that is $N(\mu, \sigma^2)$. It has been proved that $nS_n^2/(n-1)$ converges in probability to σ^2. Prove that S_n^2 converges in probability to σ^2.

5.17. Let X_n and Y_n have a bivariate normal distribution with parameters μ_1, μ_2, σ_1^2, σ_2^2 (free of n) but $\rho = 1 - 1/n$. Consider the conditional distribution of Y_n, given $X_n = x$. Investigate the limit of this conditional distribution as $n \to \infty$. What is the limiting distribution if $\rho = -1 + 1/n$? Reference to these facts was made in the Remark in Section 2.3.

5.18. Let $\bar{X}_n$ denote the mean of a random sample of size n from a Poisson distribution with parameter $\mu = 1$.

(a) Show that the m.g.f. of $Y_n = \sqrt{n}(\bar{X}_n - \mu)/\sigma = \sqrt{n}(\bar{X}_n - 1)$ is given by $\exp[-t\sqrt{n} + n(e^{t/\sqrt{n}} - 1)]$.

(b) Investigate the limiting distribution of Y_n as $n \to \infty$.

Hint: Replace, by its MacLaurin's series, the expression $e^{t/\sqrt{n}}$, which is in the exponent of the moment-generating function of Y_n.

5.19. Let $\bar{X}_n$ denote the mean of a random sample of size n from a distribution that has p.d.f. $f(x) = e^{-x}$, $0 < x < \infty$, zero elsewhere.

(a) Show that the m.g.f. $M(t; n)$ of $Y_n = \sqrt{n}(\bar{X}_n - 1)$ is equal to $[e^{t/\sqrt{n}} - (t/\sqrt{n})e^{t/\sqrt{n}}]^{-n}$, $t < \sqrt{n}$.

(b) Find the limiting distribution of Y_n as $n \to \infty$.

This exercise and the immediately preceding one are special instances of an important theorem that will be proved in the next section.

5.4 The Central Limit Theorem

It was seen (Section 4.8) that, if $X_1, X_2, \ldots, X_n$ is a random sample from a normal distribution with mean μ and variance σ^2, the random variable

$$\frac{\sum_{1}^{n} X_i - n\mu}{\sigma\sqrt{n}} = \frac{\sqrt{n}(\bar{X}_n - \mu)}{\sigma}$$

is, for every positive integer n, normally distributed with zero mean and unit variance. In probability theory there is a very elegant theorem called the *central limit theorem*. A special case of this theorem asserts the remarkable and important fact that if $X_1, X_2, \ldots, X_n$ denote the observations of a random sample of size n from any distribution having positive variance σ^2 (and hence finite mean μ), then the random variable $\sqrt{n}(\bar{X}_n - \mu)/\sigma$ has a limiting standard normal distribution. If this fact can be established, it will imply, whenever the conditions of the theorem are satisfied, that (for large n) the random variable $\sqrt{n}(\bar{X} - \mu)/\sigma$ has an approximate normal distribution with mean zero and variance 1. It will then be possible to use this approximate normal distribution to compute approximate probabilities concerning $\bar{X}$.

The more general form of the theorem is stated, but it is proved only in the modified case. However, this is exactly the proof of the theorem that would be given if we could use the characteristic function in place of the m.g.f.

Theorem 3. *Let $X_1, X_2, \ldots, X_n$ denote the observations of a random sample from a distribution that has mean μ and positive variance σ^2. Then the random variable $Y_n = \left(\sum_1^n X_i - n\mu\right)\Big/\sqrt{n}\sigma = \sqrt{n}(\bar{X}_n - \mu)/\sigma$ has a limiting distribution that is normal with mean zero and variance* 1.

Proof. In the modification of the proof, we assume the existence of the m.g.f. $M(t) = E(e^{tX})$, $-h < t < h$, of the distribution. However, this proof is essentially the same one that would be given for this theorem in a more advanced course by replacing the m.g.f. by the characteristic function $\varphi(t) = E(e^{itX})$.

The function

$$m(t) = E[e^{t(X-\mu)}] = e^{-\mu t}M(t)$$

also exists for $-h < t < h$. Since $m(t)$ is the m.g.f. for $X - \mu$, it must follow that $m(0) = 1$, $m'(0) = E(X - \mu) = 0$, and $m''(0) = E[(X - \mu)^2] = \sigma^2$. By Taylor's formula there exists a number ξ between 0 and t such that

$$\begin{aligned} m(t) &= m(0) + m'(0)t + \frac{m''(\xi)t^2}{2} \\ &= 1 + \frac{m''(\xi)t^2}{2}. \end{aligned}$$

If $\sigma^2 t^2/2$ is added and subtracted, then

$$m(t) = 1 + \frac{\sigma^2 t^2}{2} + \frac{[m''(\xi) - \sigma^2]t^2}{2}. \tag{1}$$

Next consider $M(t; n)$, where

$$\begin{aligned}
M(t; n) &= E\left[\exp\left(t\frac{\sum X_i - n\mu}{\sigma\sqrt{n}}\right)\right] \\
&= E\left[\exp\left(t\frac{X_1 - \mu}{\sigma\sqrt{n}}\right)\exp\left(t\frac{X_2 - \mu}{\sigma\sqrt{n}}\right)\cdots\exp\left(t\frac{X_n - \mu}{\sigma\sqrt{n}}\right)\right] \\
&= E\left[\exp\left(t\frac{X_1 - \mu}{\sigma\sqrt{n}}\right)\right]\cdots E\left[\exp\left(t\frac{X_n - \mu}{\sigma\sqrt{n}}\right)\right] \\
&= \left\{E\left[\exp\left(t\frac{X - \mu}{\sigma\sqrt{n}}\right)\right]\right\}^n \\
&= \left[m\left(\frac{t}{\sigma\sqrt{n}}\right)\right]^n, \quad -h < \frac{t}{\sigma\sqrt{n}} < h.
\end{aligned}$$

In Equation (1) replace t by $t/\sigma\sqrt{n}$ to obtain

$$m\left(\frac{t}{\sigma\sqrt{n}}\right) = 1 + \frac{t^2}{2n} + \frac{[m''(\xi) - \sigma^2]t^2}{2n\sigma^2},$$

where now ξ is between 0 and $t/\sigma\sqrt{n}$ with $-h\sigma\sqrt{n} < t < h\sigma\sqrt{n}$. Accordingly,

$$M(t; n) = \left\{1 + \frac{t^2}{2n} + \frac{[m''(\xi) - \sigma^2]t^2}{2n\sigma^2}\right\}^n.$$

Since $m''(t)$ is continuous at $t = 0$ and since $\xi \to 0$ as $n \to \infty$, we have

$$\lim_{n\to\infty} [m''(\xi) - \sigma^2] = 0.$$

The limit proposition cited in Section 5.3 shows that

$$\lim_{n\to\infty} M(t; n) = e^{t^2/2}$$

for all real values of t. This proves that the random variable $Y_n = \sqrt{n}(\bar{X}_n - \mu)/\sigma$ has a limiting standard normal distribution.

We interpret this theorem as saying that, when n is a large, fixed positive integer, the random variable $\bar{X}$ has an approximate normal distribution with mean μ and variance σ^2/n; and in applications we use the approximate normal p.d.f. as though it were the exact p.d.f. of $\bar{X}$.

Some illustrative examples, here and later, will help show the importance of this version of the central limit theorem.

Example 1. Let $\bar{X}$ denote the mean of a random sample of size 75 from the distribution that has the p.d.f.

$$\begin{aligned} f(x) &= 1, \qquad 0 < x < 1, \\ &= 0 \qquad \text{elsewhere.} \end{aligned}$$

It was stated in Section 5.1 that the exact p.d.f. of $\bar{X}$, say $g(\bar{x})$, is rather complicated. It can be shown that $g(\bar{x})$ has a graph when $0 < \bar{x} < 1$ that is composed of arcs of 75 different polynomials of degree 74. The computation of such a probability as $\Pr(0.45 < \bar{X} < 0.55)$ would be extremely laborious. The conditions of the theorem are satisfied, since $M(t)$ exists for all real values of t. Moreover, $\mu = \frac{1}{2}$ and $\sigma^2 = \frac{1}{12}$, so that we have approximately

$$\begin{aligned} \Pr(0.45 < \bar{X} < 0.55) &= \Pr\left[\frac{\sqrt{n}(0.45 - \mu)}{\sigma} < \frac{\sqrt{n}(\bar{X} - \mu)}{\sigma} < \frac{\sqrt{n}(0.55 - \mu)}{\sigma}\right] \\ &= \Pr[-1.5 < 30(\bar{X} - 0.5) < 1.5] \\ &= 0.866, \end{aligned}$$

from Table III in Appendix B.

Example 2. Let $X_1, X_2, \ldots, X_n$ denote a random sample from a distribution that is $b(1, p)$. Here $\mu = p$, $\sigma^2 = p(1 - p)$, and $M(t)$ exists for all real values of t. If $Y_n = X_1 + \cdots + X_n$, it is known that Y_n is $b(n, p)$. Calculation of probabilities concerning Y_n, when we do not use the Poisson approximation, can be greatly simplified by making use of the fact that $(Y_n - np)/\sqrt{np(1 - p)} = \sqrt{n}(\bar{X}_n - p)/\sqrt{p(1 - p)} = \sqrt{n}(\bar{X}_n - \mu)/\sigma$ has a limiting distribution that is normal with mean zero and variance 1. Frequently, statisticians say that Y_n, or more simply Y, has an approximate normal distribution with mean np and variance $np(1 - p)$. Even with n as small as 10, with $p = \frac{1}{2}$ so that the binomial distribution is symmetric about $np = 5$, we note in Figure 5.1 how well the normal distribution, $N(5, \frac{5}{2})$, fits the binomial distribution, $b(10, \frac{1}{2})$, where the heights of the rectangles represent the probabilities of the respective integers $0, 1, 2, \ldots, 10$. Note that the area of the rectangle whose base is $(k - 0.5, k + 0.5)$ and the area under the normal p.d.f. between $k - 0.5$ and $k + 0.5$ are *approximately* equal for each $k = 0, 1, 2, \ldots, 10$, even with $n = 10$. This example should help the reader understand Example 3.

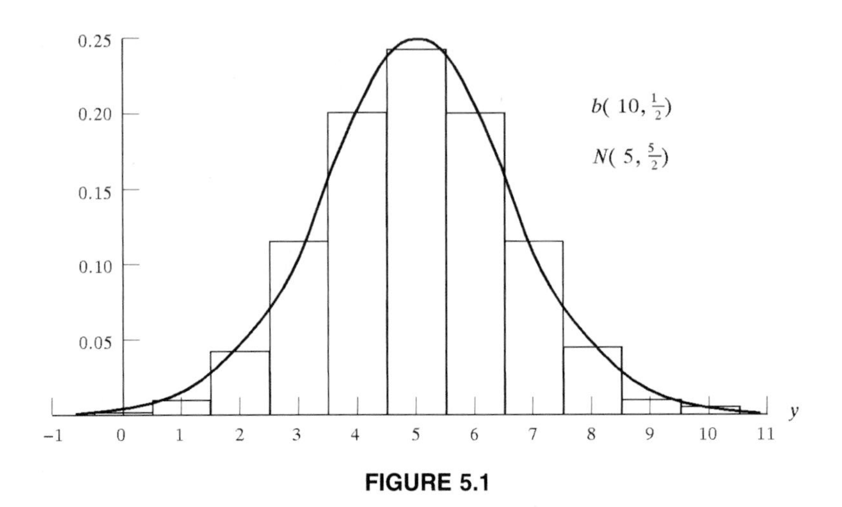

FIGURE 5.1

Example 3. With the background of Example 2, let $n = 100$ and $p = \frac{1}{2}$, and suppose that we wish to compute $\Pr(Y = 48, 49, 50, 51, 52)$. Since Y is a random variable of the discrete type, the events $Y = 48, 49, 50, 51, 52$ and $47.5 < Y < 52.5$ are equivalent. That is, $\Pr(Y = 48, 49, 50, 51, 52) = \Pr(47.5 < Y < 52.5)$. Since $np = 50$ and $np(1 - p) = 25$, the latter probability may be written

$$\Pr(47.5 < Y < 52.5) = \Pr\left(\frac{47.5 - 50}{5} < \frac{Y - 50}{5} < \frac{52.5 - 50}{5}\right)$$

$$= \Pr\left(-0.5 < \frac{Y - 50}{5} < 0.5\right).$$

Since $(Y - 50)/5$ has an approximate normal distribution with mean zero and variance 1, Table III shows this probability to be approximately 0.382.

The convention of selecting the event $47.5 < Y < 52.5$, instead of, say, $47.8 < Y < 52.3$, as the event equivalent to the event $Y = 48, 49, 50, 51, 52$ seems to have originated in the following manner: The probability, $\Pr(Y = 48, 49, 50, 51, 52)$, can be interpreted as the sum of five rectangular areas where the rectangles have bases 1 but the heights are, respectively, $\Pr(Y = 48), \ldots, \Pr(Y = 52)$. If these rectangles are so located that the midpoints of their bases are, respectively, at the points $48, 49, \ldots, 52$ on a horizontal axis, then in approximating the sum of these areas by an area bounded by the horizontal axis, the graph of a normal p.d.f., and two ordinates, it seems reasonable to take the two ordinates at the points 47.5 and 52.5.

We know that $\bar{X}$ and $\sum_{i=1}^{n} X_i$ have approximate normal distributions, provided that n is large enough. Later, we find that other statistics also have approximate normal distributions, and this is the reason that the normal distribution is so important to statisticians. That is, while not many underlying distributions are normal, the distributions of statistics calculated from random samples arising from these distributions are often very close to being normal.

Frequently, we are interested in functions of statistics that have approximate normal distributions. For illustration, Y_n of Example 2 has an approximate $N[np, np(1-p)]$. So $np(1-p)$ is an important function of p as it is the variance of Y_n. Thus, if p is unknown, we might want to estimate the variance of Y_n. Since $E(Y_n/n) = p$, we might use $n(Y_n/n)(1 - Y_n/n)$ as such an estimator and would want to know something about the latter's distribution. In particular, does it also have an approximate normal distribution? If so, what are its mean and variance? To answer questions like these, we use a procedure that is commonly called the *delta method*, which will be explained using the sample mean $\bar{X}_n$ as the statistic.

We know that $\bar{X}_n$ converges in probability to μ and $\bar{X}_n$ is approximately $N(\mu, \sigma^2/n)$. Suppose that we are interested in a function of $\bar{X}_n$, say $u(\bar{X}_n)$. Since, for large n, $\bar{X}_n$ is close to μ, we can approximate $u(\bar{X}_n)$ by the first two terms of Taylor's expansion about μ, namely

$$u(\bar{X}_n) \approx v(\bar{X}_n) = u(\mu) + (\bar{X}_n - \mu)u'(\mu),$$

where $u'(\mu)$ exists and is not zero. Since $v(\bar{X}_n)$ is a linear function of $\bar{X}_n$, it has an approximate normal distribution with mean

$$E[v(\bar{X}_n)] = u(\mu) + E[(\bar{X}_n - \mu)]u'(\mu) = u(\mu)$$

and variance

$$\text{var}\,[v(\bar{X}_n)] = [u'(\mu)]^2 \,\text{var}\,(\bar{X}_n - \mu) = [u'(\mu)]^2 \frac{\sigma^2}{n}.$$

Now, for large n, $u(\bar{X}_n)$ is approximately equal to $v(\bar{X}_n)$; so it has the same approximating distribution. That is, $u(\bar{X}_n)$ is approximately $N\{u(\mu), [u'(\mu)]^2\sigma^2/n\}$. More formally, we could say that

$$\frac{u(\bar{X}_n) - u(\mu)}{\sqrt{[u'(\mu)]^2\sigma^2/n}}$$

has a limiting standard normal distribution.

Example 4. Let Y_n (or Y for simplicity) be $b(n, p)$. Thus Y/n is approximately $N[p, p(1-p)/n)]$. Statisticians often look for functions of statistics whose variances do not depend upon the parameter. Here the variance of Y/n depends upon p. Can we find a function, say $u(Y/n)$, whose variance is essentially free of p? Since Y/n converges in probability to p, we can approximate $u(Y/n)$ by the first two terms of its Taylor's expansion about p, namely by

$$v\left(\frac{Y}{n}\right) = u(p) + \left(\frac{Y}{n} - p\right)u'(p).$$

Of course, $v(Y/n)$ is a linear function of Y/n and thus also has an approximate normal distribution; clearly, it has mean $u(p)$ and variance

$$[u'(p)]^2 \frac{p(1-p)}{n}.$$

But it is the latter that we want to be essentially free of p; thus we set it equal to a constant, obtaining the differential equation

$$u'(p) = \frac{c}{\sqrt{p(1-p)}}.$$

A solution of this is

$$u(p) = (2c)\arcsin \sqrt{p}.$$

If we take $c = \frac{1}{2}$, we have, since $u(Y/n)$ is approximately equal to $v(Y/n)$, that

$$u\left(\frac{Y}{n}\right) = \arcsin \sqrt{\frac{Y}{n}}.$$

This has an approximate normal distribution with mean $\arcsin \sqrt{p}$ and variance $1/4n$, which is free of p.

EXERCISES

5.20. Let $\bar{X}$ denote the mean of a random sample of size 100 from a distribution that is $\chi^2(50)$. Compute an approximate value of $\Pr(49 < \bar{X} < 51)$.

5.21. Let $\bar{X}$ denote the mean of a random sample of size 128 from a gamma distribution with $\alpha = 2$ and $\beta = 4$. Approximate $\Pr(7 < \bar{X} < 9)$.

5.22. Let Y be $b(72, \frac{1}{3})$. Approximate $\Pr(22 \le Y \le 28)$.

5.23. Compute an approximate probability that the mean of a random sample of size 15 from a distribution having p.d.f. $f(x) = 3x^2$, $0 < x < 1$, zero elsewhere, is between $\frac{3}{5}$ and $\frac{4}{5}$.

5.24. Let Y denote the sum of the observations of a random sample of size 12 from a distribution having p.d.f. $f(x) = \frac{1}{6}$, $x = 1, 2, 3, 4, 5, 6$, zero elsewhere. Compute an approximate value of $\Pr(36 \le Y \le 48)$.

Hint: Since the event of interest is $Y = 36, 37, \ldots, 48$, rewrite the probability as $\Pr(35.5 < Y < 48.5)$.

5.25. Let Y be $b(400, \frac{1}{5})$. Compute an approximate value of $\Pr(0.25 < Y/n)$.

5.26. If Y is $b(100, \frac{1}{2})$, approximate the value of $\Pr(Y = 50)$.

5.27. Let Y be $b(n, 0.55)$. Find the smallest value of n so that (approximately) $\Pr(Y/n > \frac{1}{2}) \geq 0.95$.

5.28. Let $f(x) = 1/x^2$, $1 < x < \infty$, zero elsewhere, be the p.d.f. of a random variable X. Consider a random sample of size 72 from the distribution having this p.d.f. Compute approximately the probability that more than 50 of the observations of the random sample are less than 3.

5.29. Forty-eight measurements are recorded to several decimal places. Each of these 48 numbers is rounded off to the nearest integer. The sum of the original 48 numbers is approximated by the sum of these integers. If we assume that the errors made by rounding off are i.i.d. and have uniform distributions over the interval $(-\frac{1}{2}, \frac{1}{2})$, compute approximately the probability that the sum of the integers is within 2 units of the true sum.

5.30. We know that $\bar{X}$ is approximately $N(\mu, \sigma^2/n)$ for large n. Find the approximate distribution of $u(\bar{X}) = \bar{X}^3$.

5.31. Let $X_1, X_2, \ldots, X_n$ be a random sample from a Poisson distribution with mean μ. Thus $Y = \sum_{i=1}^{n} X_i$ has a Poisson distribution with mean $n\mu$. Moreover, $\bar{X} = Y/n$ is approximately $N(\mu, \mu/n)$ for large n. Show that $u(Y/n) = \sqrt{Y/n}$ is a function of Y/n whose variance is essentially free of μ.

5.5 Some Theorems on Limiting Distributions

In this section, we shall present some theorems that can often be used to simplify the study of certain limiting distributions.

Theorem 4. *Let $F_n(u)$ denote the distribution function of a random variable U_n whose distribution depends upon the positive integer n. Let U_n converge in probability to the constant $c \neq 0$. The random variable U_n/c converges in probability to* 1.

The proof of this theorem is very easy and is left as an exercise.

Theorem 5. *Let $F_n(u)$ denote the distribution function of a random variable U_n whose distribution depends upon the positive integer n. Further, let U_n converge in probability to the positive constant c and let*

$\Pr(U_n < 0) = 0$ *for every n. The random variable* $\sqrt{U_n}$ *converges in probability to* $\sqrt{c}$.

Proof. We are given that the $\lim_{n\to\infty} \Pr(|U_n - c| \geq \epsilon) = 0$ for every $\epsilon > 0$.

We are to prove that the $\lim_{n\to\infty} \Pr(|\sqrt{U_n} - \sqrt{c}| \geq \epsilon') = 0$ for every $\epsilon' > 0$. Now the probability

$$\begin{aligned}\Pr(|U_n - c| \geq \epsilon) &= \Pr[|(\sqrt{U_n} - \sqrt{c})(\sqrt{U_n} + \sqrt{c})| \geq \epsilon] \\ &= \Pr\left(|\sqrt{U_n} - \sqrt{c}| \geq \frac{\epsilon}{\sqrt{U_n} + \sqrt{c}}\right) \\ &\geq \Pr\left(|\sqrt{U_n} - \sqrt{c}| \geq \frac{\epsilon}{\sqrt{c}}\right) \geq 0.\end{aligned}$$

If we let $\epsilon' = \epsilon/\sqrt{c}$, and if we take the limit, as n becomes infinite, we have

$$0 = \lim_{n\to\infty} \Pr(|U_n - c| \geq \epsilon) \geq \lim_{n\to\infty} \Pr(|\sqrt{U_n} - \sqrt{c}| \geq \epsilon') = 0$$

for every $\epsilon' > 0$. This completes the proof.

The conclusions of Theorems 4 and 5 are very natural ones and they certainly appeal to our intuition. There are many other theorems of this flavor in probability theory. As exercises, it is to be shown that if the random variables U_n and V_n converge in probability to the respective constants c and d, then U_nV_n converges in probability to the constant cd, and U_n/V_n converges in probability to the constant c/d, provided that $d \neq 0$. However, we shall accept, without proof, the following theorem, which is a modification of Slutsky's theorem.

Theorem 6. *Let* $F_n(u)$ *denote the distribution function of a random variable* U_n *whose distribution depends upon the positive integer n. Let* U_n *have a limiting distribution with distribution function* $F(u)$. *Let a random variable* V_n *converge in probability to* 1. *The limiting distribution of the random variable* $W_n = U_n/V_n$ *is the same as that of* U_n; *that is,* W_n *has a limiting distribution with distribution function* $F(w)$.

Example 1. Let Y_n denote a random variable that is $b(n, p)$, $0 < p < 1$. We know that

$$U_n = \frac{Y_n - np}{\sqrt{np(1-p)}}$$

has a limiting distribution that is $N(0, 1)$. Moreover, it has been proved that Y_n/n and $1 - Y_n/n$ converge in probability to p and $1 - p$, respectively; thus $(Y_n/n)(1 - Y_n/n)$ converges in probability to $p(1 - p)$. Then, by Theorem 4, $(Y_n/n)(1 - Y_n/n)/[p(1 - p)]$ converges in probability to 1, and Theorem 5 asserts that the following does also:

$$V_n = \left[\frac{(Y_n/n)(1 - Y_n/n)}{p(1 - p)}\right]^{1/2}.$$

Thus, in accordance with Theorem 6, the ratio $W_n = U_n/V_n$, namely

$$\frac{Y_n - np}{\sqrt{n(Y_n/n)(1 - Y_n/n)}},$$

has a limiting distribution that is $N(0, 1)$. This fact enables us to write (with n a large, fixed positive integer)

$$\Pr\left[-2 < \frac{Y - np}{\sqrt{n(Y/n)(1 - Y/n)}} < 2\right] = 0.954,$$

approximately.

Example 2. Let $\bar{X}_n$ and S_n^2 denote, respectively, the mean and the variance of a random sample of size n from a distribution that is $N(\mu, \sigma^2)$, $\sigma^2 > 0$. It has been proved that $\bar{X}_n$ converges in probability to μ and that S_n^2 converges in probability to σ^2. Theorem 5 asserts that S_n converges in probability to σ and Theorem 4 tells us that S_n/σ converges in probability to 1. In accordance with Theorem 6, the random variable $W_n = \sigma\bar{X}_n/S_n$ has the same limiting distribution as does $\bar{X}_n$. That is, $\sigma\bar{X}_n/S_n$ converges in probability to μ.

EXERCISES

5.32. Prove Theorem 4.

Hint: Note that $\Pr(|U_n/c - 1| < \epsilon) = \Pr(|U_n - c| < \epsilon|c|)$, for every $\epsilon > 0$. Then take $\epsilon' = \epsilon|c|$.

5.33. Let $\bar{X}_n$ denote the mean of a random sample of size n from a gamma distribution with parameters $\alpha = \mu > 0$ and $\beta = 1$. Show that the limiting distribution of $\sqrt{n}(\bar{X}_n - \mu)/\sqrt{\bar{X}_n}$ is $N(0, 1)$.

5.34. Let $T_n = (\bar{X}_n - \mu)/\sqrt{S_n^2/(n - 1)}$, where $\bar{X}_n$ and S_n^2 represent, respectively, the mean and the variance of a random sample of size n from a distribution that is $N(\mu, \sigma^2)$. Prove that the limiting distribution of T_n is $N(0, 1)$.

5.35. Let $X_1, \ldots, X_n$ and $Y_1, \ldots, Y_n$ be the observations of two independent random samples, each of size n, from the distributions that have the respective means μ_1 and μ_2 and the common variance σ^2. Find the limiting distribution of

$$\frac{(\bar{X}_n - \bar{Y}_n) - (\mu_1 - \mu_2)}{\sigma\sqrt{2/n}},$$

where $\bar{X}_n$ and $\bar{Y}_n$ are the respective means of the samples.

Hint: Let $\bar{Z}_n = \sum_1^n Z_i/n$, where $Z_i = X_i - Y_i$.

5.36. Let U_n and V_n converge in probability to c and d, respectively. Prove the following.

(a) The sum $U_n + V_n$ converges in probability to $c + d$.
Hint: Show that $\Pr(|U_n + V_n - c - d| \geq \epsilon) \leq \Pr(|U_n - c| + |V_n - d| \geq \epsilon) \leq \Pr(|U_n - c| \geq \epsilon/2 \text{ or } |V_n - d| \geq \epsilon/2) \leq \Pr(|U_n - c| \geq \epsilon/2) + \Pr(|V_n - d| \geq \epsilon/2)$.

(b) The product $U_n V_n$ converges in probability to cd.

(c) If $d \neq 0$, the ratio U_n/V_n converges in probability to c/d.

5.37. Let U_n converge in probability to c. If $h(u)$ is a continuous function at $u = c$, prove that $h(U_n)$ converges in probability to $h(c)$.

Hint: For each $\epsilon > 0$, there exists a $\delta > 0$ such that $\Pr[|h(U_n) - h(c)| < \epsilon] \geq \Pr[|U_n - c| < \delta]$. Why?

ADDITIONAL EXERCISES

5.38. A nail manufacturer guarantees that not more than one nail in a box of 100 nails is defective. If, in fact, the probability of each individual nail being defective is $p = 0.005$, compute the probability that:

(a) The next box of nails violates the guarantee. Use the Poisson approximation, after assuming independence.

(b) The guarantee is violated at least once in the next 25 boxes.

5.39. Let $\bar{X}_n$ and $\bar{Y}_n$ be the means of two independent random samples of size n from a distribution having variance σ^2. Determine n so that $\Pr(|\bar{X}_n - \bar{Y}_n| \leq \sigma/2) = 0.98$, approximately.

5.40. Let $X_1, X_2, \ldots, X_{25}$ be a random sample from a distribution with p.d.f. $f(x) = 6x(1 - x)$, $0 < x < 1$, zero elsewhere. Find $\Pr[0.48 < \bar{X}_n < 0.52]$ approximately.

5.41. A rolls an unbiased die 100 independent times and B rolls an unbiased die 100 independent times. What is the approximate probability that A will total at least 25 points more than B?

5.42. Compute, approximately, the probability that the sum of the observations of a random sample of size 24 from a chi-square distribution with 3 degrees of freedom is between 70 and 80.

5.43. Let X be the number of times that no heads appear on two coins when these two coins are tossed together n times. Find the smallest value of n so that $\Pr(0.24 \le X/n \le 0.26) \ge 0.954$, approximately.

5.44. Two persons have 16 and 32 dollars, respectively. They bet one dollar on each of 900 independent tosses of an unbiased coin. What is an approximation to the probability that neither person is in debt at the end of the 900 trials?

5.45. A die is rolled 720 independent times. Compute, approximately, the probability that the number of fives that appear will be between 110 and 125 inclusive.

5.46. A part is produced with a mean of 6.2 ounces and a standard deviation of 0.2 ounce. What is the probability that the weight of 100 such items is between 616 and 624 ounces?

5.47. Let $X_1, \ldots, X_{25}$ be a random sample of size 25 from a distribution having p.d.f. $f(x) = x/6$, $x = 1, 2, 3$, zero elsewhere. Approximate

$$\Pr\left(\sum_{i=1}^{25} X_i = 50, 51, \ldots, \text{ or } 60\right).$$

5.48. Say that a lot of 1000 items contains 20 defective items. A sample of size 50 is taken at random and without replacement from the lot. If 3 or fewer defective items are found in this sample, the lot is accepted. Approximate the probability of accepting this lot.

5.49. Let $X_1, X_2, \ldots, X_n$ be a random sample from a distribution having finite $E(X^m)$, $m > 0$. Show that $\sum_{i=1}^{n} X_i^m/n$ converges in probability to $E(X^m)$. Was an additional assumption needed?

5.50. It can be proved that the mean $\bar{X}_n$ of a random sample of size n from a Cauchy distribution has that same Cauchy distribution for every n. Thus $\bar{X}_n$ does not converge in probability to zero. How can this be, as earlier, under certain conditions, we proved that $\bar{X}_n$ converges in probability to the mean of the distribution?

5.51. Let Y be $\chi^2(n)$. What is the limiting distribution of $Z = \sqrt{Y} - \sqrt{n}$?

5.52. Let $\bar{X}$ be the mean of a random sample of size n from a Poisson distribution with parameter μ. Find the function $Y = u(\bar{X})$ so that Y has an approximate normal distribution with mean $u(\mu)$ and variance that is free of μ.

5.53. Let $Y_1 < Y_2 < \cdots < Y_n$ be the order statistics of a random sample $X_1, X_2, \ldots, X_n$ of size n from a distribution with distribution function $F(x)$ and p.d.f. $f(x) = F'(x)$. Say $F(\xi_p) = p$ and $f(\xi_p) > 0$. Consider the order statistic $Y_{[np]}$, where $[np]$ is the greatest integer in np.

(a) Note that the event $\sqrt{n}(Y_{[np]} - \xi_p) \le u$ is equivalent to $Z \ge [np]$, where Z is the number of X-values less than or equal to $\xi_p + u/\sqrt{n}$.

(b) Write $Z \ge np$, an approximation to $Z \ge [np]$, as

$$\frac{Z - nF(\xi_p + u/\sqrt{n})}{\sqrt{np(1-p)}} \ge \frac{-f(\xi_p)u}{\sqrt{p(1-p)}}, \quad \text{approximately,}$$

using $F(\xi_p + u/\sqrt{n}) \approx p + f(\xi_p)u/\sqrt{n}$.

(c) Since the left-hand member of the inequality in part (b) is approximately $N(0, 1)$, argue that $Y_{[np]}$ has an approximate normal distribution with mean ξ_p and variance $p(1-p)/n[f(\xi_p)]^2$.

CHAPTER 6

Introduction to Statistical Inference

6.1 Point Estimation

The first five chapters of this book deal with certain concepts and problems of probability theory. Throughout we have carefully distinguished between a sample space $\mathscr{C}$ of outcomes and the space $\mathscr{A}$ of one or more random variables defined on $\mathscr{C}$. With this chapter we begin a study of some problems in statistics and here we are more interested in the number (or numbers) by which an outcome is represented than we are in the outcome itself. Accordingly, we shall adopt a frequently used convention. We shall refer to a random variable X as the outcome of a random experiment and we shall refer to the space of X as the sample space. Were it not so awkward, we would call X the numerical outcome. Once the experiment has been performed and it is found that $X = x$, we shall call x the experimental value of X for that performance of the experiment.

This convenient terminology can be used to advantage in more general situations. To illustrate this, let a random experiment be repeated n independent times and under identical conditions. Then the random variables $X_1, X_2, \ldots, X_n$ (each of which assigns a numerical value to an outcome) constitute (Section 4.1) the observations of a random sample. If we are more concerned with the numerical representations of the outcomes than with the outcomes themselves, it seems natural to refer to $X_1, X_2, \ldots, X_n$ as the outcomes. And what more appropriate name can we give to the space of a random sample than the sample space? Once the experiment has been performed the indicated number of times and it is found that $X_1 = x_1$, $X_2 = x_2, \ldots, X_n = x_n$, we shall refer to $x_1, x_2, \ldots, x_n$ as the experimental values of $X_1, X_2, \ldots, X_n$ or as the sample data.

We shall use the terminology of the two preceding paragraphs, and in this section we shall give some examples of *statistical inference*. These examples will be built around the notion of a *point estimate* of an unknown parameter in a p.d.f.

Let a random variable X have a p.d.f. that is of known functional form but in which the p.d.f. depends upon an unknown parameter θ that may have any value in a set Ω. This will be denoted by writing the p.d.f. in the form $f(x; \theta)$, $\theta \in \Omega$. The set Ω will be called the *parameter space*. Thus we are confronted, not with one distribution of probability, but with a *family* of distributions. To each value of θ, $\theta \in \Omega$, there corresponds one member of the family. A family of probability density functions will be denoted by the symbol $\{f(x; \theta) : \theta \in \Omega\}$. Any member of this family of probability density functions will be denoted by the symbol $f(x; \theta)$, $\theta \in \Omega$. We shall continue to use the special symbols that have been adopted for the normal, the chi-square, and the binomial distributions. We may, for instance, have the family $\{N(\theta, 1) : \theta \in \Omega\}$, where Ω is the set $-\infty < \theta < \infty$. One member of this family of distributions is the distribution that is $N(0, 1)$. Any arbitrary member is $N(\theta, 1)$, $-\infty < \theta < \infty$.

Let us consider a family of probability density functions $\{f(x; \theta) : \theta \in \Omega\}$. It may be that the experimenter needs to select precisely *one* member of the family as being the p.d.f. of his random variable. That is, he needs a *point* estimate of θ. Let $X_1, X_2, \ldots, X_n$ denote a random sample from a distribution that has a p.d.f. which is one member (but which member we do not know) of the family $\{f(x; \theta) : \theta \in \Omega\}$ of probability density functions. That is, our sample arises from a distribution that has the p.d.f. $f(x; \theta): \theta \in \Omega$. Our problem is that of defining a statistic $Y_1 = u_1(X_1, X_2, \ldots, X_n)$, so that if $x_1, x_2, \ldots, x_n$ are the observed experimental values of $X_1, X_2, \ldots, X_n$, then the number $y_1 = u_1(x_1, x_2, \ldots, x_n)$ will be a good point estimate of θ.

The following illustration should help motivate one principle that is often used in finding point estimates.

Example 1. Let $X_1, X_2, \ldots, X_n$ denote a random sample from the distribution with p.d.f.

$$f(x) = \theta^x(1-\theta)^{1-x}, \qquad x = 0, 1,$$
$$= 0 \qquad \text{elsewhere},$$

where $0 \le \theta \le 1$. The probability that $X_1 = x_1, X_2 = x_2, \ldots, X_n = x_n$ is the joint p.d.f.

$$\theta^{x_1}(1-\theta)^{1-x_1}\theta^{x_2}(1-\theta)^{1-x_2}\cdots\theta^{x_n}(1-\theta)^{1-x_n} = \theta^{\Sigma x_i}(1-\theta)^{n-\Sigma x_i},$$

where x_i equals zero or 1, $i = 1, 2, \ldots, n$. This probability, which is the joint p.d.f. of $X_1, X_2, \ldots, X_n$, may be regarded as a function of θ and, when so regarded, is denoted by $L(\theta)$ and called the *likelihood function*. That is,

$$L(\theta) = \theta^{\Sigma x_i}(1-\theta)^{n-\Sigma x_i}, \qquad 0 \le \theta \le 1.$$

We might ask what value of θ would maximize the probability $L(\theta)$ of obtaining this particular observed sample $x_1, x_2, \ldots, x_n$. Certainly, this maximizing value of θ would seemingly be a good estimate of θ because it would provide the largest probability of this particular sample. Since the likelihood function $L(\theta)$ and its logarithm, $\ln L(\theta)$, are maximized for the same value θ, either $L(\theta)$ or $\ln L(\theta)$ can be used. Here

$$\ln L(\theta) = \left(\sum_1^n x_i\right)\ln\theta + \left(n - \sum_1^n x_i\right)\ln(1-\theta);$$

so we have

$$\frac{d\ln L(\theta)}{d\theta} = \frac{\sum x_i}{\theta} - \frac{n - \sum x_i}{1-\theta} = 0,$$

provided that θ is not equal to zero or 1. This is equivalent to the equation

$$(1-\theta)\sum_1^n x_i = \theta\left(n - \sum_1^n x_i\right),$$

whose solution for θ is $\sum_1^n x_i/n$. That $\sum_1^n x_i/n$ actually maximizes $L(\theta)$ and $\ln L(\theta)$ can be easily checked, even in the cases in which all of $x_1, x_2, \ldots, x_n$

equal zero together or 1 together. That is, $\sum_1^n x_i/n$ is the value of θ that maximizes $L(\theta)$. The corresponding statistic,

$$\hat{\theta} = \frac{1}{n}\sum_{i=1}^{n} X_i = \bar{X},$$

is called the *maximum likelihood estimator* of θ. The observed value of $\hat{\theta}$, namely $\sum_1^n x_i/n$, is called the *maximum likelihood estimate* of θ. For a simple example, suppose that $n = 3$, and $x_1 = 1, x_2 = 0, x_3 = 1$, then $L(\theta) = \theta^2(1 - \theta)$ and the observed $\hat{\theta} = \frac{2}{3}$ is the maximum likelihood estimate of θ.

The principle of the *method of maximum likelihood* can now be formulated easily. Consider a random sample $X_1, X_2, \ldots, X_n$ from a distribution having p.d.f. $f(x; \theta)$, $\theta \in \Omega$. The joint p.d.f. of $X_1, X_2, \ldots, X_n$ is $f(x_1; \theta)f(x_2; \theta) \cdots f(x_n; \theta)$. This joint p.d.f. may be regarded as a function of θ. When so regarded, it is called the likelihood function L of the random sample, and we write

$$L(\theta; x_1, x_2, \ldots, x_n) = f(x_1; \theta)f(x_2; \theta) \cdots f(x_n; \theta), \qquad \theta \in \Omega.$$

Suppose that we can find a nontrivial function of $x_1, x_2, \ldots, x_n$, say $u(x_1, x_2, \ldots, x_n)$, such that, when θ is replaced by $u(x_1, x_2, \ldots, x_n)$, the likelihood function L is maximized. That is, $L[u(x_1, x_2, \ldots, x_n); x_1, x_2, \ldots, x_n]$ is at least as great as $L(\theta; x_1, x_2, \ldots, x_n)$ for every $\theta \in \Omega$. Then the statistic $u(X_1, X_2, \ldots, X_n)$ will be called a *maximum likelihood estimator* (hereafter abbreviated m.l.e.) of θ and will be denoted by the symbol $\hat{\theta} = u(X_1, X_2, \ldots, X_n)$. We remark that in many instances there will be a unique m.l.e. $\hat{\theta}$ of a parameter θ, and often it may be obtained by the process of differentiation.

Example 2. Let $X_1, X_2, \ldots, X_n$ be a random sample from the normal distribution $N(\theta, 1)$, $-\infty < \theta < \infty$. Here

$$L(\theta; x_1, x_2, \ldots, x_n) = \left(\frac{1}{\sqrt{2\pi}}\right)^n \exp\left[-\sum_1^n \frac{(x_i - \theta)^2}{2}\right].$$

This function L can be maximized by setting the first derivative of L, with respect to θ, equal to zero and solving the resulting equation for θ. We note, however, that each of the functions L and $\ln L$ is maximized at the same value of θ. So it may be easier to solve

$$\frac{d \ln L(\theta; x_1, x_2, \ldots, x_n)}{d\theta} = 0.$$

For this example,

$$\frac{d \ln L(\theta; x_1, x_2, \ldots, x_n)}{d\theta} = \sum_1^n (x_i - \theta).$$

If this derivative is equated to zero, the solution for the parameter θ is $u(x_1, x_2, \ldots, x_n) = \sum_1^n x_i/n$. That $\sum_1^n x_i/n$ actually maximizes L is easily shown. Thus the statistic

$$\hat{\theta} = u(X_1, X_2, \ldots, X_n) = \frac{1}{n}\sum_1^n X_i = \bar{X}$$

is the unique m.l.e. of the mean θ.

It is interesting to note that in both Examples 1 and 2, it is true that $E(\hat{\theta}) = \theta$. That is, in each of these cases, the expected value of the estimator is equal to the corresponding parameter, which leads to the following definition.

Definition 1. Any statistic whose mathematical expectation is equal to a parameter θ is called an *unbiased* estimator of the parameter θ. Otherwise, the statistic is said to be *biased*.

Example 3. Let

$$f(x; \theta) = \frac{1}{\theta}, \quad 0 < x \le \theta, \quad 0 < \theta < \infty,$$
$$= 0 \quad \text{elsewhere},$$

and let $X_1, X_2, \ldots, X_n$ denote a random sample from this distribution. Note that we have taken $0 < x \le \theta$ instead of $0 < x < \theta$ so as to avoid a discussion of supremum versus maximum. Here

$$L(\theta; x_1, x_2, \ldots, x_n) = \frac{1}{\theta^n}, \quad 0 < x_i \le \theta,$$

which is an ever-decreasing function of θ. The maximum of such functions cannot be found by differentiation but by selecting θ as small as possible. Now $\theta \ge$ each x_i; in particular, then, $\theta \ge \max(x_i)$. Thus L can be made no larger than

$$\frac{1}{[\max(x_i)]^n}$$

and the unique m.l.e. $\hat{\theta}$ of θ in this example is the nth order statistic $\max(X_i)$. It can be shown that $E[\max(X_i)] = n\theta/(n+1)$. Thus, in this instance, the m.l.e. of the parameter θ is biased. That is, the property of unbiasedness is not in general a property of a m.l.e.

While the m.l.e. $\hat{\theta}$ of θ in Example 3 is a biased estimator, results in Chapter 5 show that the nth order statistic $\hat{\theta} = \max(X_i) = Y_n$ converges in probability to θ. Thus, in accordance with the following definition, we say that $\hat{\theta} = Y_n$ is a consistent estimator of θ.

Definition 2. Any statistic that converges in probability to a parameter θ is called a *consistent* estimator of that parameter θ.

Consistency is a desirable property of an estimator; and, in all cases of practical interest, maximum likelihood estimators are consistent.

The preceding definitions and properties are easily generalized. Let $X, Y, \ldots, Z$ denote random variables that may or may not be independent and that may or may not be identically distributed. Let the joint p.d.f. $g(x, y, \ldots, z; \theta_1, \theta_2, \ldots, \theta_m)$, $(\theta_1, \theta_2, \ldots, \theta_m) \in \Omega$, depend on m parameters. This joint p.d.f., when regarded as a function of $(\theta_1, \theta_2, \ldots, \theta_m) \in \Omega$, is called the likelihood function of the random variables. Then those functions $u_1(x, y, \ldots, z)$, $u_2(x, y, \ldots, z)$, $\ldots$, $u_m(x, y, \ldots, z)$ that maximize this likelihood function with respect to $\theta_1, \theta_2, \ldots, \theta_m$, respectively, define the maximum likelihood estimators

$$\hat{\theta}_1 = u_1(X, Y, \ldots, Z), \qquad \hat{\theta}_2 = u_2(X, Y, \ldots, Z), \ldots,$$
$$\hat{\theta}_m = u_m(X, Y, \ldots, Z)$$

of the m parameters.

Example 4. Let $X_1, X_2, \ldots, X_n$ denote a random sample from a distribution that is $N(\theta_1, \theta_2)$, $-\infty < \theta_1 < \infty, 0 < \theta_2 < \infty$. We shall find $\hat{\theta}_1$ and $\hat{\theta}_2$, the maximum likelihood estimators of θ_1 and θ_2. The logarithm of the likelihood function may be written in the form

$$\ln L(\theta_1, \theta_2; x_1, \ldots, x_n) = -\frac{\sum_1^n (x_i - \theta_1)^2}{2\theta_2} - \frac{n \ln (2\pi\theta_2)}{2}.$$

We observe that we may maximize by differentiation. We have

$$\frac{\partial \ln L}{\partial \theta_1} = \frac{\sum_1^n (x_i - \theta_1)}{\theta_2}, \qquad \frac{\partial \ln L}{\partial \theta_2} = \frac{\sum_1^n (x_i - \theta_1)^2}{2\theta_2^2} - \frac{n}{2\theta_2}.$$

If we equate these partial derivatives to zero and solve simultaneously the two equations thus obtained, the solutions for θ_1 and θ_2 are found to be $\sum_1^n x_i/n = \bar{x}$ and $\sum_1^n (x_i - \bar{x})^2/n = s^2$, respectively. It can be verified that these solutions maximize L. Thus the maximum likelihood estimators of $\theta_1 = \mu$ and $\theta_2 = \sigma^2$ are, respectively, the mean and the variance of the sample, namely $\hat{\theta}_1 = \bar{X}$ and $\hat{\theta}_2 = S^2$. Whereas $\hat{\theta}_1$ is an unbiased estimator of θ_1, the estimator $\hat{\theta}_2 = S^2$ is biased because

$$E(\hat{\theta}_2) = \frac{\sigma^2}{n} E\left(\frac{n\hat{\theta}_2}{\sigma^2}\right) = \frac{\sigma^2}{n} E\left(\frac{nS^2}{\sigma^2}\right) = \frac{(n-1)\sigma^2}{n} = \frac{(n-1)\theta_2}{n}.$$

However, in Chapter 5 it has been shown that $\hat{\theta}_1 = \bar{X}$ and $\hat{\theta}_2 = S^2$ converge in probability to θ_1 and θ_2, respectively, and thus they are consistent estimators of θ_1 and θ_2.

Suppose that we wish to estimate a function of θ, say $h(\theta)$. For convenience, let us say that $\eta = h(\theta)$ defines a one-to-one transformation. Then the value of η, say $\hat{\eta}$, that maximizes the likelihood function $L(\theta)$, or equivalently $L[\theta = h^{-1}(\eta)]$, is selected so that $\hat{\theta} = h^{-1}(\hat{\eta})$, where $\hat{\theta}$ is the m.l.e. of θ. Thus $\hat{\eta}$ is taken so that $\hat{\eta} = h(\hat{\theta})$; that is,

$$\widehat{h(\theta)} = h(\hat{\theta}).$$

This result is called the *invariance property of a maximum likelihood estimator*. For illustration, if $\eta = \theta^3$, where θ is the mean of $N(\theta, 1)$, then $\hat{\eta} = \bar{X}^3$. While there is a little complication if $h(\theta)$ is not one-to-one, we still use the fact that $\hat{\eta} = h(\hat{\theta})$. Thus if $\bar{X}$ is the mean of the sample from $b(1, \theta)$, so that $\hat{\theta} = \bar{X}$ and if $\eta = \theta(1-\theta)$, then $\hat{\eta} = \bar{X}(1-\bar{X})$. These ideas can be extended to more than one parameter. For illustration, in Example 4, if $\eta = \theta_1 + 2\sqrt{\theta_2}$, then $\hat{\eta} = \bar{X} + 2S$.

Sometimes it is impossible to find maximum likelihood estimators in a convenient closed form and numerical methods must be used to maximize the likelihood function. For illustration, suppose that $X_1, X_2, \ldots, X_n$ is a random sample from a gamma distribution with parameters $\alpha = \theta_1$ and $\beta = \theta_2$, where $\theta_1 > 0$, $\theta_2 > 0$. It is difficult to maximize

$$L(\theta_1, \theta_2; x_1, \ldots, x_n) = \left[\frac{1}{\Gamma(\theta_1)\theta_2^{\theta_1}}\right]^n (x_1 x_2 \cdots x_n)^{\theta_1 - 1} \exp\left(-\sum_1^n x_i/\theta_2\right)$$

with respect to θ_1 and θ_2, owing to the presence of the gamma function $\Gamma(\theta_1)$. Thus numerical methods must be used to maximize L once $x_1, x_2, \ldots, x_n$ are observed.

There are other ways, however, to obtain easily point estimates of

θ_1 and θ_2. For illustration, in the gamma distribution situation, let us simply equate the first two moments of the distribution to the corresponding moments of the sample. This seems like a reasonable way in which to find estimators, since the empirical distribution $F_n(x)$ converges in probability to $F(x)$, and hence corresponding moments should be about equal. Here in this illustration we have

$$\theta_1\theta_2 = \bar{X}, \quad \theta_1\theta_2^2 = S^2,$$

the solutions of which are

$$\tilde{\theta}_1 = \frac{\bar{X}^2}{S^2} \quad \text{and} \quad \tilde{\theta}_2 = \frac{S^2}{\bar{X}}.$$

We say that these latter two statistics, $\tilde{\theta}_1$ and $\tilde{\theta}_2$, are respective estimators of θ_1 and θ_2 found by the *method of moments*.

To generalize the discussion of the preceding paragraph, let $X_1, X_2, \ldots, X_n$ be a random sample of size n from a distribution with p.d.f. $f(x; \theta_1, \theta_2, \ldots, \theta_r)$, $(\theta_1, \ldots, \theta_r) \in \Omega$. The expectation $E(X^k)$ is frequently called the kth moment of the distribution, $k = 1, 2, 3, \ldots$. The sum $M_k = \sum_1^n X_i^k/n$ is the kth moment of the sample, $k = 1, 2, 3, \ldots$. The method of moments can be described as follows. Equate $E(X^k)$ to M_k, beginning with $k = 1$ and continuing until there are enough equations to provide unique solutions for $\theta_1, \theta_2, \ldots, \theta_r$, say $h_i(M_1, M_2, \ldots)$, $i = 1, 2, \ldots, r$, respectively. It should be noted that this could be done in an equivalent manner by equating $\mu = E(X)$ to $\bar{X}$ and $E[(X - \mu)^k]$ to $\sum_1^n (X_i - \bar{X})^k/n$, $k = 2, 3$, and so on until unique solutions for $\theta_1, \theta_2, \ldots, \theta_r$ are obtained. This alternative procedure was used in the preceding illustration. In most practical cases, the estimator $\tilde{\theta}_i = h_i(M_1, M_2, \ldots)$ of θ_i, found by the method of moments, is a consistent estimator of θ_i, $i = 1, 2, \ldots, r$.

EXERCISES

6.1. Let $X_1, X_2, \ldots, X_n$ represent a random sample from each of the distributions having the following probability density functions:

(a) $f(x; \theta) = \theta^x e^{-\theta}/x!$, $x = 0, 1, 2, \ldots$, $0 \le \theta < \infty$, zero elsewhere, where $f(0; 0) = 1$.

(b) $f(x; \theta) = \theta x^{\theta - 1}$, $0 < x < 1$, $0 < \theta < \infty$, zero elsewhere.

(c) $f(x; \theta) = (1/\theta)e^{-x/\theta}$, $0 < x < \infty$, $0 < \theta < \infty$, zero elsewhere.

(d) $f(x; \theta) = \frac{1}{2}e^{-|x - \theta|}$, $-\infty < x < \infty$, $-\infty < \theta < \infty$.
(e) $f(x; \theta) = e^{-(x - \theta)}$, $\theta \le x < \infty$, $-\infty < \theta < \infty$, zero elsewhere.
In each case find the m.l.e. $\hat{\theta}$ of θ.

6.2. Let $X_1, X_2, \ldots, X_n$ be i.i.d., each with the distribution having p.d.f. $f(x; \theta_1, \theta_2) = (1/\theta_2)e^{-(x - \theta_1)/\theta_2}$, $\theta_1 \le x < \infty$, $-\infty < \theta_1 < \infty$, $0 < \theta_2 < \infty$, zero elsewhere. Find the maximum likelihood estimators of θ_1 and θ_2.

6.3. Let $Y_1 < Y_2 < \cdots < Y_n$ be the order statistics of a random sample from a distribution with p.d.f. $f(x; \theta) = 1$, $\theta - \frac{1}{2} \le x \le \theta + \frac{1}{2}$, $-\infty < \theta < \infty$, zero elsewhere. Show that every statistic $u(X_1, X_2, \ldots, X_n)$ such that

$$Y_n - \tfrac{1}{2} \le u(X_1, X_2, \ldots, X_n) \le Y_1 + \tfrac{1}{2}$$

is a m.l.e. of θ. In particular, $(4Y_1 + 2Y_n + 1)/6$, $(Y_1 + Y_n)/2$, and $(2Y_1 + 4Y_n - 1)/6$ are three such statistics. Thus uniqueness is not in general a property of a m.l.e.

6.4. Let X_1, X_2, and X_3 have the multinomial distribution in which $n = 25$, $k = 4$, and the unknown probabilities are θ_1, θ_2, and θ_3, respectively. Here we can, for convenience, let $X_4 = 25 - X_1 - X_2 - X_3$ and $\theta_4 = 1 - \theta_1 - \theta_2 - \theta_3$. If the observed values of the random variables are $x_1 = 4$, $x_2 = 11$, and $x_3 = 7$, find the maximum likelihood estimates of θ_1, θ_2, and θ_3.

6.5. The *Pareto distribution* is frequently used as a model in study of incomes and has the distribution function

$$F(x; \theta_1, \theta_2) = 1 - (\theta_1/x)^{\theta_2}, \qquad \theta_1 \le x, \text{ zero elsewhere,}$$

$$\text{where } \theta_1 > 0 \quad \text{and} \quad \theta_2 > 0.$$

If $X_1, X_2, \ldots, X_n$ is a random sample from this distribution, find the maximum likelihood estimators of θ_1 and θ_2.

6.6. Let Y_n be a statistic such that $\lim_{n\to\infty} E(Y_n) = \theta$ and $\lim_{n\to\infty} \sigma^2_{Y_n} = 0$. Prove that Y_n is a consistent estimator of θ.

Hint: $\Pr(|Y_n - \theta| \ge \epsilon) \le E[(Y_n - \theta)^2]/\epsilon^2$ and $E[(Y_n - \theta)^2] = [E(Y_n - \theta)]^2 + \sigma^2_{Y_n}$. Why?

6.7. For each of the distributions in Exercise 6.1, find an estimator of θ by the method of moments and show that it is consistent.

6.8. If a random sample of size n is taken from a distribution having p.d.f. $f(x; \theta) = 2x/\theta^2$, $0 < x \le \theta$, zero elsewhere, find:
(a) The m.l.e. $\hat{\theta}$ for θ.
(b) The constant c so that $E(c\hat{\theta}) = \theta$.
(c) The m.l.e. for the median of the distribution.

6.9. Let $X_1, X_2, \ldots, X_n$ be i.i.d., each with a distribution with p.d.f. $f(x; \theta) = (1/\theta)e^{-x/\theta}$, $0 < x < \infty$, zero elsewhere. Find the m.l.e. of $\Pr(X \le 2)$.

6.10. Let X have a binomial distribution with parameters n and p. The variance of X/n is $p(1 - p)/n$; this is sometimes estimated by the m.l.e. $\dfrac{X}{n}\left(1 - \dfrac{X}{n}\right)\Big/n$. Is this an unbiased estimator of $p(1 - p)/n$? If not, can you construct one by multiplying this one by a constant?

6.11. Let the table

x	0	1	2	3	4	5
Frequency	6	10	14	13	6	1

represent a summary of a sample of size 50 from a binomial distribution having $n = 5$. Find the m.l.e. of $\Pr(X \ge 3)$.

6.12. Let $Y_1 < Y_2 < \cdots < Y_n$ be the order statistics of a random sample of size n from the uniform distribution of the continuous type over the closed interval $[\theta - \rho, \theta + \rho]$. Find the maximum likelihood estimators for θ and ρ. Are these two unbiased estimators?

6.13. Let X_1, X_2, X_3, X_4, X_5 be a random sample from a Cauchy distribution with median θ, that is, with p.d.f.

$$f(x; \theta) = \frac{1}{\pi}\frac{1}{1 + (x - \theta)^2}, \qquad -\infty < x < \infty,$$

where $-\infty < \theta < \infty$. If $x_1 = -1.94$, $x_2 = 0.59$, $x_3 = -5.98$, $x_4 = -0.08$, $x_5 = -0.77$, find by numerical methods the m.l.e. of θ.

6.2 Confidence Intervals for Means

Suppose that we are willing to accept as a fact that the (numerical) outcome X of a random experiment is a random variable that has a normal distribution with known variance σ^2 but unknown mean μ. That is, μ is some constant, but its value is unknown. To elicit some information about μ, we decide to repeat the random experiment under identical conditions n independent times, n being a fixed positive integer. Let the random variables $X_1, X_2, \ldots, X_n$ denote, respectively, the outcomes to be obtained on these n repetitions of the experiment. If our assumptions are fulfilled, we then have under consideration a random sample $X_1, X_2, \ldots, X_n$ from a distribution that is $N(\mu, \sigma^2)$, σ^2 known. Consider the maximum likelihood estimator of μ, namely $\hat{\mu} = \bar{X}$. Of course, $\bar{X}$ is $N(\mu, \sigma^2/n)$ and $(\bar{X} - \mu)/(\sigma/\sqrt{n})$ is $N(0, 1)$. Thus

$$\Pr\left(-2 < \frac{\bar{X} - \mu}{\sigma/\sqrt{n}} < 2\right) = 0.954.$$

However, the events

$$-2 < \frac{\bar{X} - \mu}{\sigma/\sqrt{n}} < 2,$$

$$\frac{-2\sigma}{\sqrt{n}} < \bar{X} - \mu < \frac{2\sigma}{\sqrt{n}},$$

and

$$\bar{X} - \frac{2\sigma}{\sqrt{n}} < \mu < \bar{X} + \frac{2\sigma}{\sqrt{n}}$$

are equivalent. Thus these events have the same probability. That is,

$$\Pr\left(\bar{X} - \frac{2\sigma}{\sqrt{n}} < \mu < \bar{X} + \frac{2\sigma}{\sqrt{n}}\right) = 0.954.$$

Since σ is a known number, each of the random variables $\bar{X} - 2\sigma/\sqrt{n}$ and $\bar{X} + 2\sigma/\sqrt{n}$ is a statistic. The interval $(\bar{X} - 2\sigma/\sqrt{n}, \bar{X} + 2\sigma/\sqrt{n})$ is a random interval. In this case, both end points of the interval are statistics. The immediately preceding probability statement can be read: Prior to the repeated independent performances of the random experiment, the probability is 0.954 that the random interval $(\bar{X} - 2\sigma/\sqrt{n}, \bar{X} + 2\sigma/\sqrt{n})$ includes the unknown fixed point (parameter) μ.

Up to this point, only probability has been involved; the determination of the p.d.f. of $\bar{X}$ and the determination of the random interval were problems of probability. Now the problem becomes statistical. Suppose the experiment yields $X_1 = x_1, X_2 = x_2, \ldots, X_n = x_n$. Then the sample value of $\bar{X}$ is $\bar{x} = (x_1 + x_2 + \cdots + x_n)/n$, a known number. Moreover, since σ is known, the interval $(\bar{x} - 2\sigma/\sqrt{n}, \bar{x} + 2\sigma/\sqrt{n})$ has known endpoints. Obviously, we cannot say that 0.954 is the probability that the particular interval $(\bar{x} - 2\sigma/\sqrt{n}, \bar{x} + 2\sigma/\sqrt{n})$ includes the parameter μ, for μ, although unknown, is some constant, and this particular interval either does or does not include μ. However, the fact that we had such a high probability, prior to the performance of the experiment, that the random interval $(\bar{X} - 2\sigma/\sqrt{n}, \bar{X} + 2\sigma/\sqrt{n})$ includes the fixed point (parameter) μ, leads us to have some reliance on the particular interval $(\bar{x} - 2\sigma/\sqrt{n}, \bar{x} + 2\sigma/\sqrt{n})$. This reliance is reflected by calling the known interval $(\bar{x} - 2\sigma/\sqrt{n}, \bar{x} + 2\sigma/\sqrt{n})$ a 95.4 percent *confidence interval* for μ. The number 0.954 is called the *confidence coefficient*. The confidence coefficient is equal to the probability that the random interval includes the parameter. One may, of course, obtain an 80, a 90, or a 99 percent confidence interval for μ by using 1.282, 1.645, or 2.576, respectively, instead of the constant 2.

A statistical inference of this sort is an example of *interval estimation* of a parameter. Note that the interval estimate of μ is found by taking a good (here maximum likelihood) estimate $\bar{x}$ of μ and adding and subtracting twice the standard deviation of $\bar{X}$, namely $2\sigma/\sqrt{n}$, which is small if n is large. If σ were not known, the end points of the random interval would not be statistics. Although the probability statement about the random interval remains valid, the sample data would not yield an interval with known end points.

Example 1. If in the preceding discussion $n = 40$, $\sigma^2 = 10$, and $\bar{x} = 7.164$, then $(7.164 - 1.282\sqrt{\frac{10}{40}}, 7.164 + 1.282\sqrt{\frac{10}{40}})$, or $(6.523, 7.805)$, is an 80 percent confidence interval for μ. Thus we have an interval estimate of μ.

In the next example we show how the central limit theorem may be used to help us find an approximate confidence interval for μ when our sample arises from a distribution that is not normal.

Example 2. Let $\bar{X}$ denote the mean of a random sample of size 25 from a distribution having variance $\sigma^2 = 100$, and mean μ. Since $\sigma/\sqrt{n} = 2$, then approximately

$$\Pr\left(-1.96 < \frac{\bar{X} - \mu}{2} < 1.96\right) = 0.95,$$

or

$$\Pr(\bar{X} - 3.92 < \mu < \bar{X} + 3.92) = 0.95.$$

Let the observed mean of the sample be $\bar{x} = 67.53$. Accordingly, the interval from $\bar{x} - 3.92 = 63.61$ to $\bar{x} + 3.92 = 71.45$ is an approximate 95 percent confidence interval for the mean μ.

Let us now turn to the problem of finding a confidence interval for the mean μ of a normal distribution when we are not so fortunate as to know the variance σ^2. From Section 4.8, we know that

$$T = \frac{\sqrt{n}(\bar{X} - \mu)/\sigma}{\sqrt{nS^2/[\sigma^2(n-1)]}} = \frac{\bar{X} - \mu}{S/\sqrt{n-1}}$$

has a t-distribution with $n - 1$ degrees of freedom, whatever the value of $\sigma^2 > 0$. For a given positive integer n and a probability of 0.95, say, we can find a number b from Table IV in Appendix B, such that

$$\Pr\left(-b < \frac{\bar{X} - \mu}{S/\sqrt{n-1}} < b\right) = 0.95,$$

which can be written in the form

$$\Pr\left(\bar{X} - \frac{bS}{\sqrt{n-1}} < \mu < \bar{X} + \frac{bS}{\sqrt{n-1}}\right) = 0.95.$$

Then the interval $[\bar{X} - (bS/\sqrt{n-1}), \bar{X} + (bS/\sqrt{n-1})]$ is a random interval having probability 0.95 of including the unknown fixed point (parameter) μ. If the experimental values of $X_1, X_2, \ldots, X_n$ are $x_1, x_2, \ldots, x_n$ with $s^2 = \sum_1^n (x_i - \bar{x})^2/n$, where $\bar{x} = \sum_1^n x_i/n$, then the interval $[\bar{x} - (bs/\sqrt{n-1}), \bar{x} + (bs/\sqrt{n-1})]$ is a 95 percent confidence interval for μ for every $\sigma^2 > 0$. Again this interval estimate of μ is found by adding and subtracting a quantity, here $bs/\sqrt{n-1}$, to the point estimate $\bar{x}$.

Example 3. If in the preceding discussion $n = 10$, $\bar{x} = 3.22$, and $s = 1.17$, then the interval $[3.22 - (2.262)(1.17)/\sqrt{9},\ 3.22 + (2.262)(1.17)/\sqrt{9}]$ or $(2.34, 4.10)$ is a 95 percent confidence interval for μ.

Remark. If one wishes to find a confidence interval for μ and if the variance σ^2 of the nonnormal distribution is unknown (unlike Example 2 of this section), he may with large samples proceed as follows. If certain weak conditions are satisfied, then S^2, the variance of a random sample of size $n \geq 2$, converges in probability to σ^2. Then in

$$\frac{\sqrt{n}(\bar{X} - \mu)/\sigma}{\sqrt{nS^2/(n-1)\sigma^2}} = \frac{\sqrt{n-1}(\bar{X} - \mu)}{S}$$

the numerator of the left-hand member has a limiting distribution that is $N(0, 1)$ and the denominator of that member converges in probability to 1. Thus $\sqrt{n-1}(\bar{X} - \mu)/S$ has a limiting distribution that is $N(0, 1)$. This fact enables us to find approximate confidence intervals for μ when our conditions are satisfied. This procedure works particularly well when the underlying nonnormal distribution is symmetric, because then $\bar{X}$ and S^2 are uncorrelated (the proof of which is beyond the level of the text). As the underlying distribution becomes more skewed, however, the sample size must be larger to achieve good approximations to the desired probabilities. A similar procedure can be followed in the next section when seeking confidence intervals for the difference of the means of two nonnormal distributions.

We shall now consider the problem of determining a confidence interval for the unknown parameter p of a binomial distribution when the parameter n is known. Let Y be $b(n, p)$, where $0 < p < 1$ and n is known. Then p is the mean of Y/n. We shall use a result of Example 1, Section 5.5, to find an approximate 95.4 percent confidence interval for the mean p. There we found that

$$\Pr\left[-2 < \frac{Y - np}{\sqrt{n(Y/n)(1 - Y/n)}} < 2\right] = 0.954,$$

approximately. Since

$$\frac{Y - np}{\sqrt{n(Y/n)(1 - Y/n)}} = \frac{(Y/n) - p}{\sqrt{(Y/n)(1 - Y/n)/n}},$$

the probability statement above can easily be written in the form

$$\Pr\left[\frac{Y}{n} - 2\sqrt{\frac{(Y/n)(1 - Y/n)}{n}} < p < \frac{Y}{n} + 2\sqrt{\frac{(Y/n)(1 - Y/n)}{n}}\right] = 0.954,$$

approximately. Thus, for large n, if the experimental value of Y is y, the interval

$$\left[\frac{y}{n} - 2\sqrt{\frac{(y/n)(1 - y/n)}{n}}, \quad \frac{y}{n} + 2\sqrt{\frac{(y/n)(1 - y/n)}{n}}\right]$$

provides an approximate 95.4 percent confidence interval for p.

A more complicated approximate 95.4 percent confidence interval can be obtained from the fact that $Z = (Y - np)/\sqrt{np(1 - p)}$ has a limiting distribution that is $N(0, 1)$, and the fact that the event $-2 < Z < 2$ is equivalent to the event

$$\frac{Y + 2 - 2\sqrt{[Y(n - Y)/n] + 1}}{n + 4} < p < \frac{Y + 2 + 2\sqrt{[Y(n - Y)/n] + 1}}{n + 4}. \tag{1}$$

The first of these facts was established in Chapter 5, and the proof of inequalities (1) is left as an exercise. Thus an experimental value y of Y may be used in inequalities (1) to determine an approximate 95.4 percent confidence interval for p.

If one wishes a 95 percent confidence interval for p that does not depend upon limiting distribution theory, he or she may use the following approach. (This approach is quite general and can be used in other instances; see Exercise 6.21.) Determine two *increasing* functions of p, say $c_1(p)$ and $c_2(p)$, such that for each value of p we have, at least approximately,

$$\Pr[c_1(p) < Y < c_2(p)] = 0.95.$$

The reason that this may be approximate is due to the fact that Y has a distribution of the discrete type and thus it is, in general, impossible to achieve the probability 0.95 exactly. With $c_1(p)$ and $c_2(p)$ increasing functions, they have single-valued inverses, say $d_1(y)$ and $d_2(y)$, respectively. Thus the events $c_1(p) < Y < c_2(p)$ and $d_2(Y) < p < d_1(Y)$ are equivalent and we have, at least approximately,

$$\Pr[d_2(Y) < p < d_1(Y)] = 0.95.$$

In the case of the binomial distribution, the functions $c_1(p)$, $c_2(p)$, $d_2(y)$, and $d_1(y)$ cannot be found explicitly, but a number of books provide tables of $d_2(y)$ and $d_1(y)$ for various values of n.

Example 4. If, in the preceding discussion, we take $n = 100$ and $y = 20$, the first approximate 95.4 percent confidence interval is given by $(0.2 - 2\sqrt{(0.2)(0.8)/100},\ 0.2 + 2\sqrt{(0.2)(0.8)/100})$ or $(0.12, 0.28)$. The approximate 95.4 percent confidence interval provided by inequalities (1) is

$$\left(\frac{22 - 2\sqrt{(1600/100) + 1}}{104}, \frac{22 + 2\sqrt{(1600/100) + 1}}{104}\right)$$

or $(0.13, 0.29)$. By referring to the appropriate tables found elsewhere, we find that an approximate 95 percent confidence interval has the limits $d_2(20) = 0.13$ and $d_1(20) = 0.29$. Thus, in this example, we see that all three methods yield results that are in substantial agreement.

Remark. The fact that the variance of Y/n is a function of p caused us some difficulty in finding a confidence interval for p. Another way of handling the problem is to try to find a function $u(Y/n)$ of Y/n, whose variance is essentially free of p. In Section 5.4, we proved that

$$u\left(\frac{Y}{n}\right) = \arcsin\sqrt{\frac{Y}{n}}$$

has an approximate normal distribution with mean $\arcsin\sqrt{p}$ and variance $1/4n$. Hence we could find an approximate 95.4 percent confidence interval by using

$$\Pr\left(-2 < \frac{\arcsin\sqrt{Y/n} - \arcsin\sqrt{p}}{\sqrt{1/4n}} < 2\right) = 0.954$$

and solving the inequalities for p.

Example 5. Suppose that we sample from a distribution with unknown mean μ and variance $\sigma^2 = 225$. We want to find the sample size n so that $\bar{x} \pm 1$ (which means $\bar{x} - 1$ to $\bar{x} + 1$) serves as a 95 percent confidence interval for μ. Using the fact that the sample mean of the observations, $\bar{X}$, is approximately $N(\mu, \sigma^2/n)$, we see that the interval given by $\bar{x} \pm 1.96(15/\sqrt{n})$ will serve as an approximate 95 percent confidence interval for μ. That is, we want

$$1.96\left(\frac{15}{\sqrt{n}}\right) = 1$$

or, equivalently,

$$\sqrt{n} = 29.4, \qquad \text{and thus} \qquad n \approx 864.36$$

or $n = 865$ because n must be an integer. Suppose, however, we could not afford to take 865 observations. In that case, the accuracy or confidence level could possibly be relaxed some. For illustration, rather than requiring $\bar{x} \pm 1$ to be a 95 percent confidence interval for μ, possibly $\bar{x} \pm 2$ would be a satisfactory 80 percent one. If this modification is acceptable, we now have

$$1.282\left(\frac{15}{\sqrt{n}}\right) = 2$$

or, equivalently,

$$\sqrt{n} = 9.615 \qquad \text{and} \qquad n \approx 92.4.$$

Since n must be an integer, we would probably use 93 in practice. Most likely, the persons involved in this project would find this is a more reasonable sample size.

EXERCISES

6.14. Let the observed value of the mean $\bar{X}$ of a random sample of size 20 from a distribution that is $N(\mu, 80)$ be 81.2. Find a 95 percent confidence interval for μ.

6.15. Let $\bar{X}$ be the mean of a random sample of size n from a distribution that is $N(\mu, 9)$. Find n such that $\Pr(\bar{X} - 1 < \mu < \bar{X} + 1) = 0.90$, approximately.

6.16. Let a random sample of size 17 from the normal distribution $N(\mu, \sigma^2)$ yield $\bar{x} = 4.7$ and $s^2 = 5.76$. Determine a 90 percent confidence interval for μ.

6.17. Let $\bar{X}$ denote the mean of a random sample of size n from a distribution that has mean μ and variance $\sigma^2 = 10$. Find n so that the probability is approximately 0.954 that the random interval $(\bar{X} - \frac{1}{2}, \bar{X} + \frac{1}{2})$ includes μ.

6.18. Let $X_1, X_2, \ldots, X_9$ be a random sample of size 9 from a distribution that is $N(\mu, \sigma^2)$.

(a) If σ is known, find the length of a 95 percent confidence interval for μ if this interval is based on the random variable $\sqrt{9}(\bar{X} - \mu)/\sigma$.

(b) If σ is unknown, find the expected value of the length of a 95 percent confidence interval for μ if this interval is based on the random variable $\sqrt{8}(\bar{X} - \mu)/S$.
Hint: Write $E(S) = (\sigma/\sqrt{n})E[(nS^2/\sigma^2)^{1/2}]$.

(c) Compare these two answers.

6.19. Let $X_1, X_2, \ldots, X_n, X_{n+1}$ be a random sample of size $n + 1, n > 1$, from a distribution that is $N(\mu, \sigma^2)$. Let $\bar{X} = \sum_1^n X_i/n$ and $S^2 = \sum_1^n (X_i - \bar{X})^2/n$. Find the constant c so that the statistic $c(\bar{X} - X_{n+1})/S$ has a t-distribution. If $n = 8$, determine k such that $\Pr(\bar{X} - kS < X_9 < \bar{X} + kS) = 0.80$. The observed interval $(\bar{x} - ks, \bar{x} + ks)$ is often called an 80 percent *prediction interval* for X_9.

6.20. Let Y be $b(300, p)$. If the observed value of Y is $y = 75$, find an approximate 90 percent confidence interval for p.

6.21. Let $\bar{X}$ be the mean of a random sample of size n from a distribution that is $N(\mu, \sigma^2)$, where the positive variance σ^2 is known. Use the fact that $\Phi(2) - \Phi(-2) = 0.954$ to find, for each μ, $c_1(\mu)$ and $c_2(\mu)$ such that $\Pr[c_1(\mu) < \bar{X} < c_2(\mu)] = 0.954$. Note that $c_1(\mu)$ and $c_2(\mu)$ are increasing functions of μ. Solve for the respective functions $d_1(\bar{x})$ and $d_2(\bar{x})$; thus we also have that $\Pr[d_2(\bar{X}) < \mu < d_1(\bar{X})] = 0.954$. Compare this with the answer obtained previously in the text.

6.22. In the notation of the discussion of the confidence interval for p, show that the event $-2 < Z < 2$ is equivalent to inequalities (1).

Hint: First observe that $-2 < Z < 2$ is equivalent to $Z^2 < 4$, which can be written as an inequality involving a quadratic expression in p.

6.23. Let $\bar{X}$ denote the mean of a random sample of size 25 from a gamma-type distribution with $\alpha = 4$ and $\beta > 0$. Use the central limit theorem to find an approximate 0.954 confidence interval for μ, the mean of the gamma distribution.

Hint: Base the confidence interval on the random variable $(\bar{X} - 4\beta)/(4\beta^2/25)^{1/2} = 5\bar{X}/2\beta - 10$.

6.24. Let $\bar{x}$ be the observed mean of a random sample of size n from a distribution having mean μ and known variance σ^2. Find n so that $\bar{x} - \sigma/4$ to $\bar{x} + \sigma/4$ is an approximate 95 percent confidence interval for μ.

6.25. Assume a binomial model for a certain random variable. If we desire a 90 percent confidence interval for p that is at most 0.02 in length, find n.

Hint: Note that $\sqrt{(y/n)(1 - y/n)} \le \sqrt{(\frac{1}{2})(1 - \frac{1}{2})}$.

6.26. It is known that a random variable X has a Poisson distribution with parameter μ. A sample of 200 observations from this population has a mean equal to 3.4. Construct an approximate 90 percent confidence interval for μ.

6.27. Let $Y_1 < Y_2 < \cdots < Y_n$ denote the order statistics of a random sample of size n from a distribution that has p.d.f. $f(x) = 3x^2/\theta^3$, $0 < x < \theta$, zero elsewhere.
(a) Show that $\Pr(c < Y_n/\theta < 1) = 1 - c^{3n}$, where $0 < c < 1$.
(b) If n is 4 and if the observed value of Y_4 is 2.3, what is a 95 percent confidence interval for θ?

6.28. Let $X_1, X_2, \ldots, X_n$ be a random sample from $N(\mu, \sigma^2)$, where both parameters μ and σ^2 are unknown. A *confidence interval* for σ^2 can be found as follows. We know that nS^2/σ^2 is $\chi^2(n-1)$. Thus we can find constants a and b so that $\Pr(nS^2/\sigma^2 < b) = 0.975$ and $\Pr(a < nS^2/\sigma^2 < b) = 0.95$.
(a) Show that this second probability statement can be written as $\Pr(nS^2/b < \sigma^2 < nS^2/a) = 0.95$.
(b) If $n = 9$ and $s^2 = 7.63$, find a 95 percent confidence interval for σ^2.
(c) If μ is known, how would you modify the preceding procedure for finding a confidence interval for σ^2?

6.29. Let $X_1, X_2, \ldots, X_n$ be a random sample from a gamma distribution with known parameter $\alpha = 3$ and unknown $\beta > 0$. Discuss the construction of a confidence interval for β.

Hint: What is the distribution of $2 \sum_{i=1}^{n} X_i/\beta$? Follow the procedure outlined in Exercise 6.28.

6.3 Confidence Intervals for Differences of Means

The random variable T may also be used to obtain a confidence interval for the difference $\mu_1 - \mu_2$ between the means of two normal distributions, say $N(\mu_1, \sigma^2)$ and $N(\mu_2, \sigma^2)$, when the distributions have the same, but unknown, variance σ^2.

Remark. Let X have a normal distribution with unknown parameters μ_1 and σ^2. A modification can be made in conducting the experiment so that the variance of the distribution will remain the same but the mean of the distribution will be changed; say, increased. After the modification has been effected, let the random variable be denoted by Y, and let Y have a normal distribution with unknown parameters μ_2 and σ^2. Naturally, it is hoped that

μ_2 is greater than μ_1, that is, that $\mu_1 - \mu_2 < 0$. Accordingly, one seeks a confidence interval for $\mu_1 - \mu_2$ in order to make a statistical inference.

A confidence interval for $\mu_1 - \mu_2$ may be obtained as follows: Let $X_1, X_2, \ldots, X_n$ and $Y_1, Y_2, \ldots, Y_m$ denote, respectively, independent random samples from the two distributions, $N(\mu_1, \sigma^2)$ and $N(\mu_2, \sigma^2)$, respectively. Denote the means of the samples by $\bar{X}$ and $\bar{Y}$ and the variances of the samples by S_1^2 and S_2^2, respectively. It should be noted that these four statistics are independent. The independence of $\bar{X}$ and S_1^2 (and, inferentially that of $\bar{Y}$ and S_2^2) was established in Section 4.8; the assumption that the two samples are independent accounts for the independence of the others. Thus $\bar{X}$ and $\bar{Y}$ are normally and independently distributed with means μ_1 and μ_2 and variances σ^2/n and σ^2/m, respectively. In accordance with Section 4.7, their difference $\bar{X} - \bar{Y}$ is normally distributed with mean $\mu_1 - \mu_2$ and variance $\sigma^2/n + \sigma^2/m$. Then the random variable

$$\frac{(\bar{X} - \bar{Y}) - (\mu_1 - \mu_2)}{\sqrt{\sigma^2/n + \sigma^2/m}}$$

is normally distributed with zero mean and unit variance. This random variable may serve as the numerator of a T random variable. Further, nS_1^2/σ^2 and mS_2^2/σ^2 have independent chi-square distributions with $n - 1$ and $m - 1$ degrees of freedom, respectively, so that their sum $(nS_1^2 + mS_2^2)/\sigma^2$ has a chi-square distribution with $n + m - 2$ degrees of freedom, provided that $m + n - 2 > 0$. Because of the independence of $\bar{X}$, $\bar{Y}$, S_1^2, and S_2^2, it is seen that

$$\sqrt{\frac{nS_1^2 + mS_2^2}{\sigma^2(n + m - 2)}}$$

may serve as the denominator of a T random variable. That is, the random variable

$$T = \frac{(\bar{X} - \bar{Y}) - (\mu_1 - \mu_2)}{\sqrt{\dfrac{nS_1^2 + mS_2^2}{n + m - 2}\left(\dfrac{1}{n} + \dfrac{1}{m}\right)}}$$

has a t-distribution with $n + m - 2$ degrees of freedom. As in the previous section, we can (once n and m are specified positive integers with $n + m - 2 > 0$) find a positive number b from Table IV of Appendix B such that

$$\Pr(-b < T < b) = 0.95.$$

If we set

$$R = \sqrt{\frac{nS_1^2 + mS_2^2}{n + m - 2}\left(\frac{1}{n} + \frac{1}{m}\right)},$$

this probability may be written in the form

$$\Pr [(\bar{X} - \bar{Y}) - bR < \mu_1 - \mu_2 < (\bar{X} - \bar{Y}) + bR] = 0.95.$$

It follows that the random interval

$$\left[(\bar{X} - \bar{Y}) - b\sqrt{\frac{nS_1^2 + mS_2^2}{n + m - 2}\left(\frac{1}{n} + \frac{1}{m}\right)},\right.$$

$$\left.(\bar{X} - \bar{Y}) + b\sqrt{\frac{nS_1^2 + mS_2^2}{n + m - 2}\left(\frac{1}{n} + \frac{1}{m}\right)}\right]$$

has probability 0.95 of including the unknown fixed point $(\mu_1 - \mu_2)$. As usual, the experimental values of $\bar{X}$, $\bar{Y}$, S_1^2, and S_2^2, namely $\bar{x}$, $\bar{y}$, s_1^2, and s_2^2, will provide a 95 percent confidence interval for $\mu_1 - \mu_2$ when the variances of the two normal distributions are unknown but equal. A consideration of the difficulty encountered when the unknown variances of the two normal distributions are not equal is assigned to one of the exercises.

Example 1. It may be verified that if in the preceding discussion $n = 10$, $m = 7$, $\bar{x} = 4.2$, $\bar{y} = 3.4$, $s_1^2 = 49$, $s_2^2 = 32$, then the interval $(-5.16, 6.76)$ is a 90 percent confidence interval for $\mu_1 - \mu_2$.

Let Y_1 and Y_2 be two independent random variables with binomial distributions $b(n_1, p_1)$ and $b(n_2, p_2)$, respectively. Let us now turn to the problem of finding a confidence interval for the difference $p_1 - p_2$ of the means of Y_1/n_1 and Y_2/n_2 when n_1 and n_2 are known. Since the mean and the variance of $Y_1/n_1 - Y_2/n_2$ are, respectively, $p_1 - p_2$ and $p_1(1 - p_1)/n_1 + p_2(1 - p_2)/n_2$, then the random variable given by the ratio

$$\frac{(Y_1/n_1 - Y_2/n_2) - (p_1 - p_2)}{\sqrt{p_1(1 - p_1)/n_1 + p_2(1 - p_2)/n_2}}$$

has mean zero and variance 1 for all positive integers n_1 and n_2. Moreover, since both Y_1 and Y_2 have approximate normal distributions for large n_1 and n_2, one suspects that the ratio has an approximate normal distribution. This is actually the case, but it will not be proved here. Moreover, if $n_1/n_2 = c$, where c is a fixed positive constant, the result of Exercise 6.36 shows that the random variable

$$\frac{(Y_1/n_1)(1 - Y_1/n_1)/n_1 + (Y_2/n_2)(1 - Y_2/n_2)/n_2}{p_1(1 - p_1)/n_1 + p_2(1 - p_2)/n_2} \tag{1}$$

converges in probability to 1 as $n_2 \to \infty$ (and thus $n_1 \to \infty$, since $n_1/n_2 = c$, $c > 0$). In accordance with Theorem 6, Section 5.5, the random variable

$$W = \frac{(Y_1/n_1 - Y_2/n_2) - (p_1 - p_2)}{U},$$

where

$$U = \sqrt{(Y_1/n_1)(1 - Y_1/n_1)/n_1 + (Y_2/n_2)(1 - Y_2/n_2)/n_2},$$

has a limiting distribution that is $N(0, 1)$. The event $-2 < W < 2$, the probability of which is approximately equal to 0.954, is equivalent to the event

$$\frac{Y_1}{n_1} - \frac{Y_2}{n_2} - 2U < p_1 - p_2 < \frac{Y_1}{n_1} - \frac{Y_2}{n_2} + 2U.$$

Accordingly, the experimental values y_1 and y_2 of Y_1 and Y_2, respectively, will provide an approximate 95.4 percent confidence interval for $p_1 - p_2$.

Example 2. If, in the preceding discussion, we take $n_1 = 100$, $n_2 = 400$, $y_1 = 30$, $y_2 = 80$, then the experimental values of $Y_1/n_1 - Y_2/n_2$ and U are 0.1 and $\sqrt{(0.3)(0.7)/100 + (0.2)(0.8)/400} = 0.05$, respectively. Thus the interval $(0, 0.2)$ is an approximate 95.4 percent confidence interval for $p_1 - p_2$.

EXERCISES

6.30. Let two independent random samples, each of size 10, from two normal distributions $N(\mu_1, \sigma^2)$ and $N(\mu_2, \sigma^2)$ yield $\bar{x} = 4.8$, $s_1^2 = 8.64$, $\bar{y} = 5.6$, $s_2^2 = 7.88$. Find a 95 percent confidence interval for $\mu_1 - \mu_2$.

6.31. Let two independent random variables Y_1 and Y_2, with binomial distributions that have parameters $n_1 = n_2 = 100$, p_1, and p_2, respectively, be observed to be equal to $y_1 = 50$ and $y_2 = 40$. Determine an approximate 90 percent confidence interval for $p_1 - p_2$.

6.32. Discuss the problem of finding a confidence interval for the difference $\mu_1 - \mu_2$ between the two means of two normal distributions if the variances σ_1^2 and σ_2^2 are known but not necessarily equal.

6.33. Discuss Exercise 6.32 when it is assumed that the variances are unknown and unequal. This is a very difficult problem, and the discussion should point out exactly where the difficulty lies. If, however, the variances are unknown but their ratio σ_1^2/σ_2^2 is a known constant k, then a statistic that is a T random variable can again be used. Why?

6.34. As an illustration of Exercise 6.33, one can let $X_1, X_2, \ldots, X_9$ and $Y_1, Y_2, \ldots, Y_{12}$ represent two independent random samples from the respective normal distributions $N(\mu_1, \sigma_1^2)$ and $N(\mu_2, \sigma_2^2)$. It is given that $\sigma_1^2 = 3\sigma_2^2$, but σ_2^2 is unknown. Define a random variable which has a t-distribution that can be used to find a 95 percent interval for $\mu_1 - \mu_2$.

6.35. Let $\bar{X}$ and $\bar{Y}$ be the means of two independent random samples, each of size n, from the respective distributions $N(\mu_1, \sigma^2)$ and $N(\mu_2, \sigma^2)$, where the common variance is known. Find n such that

$$\Pr(\bar{X} - \bar{Y} - \sigma/5 < \mu_1 - \mu_2 < \bar{X} - \bar{Y} + \sigma/5) = 0.90$$

6.36. Under the conditions given, show that the random variable defined by ratio (1) of the text converges in probability to 1.

6.37. Let $X_1, X_2, \ldots, X_n$ and $Y_1, Y_2, \ldots, Y_m$ be two independent random samples from the respective normal distributions $N(\mu_1, \sigma_1^2)$ and $N(\mu_2, \sigma_2^2)$, where the four parameters are unknown. To construct a *confidence interval for the ratio*, σ_1^2/σ_2^2, of the variances, form the quotient of the two independent chi-square variables, each divided by its degrees of freedom, namely

$$F = \frac{\dfrac{mS_2^2}{\sigma_2^2}\Big/(m-1)}{\dfrac{nS_1^2}{\sigma_1^2}\Big/(n-1)},$$

where S_1^2 and S_2^2 are the respective sample variances.

(a) What kind of distribution does F have?

(b) From the appropriate table, a and b can be found so that $\Pr(F < b) = 0.975$ and $\Pr(a < F < b) = 0.95$.

(c) Rewrite the second probability statement as

$$\Pr\left[a\frac{nS_1^2/(n-1)}{mS_2^2/(m-1)} < \frac{\sigma_1^2}{\sigma_2^2} < b\frac{nS_1^2/(n-1)}{mS_2^2/(m-1)}\right] = 0.95.$$

The observed values, s_1^2 and s_2^2, can be inserted in these inequalities to provide a 95 percent confidence interval for σ_1^2/σ_2^2.

6.4 Tests of Statistical Hypotheses

The two principal areas of statistical inference are the areas of estimation of parameters and of tests of statistical hypotheses. The problem of estimation of parameters, both point and interval estimation, has been treated. In Sections 6.4 and 6.5 some aspects of statistical hypotheses and tests of statistical hypotheses will be considered. The subject will be introduced by way of example.

Example 1. Let it be known that the outcome X of a random experiment is $N(\theta, 100)$. For instance, X may denote a score on a test, which score we assume to be normally distributed with mean θ and variance 100. Let us say the past experience with this random experiment indicates that $\theta = 75$. Suppose, owing possibly to some research in the area pertaining to this experiment, some changes are made in the method of performing this random experiment. It is then suspected that no longer does $\theta = 75$ but that now $\theta > 75$. There is as yet no formal experimental evidence that $\theta > 75$; hence the statement $\theta > 75$ is a conjecture or a *statistical hypothesis*. In admitting that the statistical hypothesis $\theta > 75$ may be false, we allow, in effect, the possibility that $\theta \le 75$. Thus there are actually two statistical hypotheses. First, that the unknown parameter $\theta \le 75$; that is, there has been no increase in θ. Second, that the unknown parameter $\theta > 75$. Accordingly, the parameter space is $\Omega = \{\theta : -\infty < \theta < \infty\}$. We denote the first of these hypotheses by the symbols $H_0 : \theta \le 75$ and the second by the symbols $H_1 : \theta > 75$. Since the values $\theta > 75$ are alternatives to those where $\theta \le 75$, the hypothesis $H_1 : \theta > 75$ is called the *alternative hypothesis*. Needless to say, H_0 could be called the alternative to H_1; however, the conjecture, here $\theta > 75$, that is made by the research worker is usually taken to be the alternative hypothesis. In any case the problem is to decide which of these hypotheses is to be accepted. To reach a decision, the random experiment is to be repeated a number of independent times, say n, and the results observed. That is, we consider a random sample $X_1, X_2, \ldots, X_n$ from a distribution that is $N(\theta, 100)$, and we devise a rule that will tell us what decision to make once the experimental values, say $x_1, x_2, \ldots, x_n$, have been determined. Such a rule is called a *test* of the hypothesis $H_0 : \theta \le 75$ against the alternative hypothesis $H_1 : \theta > 75$. There is no bound on the number of rules or tests that can be constructed. We shall consider three such tests. Our tests will be constructed around the following notion. We shall partition the sample space $\mathscr{A}$ into a subset C and its complement C^*. If the experimental values of $X_1, X_2, \ldots, X_n$, say $x_1, x_2, \ldots, x_n$, are such that the point $(x_1, x_2, \ldots, x_n) \in C$, we shall reject the hypothesis H_0 (accept the hypothesis H_1). If we have $(x_1, x_2, \ldots, x_n) \in C^*$, we shall accept the hypothesis H_0 (reject the hypothesis H_1).

Test 1. Let $n = 25$. The sample space $\mathscr{A}$ is the set

$$\{(x_1, x_2, \ldots, x_{25}) : -\infty < x_i < \infty, i = 1, 2, \ldots, 25\}.$$

Let the subset C of the sample space be

$$C = \{(x_1, x_2, \ldots, x_{25}) : x_1 + x_2 + \cdots + x_{25} > (25)(75)\}.$$

We shall reject the hypothesis H_0 if and only if our 25 experimental values are such that $(x_1, x_2, \ldots, x_{25}) \in C$. If $(x_1, x_2, \ldots, x_{25})$ is not an element of C, we shall accept the hypothesis H_0. This subset C of the sample space that leads to the rejection of the hypothesis $H_0 : \theta \le 75$ is called the *critical region* of Test 1. Now $\sum_1^{25} x_i > (25)(75)$ if and only if $\bar{x} > 75$, where $\bar{x} = \sum_1^{25} x_i/25$. Thus we can much more conveniently say that we shall reject the hypothesis $H_0 : \theta \le 75$ and accept the hypothesis $H_1 : \theta > 75$ if and only if the experimentally determined value of the sample mean $\bar{x}$ is greater than 75. If $\bar{x} \le 75$, we accept the hypothesis $H_0 : \theta \le 75$. Our test then amounts to this: We shall reject the hypothesis $H_0 : \theta \le 75$ if the mean of the sample exceeds the maximum value of the mean of the distribution when the hypothesis H_0 is true.

It would help us to evaluate a test of a statistical hypothesis if we knew the probability of rejecting that hypothesis (and hence of accepting the alternative hypothesis). In our Test 1, this means that we want to compute the probability

$$\Pr[(X_1, \ldots, X_{25}) \in C] = \Pr(\bar{X} > 75).$$

Obviously, this probability is a function of the parameter θ and we shall denote it by $K_1(\theta)$. The function $K_1(\theta) = \Pr(\bar{X} > 75)$ is called the *power function* of Test 1, and the value of the power function at a parameter point is called the *power* of Test 1 at that point. Because $\bar{X}$ is $N(\theta, 4)$, we have

$$K_1(\theta) = \Pr\left(\frac{\bar{X} - \theta}{2} > \frac{75 - \theta}{2}\right) = 1 - \Phi\left(\frac{75 - \theta}{2}\right).$$

So, for illustration, we have, by Table III of Appendix B, that the power at $\theta = 75$ is $K_1(75) = 0.500$. Other powers are $K_1(73) = 0.159$, $K_1(77) = 0.841$, and $K_1(79) = 0.977$. The graph of $K_1(\theta)$ of Test 1 is depicted in Figure 6.1. Among other things, this means that, if $\theta = 75$, the probability of rejecting the hypothesis $H_0 : \theta \le 75$ is $\frac{1}{2}$. That is, if $\theta = 75$ so that H_0 is true, the

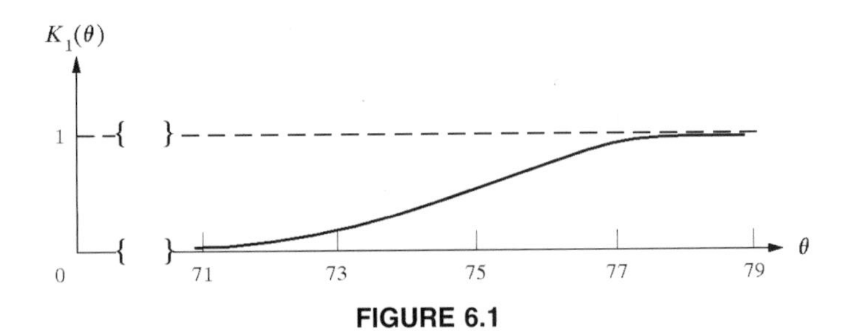

FIGURE 6.1

probability of rejecting this true hypothesis H_0 is $\frac{1}{2}$. Many statisticians and research workers find it very undesirable to have such a high probability as $\frac{1}{2}$ assigned to this kind of mistake: namely the rejection of H_0 when H_0 is a true hypothesis. Thus Test 1 does not appear to be a very satisfactory test. Let us try to devise another test that does not have this objectionable feature. We shall do this by making it more difficult to reject the hypothesis H_0, with the hope that this will give a smaller probability of rejecting H_0 when that hypothesis is true.

Test 2. Let $n = 25$. We shall reject the hypothesis $H_0 : \theta \le 75$ and accept the hypothesis $H_1 : \theta > 75$ if and only if $\bar{x} > 78$. Here the critical region is $C = \{(x_1, \ldots, x_{25}) : x_1 + \cdots + x_{25} > (25)(78)\}$. The power function of Test 2 is, because $\bar{X}$ is $N(\theta, 4)$,

$$K_2(\theta) = \Pr(\bar{X} > 78) = 1 - \Phi\left(\frac{78 - \theta}{2}\right).$$

Some values of the power function of Test 2 are $K_2(73) = 0.006$, $K_2(75) = 0.067$, $K_2(77) = 0.309$, and $K_2(79) = 0.691$. That is, if $\theta = 75$, the probability of rejecting $H_0 : \theta \le 75$ is 0.067; this is much more desirable than the corresponding probability $\frac{1}{2}$ that resulted from Test 1. However, if H_0 is false and, in fact, $\theta = 77$, the probability of rejecting $H_0 : \theta \le 75$ (and hence of accepting $H_1 : \theta > 75$) is only 0.309. In certain instances, this low probability 0.309 of a correct decision (the acceptance of H_1 when H_1 is true) is objectionable. That is, Test 2 is not wholly satisfactory. Perhaps we can overcome the undesirable features of Tests 1 and 2 if we proceed as in Test 3.

Test 3. Let us first select a power function $K_3(\theta)$ that has the features of a small value at $\theta = 75$ and a large value at $\theta = 77$. For instance, take $K_3(75) = 0.159$ and $K_3(77) = 0.841$. To determine a test with such a power function, let us reject $H_0 : \theta \le 75$ if and only if the experimental value $\bar{x}$ of the mean of a random sample of size n is greater than some constant c. Thus the critical region is $C = \{(x_1, x_2, \ldots, x_n) : x_1 + x_2 + \ + \cdots + x_n > nc\}$. It should be noted that the sample size n and the constant c have not been determined as yet. However, since $\bar{X}$ is $N(\theta, 100/n)$, the power function is

$$K_3(\theta) = \Pr(\bar{X} > c) = 1 - \Phi\left(\frac{c - \theta}{10/\sqrt{n}}\right).$$

The conditions $K_3(75) = 0.159$ and $K_3(77) = 0.841$ require that

$$1 - \Phi\left(\frac{c - 75}{10/\sqrt{n}}\right) = 0.159, \qquad 1 - \Phi\left(\frac{c - 77}{10/\sqrt{n}}\right) = 0.841.$$

Equivalently, from Table III of Appendix B, we have

$$\frac{c - 75}{10/\sqrt{n}} = 1, \qquad \frac{c - 77}{10/\sqrt{n}} = -1.$$

The solution to these two equations in n and c is $n = 100$, $c = 76$. With these values of n and c, other powers of Test 3 are $K_3(73) = 0.001$ and $K_3(79) = 0.999$. It is important to observe that although Test 3 has a more desirable power function than those of Tests 1 and 2, a certain "price" has been paid—a sample size of $n = 100$ is required in Test 3, whereas we had $n = 25$ in the earlier tests.

Remark. Throughout the text we frequently say that we accept the hypothesis H_0 if we do not reject H_0 in favor of H_1. If this decision is made, it certainly does not mean that H_0 is true or that we even believe that it is true. All it means is, based upon the data at hand, that we are not convinced that the hypothesis H_0 is wrong. Accordingly, the statement "We accept H_0" would possibly be better read as "We do not reject H_0." However, because it is in fairly common use, we use the statement "We accept H_0," but read it with this remark in mind.

We have now illustrated the following concepts:

1. A statistical hypothesis.
2. A test of a hypothesis against an alternative hypothesis and the associated concept of the critical region of the test.
3. The power of a test.

These concepts will now be formally defined.

Definition 3. A *statistical hypothesis* is an assertion about the distribution of one or more random variables. If the statistical hypothesis completely specifies the distribution, it is called a *simple statistical hypothesis*; if it does not, it is called a *composite statistical hypothesis*.

If we refer to Example 1, we see that both $H_0 : \theta \le 75$ and $H_1 : \theta > 75$ are composite statistical hypotheses, since neither of them completely specifies the distribution. If there, instead of $H_0 : \theta \le 75$, we had $H_0 : \theta = 75$, then H_0 would have been a simple statistical hypothesis.

Definition 4. A *test* of a statistical hypothesis is a rule which, when the experimental sample values have been obtained, leads to a decision to accept or to reject the hypothesis under consideration.

Definition 5. Let C be that subset of the sample space which, in accordance with a prescribed test, leads to the rejection of the hypothesis under consideration. Then C is called the *critical region* of the test.

Definition 6. The *power function* of a test of a statistical hypothesis H_0 against an alternative hypothesis H_1 is that function, defined for all distributions under consideration, which yields the probability that the sample point falls in the critical region C of the test, that is, a function that yields the probability of rejecting the hypothesis under consideration. The value of the power function at a parameter point is called the *power* of the test at that point.

Definition 7. Let H_0 denote a hypothesis that is to be tested against an alternative hypothesis H_1 in accordance with a prescribed test. The *significance level* of the test (or the *size* of the critical region C) is the maximum value (actually supremum) of the power function of the test when H_0 is true.

If we refer again to Example 1, we see that the significance levels of Tests 1, 2, and 3 of that example are 0.500, 0.067, and 0.159, respectively. An additional example may help clarify these definitions.

Example 2. It is known that the random variable X has a p.d.f. of the form

$$f(x; \theta) = \frac{1}{\theta} e^{-x/\theta}, \qquad 0 < x < \infty,$$
$$= 0 \qquad \text{elsewhere.}$$

It is desired to test the simple hypothesis $H_0 : \theta = 2$ against the alternative simple hypothesis $H_1 : \theta = 4$. Thus $\Omega = \{\theta : \theta = 2, 4\}$. A random sample X_1, X_2 of size $n = 2$ will be used. The test to be used is defined by taking the critical region to be $C = \{(x_1, x_2) : 9.5 \le x_1 + x_2 < \infty\}$. The power function of the test and the significance level of the test will be determined.

There are but two probability density functions under consideration, namely, $f(x; 2)$ specified by H_0 and $f(x; 4)$ specified by H_1. Thus the power function is defined at but two points $\theta = 2$ and $\theta = 4$. The power function of the test is given by $\Pr[(X_1, X_2) \in C]$. If H_0 is true, that is, $\theta = 2$, the joint p.d.f. of X_1 and X_2 is

$$f(x_1; 2)f(x_2; 2) = \tfrac{1}{4} e^{-(x_1 + x_2)/2}, \qquad 0 < x_1 < \infty, \quad 0 < x_2 < \infty,$$
$$= 0 \qquad \text{elsewhere,}$$

and

$$\Pr[(X_1, X_2) \in C] = 1 - \Pr[(X_1, X_2) \in C^*]$$
$$= 1 - \int_0^{9.5} \int_0^{9.5 - x_2} \tfrac{1}{4} e^{-(x_1 + x_2)/2}\, dx_1\, dx_2$$
$$= 0.05, \quad \text{approximately.}$$

If H_1 is true, that is, $\theta = 4$, the joint p.d.f. of X_1 and X_2 is

$$f(x_1; 4)f(x_2; 4) = \tfrac{1}{16}e^{-(x_1+x_2)/4}, \qquad 0 < x_1 < \infty, \quad 0 < x_2 < \infty,$$
$$= 0 \qquad \text{elsewhere,}$$

and

$$\Pr[(X_1, X_2) \in C] = 1 - \int_0^{9.5}\int_0^{9.5-x_2} \tfrac{1}{16}e^{-(x_1+x_2)/4}\,dx_1\,dx_2$$
$$= 0.31, \quad \text{approximately.}$$

Thus the power of the test is given by 0.05 for $\theta = 2$ and by 0.31 for $\theta = 4$. That is, the probability of rejecting H_0 when H_0 is true is 0.05, and the probability of rejecting H_0 when H_0 is false is 0.31. Since the significance level of this test (or the size of the critical region) is the power of the test when H_0 is true, the significance level of this test is 0.05.

The fact that the power of this test, when $\theta = 4$, is only 0.31 immediately suggests that a search be made for another test which, with the same power when $\theta = 2$, would have a power greater than 0.31 when $\theta = 4$. However later, it will be clear that such a search would be fruitless. That is, there is no test with a significance level of 0.05 and based on a random sample of size $n = 2$ that has greater power at $\theta = 4$. The only manner in which the situation may be improved is to have recourse to a random sample of size n greater than 2.

Our computations of the powers of this test at the two points $\theta = 2$ and $\theta = 4$ were purposely done the hard way to focus attention on fundamental concepts. A procedure that is computationally simpler is the following. When the hypothesis H_0 is true, the random variable X is $\chi^2(2)$. Thus the random variable $X_1 + X_2 = Y$, say, is $\chi^2(4)$. Accordingly, the power of the test when H_0 is true is given by

$$\Pr(Y \ge 9.5) = 1 - \Pr(Y < 9.5) = 1 - 0.95 = 0.05,$$

from Table II of Appendix B. When the hypothesis H_1 is true, the random variable $X/2$ is $\chi^2(2)$; so the random variable $(X_1 + X_2)/2 = Z$, say, is $\chi^2(4)$. Accordingly, the power of the test when H_1 is true is given by

$$\Pr(X_1 + X_2 \ge 9.5) = \Pr(Z \ge 4.75)$$
$$= \int_{4.75}^{\infty} \tfrac{1}{4}ze^{-z/2}\,dz,$$

which is equal to 0.31, approximately.

Remark. The rejection of the hypothesis H_0 when that hypothesis is true is, of course, an incorrect decision or an error. This incorrect decision is often called a type I error; accordingly, the significance level of the test is the probability of committing an error of type I. The acceptance of H_0 when H_0

is false (H_1 is true) is called an error of type II. Thus the probability of a type II error is 1 minus the power of the test when H_1 is true. Frequently, it is disconcerting to the student to discover that there are so many names for the same thing. However, since all of them are used in the statistical literature, we feel obligated to point out that "significance level," "size of the critical region," "power of the test when H_0 is true," and "the probability of committing an error of type I" are all equivalent.

EXERCISES

6.38. Let X have a p.d.f. of the form $f(x; \theta) = \theta x^{\theta - 1}, 0 < x < 1$, zero elsewhere, where $\theta \in \{\theta : \theta = 1, 2\}$. To test the simple hypothesis $H_0 : \theta = 1$ against the alternative simple hypothesis $H_1 : \theta = 2$, use a random sample X_1, X_2 of size $n = 2$ and define the critical region to be $C = \{(x_1, x_2) : \frac{3}{4} \le x_1 x_2\}$. Find the power function of the test.

6.39. Let X have a binomial distribution with parameters $n = 10$ and $p \in \{p : p = \frac{1}{4}, \frac{1}{2}\}$. The simple hypothesis $H_0 : p = \frac{1}{2}$ is rejected, and the alternative simple hypothesis $H_1 : p = \frac{1}{4}$ is accepted, if the observed value of X_1, a random sample of size 1, is less than or equal to 3. Find the power function of the test.

6.40. Let X_1, X_2 be a random sample of size $n = 2$ from the distribution having p.d.f. $f(x; \theta) = (1/\theta)e^{-x/\theta}, 0 < x < \infty$, zero elsewhere. We reject $H_0 : \theta = 2$ and accept $H_1 : \theta = 1$ if the observed values of X_1, X_2, say x_1, x_2, are such that

$$\frac{f(x_1; 2)f(x_2; 2)}{f(x_1; 1)f(x_2;1)} \le \frac{1}{2}.$$

Here $\Omega = \{\theta : \theta = 1, 2\}$. Find the significance level of the test and the power of the test when H_0 is false.

6.41. Sketch, as in Figure 6.1, the graphs of the power functions of Tests 1, 2, and 3 of Example 1 of this section.

6.42. Let us assume that the life of a tire in miles, say X, is normally distributed with mean θ and standard deviation 5000. Past experience indicates that $\theta = 30{,}000$. The manufacturer claims that the tires made by a new process have mean $\theta > 30{,}000$, and it is very possible that $\theta = 35{,}000$. Let us check his claim by testing $H_0 : \theta = 30{,}000$ against $H_1 : \theta > 30{,}000$. We shall observe n independent values of X, say $x_1, \ldots, x_n$, and we shall reject H_0 (thus accept H_1) if and only if $\bar{x} \ge c$. Determine n and c so that the power function $K(\theta)$ of the test has the values $K(30{,}000) = 0.01$ and $K(35{,}000) = 0.98$.

6.43. Let X have a Poisson distribution with mean θ. Consider the simple hypothesis $H_0 : \theta = \frac{1}{2}$ and the alternative composite hypothesis $H_1 : \theta < \frac{1}{2}$.

Thus $\Omega = \{\theta : 0 < \theta \leq \frac{1}{2}\}$. Let $X_1, \ldots, X_{12}$ denote a random sample of size 12 from this distribution. We reject H_0 if and only if the observed value of $Y = X_1 + \cdots + X_{12} \leq 2$. If $K(\theta)$ is the power function of the test, find the powers $K(\frac{1}{2})$, $K(\frac{1}{3})$, $K(\frac{1}{4})$, $K(\frac{1}{6})$, and $K(\frac{1}{12})$. Sketch the graph of $K(\theta)$. What is the significance level of the test?

6.44. Let Y have a binomial distribution with parameters n and p. We reject $H_0 : p = \frac{1}{2}$ and accept $H_1 : p > \frac{1}{2}$ if $Y \geq c$. Find n and c to give a power function $K(p)$ which is such that $K(\frac{1}{2}) = 0.10$ and $K(\frac{2}{3}) = 0.95$, approximately.

6.45. Let $Y_1 < Y_2 < Y_3 < Y_4$ be the order statistics of a random sample of size $n = 4$ from a distribution with p.d.f. $f(x; \theta) = 1/\theta$, $0 < x < \theta$, zero elsewhere, where $0 < \theta$. The hypothesis $H_0 : \theta = 1$ is rejected and $H_1 : \theta > 1$ accepted if the observed $Y_4 \geq c$.

(a) Find the constant c so that the significance level is $\alpha = 0.05$.
(b) Determine the power function of the test.

6.5 Additional Comments About Statistical Tests

All of the alternative hypotheses considered in Section 6.4 were *one-sided hypotheses*. For illustration, in Exercise 6.42 we tested $H_0 : \theta = 30{,}000$ against the one-sided alternative $H_1 : \theta > 30{,}000$, where θ is the mean of a normal distribution having standard deviation $\sigma = 5000$. The test associated with this situation, namely reject H_0 if and only if the sample mean $\bar{X} \geq c$, is a *one-sided test*. For convenience, we often call $H_0 : \theta = 30{,}000$ the *null hypothesis* because, as in this exercise, it suggests that the new process has not changed the mean of the distribution. That is, the new process has been used without consequence if in fact the mean still equals 30,000; hence the terminology null hypothesis is appropriate. So in Exercise 6.42 we are testing a simple null hypothesis against a composite one-sided alternative with a one-sided test.

This does suggest that there could be two-sided alternative hypotheses. For illustration, in Exercise 6.42, suppose there is the possibility that the new process might decrease the mean. That is, say that we simply do not know whether with the new process $\theta > 30{,}000$ or $\theta < 30{,}000$; or there has been no change and the null hypothesis $H_0 : \theta = 30{,}000$ is still true. Then we would want to test $H_0 : \theta = 30{,}000$ against the two-sided alternative $H_1 : \theta \neq 30{,}000$. To help see how to construct a *two-sided test* for H_0 against H_1, consider the following argument.

In dealing with a test of $H_0 : \theta = 30{,}000$ against the one-sided alternative $\theta > 30{,}000$, we used $\bar{X} \geq c$ or, equivalently,

$$Z = \frac{\bar{X} - 30{,}000}{\sigma/\sqrt{n}} \geq \frac{c - 30{,}000}{\sigma/\sqrt{n}} = c_1,$$

where since $\bar{X}$ is $N(\theta = 30{,}000, \sigma^2/n)$ under H_0, Z is $N(0, 1)$; and we could select $c_1 = 1.645$ to have a test of significance level $\alpha = 0.05$. That is, if $\bar{X}$ is $1.645\sigma/\sqrt{n}$ greater than the mean $\theta = 30{,}000$, we would reject H_0 and accept H_1 and the significance level would be equal to $\alpha = 0.05$. To test $H_0 : \theta = 30{,}000$ against $H_1 : \theta \neq 30{,}000$, let us again use $\bar{X}$ through Z and reject H_0 if $\bar{X}$ or Z is too large or too small. Namely, if we reject H_0 and accept H_1 when

$$|Z| = \left| \frac{\bar{X} - 30{,}000}{\sigma/\sqrt{n}} \right| \geq 1.96,$$

the significance level $\alpha = 0.05$ because this is the probability of $|Z| \geq 1.96$ when H_0 is true.

It is interesting to note that the latter test is the equivalent of saying that we reject H_0 and accept H_1 if 30,000 is not in the (two-sided) confidence interval for the mean θ. Or equivalently, if

$$\bar{X} - 1.96\frac{\sigma}{\sqrt{n}} < 30{,}000 < \bar{X} + 1.96\frac{\sigma}{\sqrt{n}},$$

then we accept $H_0 : \theta = 30{,}000$ because those two inequalities are equivalent to

$$\left| \frac{\bar{X} - 30{,}000}{\sigma/\sqrt{n}} \right| < 1.96,$$

which leads to the acceptance of $H_0 : \theta = 30{,}000$.

Once we recognize this relationship between confidence intervals and tests of hypotheses, we can use all those statistics that we used to construct confidence intervals to test hypotheses, not only against two-sided alternatives but one-sided ones as well. Without listing all of these in a table, we give enough of them so that the principle can be understood.

Example 1. Let $\bar{X}$ and S^2 be the mean and the variance of a random sample of size n coming from $N(\mu, \sigma^2)$. To test, at significance level $\alpha = 0.05$, $H_0 : \mu = \mu_0$ against the two-sided alternative $H_1 : \mu \neq \mu_0$, reject if

$$|T| = \left| \frac{\bar{X} - \mu_0}{S/\sqrt{n - 1}} \right| \geq b,$$

where b is the 97.5th percentile of the t-distribution with $n - 1$ degrees of freedom.

Example 2. Let independent random samples be taken from $N(\mu_1, \sigma^2)$ and $N(\mu_2, \sigma^2)$, respectively. Say these have the respective sample characteristics n, $\bar{X}$, S_1^2 and m, $\bar{Y}$, S_2^2. At $\alpha = 0.05$, reject $H_0 : \mu_1 = \mu_2$ and accept the one-sided alternative $H_1 : \mu_1 > \mu_2$ if

$$T = \frac{\bar{X} - \bar{Y} - 0}{\sqrt{\dfrac{nS_1^2 + mS_2^2}{n + m - 2}\left(\dfrac{1}{n} + \dfrac{1}{m}\right)}} \geq c.$$

Note that $\bar{X} - \bar{Y}$ has a normal distribution with mean zero under H_0. So c is taken as the 95th percentile of a t-distribution with $n + m - 2$ degrees of freedom to provide $\alpha = 0.05$.

Example 3. Say Y is $b(n, p)$. To test $H_0 : p = p_0$ against $H_1 : p < p_0$, we use either

$$Z_1 = \frac{(Y/n) - p_0}{\sqrt{p_0(1 - p_0)/n}} \leq c \qquad \text{or} \qquad Z_2 = \frac{(Y/n) - p_0}{\sqrt{(Y/n)(1 - Y/n)/n}} \leq c.$$

If n is large, both Z_1 and Z_2 have approximate standard normal distributions provided that $H_0 : p = p_0$ is true. Hence c is taken to be -1.645 to give an approximate significance level of $\alpha = 0.05$. Some statisticians use Z_1 and others Z_2. We do not have strong preference one way or the other because the two methods provide about the same numerical result. As one might suspect, using Z_1 provides better probabilities for power calculations if the true p is close to p_0 while Z_2 is better if H_0 is clearly false. However, with a two-sided alternative hypothesis, Z_2 does provide a better relationship with the confidence interval for p. That is, $|Z_2| < 2$ is equivalent to p_0 being in the interval from

$$\frac{Y}{n} - 2\sqrt{\frac{(Y/n)(1 - Y/n)}{n}} \quad \text{to} \quad \frac{Y}{n} + 2\sqrt{\frac{(Y/n)(1 - Y/n)}{n}},$$

which is the interval that provides a 95.4 percent confidence interval for p as considered in Section 6.2.

In closing this section, we introduce the concepts of *randomized tests* and *p-values* through an example and remarks that follow the example.

Example 4. Let $X_1, X_2, \ldots, X_{10}$ be a random sample of size $n = 10$ from a Poisson distribution with mean θ. A critical region for testing $H_0 : \theta = 0.1$ against $H_1 : \theta > 0.1$ is given by $Y = \sum_{i=1}^{10} X_i \geq 3$. The statistic Y has a Poisson distribution with mean 10θ. Thus, with $\theta = 0.1$ so that the mean of Y is 1, the significance level of the test is

$$\Pr(Y \geq 3) = 1 - \Pr(Y \leq 2) = 1 - 0.920 = 0.080.$$

If the critical region defined by $\sum_{1}^{10} x_i \geq 4$ is used, the significance level is

$$\alpha = \Pr(Y \geq 4) = 1 - \Pr(Y \leq 3) = 1 - 0.981 = 0.019.$$

If a significance level of about $\alpha = 0.05$, say, is desired, most statisticians would use one of these tests; that is, they would adjust the significance level to that of one of these convenient tests. However, a significance level of $\alpha = 0.05$ can be achieved exactly by rejecting H_0 if $\sum_{1}^{10} x_i \geq 4$ or if $\sum_{1}^{10} x_i = 3$ and an auxiliary independent random experiment resulted in "success," where the probability of success is selected to be equal to

$$\frac{0.050 - 0.019}{0.080 - 0.019} = \frac{31}{61}.$$

This is due to the fact that, when $\theta = 0.1$ so that the mean of Y is 1,

$$\Pr(Y \geq 4) + \Pr(Y = 3 \text{ and success}) = 0.019 + \Pr(Y = 3)\Pr(\text{success})$$
$$= 0.019 + (0.061)\tfrac{31}{61} = 0.05.$$

The process of performing the auxiliary experiment to decide whether to reject or not when $Y = 3$ is sometimes referred to as a *randomized test.*

Remarks. Not many statisticians like randomized tests in practice, because the use of them means that two statisticians could make the same assumptions, observe the same data, apply the same test, and yet make different decisions. Hence they usually adjust their significance level so as not to randomize. As a matter of fact, many statisticians report what are commonly called *p-values* (for *probability values*). For illustration, if in Example 4 the observed Y is $y = 4$, the p-value is 0.019; and if it is $y = 3$, the p-value is 0.080. That is, the p-value is the observed "tail" probability of a statistic being at least as extreme as the particular observed value when H_0 is true. Hence, more generally, if $Y = u(X_1, X_2, \ldots, X_n)$ is the statistic to be used in a test of H_0 and if the critical region is of the form

$$u(x_1, x_2, \ldots, x_n) \leq c,$$

an observed value $u(x_1, x_2, \ldots, x_n) = d$ would mean that the

$$p\text{-value} = \Pr(Y \leq d; H_0).$$

That is, if $G(y)$ is the distribution function of $Y = u(X_1, X_2, \ldots, X_n)$, provided that H_0 is true, the p-value is equal to $G(d)$ in this case. However,

$G(Y)$, in the continuous case, is uniformly distributed on the unit interval, so an observed value $G(d) \le 0.05$ would be equivalent to selecting c, so that

$$\Pr[u(X_1, X_2, \ldots, X_n) \le c; H_0] = 0.05$$

and observing that $d \le c$. Most computer programs automatically print out the p-value of a test.

Example 5. Let $X_1, X_2, \ldots, X_{25}$ be a random sample from $N(\mu, \sigma^2 = 4)$. To test $H_0 : \mu = 77$ against the one-sided alternative hypothesis $H_1 : \mu < 77$, say we observe the 25 values and determine that $\bar{x} = 76.1$. The variance of $\bar{X}$ is $\sigma^2/n = 4/25 = 0.16$; so we know that $Z = (\bar{X} - 77)/0.4$ is $N(0, 1)$ provided that $\mu = 77$. Since the observed value of this test statistic is $z = (76.1 - 77)/0.4 = -2.25$, the p-value of the test is $\Phi(-2.25) = 1 - 0.988 = 0.012$. Accordingly, if we were using a significance level of $\alpha = 0.05$, we would reject H_0 and accept $H_1 : \mu < 77$ because $0.012 < 0.05$.

EXERCISES

6.46. Assume that the weight of cereal in a "10-ounce box" is $N(\mu, \sigma^2)$. To test $H_0 : \mu = 10.1$ against $H_1 : \mu > 10.1$, we take a random sample of size $n = 16$ and observe that $\bar{x} = 10.4$ and $s = 0.4$.
(a) Do we accept or reject H_0 at the 5 percent significance level?
(b) What is the approximate p-value of this test?

6.47. Each of 51 golfers hit three golf balls of brand X and three golf balls of brand Y in a random order. Let X_i and Y_i equal the averages of the distances traveled by the brand X and brand Y golf balls hit by the ith golfer, $i = 1, 2, \ldots, 51$. Let $W_i = X_i - Y_i$, $i = 1, 2, \ldots, 51$. Test $H_0 : \mu_W = 0$ against $H_1 : \mu_W > 0$, where μ_W is the mean of the differences. If $\bar{w} = 2.07$ and $s_w^2 = 84.63$, would H_0 be accepted or rejected at an $\alpha = 0.05$ significance level? What is the p-value of this test?

6.48. Among the data collected for the World Health Organization air quality monitoring project is a measure of suspended particles in $\mu g/m^3$. Let X and Y equal the concentration of suspended particles in $\mu g/m^3$ in the city center (commercial district) for Melbourne and Houston, respectively. Using $n = 13$ observations of X and $m = 16$ observations of Y, we shall test $H_0 : \mu_X = \mu_Y$ against $H_1 : \mu_X < \mu_Y$.
(a) Define the test statistic and critical region, assuming that the variances are equal. Let $\alpha = 0.05$.
(b) If $\bar{x} = 72.9$, $s_x = 25.6$, $\bar{y} = 81.7$, and $s_y = 28.3$, calculate the value of the test statistic and state your conclusion.

6.49. Let p equal the proportion of drivers who use a seat belt in a state that does not have a mandatory seat belt law. It was claimed that $p = 0.14$. An advertising campaign was conducted to increase this proportion. Two months after the campaign, $y = 104$ out of a random sample of $n = 590$ drivers were wearing their seat belts. Was the campaign successful?

(a) Define the null and alternative hypotheses.

(b) Define a critical region with an $\alpha = 0.01$ significance level.

(c) Determine the approximate p-value and state your conclusion.

6.50. A machine shop that manufactures toggle levers has both a day and a night shift. A toggle lever is defective if a standard nut cannot be screwed onto the threads. Let p_1 and p_2 be the proportion of defective levers among those manufactured by the day and night shifts, respectively. We shall test the null hypothesis, $H_0 : p_1 = p_2$, against a two-sided alternative hypothesis based on two random samples, each of 1000 levers taken from the production of the respective shifts.

(a) Define the test statistic which has an approximate $N(0, 1)$ distribution. Sketch a standard normal p.d.f. illustrating the critical region having $\alpha = 0.05$.

(b) If $y_1 = 37$ and $y_2 = 53$ defectives were observed for the day and night shifts, respectively, calculate the value of the test statistic and the approximate p-value (note that this is a two-sided test). Locate the calculated test statistic on your figure in part (a) and state your conclusion.

6.51. In Exercise 6.28 we found a confidence interval for the variance σ^2 using the variance S^2 of a random sample of size n arising from $N(\mu, \sigma^2)$, where the mean μ is unknown. In testing $H_0 : \sigma^2 = \sigma_0^2$ against $H_1 : \sigma^2 > \sigma_0^2$, use the critical region defined by $nS^2/\sigma_0^2 \ge c$. That is, reject H_0 and accept H_1 if $S^2 \ge c\sigma_0^2/n$. If $n = 13$ and the significance level $\alpha = 0.025$, determine c.

6.52. In Exercise 6.37, in finding a confidence interval for the ratio of the variances of two normal distributions, we used a statistic $[nS_1^2/(n-1)]/[mS_2^2/(m-1)]$, which has an F-distribution when those two variances are equal. If we denote that statistic by F, we can test $H_0 : \sigma_1^2 = \sigma_2^2$ against $H_1 : \sigma_1^2 > \sigma_2^2$ using the critical region $F \ge c$. If $n = 13$, $m = 11$, and $\alpha = 0.05$, find c.

6.6 Chi-Square Tests

In this section we introduce tests of statistical hypotheses called *chi-square tests*. A test of this sort was originally proposed by Karl Pearson in 1900, and it provided one of the earlier methods of statistical inference.

Let the random variable X_i be $N(\mu_i, \sigma_i^2)$, $i = 1, 2, \ldots, n$, and let $X_1, X_2, \ldots, X_n$ be mutually independent. Thus the joint p.d.f. of these variables is

$$\frac{1}{\sigma_1\sigma_2\cdots\sigma_n(2\pi)^{n/2}}\exp\left[-\frac{1}{2}\sum_1^n\left(\frac{x_i-\mu_i}{\sigma_i}\right)^2\right], \qquad -\infty < x_i < \infty.$$

The random variable that is defined by the exponent (apart from the coefficient $-\frac{1}{2}$) is $\sum_1^n (X_i - \mu_i)^2/\sigma_i^2$, and this random variable is $\chi^2(n)$. In Section 4.10 we generalized this joint normal distribution of probability to n random variables that are *dependent* and we call the distribution a *multivariate normal distribution*. In Section 10.8, it will be shown that a certain exponent in the joint p.d.f. (apart from a coefficient of $-\frac{1}{2}$) defines a random variable that is $\chi^2(n)$. This fact is the mathematical basis of the chi-square tests.

Let us now discuss some random variables that have approximate chi-square distributions. Let X_1 be $b(n, p_1)$. Since the random variable $Y = (X_1 - np_1)/\sqrt{np_1(1-p_1)}$ has, as $n \to \infty$, a limiting distribution that is $N(0, 1)$, we would strongly suspect that the limiting distribution of $Z = Y^2$ is $\chi^2(1)$. This is, in fact, the case, as will now be shown. If $G_n(y)$ represents the distribution function of Y, we know that

$$\lim_{n\to\infty} G_n(y) = \Phi(y), \qquad -\infty < y < \infty,$$

where $\Phi(y)$ is the distribution function of a distribution that is $N(0, 1)$. Let $H_n(z)$ represent, for each positive integer n, the distribution function of $Z = Y^2$. Thus, if $z \geq 0$,

$$\begin{aligned} H_n(z) = \Pr(Z \leq z) &= \Pr(-\sqrt{z} \leq Y \leq \sqrt{z}) \\ &= G_n(\sqrt{z}) - G_n[(-\sqrt{z})-]. \end{aligned}$$

Accordingly, since $\Phi(y)$ is everywhere continuous,

$$\begin{aligned} \lim_{n\to\infty} H_n(z) &= \Phi(\sqrt{z}) - \Phi(-\sqrt{z}) \\ &= 2\int_0^{\sqrt{z}} \frac{1}{\sqrt{2\pi}} e^{-w^2/2}\, dw. \end{aligned}$$

If we change the variable of integration in this last integral by writing $w^2 = v$, then

$$\lim_{n \to \infty} H_n(z) = \int_0^z \frac{1}{\Gamma(\frac{1}{2})2^{1/2}} v^{1/2 - 1} e^{-v/2}\, dv,$$

provided that $z \geq 0$. If $z < 0$, then $\lim_{n \to \infty} H_n(z) = 0$. Thus $\lim_{n \to \infty} H_n(z)$ is equal to the distribution function of a random variable that is $\chi^2(1)$. This is the desired result.

Let us now return to the random variable X_1 which is $b(n, p_1)$. Let $X_2 = n - X_1$ and let $p_2 = 1 - p_1$. If we denote Y^2 by Q_1 instead of Z, we see that Q_1 may be written as

$$Q_1 = \frac{(X_1 - np_1)^2}{np_1(1 - p_1)} = \frac{(X_1 - np_1)^2}{np_1} + \frac{(X_1 - np_1)^2}{n(1 - p_1)}$$

$$= \frac{(X_1 - np_1)^2}{np_1} + \frac{(X_2 - np_2)^2}{np_2}$$

because $(X_1 - np_1)^2 = (n - X_2 - n + np_2)^2 = (X_2 - np_2)^2$. Since Q_1 has a limiting chi-square distribution with 1 degree of freedom, we say, when n is a positive integer, that Q_1 has an approximate chi-square distribution with 1 degree of freedom. This result can be generalized as follows.

Let $X_1, X_2, \ldots, X_{k-1}$ have a multinomial distribution with the parameters $n, p_1, \ldots, p_{k-1}$, as in Section 3.1. As a convenience, let $X_k = n - (X_1 + \cdots + X_{k-1})$ and let $p_k = 1 - (p_1 + \cdots + p_{k-1})$. Define Q_{k-1} by

$$Q_{k-1} = \sum_{i=1}^{k} \frac{(X_i - np_i)^2}{np_i}.$$

It is proved in a more advanced course that, as $n \to \infty$, Q_{k-1} has a limiting distribution that is $\chi^2(k - 1)$. If we accept this fact, we can say that Q_{k-1} has an approximate chi-square distribution with $k - 1$ degrees of freedom when n is a positive integer. Some writers caution the user of this approximation to be certain that n is large enough that each np_i, $i = 1, 2, \ldots, k$, is at least equal to 5. In any case it is important to realize that Q_{k-1} does not have a chi-square distribution, only an approximate chi-square distribution.

The random variable Q_{k-1} may serve as the basis of the tests of certain statistical hypotheses which we now discuss. Let the sample space $\mathscr{A}$ of a random experiment be the union of a finite number k of mutually disjoint sets $A_1, A_2, \ldots, A_k$. Furthermore, let $P(A_i) = p_i$, $i = 1, 2, \ldots, k$, where $p_k = 1 - p_1 - \cdots - p_{k-1}$, so that p_i is the probability that the outcome of the random experiment is an element of the set A_i. The random experiment is to be repeated n independent times and X_i will represent the number of times the outcome is an element of the set A_i. That is, $X_1, X_2, \ldots, X_k = n - X_1 - \cdots - X_{k-1}$ are the frequencies with which the outcome is, respectively, an element of $A_1, A_2, \ldots, A_k$. Then the joint p.d.f. of $X_1, X_2, \ldots, X_{k-1}$ is the multinomial p.d.f. with the parameters $n, p_1, \ldots, p_{k-1}$. Consider the simple hypothesis (concerning this multinomial p.d.f.) $H_0 : p_1 = p_{10}$, $p_2 = p_{20}, \ldots, p_{k-1} = p_{k-1,0}$ $(p_k = p_{k0} = 1 - p_{10} - \cdots - p_{k-1,0})$, where $p_{10}, \ldots, p_{k-1,0}$ are specified numbers. It is desired to test H_0 against all alternatives.

If the hypothesis H_0 is true, the random variable

$$Q_{k-1} = \sum_1^k \frac{(X_i - np_{i0})^2}{np_{i0}}$$

has an approximate chi-square distribution with $k - 1$ degrees of freedom. Since, when H_0 is true, np_{i0} is the expected value of X_i, one would feel intuitively that experimental values of Q_{k-1} should not be too large if H_0 is true. With this in mind, we may use Table II of Appendix B, with $k - 1$ degrees of freedom, and find c so that $\Pr(Q_{k-1} \geq c) = \alpha$, where α is the desired significance level of the test. If, then, the hypothesis H_0 is rejected when the observed value of Q_{k-1} is at least as great as c, the test of H_0 will have a significance level that is approximately equal to α.

Some illustrative examples follow.

Example 1. One of the first six positive integers is to be chosen by a random experiment (perhaps by the cast of a die). Let $A_i = \{x : x = i\}$, $i = 1, 2, \ldots, 6$. The hypothesis $H_0 : P(A_i) = p_{i0} = \frac{1}{6}$, $i = 1, 2, \ldots, 6$, will be tested, at the approximate 5 percent significance level, against all alternatives. To make the test, the random experiment will be repeated, under the same conditions, 60 independent times. In this example $k = 6$ and $np_{i0} = 60(\frac{1}{6}) = 10$, $i = 1, 2, \ldots, 6$. Let X_i denote the frequency with which the random experiment terminates with the outcome in A_i, $i = 1, 2, \ldots, 6$, and let $Q_5 = \sum_1^6 (X_i - 10)^2/10$. If H_0 is true, Table II, with $k - 1 = 6 - 1 = 5$ degrees of freedom, shows that we have $\Pr(Q_5 \geq 11.1) = 0.05$. Now suppose that the experimental frequencies of $A_1, A_2, \ldots, A_6$ are, respectively, 13, 19, 11, 8, 5, and 4. The observed value of Q_5 is

$$\frac{(13-10)^2}{10}+\frac{(19-10)^2}{10}+\frac{(11-10)^2}{10}+\frac{(8-10)^2}{10}+\frac{(5-10)^2}{10}+\frac{(4-10)^2}{10}=15.6$$

Since $15.6 > 11.1$, the hypothesis $P(A_i) = \frac{1}{6}$, $i = 1, 2, \ldots, 6$, is rejected at the (approximate) 5 percent significance level.

Example 2. A point is to be selected from the unit interval $\{x : 0 < x < 1\}$ by a random process. Let $A_1 = \{x : 0 < x \le \frac{1}{4}\}$, $A_2 = \{x : \frac{1}{4} < x \le \frac{1}{2}\}$, $A_3 = \{x : \frac{1}{2} < x \le \frac{3}{4}\}$, and $A_4 = \{x : \frac{3}{4} < x < 1\}$. Let the probabilities p_i, $i = 1, 2, 3, 4$, assigned to these sets under the hypothesis be determined by the p.d.f. $2x$, $0 < x < 1$, zero elsewhere. Then these probabilities are, respectively,

$$p_{10} = \int_0^{1/4} 2x\,dx = \frac{1}{16}, \qquad p_{20} = \frac{3}{16}, \qquad p_{30} = \frac{5}{16}, \qquad p_{40} = \frac{7}{16}.$$

Thus the hypothesis to be tested is that p_1, p_2, p_3, and $p_4 = 1 - p_1 - p_2 - p_3$ have the preceding values in a multinomial distribution with $k = 4$. This hypothesis is to be tested at an approximate 0.025 significance level by repeating the random experiment $n = 80$ independent times under the same conditions. Here the np_{i0}, $i = 1, 2, 3, 4$, are, respectively, 5, 15, 25, and 35. Suppose the observed frequencies of A_1, A_2, A_3, and A_4 are 6, 18, 20, and 36, respectively. Then the observed value of $Q_3 = \sum_1^4 (X_i - np_{i0})^2/(np_{i0})$ is

$$\frac{(6-5)^2}{5}+\frac{(18-15)^2}{15}+\frac{(20-25)^2}{25}+\frac{(36-35)^2}{35}=\frac{64}{35}=1.83,$$

approximately. From Table II, with $4 - 1 = 3$ degrees of freedom, the value corresponding to a 0.025 significance level is $c = 9.35$. Since the observed value of Q_3 is less than 9.35, the hypothesis is accepted at the (approximate) 0.025 level of significance.

Thus far we have used the chi-square test when the hypothesis H_0 is a simple hypothesis. More often we encounter hypotheses H_0 in which the multinomial probabilities $p_1, p_2, \ldots, p_k$ are not completely specified by the hypothesis H_0. That is, under H_0, these probabilities are functions of unknown parameters. For illustration, suppose that a certain random variable Y can take on any real value. Let us partition the space $\{y : -\infty < y < \infty\}$ into k mutually disjoint sets $A_1, A_2, \ldots, A_k$ so that the events $A_1, A_2, \ldots, A_k$ are mutually exclusive and exhaustive. Let H_0 be the hypothesis that Y is $N(\mu, \sigma^2)$ with μ and σ^2 unspecified. Then each

$$p_i = \int_{A_i} \frac{1}{\sqrt{2\pi}\,\sigma} \exp\left[-(y-\mu)^2/2\sigma^2\right] dy, \qquad i = 1, 2, \ldots, k,$$

is a function of the unknown parameters μ and σ^2. Suppose that we take a random sample $Y_1, \ldots, Y_n$ of size n from this distribution. If we let X_i denote the frequency of A_i, $i = 1, 2, \ldots, k$, so that $X_1 + \cdots + X_k = n$, the random variable

$$Q_{k-1} = \sum_{i=1}^{k} \frac{(X_i - np_i)^2}{np_i}$$

cannot be computed once $X_1, \ldots, X_k$ have been observed, since each p_i, and hence Q_{k-1}, is a function of the unknown parameters μ and σ^2.

There is a way out of our trouble, however. We have noted that Q_{k-1} is a function of μ and σ^2. Accordingly, choose the values of μ and σ^2 that minimize Q_{k-1}. Obviously, these values depend upon the observed $X_1 = x_1, \ldots, X_k = x_k$ and are called *minimum chi-square estimates* of μ and σ^2. These point estimates of μ and σ^2 enable us to compute numerically the estimates of each p_i. Accordingly, if these values are used, Q_{k-1} can be computed once $Y_1, Y_2, \ldots, Y_n$, and hence $X_1, X_2, \ldots, X_k$, are observed. However, a very important aspect of the fact, which we accept without proof, is that now Q_{k-1} is approximately $\chi^2(k-3)$. That is, the number of degrees of freedom of the limiting chi-square distribution of Q_{k-1} is reduced by one for each parameter estimated by the experimental data. This statement applies not only to the problem at hand but also to more general situations. Two examples will now be given. The first of these examples will deal with the test of the hypothesis that two multinominal distributions are the same.

Remark. In many instances, such as that involving the mean μ and the variance σ^2 of a normal distribution, minimum chi-square estimates are difficult to compute. Hence other estimates, such as the maximum likelihood estimates $\hat{\mu} = \bar{Y}$ and $\widehat{\sigma^2} = S^2$, are used to evaluate p_i and Q_{k-1}. In general, Q_{k-1} is not minimized by maximum likelihood estimates, and thus its computed value is somewhat greater than it would be if minimum chi-square estimates were used. Hence, when comparing it to a critical value listed in the chi-square table with $k-3$ degrees of freedom, there is a greater chance of rejecting than there would be if the actual minimum of Q_{k-1} is used.

Accordingly, the approximate significance level of such a test will be somewhat higher than that value found in the table. This modification should be kept in mind and, if at all possible, each p_i should be estimated using the frequencies $X_1, \ldots, X_k$ rather than using directly the observations $Y_1, Y_2, \ldots, Y_n$ of the random sample.

Example 3. Let us consider two multinomial distributions with parameters $n_j, p_{1j}, p_{2j}, \ldots, p_{kj}$, $j = 1, 2$, respectively. Let X_{ij}, $i = 1, 2, \ldots, k$, $j = 1, 2$, represent the corresponding frequencies. If n_1 and n_2 are large and the observations from one distribution are independent of those from the other, the random variable

$$\sum_{j=1}^{2} \sum_{i=1}^{k} \frac{(X_{ij} - n_j p_{ij})^2}{n_j p_{ij}}$$

is the sum of two independent random variables, each of which we treat as though it were $\chi^2(k-1)$; that is, the random variable is approximately $\chi^2(2k-2)$. Consider the hypothesis

$$H_0: p_{11} = p_{12}, p_{21} = p_{22}, \ldots, p_{k1} = p_{k2},$$

where each $p_{i1} = p_{i2}$, $i = 1, 2, \ldots, k$, is unspecified. Thus we need point estimates of these parameters. The maximum likelihood estimator of $p_{i1} = p_{i2}$, based upon the frequencies X_{ij}, is $(X_{i1} + X_{i2})/(n_1 + n_2)$, $i = 1, 2, \ldots, k$. Note that we need only $k - 1$ point estimates, because we have a point estimate of $p_{k1} = p_{k2}$ once we have point estimates of the first $k - 1$ probabilities. In accordance with the fact that has been stated, the random variable

$$\sum_{j=1}^{2} \sum_{i=1}^{k} \frac{\{X_{ij} - n_j[(X_{i1} + X_{i2})/(n_1 + n_2)]\}^2}{n_j[(X_{i1} + X_{i2})/(n_1 + n_2)]}$$

has an approximate χ^2 distribution with $2k - 2 - (k - 1) = k - 1$ degrees of freedom. Thus we are able to test the hypothesis that two multinomial distributions are the same; this hypothesis is rejected when the computed value of this random variable is at least as great as an appropriate number from Table II, with $k - 1$ degrees of freedom.

The second example deals with the subject of *contingency tables.*

Example 4. Let the result of a random experiment be classified by two attributes (such as the color of the hair and the color of the eyes). That is, one attribute of the outcome is one and only one of certain mutually exclusive and exhaustive events, say $A_1, A_2, \ldots, A_a$; and the other attribute of the outcome is also one and only one of certain mutually exclusive and exhaustive events, say $B_1, B_2, \ldots, B_b$. Let $p_{ij} = P(A_i \cap B_j)$, $i = 1, 2, \ldots, a$; $j = 1, 2, \ldots, b$. The random experiment is to be repeated n independent times and X_{ij} will denote the frequency of the event $A_i \cap B_j$. Since there are $k = ab$ such events as $A_i \cap B_j$, the random variable

$$Q_{ab-1} = \sum_{j=1}^{b} \sum_{i=1}^{a} \frac{(X_{ij} - np_{ij})^2}{np_{ij}}$$

has an approximate chi-square distribution with $ab - 1$ degrees of freedom, provided that n is large. Suppose that we wish to test the independence of the A attribute and the B attribute; that is, we wish to test the hypothesis $H_0 : P(A_i \cap B_j) = P(A_i)P(B_j)$, $i = 1, 2, \ldots, a$; $j = 1, 2, \ldots, b$. Let us denote $P(A_i)$ by $p_{i.}$ and $P(B_j)$ by $p_{.j}$; thus

$$p_{i.} = \sum_{j=1}^{b} p_{ij}, \qquad p_{.j} = \sum_{i=1}^{a} p_{ij},$$

and

$$1 = \sum_{j=1}^{b} \sum_{i=1}^{a} p_{ij} = \sum_{j=1}^{b} p_{.j} = \sum_{i=1}^{a} p_{i.}.$$

Then the hypothesis can be formulated as $H_0 : p_{ij} = p_{i.}p_{.j}$, $i = 1, 2, \ldots, a$; $j = 1, 2, \ldots, b$. To test H_0, we can use Q_{ab-1} with p_{ij} replaced by $p_{i.}p_{.j}$. But if $p_{i.}$, $i = 1, 2, \ldots, a$, and $p_{.j}$, $j = 1, 2, \ldots, b$, are unknown, as they frequently are in applications, we cannot compute Q_{ab-1} once the frequencies are observed. In such a case we estimate these unknown parameters by

$$\hat{p}_{i.} = \frac{X_{i.}}{n}, \quad \text{where} \quad X_{i.} = \sum_{j=1}^{b} X_{ij}, \qquad i = 1, 2, \ldots, a,$$

and

$$\hat{p}_{.j} = \frac{X_{.j}}{n}, \quad \text{where} \quad X_{.j} = \sum_{i=1}^{a} X_{ij}, \qquad j = 1, 2, \ldots, b.$$

Since $\sum_i p_{i.} = \sum_j p_{.j} = 1$, we have estimated only $a - 1 + b - 1 = a + b - 2$ parameters. So if these estimates are used in Q_{ab-1}, with $p_{ij} = p_{i.}p_{.j}$, then, according to the rule that has been stated in this section, the random variable

$$\sum_{j=1}^{b} \sum_{i=1}^{a} \frac{[X_{ij} - n(X_{i.}/n)(X_{.j}/n)]^2}{n(X_{i.}/n)(X_{.j}/n)}$$

has an approximate chi-square distribution with $ab - 1 - (a + b - 2) = (a - 1)(b - 1)$ degrees of freedom provided that H_0 is true. The hypothesis H_0 is then rejected if the computed value of this statistic exceeds the constant c, where c is selected from Table II so that the test has the desired significance level α.

In each of the four examples of this section, we have indicated that the statistic used to test the hypothesis H_0 has an approximate chi-square distribution, provided that n is sufficiently large and H_0 is true. To compute the power of any of these tests for values of the parameters not described by H_0, we need the distribution of the statistic when H_0 is not true. In each of these cases, the statistic has an approximate distribution called a *noncentral chi-square distribution*. The noncentral chi-square distribution will be discussed in Section 10.3.

EXERCISES

6.53. A number is to be selected from the interval $\{x : 0 < x < 2\}$ by a random process. Let $A_i = \{x : (i-1)/2 < x \le i/2\}$, $i = 1, 2, 3$, and let $A_4 = \{x : \frac{3}{2} < x < 2\}$. A certain hypothesis assigns probabilities p_{i0} to these sets in accordance with $p_{i0} = \int_{A_i} (\frac{1}{2})(2 - x)\, dx$, $i = 1, 2, 3, 4$. This hypothesis (concerning the multinomial p.d.f. with $k = 4$) is to be tested, at the 5 percent level of significance, by a chi-square test. If the observed frequencies of the sets A_i, $i = 1, 2, 3, 4$, are, respectively, 30, 30, 10, 10, would H_0 be accepted at the (approximate) 5 percent level of significance?

6.54. Let the following sets be defined: $A_1 = \{x : -\infty < x \le 0\}$, $A_i = \{x : i - 2 < x \le i - 1\}$, $i = 2, \ldots, 7$, and $A_8 = \{x : 6 < x < \infty\}$. A certain hypothesis assigns probabilities p_{i0} to these sets A_i in accordance with

$$p_{i0} = \int_{A_i} \frac{1}{2\sqrt{2\pi}} \exp\left[-\frac{(x-3)^2}{2(4)}\right] dx, \qquad i = 1, 2, \ldots, 7, 8.$$

This hypothesis (concerning the multinomial p.d.f. with $k = 8$) is to be tested, at the 5 percent level of significance, by a chi-square test. If the observed frequencies of the sets A_i, $i = 1, 2, \ldots, 8$, are, respectively, 60, 96, 140, 210, 172, 160, 88, and 74, would H_0 be accepted at the (approximate) 5 percent level of significance?

6.55. A die was cast $n = 120$ independent times and the following data resulted:

Spots up	1	2	3	4	5	6
Frequency	b	20	20	20	20	40-b

If we use a chi-square test, for what values of b would the hypothesis that the die is unbiased be rejected at the 0.025 significance level?

6.56. Consider the problem from genetics of crossing two types of peas. The Mendelian theory states that the probabilities of the classifications (a) round and yellow, (b) wrinkled and yellow, (c) round and green, and (d) wrinkled and green are $\frac{9}{16}$, $\frac{3}{16}$, $\frac{3}{16}$, and $\frac{1}{16}$, respectively. If, from 160 independent observations, the observed frequencies of these respective classifications are 86, 35, 26, and 13, are these data consistent with the Mendelian theory? That is, test, with $\alpha = 0.01$, the hypothesis that the respective probabilities are $\frac{9}{16}$, $\frac{3}{16}$, $\frac{3}{16}$, and $\frac{1}{16}$.

6.57. Two different teaching procedures were used on two different groups of students. Each group contained 100 students of about the same ability. At the end of the term, an evaluating team assigned a letter grade to each student. The results were tabulated as follows.

	Grade					
Group	A	B	C	D	F	Total
I	15	25	32	17	11	100
II	9	18	29	28	16	100

If we consider these data to be independent observations from two respective multinomial distributions with $k = 5$, test, at the 5 percent significance level, the hypothesis that the two distributions are the same (and hence the two teaching procedures are equally effective).

6.58. Let the result of a random experiment be classified as one of the mutually exclusive and exhaustive ways A_1, A_2, A_3 and also as one of the mutually exclusive and exhaustive ways B_1, B_2, B_3, B_4. Two hundred independent trials of the experiment result in the following data:

	B_1	B_2	B_3	B_4
A_1	10	21	15	6
A_2	11	27	21	13
A_3	6	19	27	24

Test, at the 0.05 significance level, the hypothesis of independence of the A attribute and the B attribute, namely $H_0: P(A_i \cap B_j) = P(A_i)P(B_j)$, $i = 1, 2, 3$ and $j = 1, 2, 3, 4$, against the alternative of dependence.

6.59. A certain genetic model suggests that the probabilities of a particular trinomial distribution are, respectively, $p_1 = p^2$, $p_2 = 2p(1 - p)$, and $p_3 = (1 - p)^2$, where $0 < p < 1$. If X_1, X_2, X_3 represent the respective frequencies in n independent trials, explain how we could check on the adequacy of the genetic model.

6.60. Let the result of a random experiment be classified as one of the mutually exclusive and exhaustive ways A_1, A_2, A_3 and also as one of the mutually and exhaustive ways B_1, B_2, B_3, B_4. Say that 180 independent trials of the experiment result in the following frequencies:

	B_1	B_2	B_3	B_4
A_1	$15 - 3k$	$15 - k$	$15 + k$	$15 + 3k$
A_2	15	15	15	15
A_3	$15 + 3k$	$15 + k$	$15 - k$	$15 - 3k$

where k is one of the integers 0, 1, 2, 3, 4, 5. What is the smallest value of k that will lead to the rejection of the independence of the A attribute and the B attribute at the $\alpha = 0.05$ significance level?

6.61. It is proposed to fit the Poisson distribution to the following data

x	0	1	2	3	$3 < x$
Frequency	20	40	16	18	6

(a) Compute the corresponding chi-square goodness-of-fit statistic. *Hint:* In computing the mean, treat $3 < x$ as $x = 4$.
(b) How many degrees of freedom are associated with this chi-square?
(c) Do these data result in the rejection of the Poisson model at the $\alpha = 0.05$ significance level?

ADDITIONAL EXERCISES

6.62. Let $Y_1 < Y_2 < \cdots < Y_n$ be the order statistics of a random sample of size n from the distribution having p.d.f. $f(x) = 2x/\theta^2$, $0 < x < \theta$, zero elsewhere.
(a) If $0 < c < 1$, show that $\Pr(c < Y_n/\theta < 1) = 1 - c^{2n}$.
(b) If $n = 5$ and if the observed value of Y_n is 1.8, find a 99 percent confidence interval for θ.

6.63. If 0.35, 0.92, 0.56, and 0.71 are the four observed values of a random sample from a distribution having p.d.f. $f(x; \theta) = \theta x^{\theta - 1}$, $0 < x < 1$, zero elsewhere, find an estimate for θ.

6.64. Let the table

x	0	1	2	3	4	5
Frequency	6	10	14	13	6	1

represent a summary of a random sample of size 50 from a Poisson distribution. Find the maximum likelihood estimate of Pr $(X = 2)$.

6.65. Let X be $N(\mu, 100)$. To test $H_0 : \mu = 80$ against $H_1 : \mu > 80$, let the critical region be defined by $C = \{(x_1, x_2, \ldots, x_{25}) : \bar{x} \geq 83\}$, where $\bar{x}$ is the sample mean of a random sample of size $n = 25$ from this distribution.
(a) How is the power function $K(\mu)$ defined for this test?
(b) What is the significance level of this test?
(c) What are the values of $K(80)$, $K(83)$, and $K(86)$?
(d) Sketch the graph of the power function.
(e) What is the p-value corresponding to $\bar{x} = 83.41$?

6.66. Let X equal the yield of alfalfa in tons per acre per year. Assume that X is $N(1.5, 0.09)$. It is hoped that new fertilizer will increase the average yield. We shall test the null hypothesis $H_0 : \mu = 1.5$ against the alternative hypothesis $H_1 : \mu > 1.5$. Assume that the variance continues to equal $\sigma^2 = 0.09$ with the new fertilizer. Using $\bar{X}$, the mean of a random sample of size n, as the test statistic, reject H_0 if $\bar{x} \geq c$. Find n and c so that the power function $K(\mu) = \Pr(\bar{X} \geq c : \mu)$ is such that $\alpha = K(1.5) = 0.05$ and $K(1.7) = 0.95$.

6.67. A random sample of 100 observations from a Poisson distribution has a mean equal to 6.25. Construct an approximate 95 percent confidence interval for the mean of the distribution.

6.68. Say that a random sample of size 25 is taken from a binomial distribution with parameters $n = 5$ and p. These data are then lost, but we recall that the relative frequency of the value 5 was $\frac{6}{25}$. Under these conditions, how would you estimate p? Is this suggested estimate unbiased?

6.69. When 100 tacks were thrown on a table, 60 of them landed point up. Obtain a 95 percent confidence interval for the probability that a tack of this type will land point up. Assume independence.

6.70. Let $X_1, X_2, \ldots, X_8$ be a random sample of size $n = 8$ from a Poisson distribution with mean μ. Reject the simple null hypothesis $H_0 : \mu = 0.5$ and accept $H_1 : \mu > 0.5$ if the observed sum $\sum_{i=1}^{8} x_i \geq 8$.
(a) Compute the significance level α of the test.
(b) Find the power function $K(\mu)$ of the test as a sum of Poisson probabilities.
(c) Using the Appendix, determine $K(0.75)$, $K(1)$, and $K(1.25)$.

6.71. Let p denote the probability that, for a particular tennis player, the first serve is good. Since $p = 0.40$, this player decided to take lessons in order to increase p. When the lessons are completed, the hypothesis

$H_0 : p = 0.40$ will be tested against $H_1 : p > 0.40$ based on $n = 25$ trials. Let y equal the number of first serves that are good, and let the critical region be defined by $C = \{y : y \geq 13\}$.
(a) Determine $\alpha = \Pr(Y \geq 13; p = 0.40)$.
(b) Find $\beta = \Pr(Y < 13)$ when $p = 0.60$; that is, $\beta = \Pr(Y \leq 12; p = 0.60)$.

6.72. The mean birth weight in the United States is $\mu = 3315$ grams with a standard deviation of $\sigma = 575$. Let X equal the birth weight in grams in Jerusalem. Assume that the distribution of X is $N(\mu, \sigma^2)$. We shall test the null hypothesis $H_0 : \mu = 3315$ against the alternative hypothesis $H_1 : \mu < 3315$ using a random sample of size $n = 30$.
(a) Define a critical region that has a significance level of $\alpha = 0.05$.
(b) If the random sample of $n = 30$ yielded $\bar{x} = 3189$ and $s = 488$, what is your conclusion?
(c) What is the approximate p-value of your test?

6.73. Let $Y_1 < Y_2 < \cdots < Y_5$ be the order statistics of a random sample of size 5 from the distribution having p.d.f. $f(x) = \exp[-(x - \theta)/\beta]/\beta$, $\theta < x < \infty$, zero elsewhere. Discuss the construction of a 90 percent confidence interval for β if θ is known.

6.74. Three independent random samples, each of size 6, are drawn from three normal distributions having common unknown variance. We find the three sample variances to be 10, 14, and 8, respectively.
(a) Compute an unbiased estimate of the common variance.
(b) Determine a 90 percent confidence interval for the common variance.

6.75. Let $X_1, X_2, \ldots, X_n$ be a random sample from $N(\mu, \sigma^2)$.
(a) If the constant b is defined by the equation $\Pr(X \leq b) = 0.90$, find the m.l.e. of b.
(b) If c is given constant, find the m.l.e. of $\Pr(X \leq c)$.

6.76. Let $\bar{X}_1, \bar{X}_2$, and $\bar{X}_3$ and S_1^2, S_2^2, and S_3^2 denote the means and the variances of three independent random samples, each of size 10, from a normal distribution with mean μ and variance σ^2. Find the constant c so that

$$\Pr\left(\frac{\bar{X}_1 + \bar{X}_2 - 2\bar{X}_3}{\sqrt{10S_1^2 + 10S_2^2 + 10S_3^2}} \leq c\right) = 0.95.$$

6.77. Let Y be $b(192, p)$. We reject $H_0 : p = 0.75$ and accept $H_1 : p > 0.75$ if and only if $Y \geq 152$. Use the normal approximation to determine:
(a) $\alpha = \Pr(Y \geq 152; p = 0.75)$.
(b) $\beta = \Pr(Y < 152)$ when $p = 0.80$.

6.78. Let Y be $b(100, p)$. To test $H_0 : p = 0.08$ against $H_1 : p < 0.08$, we reject H_0 and accept H_1 if and only if $Y \leq 6$.

(a) Determine the significance level α of the test.
(b) Find the probability of the type II error if in fact $p = 0.04$.

6.79. Let $X_1, X_2, \ldots, X_n$ be a random sample from a Bernoulli distribution with parameter p. If p is restricted so that we know that $\frac{1}{2} \leq p \leq 1$, find the m.l.e. of this parameter.

6.80. Consider two Bernoulli distributions with unknown parameters p_1 and p_2, respectively. If Y and Z equal the numbers of successes in two independent random samples, each of sample size n, from the respective distributions, determine the maximum likelihood estimators of p_1 and p_2 if we know that $0 \leq p_1 \leq p_2 \leq 1$.

6.81. Let $(X_1, Y_1), (X_2, Y_2), \ldots, (X_n, Y_n)$ be n i.i.d. pairs of random variables, each with the bivariate normal distribution having five parameters $\mu_1, \mu_2, \sigma_1^2, \sigma_2^2$, and ρ.
(a) Show that $Z_i = X_i - Y_i$ is $N(\mu, \sigma^2)$, where $\mu = \mu_1 - \mu_2$ and $\sigma^2 = \sigma_1^2 - 2\rho\sigma_1\sigma_2 + \sigma_2^2$, $i = 1, 2, \ldots, n$.
(b) Since all five parameters are unknown, μ and σ^2 are unknown. To test $H_0: \mu = 0$ $(H_0: \mu_1 = \mu_2)$ against $H_1: \mu > 0$ $(H_1: \mu_1 > \mu_2)$, construct a t-test based upon the mean and the variance of the n differences $Z_1, Z_2, \ldots, Z_n$. This is often called a *paired t-test*.

CHAPTER 7

Sufficient Statistics

7.1 Measures of Quality of Estimators

In Chapter 6 we presented some procedures for finding point estimates, interval estimates, and tests of statistical hypotheses. In this and the next two chapters, we provide reasons why certain statistics are used in these various statistical inferences. We begin by considering desirable properties of a point estimate.

Now it would seem that if $y = u(x_1, x_2, \ldots, x_n)$ is to qualify as a good point estimate of θ, there should be a great probability that the statistic $Y = u(X_1, X_2, \ldots, X_n)$ will be close to θ; that is, θ should be a sort of rallying point for the numbers $y = u(x_1, x_2, \ldots, x_n)$. This can be achieved in one way by selecting $Y = u(X_1, X_2, \ldots, X_n)$ in such a way that not only is Y an unbiased estimator of θ, but also the variance of Y is as small as it can be made. We do this because the variance of Y is a measure of the intensity of the concentration of the probability for Y in the neighborhood of the point $\theta = E(Y)$. Accordingly, we define an unbiased minimum variance estimator of the parameter θ in the following manner.

Definition 1. For a given positive integer n, $Y = u(X_1, X_2, \ldots, X_n)$ will be called an *unbiased minimum variance* estimator of the parameter θ if Y is unbiased, that is, $E(Y) = \theta$, and if the variance of Y is less than or equal to the variance of every other unbiased estimator of θ.

For illustration, let $X_1, X_2, \ldots, X_9$ denote a random sample from a distribution that is $N(\theta, 1)$, $-\infty < \theta < \infty$. Since the statistic $\bar{X} = (X_1 + X_2 + \cdots + X_9)/9$ is $N(\theta, \frac{1}{9})$, $\bar{X}$ is an unbiased estimator of θ. The statistic X_1 is $N(\theta, 1)$, so X_1 is also an unbiased estimator of θ. Although the variance $\frac{1}{9}$ of $\bar{X}$ is less than the variance 1 of X_1, we cannot say, with $n = 9$, that $\bar{X}$ is the unbiased minimum variance estimator of θ; that definition requires that the comparison be made with every unbiased estimator of θ. To be sure, it is quite impossible to tabulate all other unbiased estimators of this parameter θ, so other methods must be developed for making the comparisons of the variances. A beginning on this problem will be made in this chapter.

Let us now discuss the problem of point estimation of a parameter from a slightly different standpoint. Let $X_1, X_2, \ldots, X_n$ denote a random sample of size n from a distribution that has the p.d.f. $f(x; \theta)$, $\theta \in \Omega$. The distribution may be either of the continuous or the discrete type. Let $Y = u(X_1, X_2, \ldots, X_n)$ be a statistic on which we wish to base a point estimate of the parameter θ. Let $\delta(y)$ be that function of the observed value of the statistic Y which is the point estimate of θ. Thus the function δ *decides* the value of our point estimate of θ and δ is called a *decision function* or a *decision rule*. One value of the decision function, say $\delta(y)$, is called a *decision*. Thus a numerically determined point estimate of a parameter θ is a decision. Now a decision may be correct or it may be wrong. It would be useful to have a measure of the seriousness of the difference, if any, between the true value of θ and the point estimate $\delta(y)$. Accordingly, with each pair, $[\theta, \delta(y)]$, $\theta \in \Omega$, we will associate a nonnegative number $\mathscr{L}[\theta, \delta(y)]$ that reflects this seriousness. We call the function $\mathscr{L}$ the *loss function*. The expected (mean) value of the loss function is called the *risk function*. If $g(y; \theta)$, $\theta \in \Omega$, is the p.d.f. of Y, the risk function $R(\theta, \delta)$ is given by

$$R(\theta, \delta) = E\{\mathscr{L}[\theta, \delta(Y)]\} = \int_{-\infty}^{\infty} \mathscr{L}[\theta, \delta(y)]g(y; \theta)\, dy$$

if Y is a random variable of the continuous type. It would be desirable to select a decision function that minimizes the risk $R(\theta, \delta)$ for all values of θ, $\theta \in \Omega$. But this is usually impossible because the decision function δ that minimizes $R(\theta, \delta)$ for one value of θ may not minimize

$R(\theta, \delta)$ for another value of θ. Accordingly, we need either to restrict our decision function to a certain class or to consider methods of ordering the risk functions. The following example, while very simple, dramatizes these difficulties.

Example 1. Let $X_1, X_2, \ldots, X_{25}$ be a random sample from a distribution that is $N(\theta, 1)$, $-\infty < \theta < \infty$. Let $Y = \bar{X}$, the mean of the random sample, and let $\mathscr{L}[\theta, \delta(y)] = [\theta - \delta(y)]^2$. We shall compare the two decision functions given by $\delta_1(y) = y$ and $\delta_2(y) = 0$ for $-\infty < y < \infty$. The corresponding risk functions are

$$R(\theta, \delta_1) = E[(\theta - Y)^2] = \tfrac{1}{25}$$

and

$$R(\theta, \delta_2) = E[(\theta - 0)^2] = \theta^2.$$

Obviously, if, in fact, $\theta = 0$, then $\delta_2(y) = 0$ is an excellent decision and we have $R(0, \delta_2) = 0$. However, if θ differs from zero by very much, it is equally clear that $\delta_2(y) = 0$ is a poor decision. For example, if, in fact, $\theta = 2$, $R(2, \delta_2) = 4 > R(2, \delta_1) = \frac{1}{25}$. In general, we see that $R(\theta, \delta_2) < R(\theta, \delta_1)$, provided that $-\frac{1}{5} < \theta < \frac{1}{5}$ and that otherwise $R(\theta, \delta_2) \geq R(\theta, \delta_1)$. That is, one of these decision functions is better than the other for some values of θ and the other decision functions are better for other values of θ. If, however, we had restricted our consideration to decision functions δ such that $E[\delta(Y)] = \theta$ for all values of θ, $\theta \in \Omega$, then the decision $\delta_2(y) = 0$ is not allowed. Under this restriction and with the given $\mathscr{L}[\theta, \delta(y)]$, the risk function is the variance of the unbiased estimator $\delta(Y)$, and we are confronted with the problem of finding the unbiased minimum variance estimator. Later in this chapter we show that the solution is $\delta(y) = y = \bar{x}$.

Suppose, however, that we do not want to restrict ourselves to decision functions δ, such that $E[\delta(Y)] = \theta$ for all values of θ, $\theta \in \Omega$. Instead, let us say that the decision function that minimizes the maximum of the risk function is the best decision function. Because, in this example, $R(\theta, \delta_2) = \theta^2$ is unbounded, $\delta_2(y) = 0$ is not, in accordance with this criterion, a good decision function. On the other hand, with $-\infty < \theta < \infty$, we have

$$\max_\theta R(\theta, \delta_1) = \max_\theta \left(\tfrac{1}{25}\right) = \tfrac{1}{25}.$$

Accordingly, $\delta_1(y) = y = \bar{x}$ seems to be a very good decision in accordance with this criterion because $\frac{1}{25}$ is small. As a matter of fact, it can be proved that δ_1 is the best decision function, as measured by the *minimax criterion*, when the loss function is $\mathscr{L}[\theta, \delta(y)] = [\theta - \delta(y)]^2$.

In this example we illustrated the following:

1. Without some restriction on the decision function, it is difficult to

find a decision function that has a risk function which is uniformly less than the risk function of another decision function.

2. A principle of selecting a best decision function, called the *minimax principle*. This principle may be stated as follows: If the decision function given by $\delta_0(y)$ is such that, for all $\theta \in \Omega$,

$$\max_{\theta} R[\theta, \delta_0(y)] \leq \max_{\theta} R[\theta, \delta(y)]$$

for every other decision function $\delta(y)$, then $\delta_0(y)$ is called a *minimax decision function*.

With the restriction $E[\delta(Y)] = \theta$ and the loss function $\mathscr{L}[\theta, \delta(y)] = [\theta - \delta(y)]^2$, the decision function that minimizes the risk function yields an unbiased estimator with minimum variance. If, however, the restriction $E[\delta(Y)] = \theta$ is replaced by some other condition, the decision function $\delta(Y)$, if it exists, which minimizes $E\{[\theta - \delta(Y)]^2\}$ uniformly in θ is sometimes called the *minimum mean-square-error estimator*. Exercises 7.6, 7.7, and 7.8 provide examples of this type of estimator.

There are two additional observations about decision rules and loss functions that should be made at this point. First, since Y is a statistic, the decision rule $\delta(Y)$ is also a statistic, and we could have started directly with a decision rule based on the observations in a random sample, say $\delta_1(X_1, X_2, \ldots, X_n)$. The risk function is then given by

$$\begin{aligned} R(\theta, \delta_1) &= E\{\mathscr{L}[\theta, \delta_1(X_1, X_2, \ldots, X_n)]\} \\ &= \int_{-\infty}^{\infty}\int_{-\infty}^{\infty}\cdots\int_{-\infty}^{\infty} \mathscr{L}[\theta, \delta_1(x_1, x_2, \ldots, x_n)] \\ &\qquad \times f(x_1; \theta)\cdots f(x_n; \theta)\, dx_1 \cdots dx_n \end{aligned}$$

if the random sample arises from a continuous-type distribution. We did not do this because, as you will see in this chapter, it is rather easy to find a good statistic, say Y, upon which to base all of the statistical inferences associated with a particular model. Thus we thought it more appropriate to start with a statistic that would be familiar, like the m.l.e. $Y = \bar{X}$ in Example 1. The second decision rule of that example could be written $\delta_2(X_1, X_2, \ldots, X_n) = 0$, a constant no matter what values of $X_1, X_2, \ldots, X_n$ are observed.

The second observation is that we have only used one loss function, namely the *square-error loss function* $\mathscr{L}(\theta, \delta) = (\theta - \delta)^2$.

The *absolute-error loss function* $\mathscr{L}(\theta, \delta) = |\theta - \delta|$ is another popular one. The loss function defined by

$$\begin{aligned} \mathscr{L}(\theta, \delta) &= 0, \qquad |\theta - \delta| \le a, \\ &= b, \qquad |\theta - \delta| > a, \end{aligned}$$

where a and b are positive constants, is sometimes referred to as the *goal post loss function*. The reason for this terminology is that football fans recognize it is like kicking a field goal: There is no loss (actually a three-point gain) if within a units of the middle but b units of loss (zero points awarded) if outside that restriction. In addition, loss functions can be asymmetric as well as symmetric as the three previous ones have been. That is, for example, it might be more costly to underestimate the value of θ than to overestimate it. (Many of us think about this type of loss function when estimating the time it takes us to reach an airport to catch a plane.) Some of these loss functions are considered when studying Bayesian estimates in Chapter 8.

Let us close this section with an interesting illustration that raises a question leading to the likelihood principle which many statisticians believe is a quality characteristic that estimators should enjoy. Suppose that two statisticians, A and B, observe 10 independent trials of a random experiment ending in success or failure. Let the probability of success on each trial be θ, where $0 < \theta < 1$. Let us say that each statistician observes one success in these 10 trials. Suppose, however, that A had decided to take $n = 10$ such observations in advance and found only one success while B had decided to take as many observations as needed to get the first success, which happened on the 10th trial. The model of A is that Y is $b(n = 10, \theta)$ and $y = 1$ is observed. On the other hand, B is considering the random variable Z that has a geometric p.d.f. $g(z) = (1 - \theta)^{z-1}\theta$, $z = 1, 2, 3, \ldots$, and $z = 10$ is observed. In either case, the relative frequency of success is

$$\frac{y}{n} = \frac{1}{z} = \frac{1}{10},$$

which could be used as an estimate of θ.

Let us observe, however, that one of the corresponding estimators, Y/n and $1/Z$, is biased. We have

$$E\left(\frac{Y}{10}\right) = \frac{1}{10} E(Y) = \frac{1}{10}(10\theta) = \theta$$

while

$$E\left(\frac{1}{Z}\right) = \sum_{z=1}^{\infty} \frac{1}{z}(1-\theta)^{z-1}\theta$$

$$= \theta + \tfrac{1}{2}(1-\theta)\theta + \tfrac{1}{3}(1-\theta)^2\theta + \cdots > \theta.$$

That is, $1/Z$ is a biased estimator while $Y/10$ is unbiased. Thus A is using an unbiased estimator while B is not. Should we adjust B's estimator so that it too is unbiased?

It is interesting to note that if we maximize the two respective likelihood functions, namely

$$L_1(\theta) = \binom{10}{y}\theta^y(1-\theta)^{10-y}$$

and

$$L_2(\theta) = (1-\theta)^{z-1}\theta,$$

with $n = 10$, $y = 1$, and $z = 10$, we get exactly the same answer, $\hat{\theta} = \frac{1}{10}$. This must be the case, because in each situation we are maximizing $(1-\theta)^9\theta$. Many statisticians believe that this is the way it should be and accordingly adopt the *likelihood principle:*

Suppose two different sets of data from possibly two different random experiments lead to respective likelihood ratios, $L_1(\theta)$ and $L_2(\theta)$, that are proportional to each other. These two data sets provide the same information about the parameter θ and a statistician should obtain the same estimate of θ from either.

In our special illustration, we note that $L_1(\theta) \propto L_2(\theta)$, and the likelihood principle states that statisticians A and B should make the same inference. Thus believers in the likelihood principle would not adjust the second estimator to make it unbiased.

EXERCISES

7.1. Show that the mean $\bar{X}$ of a random sample of size n from a distribution having p.d.f. $f(x; \theta) = (1/\theta)e^{-(x/\theta)}$, $0 < x < \infty$, $0 < \theta < \infty$, zero elsewhere, is an unbiased estimator of θ and has variance θ^2/n.

7.2. Let $X_1, X_2, \ldots, X_n$ denote a random sample from a normal distribution with mean zero and variance θ, $0 < \theta < \infty$. Show that $\sum_1^n X_i^2/n$ is an unbiased estimator of θ and has variance $2\theta^2/n$.

7.3. Let $Y_1 < Y_2 < Y_3$ be the order statistics of a random sample of size 3 from the uniform distribution having p.d.f. $f(x; \theta) = 1/\theta, 0 < x < \theta, 0 < \theta < \infty$, zero elsewhere. Show that $4Y_1$, $2Y_2$, and $\frac{4}{3}Y_3$ are all unbiased estimators of θ. Find the variance of each of these unbiased estimators.

7.4. Let Y_1 and Y_2 be two independent unbiased estimators of θ. Say the variance of Y_1 is twice the variance of Y_2. Find the constants k_1 and k_2 so that $k_1 Y_1 + k_2 Y_2$ is an unbiased estimator with smallest possible variance for such a linear combination.

7.5. In Example 1 of this section, take $\mathscr{L}[\theta, \delta(y)] = |\theta - \delta(y)|$. Show that $R(\theta, \delta_1) = \frac{1}{5}\sqrt{2/\pi}$ and $R(\theta, \delta_2) = |\theta|$. Of these two decision functions δ_1 and δ_2, which yields the smaller maximum risk?

7.6. Let $X_1, X_2, \ldots, X_n$ denote a random sample from a Poisson distribution with parameter θ, $0 < \theta < \infty$. Let $Y = \sum_1^n X_i$ and let $\mathscr{L}[\theta, \delta(y)] = [\theta - \delta(y)]^2$. If we restrict our considerations to decision functions of the form $\delta(y) = b + y/n$, where b does not depend upon y, show that $R(\theta, \delta) = b^2 + \theta/n$. What decision function of this form yields a uniformly smaller risk than every other decision function of this form? With this solution, say δ, and $0 < \theta < \infty$, determine $\max_\theta R(\theta, \delta)$ if it exists.

7.7. Let $X_1, X_2, \ldots, X_n$ denote a random sample from a distribution that is $N(\mu, \theta)$, $0 < \theta < \infty$, where μ is unknown. Let $Y = \sum_1^n (X_i - \bar{X})^2/n = S^2$ and let $\mathscr{L}[\theta, \delta(y)] = [\theta - \delta(y)]^2$. If we consider decision functions of the form $\delta(y) = by$, where b does not depend upon y, show that $R(\theta, \delta) = (\theta^2/n^2)[(n^2 - 1)b^2 - 2n(n - 1)b + n^2]$. Show that $b = n/(n + 1)$ yields a minimum risk for decision functions of this form. Note that $nY/(n + 1)$ is not an unbiased estimator of θ. With $\delta(y) = ny/(n + 1)$ and $0 < \theta < \infty$, determine $\max_\theta R(\theta, \delta)$ if it exists.

7.8. Let $X_1, X_2, \ldots, X_n$ denote a random sample from a distribution that is $b(1, \theta)$, $0 \le \theta \le 1$. Let $Y = \sum_1^n X_i$ and let $\mathscr{L}[\theta, \delta(y)] = [\theta - \delta(y)]^2$. Consider decision functions of the form $\delta(y) = by$, where b does not depend upon y. Prove that $R(\theta, \delta) = b^2 n\theta(1 - \theta) + (bn - 1)^2\theta^2$. Show that

$$\max_\theta R(\theta, \delta) = \frac{b^4 n^2}{4[b^2 n - (bn - 1)^2]},$$

provided that the value b is such that $b^2 n \ge 2(bn - 1)^2$. Prove that $b = 1/n$ does not minimize $\max_\theta R(\theta, \delta)$.

7.9. Let $X_1, X_2, \ldots, X_n$ be a random sample from a Poisson distribution with mean $\theta > 0$.

(a) Statistician A observes the sample to be the values $x_1, x_2, \ldots, x_n$ with sum $y_1 = \Sigma\, x_i$. Find the m.l.e. of θ.

(b) Statistician B loses the sample values $x_1, x_2, \ldots, x_n$ but remembers the sum y_1 and the fact that the sample arose from a Poisson distribution. Thus B decides to create some fake observations which he calls $z_1, z_2, \ldots, z_n$ (as he knows they will probably not equal the original x-values) as follows. He notes that the conditional probability of independent Poisson random variables $Z_1, Z_2, \ldots, Z_n$ being equal to $z_1, z_2, \ldots, z_n$, given $\Sigma\, z_i = y_1$, is

$$\frac{\dfrac{\theta^{z_1}e^{-\theta}}{z_1!}\,\dfrac{\theta^{z_2}e^{-\theta}}{z_2!}\cdots\dfrac{\theta^{z_n}e^{-\theta}}{z_n!}}{\dfrac{(n\theta)^{y_1}e^{-n\theta}}{y_1!}} = \frac{y_1!}{z_1!\,z_2!\cdots z_n!}\left(\frac{1}{n}\right)^{z_1}\left(\frac{1}{n}\right)^{z_2}\cdots\left(\frac{1}{n}\right)^{z_n}$$

since $Y_1 = \Sigma\, Z_i$ has a Poisson distribution with mean $n\theta$. The latter distribution is multinomial with y_1 independent trials, each terminating in one of n mutually exclusive and exhaustive ways, each of which has the same probability $1/n$. Accordingly, B runs such a multinomial experiment y_1 independent trials and obtains $z_1, z_2, \ldots, z_n$. Find the likelihood function using these z-values. Is it proportional to that of statistician A?

Hint: Here the likelihood function is the product of this conditional p.d.f. and the p.d.f. of $Y_1 = \Sigma\, Z_i$.

7.2 A Sufficient Statistic for a Parameter

Suppose that $X_1, X_2, \ldots, X_n$ is a random sample from a distribution that has p.d.f. $f(x; \theta)$, $\theta \in \Omega$. In Chapter 6 and Section 7.1 we constructed statistics to make statistical inferences as illustrated by point and interval estimation and tests of statistical hypotheses. We note that a statistic, say $Y = u(X_1, X_2, \ldots, X_n)$, is a form of data reduction. For illustration, instead of listing all of the individual observations $X_1, X_2, \ldots, X_n$, we might prefer to give only the sample mean $\bar{X}$ or the sample variance S^2. Thus statisticians look for ways of reducing a set of data so that these data can be more easily understood without losing the meaning associated with the entire set of observations.

It is interesting to note that a statistic $Y = u(X_1, X_2, \ldots, X_n)$ really partitions the sample space of $X_1, X_2, \ldots, X_n$. For illustration, suppose we say that the sample was observed and $\bar{x} = 8.32$. There are many points in the sample space which have that same mean of 8.32, and we can consider them as belonging to the set $\{(x_1, x_2, \ldots, x_n) : \bar{x} = 8.32\}$. As a matter of fact, all points on the hyperplane

$$x_1 + x_2 + \cdots + x_n = (8.32)n$$

yield the mean of $\bar{x} = 8.32$, so this hyperplane is that set. However, there are many values that $\bar{X}$ can take and thus there are many such sets. So, in this sense, the sample mean $\bar{X}$—or any statistic $Y = u(X_1, X_2, \ldots, X_n)$—partitions the sample space into a collection of sets.

Often in the study of statistics the parameter θ of the model is unknown; thus we desire to make some statistical inference about it. In this section we consider a statistic denoted by $Y_1 = u_1(X_1, X_2, \ldots, X_n)$, which we call a *sufficient statistic* and which we find is good for making those inferences. This sufficient statistic partitions the sample space in such a way that, given

$$(X_1, X_2, \ldots, X_n) \in \{(x_1, x_2, \ldots, x_n) : u_1(x_1, x_2, \ldots, x_n) = y_1\},$$

the conditional probability of $X_1, X_2, \ldots, X_n$ does not depend upon θ. Intuitively, this means that once the set determined by $Y_1 = y_1$ is fixed, the distribution of another statistic, say $Y_2 = u_2(X_1, X_2, \ldots, X_n)$, does not depend upon the parameter θ because the conditional distribution of $X_1, X_2, \ldots, X_n$ does not depend upon θ. Hence it is impossible to use Y_2, given $Y_1 = y_1$, to make a statistical inference about θ. So, in a sense, Y_1 *exhausts* all the information about θ that is contained in the sample. This is why we call $Y_1 = u_1(X_1, X_2, \ldots, X_n)$ a sufficient statistic.

To understand clearly the definition of a sufficient statistic for a parameter θ, we start with an illustration.

Example 1. Let $X_1, X_2, \ldots, X_n$ denote a random sample from the distribution that has p.d.f.

$$\begin{aligned} f(x; \theta) &= \theta^x(1-\theta)^{1-x}, \qquad x = 0, 1; \quad 0 < \theta < 1; \\ &= 0 \qquad \text{elsewhere.} \end{aligned}$$

The statistic $Y_1 = X_1 + X_2 + \cdots + X_n$ has the p.d.f.

$$\begin{aligned} g_1(y_1; \theta) &= \binom{n}{y_1}\theta^{y_1}(1-\theta)^{n-y_1}, \qquad y_1 = 0, 1, \ldots, n, \\ &= 0 \qquad \text{elsewhere.} \end{aligned}$$

What is the conditional probability

$$\Pr(X_1 = x_1, X_2 = x_2, \ldots, X_n = x_n | Y_1 = y_1) = P(A|B),$$

say, where $y_1 = 0, 1, 2, \ldots, n$? Unless the sum of the integers $x_1, x_2, \ldots, x_n$ (each of which equals zero or 1) is equal to y_1, the conditional probability obviously equals zero because $A \cap B = \varnothing$. But in the case $y_1 = \Sigma x_i$, we have that $A \subset B$ so that $A \cap B = A$ and $P(A|B) = P(A)/P(B)$; thus the conditional probability equals

$$\frac{\theta^{x_1}(1-\theta)^{1-x_1}\theta^{x_2}(1-\theta)^{1-x_2}\cdots\theta^{x_n}(1-\theta)^{1-x_n}}{\binom{n}{y_1}\theta^{y_1}(1-\theta)^{n-y_1}} = \frac{\theta^{\Sigma x_i}(1-\theta)^{n-\Sigma x_i}}{\binom{n}{\sum x_i}\theta^{\Sigma x_i}(1-\theta)^{n-\Sigma x_i}}$$

$$= \frac{1}{\binom{n}{\sum x_i}}.$$

Since $y_1 = x_1 + x_2 + \cdots + x_n$ equals the number of 1's in the n independent trials, this conditional probability is the probability of selecting a particular arrangement of y_1 1's and $(n - y_1)$ zeros. Note that this conditional probability does *not* depend upon the value of the parameter θ.

In general, let $g_1(y_1; \theta)$ be the p.d.f. of the statistic $Y_1 = u_1(X_1, X_2, \ldots, X_n)$, where $X_1, X_2, \ldots, X_n$ is a random sample arising from a distribution of the discrete type having p.d.f. $f(x; \theta)$, $\theta \in \Omega$. The conditional probability of $X_1 = x_1, X_2 = x_2, \ldots, X_n = x_n$, given $Y_1 = y_1$, equals

$$\frac{f(x_1; \theta)f(x_2; \theta)\cdots f(x_n; \theta)}{g_1[u_1(x_1, x_2, \ldots, x_n); \theta]},$$

provided that $x_1, x_2, \ldots, x_n$ are such that the fixed $y_1 = u_1(x_1, x_2, \ldots, x_n)$, and equals zero otherwise. We say that $Y_1 = u_1(X_1, X_2, \ldots, X_n)$ is a *sufficient statistic* for θ if and only if this ratio does not depend upon θ. While, with distributions of the continuous type, we cannot use the same argument, we do, in this case, accept the fact that if this ratio does not depend upon θ, then the conditional distribution of $X_1, X_2, \ldots, X_n$, given $Y_1 = y_1$, does not depend upon θ. Thus, in both cases, we use the same definition of a sufficient statistic for θ.

Definition 2. Let $X_1, X_2, \ldots, X_n$ denote a random sample of size n from a distribution that has p.d.f. $f(x; \theta)$, $\theta \in \Omega$. Let

$Y_1 = u_1(X_1, X_2, \ldots, X_n)$ be a statistic whose p.d.f. is $g_1(y_1; \theta)$. Then Y_1 is a *sufficient statistic* for θ if and only if

$$\frac{f(x_1; \theta)f(x_2; \theta) \cdots f(x_n; \theta)}{g_1[u_1(x_1, x_2, \ldots, x_n); \theta]} = H(x_1, x_2, \ldots, x_n),$$

where $H(x_1, x_2, \ldots, x_n)$ does not depend upon $\theta \in \Omega$.

Remark. In most cases in this book, $X_1, X_2, \ldots, X_n$ do represent the observations of a random sample; that is, they are i.i.d. It is not necessary, however, in more general situations, that these random variables be independent; as a matter of fact, they do not need to be identically distributed. Thus, more generally, the definition of sufficiency of a statistic $Y_1 = u_1(X_1, X_2, \ldots, X_n)$ would be extended to read that

$$\frac{f(x_1, x_2, \ldots, x_n; \theta)}{g_1[u_1(x_1, x_2, \ldots, x_n); \theta]} = H(x_1, x_2, \ldots, x_n)$$

does not depend upon $\theta \in \Omega$, where $f(x_1, x_2, \ldots, x_n; \theta)$ is the joint p.d.f. of $X_1, X_2, \ldots, X_n$. There are even a few situations in which we need an extension like this one in this book.

We now give two examples that are illustrative of the definition.

Example 2. Let $X_1, X_2, \ldots, X_n$ be a random sample from a gamma distribution with $\alpha = 2$ and $\beta = \theta > 0$. Since the m.g.f. associated with this distribution is $M(t) = (1 - \theta t)^{-2}$, $t < 1/\theta$, the m.g.f. of $Y_1 = \sum_{i=1}^{n} X_i$ is

$$\begin{aligned} E[e^{t(X_1 + X_2 + \cdots + X_n)}] &= E(e^{tX_1})E(e^{tX_2}) \cdots E(e^{tX_n}) \\ &= [(1 - \theta t)^{-2}]^n = (1 - \theta t)^{-2n}. \end{aligned}$$

Thus Y_1 has a gamma distribution with $\alpha = 2n$ and $\beta = \theta$, so that its p.d.f. is

$$\begin{aligned} g_1(y_1; \theta) &= \frac{1}{\Gamma(2n)\theta^{2n}} y_1^{2n-1} e^{-y_1/\theta}, \qquad 0 < y_1 < \infty, \\ &= 0 \qquad \text{elsewhere.} \end{aligned}$$

Thus we have that the ratio in Definition 2 equals

$$\frac{\left[\dfrac{x_1^{2-1}e^{-x_1/\theta}}{\Gamma(2)\theta^2}\right]\left[\dfrac{x_2^{2-1}e^{-x_2/\theta}}{\Gamma(2)\theta^2}\right]\cdots\left[\dfrac{x_n^{2-1}e^{-x_n/\theta}}{\Gamma(2)\theta^2}\right]}{\dfrac{(x_1 + x_2 + \cdots + x_n)^{2n-1}e^{-(x_1 + x_2 + \cdots + x_n)/\theta}}{\Gamma(2n)\theta^{2n}}} = \frac{\Gamma(2n)}{[\Gamma(2)]^n}\frac{x_1 x_2 \cdots x_n}{(x_1 + x_2 + \cdots x_n)^{2n-1}},$$

where $0 < x_i < \infty$, $i = 1, 2, \ldots, n$. Since this ratio does not depend upon θ, the sum Y_1 is a sufficient statistic for θ.

Example 3. Let $Y_1 < Y_2 < \cdots < Y_n$ denote the order statistics of a random sample of size n from the distribution with p.d.f.

$$f(x; \theta) = e^{-(x-\theta)} I_{(\theta, \infty)}(x).$$

Here we use the indicator function of set A defined by

$$\begin{aligned} I_A(x) &= 1, \qquad x \in A, \\ &= 0, \qquad x \notin A. \end{aligned}$$

This means, of course, that $f(x; \theta) = e^{-(x-\theta)}$, $\theta < x < \infty$, and zero elsewhere. The p.d.f. of $Y_1 = \min (X_i)$ is

$$g_1(y_1; \theta) = ne^{-n(y_1-\theta)} I_{(\theta,\infty)}(y_1).$$

Thus we have that

$$\frac{\prod_{i=1}^{n} e^{-(x_i-\theta)} I_{(\theta,\infty)}(x_i)}{ne^{-n(\min x_i - \theta)} I_{(\theta,\infty)}(\min x_i)} = \frac{e^{-x_1 - x_2 - \cdots - x_n}}{ne^{-n \min x_i}}$$

since $\prod_{i=1}^{n} I_{(\theta,\infty)}(x_i) = I_{(\theta,\infty)}(\min x_i)$, because when $\theta < \min x_i$, then $\theta < x_i$, $i = 1, 2, \ldots, n$, and at least one x-value is less than or equal to θ when $\min x_i \le \theta$. Since this ratio does not depend upon θ, the first order statistic Y_1 is a sufficient statistic for θ.

If we are to show, by means of the definition, that a certain statistic Y_1 is or is not a sufficient statistic for a parameter θ, we must first of all know the p.d.f. of Y_1, say $g_1(y_1; \theta)$. In some instances it may be quite tedious to find this p.d.f. Fortunately, this problem can be avoided if we will but prove the following *factorization theorem* of Neyman.

Theorem 1. *Let $X_1, X_2, \ldots, X_n$ denote a random sample from a distribution that has* p.d.f. *$f(x; \theta)$, $\theta \in \Omega$. The statistic $Y_1 = u_1(X_1, X_2, \ldots, X_n)$ is a sufficient statistic for θ if and only if we can find two nonnegative functions, k_1 and k_2, such that*

$$f(x_1; \theta) f(x_2; \theta) \cdots f(x_n; \theta) = k_1[u_1(x_1, x_2, \ldots, x_n); \theta] k_2(x_1, x_2, \ldots, x_n),$$

where $k_2(x_1, x_2, \ldots, x_n)$ does not depend upon θ.

Proof. We shall prove the theorem when the random variables are of the continuous type. Assume that the factorization is as stated in the theorem. In our proof we shall make the one-to-one transformation $y_1 = u_1(x_1, \ldots, x_n)$, $y_2 = u_2(x_1, \ldots, x_n), \ldots, y_n = u_n(x_1, \ldots, x_n)$ having the inverse functions $x_1 = w_1(y_1, \ldots, y_n)$, $x_2 = w_2(y_1, \ldots, y_n), \ldots, x_n = w_n(y_1, \ldots, y_n)$ and Jacobian J. The joint p.d.f. of the statistics $Y_1, Y_2, \ldots, Y_n$ is then given by

$$g(y_1, y_2, \ldots, y_n; \theta) = k_1(y_1; \theta)k_2(w_1, w_2, \ldots, w_n)|J|,$$

where $w_i = w_i(y_1, y_2, \ldots, y_n)$, $i = 1, 2, \ldots, n$. The p.d.f. of Y_1, say $g_1(y_1; \theta)$, is given by

$$\begin{aligned} g_1(y_1; \theta) &= \int_{-\infty}^{\infty} \cdots \int_{-\infty}^{\infty} g(y_1, y_2, \ldots, y_n; \theta)\, dy_2 \cdots dy_n \\ &= k_1(y_1; \theta) \int_{-\infty}^{\infty} \cdots \int_{-\infty}^{\infty} |J| k_2(w_1, w_2, \ldots, w_n)\, dy_2 \cdots dy_n. \end{aligned}$$

Now the function k_2 does not depend upon θ. Nor is θ involved in either the Jacobian J or the limits of integration. Hence the $(n - 1)$-fold integral in the right-hand member of the preceding equation is a function of y_1 alone, say $m(y_1)$. Thus

$$g_1(y_1; \theta) = k_1(y_1; \theta)m(y_1).$$

If $m(y_1) = 0$, then $g_1(y_1; \theta) = 0$. If $m(y_1) > 0$, we can write

$$k_1[u_1(x_1, \ldots, x_n); \theta] = \frac{g_1[u_1(x_1, \ldots, x_n); \theta]}{m[u_1(x_1, \ldots, x_n)]},$$

and the assumed factorization becomes

$$f(x_1; \theta) \cdots f(x_n; \theta) = g_1[u_1(x_1, \ldots, x_n); \theta] \frac{k_2(x_1, \ldots, x_n)}{m[u_1(x_1, \ldots, x_n)]}.$$

Since neither the function k_2 nor the function m depends upon θ, then in accordance with the definition, Y_1 is a sufficient statistic for the parameter θ.

Conversely, if Y_1 is a sufficient statistic for θ, the factorization can be realized by taking the function k_1 to be the p.d.f. of Y_1, namely the function g_1. This completes the proof of the theorem.

Example 4. Let $X_1, X_2, \ldots, X_n$ denote a random sample from a distribution that is $N(\theta, \sigma^2)$, $-\infty < \theta < \infty$, where the variance $\sigma^2 > 0$ is known. If $\bar{x} = \sum_1^n x_i/n$, then

$$\sum_{i=1}^n (x_i - \theta)^2 = \sum_{i=1}^n [(x_i - \bar{x}) + (\bar{x} - \theta)]^2 = \sum_{i=1}^n (x_i - \bar{x})^2 + n(\bar{x} - \theta)^2$$

because

$$2 \sum_{i=1}^n (x_i - \bar{x})(\bar{x} - \theta) = 2(\bar{x} - \theta) \sum_{i=1}^n (x_i - \bar{x}) = 0.$$

Thus the joint p.d.f. of $X_1, X_2, \ldots, X_n$ may be written

$$\left(\frac{1}{\sigma\sqrt{2\pi}}\right)^n \exp\left[-\sum_{i=1}^n (x_i - \theta)^2/2\sigma^2\right]$$

$$= \{\exp[-n(\bar{x} - \theta)^2/2\sigma^2]\}\left\{\frac{\exp\left[-\sum_{i=1}^n (x_i - \bar{x})^2/2\sigma^2\right]}{(\sigma\sqrt{2\pi})^n}\right\}.$$

Since the first factor of the right-hand member of this equation depends upon $x_1, x_2, \ldots, x_n$ only through $\bar{x}$, and since the second factor does not depend upon θ, the factorization theorem implies that the mean $\bar{X}$ of the sample is, for any particular value of σ^2, a sufficient statistic for θ, the mean of the normal distribution.

We could have used the definition in the preceding example because we know that $\bar{X}$ is $N(\theta, \sigma^2/n)$. Let us now consider an example in which the use of the definition is inappropriate.

Example 5. Let $X_1, X_2, \ldots, X_n$ denote a random sample from a distribution with p.d.f.

$$\begin{aligned} f(x; \theta) &= \theta x^{\theta - 1}, \qquad 0 < x < 1, \\ &= 0 \qquad \text{elsewhere,} \end{aligned}$$

where $0 < \theta$. We shall use the factorization theorem to prove that the product $u_1(X_1, X_2, \ldots, X_n) = X_1 X_2 \cdots X_n$ is a sufficient statistic for θ. The joint p.d.f. of $X_1, X_2, \ldots, X_n$ is

$$\theta^n (x_1 x_2 \cdots x_n)^{\theta - 1} = [\theta^n (x_1 x_2 \cdots x_n)^\theta]\left(\frac{1}{x_1 x_2 \cdots x_n}\right),$$

where $0 < x_i < 1$, $i = 1, 2, \ldots, n$. In the factorization theorem let

$$k_1[u_1(x_1, x_2, \ldots, x_n); \theta] = \theta^n (x_1 x_2 \cdots x_n)^\theta$$

and

$$k_2(x_1, x_2, \ldots, x_n) = \frac{1}{x_1 x_2 \cdots x_n}.$$

Since $k_2(x_1, x_2, \ldots, x_n)$ does not depend upon θ, the product $X_1 X_2 \cdots X_n$ is a sufficient statistic for θ.

There is a tendency for some readers to apply incorrectly the factorization theorem in those instances in which the domain of positive probability density depends upon the parameter θ. This is due to the fact that they do not give proper consideration to the domain of the function $k_2(x_1, x_2, \ldots, x_n)$. This will be illustrated in the next example.

Example 6. In Example 3 with $f(x; \theta) = e^{-(x-\theta)} I_{(\theta,\infty)}(x)$, it was found that the first order statistic Y_1 is a sufficient statistic for θ. To illustrate our point about not considering the domain of the function, take $n = 3$ and note that

$$e^{-(x_1-\theta)} e^{-(x_2-\theta)} e^{-(x_3-\theta)} = [e^{-3\max x_i + 3\theta}][e^{-x_1 - x_2 - x_3 + 3\max x_i}]$$

or a similar expression. Certainly, in the latter formula, there is no θ in the second factor and it might be assumed that $Y_3 = \max X_i$ is a sufficient statistic for θ. Of course, this is incorrect because we should have written the joint p.d.f. of X_1, X_2, X_3 as

$$[e^{-(x_1-\theta)} I_{(\theta,\infty)}(x_1)][e^{-(x_2-\theta)} I_{(\theta,\infty)}(x_2)][e^{-(x_3-\theta)} I_{(\theta,\infty)}(x_3)]$$

$$= [e^{3\theta} I_{(\theta,\infty)}(\min x_i)][e^{-x_1 - x_2 - x_3}]$$

because $I_{(\theta,\infty)}(\min x_i) = I_{(\theta,\infty)}(x_1) I_{(\theta,\infty)}(x_2) I_{(\theta,\infty)}(x_3)$. A similar statement cannot be made with $\max x_i$. Thus $Y_1 = \min X_i$ is the sufficient statistic for θ, not $Y_3 = \max X_i$.

EXERCISES

7.10. Let $X_1, X_2, \ldots, X_n$ be a random sample from the normal distribution $N(0, \theta)$, $0 < \theta < \infty$. Show that $\sum_1^n X_i^2$ is a sufficient statistic for θ.

7.11. Prove that the sum of the observations of a random sample of size n from a Poisson distribution having parameter θ, $0 < \theta < \infty$, is a sufficient statistic for θ.

7.12. Show that the nth order statistic of a random sample of size n from the uniform distribution having p.d.f. $f(x; \theta) = 1/\theta$, $0 < x < \theta$, $0 < \theta < \infty$, zero elsewhere, is a sufficient statistic for θ. Generalize this result by considering the p.d.f. $f(x; \theta) = Q(\theta)M(x)$, $0 < x < \theta$, $0 < \theta < \infty$, zero elsewhere. Here, of course,

$$\int_0^\theta M(x)\,dx = \frac{1}{Q(\theta)}.$$

7.13. Let $X_1, X_2, \ldots, X_n$ be a random sample of size n from a geometric distribution that has p.d.f. $f(x; \theta) = (1 - \theta)^x\theta$, $x = 0, 1, 2, \ldots, 0 < \theta < 1$, zero elsewhere. Show that $\sum_1^n X_i$ is a sufficient statistic for θ.

7.14. Show that the sum of the observations of a random sample of size n from a gamma distribution that has p.d.f. $f(x; \theta) = (1/\theta)e^{-x/\theta}$, $0 < x < \infty$, $0 < \theta < \infty$, zero elsewhere, is a sufficient statistic for θ.

7.15. Let $X_1, X_2, \ldots, X_n$ be a random sample of size n from a beta distribution with parameters $\alpha = \theta > 0$ and $\beta = 2$. Show that the product $X_1 X_2 \cdots X_n$ is a sufficient statistic for θ.

7.16. Show that the product of the sample observations is a sufficient statistic for $\theta > 0$ if the random sample is taken from a gamma distribution with parameters $\alpha = \theta$ and $\beta = 6$.

7.17. What is the sufficient statistic for θ if the sample arises from a beta distribution in which $\alpha = \beta = \theta > 0$?

7.3 Properties of a Sufficient Statistic

Suppose that a random sample $X_1, X_2, \ldots, X_n$ is taken from a distribution with p.d.f. $f(x; \theta)$ that depends upon one parameter $\theta \in \Omega$. Say that a sufficient statistic $Y_1 = u_1(X_1, X_2, \ldots, X_n)$ for θ exists and has p.d.f. $g_1(y_1; \theta)$. Now consider two statisticians, A and B. The first statistician, A, has all of the observed data $x_1, x_2, \ldots, x_n$; but the second, B, has only the value y_1 of the sufficient statistic. Clearly, A has as much information as does B. However, it turns out that B is as well off as A in making statistical inferences about θ in the following sense. Since the conditional probability of $X_1, X_2, \ldots, X_n$, given $Y_1 = y_1$, does not depend upon θ, statistician B can create some pseudo observations, say $Z_1, Z_2, \ldots, Z_n$, that provide a likelihood function that is proportional to that based on $X_1, X_2, \ldots, X_n$ with the factor $g_1(y_1; \theta)$ being common to each likelihood. The other factors of the two likelihood functions do not depend upon θ. Hence, in either case, inferences, like the m.l.e. of θ, would be based upon the sufficient statistic Y_1.

To make this clear, we provide two illustrations. The first is based upon Example 1 of Section 7.2. There the ratio of the likelihood function and the p.d.f. of Y_1 is

$$\frac{L(\theta)}{g_1(y_1;\theta)}=\frac{1}{\binom{n}{y_1}},$$

where $y_1=\sum_{i=1}^{n} x_i$. Recall that each x_i is equal to zero or 1, and thus y_1 is the sum of y_1 *ones* and $(n-y_1)$ *zeros*. Say we know only the value y_1 and not $x_1, x_2, \ldots, x_n$; so we create pseudovalues $z_1, z_2, \ldots, z_n$ by arranging at random y_1 *ones* and $(n-y_1)$ *zeros* so that the probability of each arrangement is $p=1\Big/\binom{n}{y_1}$. Thus the probability that these z-values equal the original x-values is p, and hence it is highly unlikely, namely with probability p, that those two sets of values would be equal. Yet the two likelihood functions are proportional, namely

$$\left[\binom{n}{y_1}\theta^{y_1}(1-\theta)^{n-y_1}\right]\frac{1}{\binom{n}{\sum x_i}} \propto \left[\binom{n}{y_1}\theta^{y_1}(1-\theta)^{n-y_1}\right]\frac{1}{\binom{n}{\sum z_i}}$$

because $y_1=\sum_{i=1}^{n} x_i=\sum_{i=1}^{n} z_i$. Clearly, the m.l.e. of θ, using either expression, is y_1/n.

The next illustration refers back to Exercise 7.9. There the sample arose from a Poisson distribution with parameter $\theta>0$. It turns out that $Y_1=\sum_{i=1}^{n} X_i$ is a sufficient statistic for θ (see Exercise 7.11). In Exercise 7.9 we found that

$$\frac{L(\theta)}{g_1(y_1;\theta)}=\frac{y_1!}{x_1!\,x_2!\cdots x_n!}\left(\frac{1}{n}\right)^{x_1}\left(\frac{1}{n}\right)^{x_2}\cdots\left(\frac{1}{n}\right)^{x_n},$$

when $L(\theta)$ is the likelihood function based upon $x_1, x_2, \ldots, x_n$. Since this is a multinomial distribution that does not depend upon θ, we can generate some values of $Z_1, Z_2, \ldots, Z_n$, say $z_1, z_2, \ldots, z_n$, that have this multinomial distribution. It is interesting to note that while in the previous examples the z-values provided an arrangement of the x-values, here the z-values do not need to equal those x-values. That is, the values $z_1, z_2, \ldots, z_n$ do not necessarily provide an arrangement of $x_1, x_2, \ldots, x_n$. It is, however, true that $\sum z_i=\sum x_i=y_1$. Of course, from the way the z-values were obtained, the two likelihood functions enjoy the property of being proportional, namely

$$g_1(y_1; \theta) \frac{y_1!}{x_1!\, x_2! \cdots x_n!} \left(\frac{1}{n}\right)^{x_1} \left(\frac{1}{n}\right)^{x_2} \cdots \left(\frac{1}{n}\right)^{x_n}$$

$$\propto g_1(y_1; \theta) \frac{y_1!}{z_1!\, z_2! \cdots z_n!} \left(\frac{1}{n}\right)^{z_1} \left(\frac{1}{n}\right)^{z_2} \cdots \left(\frac{1}{n}\right)^{z_n}.$$

Thus, for illustration, using either of these likelihood functions, the m.l.e. of θ is y_1/n because this is the value of θ that maximizes $g_1(y_1; \theta)$.

Since we have considered how the statistician knowing only the value of the sufficient statistic can create a sample that satisfies the likelihood principle; and thus, in this sense, she is as well off as the statistician that has knowledge of all the data. So let us now state a fairly obvious theorem that relates the m.l.e. of θ to a sufficient statistic.

Theorem 2. *Let $X_1, X_2, \ldots, X_n$ denote a random sample from a distribution that has* p.d.f. $f(x; \theta)$, $\theta \in \Omega$. *If a sufficient statistic $Y_1 = u_1(X_1, X_2, \ldots, X_n)$ for θ exists and if a maximum likelihood estimator $\hat{\theta}$ of θ also exists uniquely, then $\hat{\theta}$ is a function of $Y_1 = u_1(X_1, X_2, \ldots, X_n)$.*

Proof. Let $g_1(y_1; \theta)$ be the p.d.f. of Y_1. Then by the definition of sufficiency, the likelihood function

$$\begin{aligned} L(\theta; x_1, x_2, \ldots, x_n) &= f(x_1; \theta) f(x_2; \theta) \cdots f(x_n; \theta) \\ &= g_1[u_1(x_1, \ldots, x_n); \theta] H(x_1, \ldots, x_n), \end{aligned}$$

where $H(x_1, \ldots, x_n)$ does not depend upon θ. Thus L and g_1, as functions of θ, are maximized simultaneously. Since there is one and only one value of θ that maximizes L and hence $g_1[u_1(x_1, \ldots, x_n); \theta]$, that value of θ must be a function of $u_1(x_1, x_2, \ldots, x_n)$. Thus the m.l.e. $\hat{\theta}$ is a function of the sufficient statistic $Y_1 = u_1(X_1, X_2, \ldots, X_n)$.

Let us consider another important property possessed by a sufficient statistic $Y_1 = u_1(X_1, X_2, \ldots, X_n)$ for θ. The conditional p.d.f. of a second statistic, say $Y_2 = u_2(X_1, X_2, \ldots, X_n)$, given $Y_1 = y_1$, does not depend upon θ. On intuitive grounds, we might surmise that the conditional p.d.f. of Y_2, given some linear function $aY_1 + b$, $a \neq 0$, of Y_1, does not depend upon θ. That is, it seems as though the random variable $aY_1 + b$ is also a sufficient statistic for θ. This conjecture is correct. In fact, every function $Z = u(Y_1)$, or $Z = u[u_1(X_1, X_2, \ldots, X_n)] = v(X_1, X_2, \ldots, X_n)$, not involving θ, with a single-valued inverse $Y_1 = w(Z)$, is also a sufficient statistic for θ. To prove this, we write, in accordance with the factorization theorem,

$$f(x_1; \theta) \cdots f(x_n; \theta) = k_1[u_1(x_1, x_2, \ldots, x_n); \theta]k_2(x_1, x_2, \ldots, x_n).$$

However, we find that $y_1 = w(z)$ or, equivalently, $u_1(x_1, x_2, \ldots, x_n) = w[v(x_1, x_2, \ldots, x_n)]$, which is not a function of θ. Hence

$$f(x_1; \theta) \cdots f(x_n; \theta) = k_1\{w[v(x_1, \ldots, x_n)]; \theta\}k_2(x_1, x_2, \ldots, x_n).$$

Since the first factor of the right-hand member of this equation is a function of $z = v(x_1, \ldots, x_n)$ and θ, while the second factor does not depend upon θ, the factorization theorem implies that $Z = u(Y_1)$ is also a sufficient statistic for θ.

Possibly, the preceding observation is obvious if we think about the sufficient statistic Y_1 partitioning the sample space in such a way that the conditional probability of $X_1, X_2, \ldots, X_n$, given $Y_1 = y_1$, does not depend upon θ. We say this because every function $Z = u(Y_1)$ with a single-valued inverse $Y_1 = w(Z)$ would partition the sample space in exactly the same way, that is, the set of points

$$\{(x_1, x_2, \ldots, x_n) : u_1(x_1, x_2, \ldots, x_n) = y_1\},$$

for each y_1, is exactly the same as

$$\{(x_1, x_2, \ldots, x_n) : v(x_1, x_2, \ldots, x_n) = u(y_1)\}$$

because $w[v(x_1, x_2, \ldots, x_n)] = u_1(x_1, x_2, \ldots, x_n) = y_1$.

Remark. Throughout the discussion of sufficient statistics, as a matter of fact throughout much of the mathematics of statistical inference, we hope the reader recognizes the importance of the assumption of having a certain model. Clearly, when we say that a statistician having the value of a certain statistic (here sufficient) is as well off in making statistical inferences as the statistician who has all of the data, we depend upon the fact that a certain model is true. For illustration, knowing that we have i.i.d. variables, each with p.d.f. $f(x; \theta)$, is extremely important; because if that $f(x; \theta)$ is incorrect or if the independence assumption does not hold, our resulting inferences could be very bad. The statistician with all the data could—and should—check to see if the model is reasonably good. Such procedures checking the model are often called *model diagnostics*, the discussion of which we leave to a more applied course in statistics.

We now consider a result of Rao and Blackwell from which we see that we need consider only functions of the sufficient statistic in finding the unbiased point estimates of parameters. In showing this, we can refer back to a result of Section 2.2: If X_1 and X_2 are random variables and certain expectations exist, then

$$E[X_2] = E[E(X_2|X_1)]$$

and

$$\text{var}\,(X_2) \geq \text{var}\,[E(X_2|X_1)].$$

For the adaptation in context of sufficient statistics, we let the sufficient statistic Y_1 be X_1 and Y_2, an unbiased statistic of θ, be X_2. Thus, with $E(Y_2|y_1) = \varphi(y_1)$, we have

$$\theta = E(Y_2) = E[\varphi(Y_1)]$$

and

$$\text{var}\,(Y_2) \geq \text{var}\,[\varphi(Y_1)].$$

That is, through this conditioning, the function $\varphi(Y_1)$ of the sufficient statistic Y_1 is an unbiased estimator of θ having smaller variance than that of the unbiased estimator Y_2. We summarize this discussion more formally in the following theorem, which can be attributed to Rao and Blackwell.

Theorem 3. *Let $X_1, X_2, \ldots, X_n$, n a fixed positive integer, denote a random sample from a distribution (continuous or discrete) that has* p.d.f. *$f(x; \theta)$, $\theta \in \Omega$. Let $Y_1 = u_1(X_1, X_2, \ldots, X_n)$ be a sufficient statistic for θ, and let $Y_2 = u_2(X_1, X_2, \ldots, X_n)$, not a function of Y_1 alone, be an unbiased estimator of θ. Then $E(Y_2|y_1) = \varphi(y_1)$ defines a statistic $\varphi(Y_1)$. This statistic $\varphi(Y_1)$ is a function of the sufficient statistic for θ; it is an unbiased estimator of θ; and its variance is less than that of Y_2.*

This theorem tells us that in our search for an unbiased minimum variance estimator of a parameter, we may, if a sufficient statistic for the parameter exists, restrict that search to functions of the sufficient statistic. For if we begin with an unbiased estimator Y_2 that is not a function of the sufficient statistic Y_1 alone, then we can always improve on this by computing $E(Y_2|y_1) = \varphi(y_1)$ so that $\varphi(Y_1)$ is an unbiased estimator with smaller variance than that of Y_2.

After Theorem 3 many students believe that it is necessary to find first some unbiased estimator Y_2 in their search for $\varphi(Y_1)$, an unbiased estimator of θ based upon the sufficient statistic Y_1. This is not the case at all, and Theorem 3 simply convinces us that we can restrict our search for a best estimator to functions of Y_1. It frequently happens that $E(Y_1) = a\theta + b$, where $a \neq 0$ and b are constants, and thus $(Y_1 - b)/a$ is a function of Y_1 that is an unbiased estimator of θ. That is, we can usually find an unbiased estimator based on Y_1 without first finding an estimator Y_2. In the next two sections we discover that, in most instances, if there is one function $\varphi(Y_1)$ that is unbiased, $\varphi(Y_1)$ is the only unbiased estimator based on the sufficient statistic Y_1.

Remark. Since the unbiased estimator $\varphi(Y_1)$, where $\varphi(y_1) = E(Y_2|y_1)$, has variance smaller than that of the unbiased estimator Y_2 of θ, students sometimes reason as follows. Let the function $\Upsilon(y_3) = E[\varphi(Y_1)|Y_3 = y_3]$, where Y_3 is another statistic, which is not sufficient for θ. By the Rao–Blackwell theorem, we have that $E[\Upsilon(Y_3)] = \theta$ and $\Upsilon(Y_3)$ has a smaller variance than does $\varphi(Y_1)$. Accordingly, $\Upsilon(Y_3)$ must be better than $\varphi(Y_1)$ as an unbiased estimator of θ. But this is *not* true because Y_3 is not sufficient; thus θ is present in the conditional distribution of Y_1, given $Y_3 = y_3$, and the conditional mean $\Upsilon(y_3)$. So although indeed $E[\Upsilon(Y_3)] = \theta$, $\Upsilon(Y_3)$ is not even a statistic because it involves the unknown parameter θ and hence cannot be used as an estimator.

Example 1. Let X_1, X_2, X_3 be a random sample from an exponential distribution with mean $\theta > 0$, so that the joint p.d.f. is

$$\left(\frac{1}{\theta}\right)^3 e^{-(x_1+x_2+x_3)/\theta}, \qquad 0 < x_i < \infty,$$

$i = 1, 2, 3$, zero elsewhere. From the factorization theorem, we see that $Y_1 = X_1 + X_2 + X_3$ is a sufficient statistic for θ. Of course,

$$E(Y_1) = E(X_1 + X_2 + X_3) = 3\theta,$$

and thus $Y_1/3 = \bar{X}$ is a function of the sufficient statistic that is an unbiased estimator of θ.

In addition, let $Y_2 = X_2 + X_3$ and $Y_3 = X_3$. The one-to-one transformation defined by

$$x_1 = y_1 - y_2, \qquad x_2 = y_2 - y_3, \qquad x_3 = y_3$$

has Jacobian equal to 1 and the joint p.d.f. of Y_1, Y_2, Y_3 is

$$g(y_1, y_2, y_3; \theta) = \left(\frac{1}{\theta}\right)^3 e^{-y_1/\theta}, \qquad 0 < y_3 < y_2 < y_1 < \infty,$$

zero elsewhere. The marginal p.d.f. of Y_1 and Y_3 is found by integrating out y_2 to obtain

$$g_{13}(y_1, y_3; \theta) = \left(\frac{1}{\theta}\right)^3 (y_1 - y_3)e^{-y_1/\theta}, \qquad 0 < y_3 < y_1 < \infty,$$

zero elsewhere. The p.d.f. of Y_3 alone is

$$g_3(y_3; \theta) = \frac{1}{\theta} e^{-y_3/\theta}, \qquad 0 < y_3 < \infty,$$

zero elsewhere, since $Y_3 = X_3$ is an observation of a random sample from this exponential distribution.

Accordingly, the conditional p.d.f. of Y_1, given $Y_3 = y_3$, is

$$\begin{aligned} g_{1|3}(y_1|y_3) &= \frac{g_{13}(y_1, y_3; \theta)}{g_3(y_3; \theta)} \\ &= \left(\frac{1}{\theta}\right)^2 (y_1 - y_3)e^{-(y_1 - y_3)/\theta}, \qquad 0 < y_3 < y_1 < \infty, \end{aligned}$$

zero elsewhere. Thus

$$\begin{aligned} E\left(\frac{Y_1}{3}|y_3\right) &= E\left(\frac{Y_1 - Y_3}{3}|y_3\right) + E\left(\frac{Y_3}{3}|y_3\right) \\ &= \left(\frac{1}{3}\right)\int_{y_3}^{\infty} \left(\frac{1}{\theta}\right)^2 (y_1 - y_3)^2 e^{-(y_1 - y_3)/\theta}\, dy_1 + \frac{y_3}{3} \\ &= \left(\frac{1}{3}\right)\frac{\Gamma(3)\theta^3}{\theta^2} + \frac{y_3}{3} = \frac{2\theta}{3} + \frac{y_3}{3} = \Upsilon(y_3). \end{aligned}$$

Of course, $E[\Upsilon(Y_3)] = \theta$ and $\operatorname{var}[\Upsilon(Y_3)] \le \operatorname{var}(Y_1/3)$, but $\Upsilon(Y_3)$ is not a statistic as it involves θ and cannot be used as an estimator of θ. This illustrates the preceding remark.

EXERCISES

7.18. In each of the Exercises 7.10, 7.11, 7.13, and 7.14, show that the m.l.e. of θ is a function of the sufficient statistic for θ.

7.19. Let $Y_1 < Y_2 < Y_3 < Y_4 < Y_5$ be the order statistics of a random sample of size 5 from the uniform distribution having p.d.f. $f(x; \theta) = 1/\theta$, $0 < x < \theta$, $0 < \theta < \infty$, zero elsewhere. Show that $2Y_3$ is an unbiased estimator of θ. Determine the joint p.d.f. of Y_3 and the sufficient statistic Y_5 for θ. Find the conditional expectation $E(2Y_3|y_5) = \varphi(y_5)$. Compare the variances of $2Y_3$ and $\varphi(Y_5)$.

Hint: All of the integrals needed in this exercise can be evaluated by making a change of variable such as $z = y/\theta$ and using the results associated with the beta p.d.f.; see Section 4.4.

7.20. If X_1, X_2 is a random sample of size 2 from a distribution having p.d.f. $f(x; \theta) = (1/\theta)e^{-x/\theta}$, $0 < x < \infty$, $0 < \theta < \infty$, zero elsewhere, find the joint p.d.f. of the sufficient statistic $Y_1 = X_1 + X_2$ for θ and $Y_2 = X_2$. Show that Y_2 is an unbiased estimator of θ with variance θ^2. Find $E(Y_2|y_1) = \varphi(y_1)$ and the variance of $\varphi(Y_1)$.

7.21. Let the random variables X and Y have the joint p.d.f. $f(x, y) = (2/\theta^2)e^{-(x+y)/\theta}$, $0 < x < y < \infty$, zero elsewhere.

(a) Show that the mean and the variance of Y are, respectively, $3\theta/2$ and $5\theta^2/4$.

(b) Show that $E(Y|x) = x + \theta$. In accordance with the theory, the expected value of $X + \theta$ is that of Y, namely, $3\theta/2$, and the variance of $X + \theta$ is less than that of Y. Show that the variance of $X + \theta$ is in fact $\theta^2/4$.

7.22. In each of Exercises 7.10, 7.11, and 7.12, compute the expected value of the given sufficient statistic and, in each case, determine an unbiased estimator of θ that is a function of that sufficient statistic alone.

7.4 Completeness and Uniqueness

Let $X_1, X_2, \ldots, X_n$ be a random sample from the Poisson distribution that has p.d.f.

$$f(x; \theta) = \frac{\theta^x e^{-\theta}}{x!}, \qquad x = 0, 1, 2, \ldots; \quad 0 < \theta;$$
$$= 0 \qquad \text{elsewhere.}$$

From Exercise 7.11 of Section 7.2 we know that $Y_1 = \sum_{i=1}^{n} X_i$ is a sufficient statistic for θ and its p.d.f. is

$$g_1(y_1; \theta) = \frac{(n\theta)^{y_1} e^{-n\theta}}{y_1!}, \qquad y_1 = 0, 1, 2, \ldots,$$
$$= 0 \qquad \text{elsewhere.}$$

Let us consider the family $\{g_1(y_1; \theta) : 0 < \theta\}$ of probability density functions. Suppose that the function $u(Y_1)$ of Y_1 is such that $E[u(Y_1)] = 0$ for every $\theta > 0$. We shall show that this requires $u(y_1)$ to be zero at every point $y_1 = 0, 1, 2, \ldots$. That is,

$$E[u(Y_1)] = 0, \qquad 0 < \theta,$$

implies that

$$0 = u(0) = u(1) = u(2) = u(3) = \cdots.$$

We have for all $\theta > 0$ that

$$0 = E[u(Y_1)] = \sum_{y_1=0}^{\infty} u(y_1) \frac{(n\theta)^{y_1} e^{-n\theta}}{y_1!}$$

$$= e^{-n\theta}\left[u(0) + u(1)\frac{n\theta}{1!} + u(2)\frac{(n\theta)^2}{2!} + \cdots\right].$$

Since $e^{-n\theta}$ does not equal zero, we have that

$$0 = u(0) + [nu(1)]\theta + \left[\frac{n^2u(2)}{2}\right]\theta^2 + \cdots.$$

However, if such an infinite series converges to zero for all $\theta > 0$, then each of the coefficients must equal zero. That is,

$$u(0) = 0, \qquad nu(1) = 0, \qquad \frac{n^2u(2)}{2} = 0, \ldots$$

and thus $0 = u(0) = u(1) = u(2) = \cdots$, as we wanted to show. Of course, the condition $E[u(Y_1)] = 0$ for all $\theta > 0$ does not place any restriction on $u(y_1)$ when y_1 is not a nonnegative integer. So we see that, in this illustration, $E[u(Y_1)] = 0$ for all $\theta > 0$ requires that $u(y_1)$ equals zero except on a set of points that has probability zero for each p.d.f. $g_1(y_1; \theta)$, $0 < \theta$. From the following definition we observe that the family $\{g_1(y_1; \theta) : 0 < \theta\}$ is complete.

Definition 3. Let the random variable Z of either the continuous type or the discrete type have a p.d.f. that is one member of the family $\{h(z; \theta) : \theta \in \Omega\}$. If the condition $E[u(Z)] = 0$, for every $\theta \in \Omega$, requires that $u(z)$ be zero except on a set of points that has probability zero for each p.d.f. $h(z; \theta)$, $\theta \in \Omega$, then the family $\{h(z; \theta) : \theta \in \Omega\}$ is called a *complete family* of probability density functions.

Remark. In Section 1.9 it was noted that the existence of $E[u(X)]$ implies that the integral (or sum) converges absolutely. This absolute convergence was tacitly assumed in our definition of completeness and it is needed to prove that certain families of probability density functions are complete.

In order to show that certain families of probability density functions of the continuous type are complete, we must appeal to the same type of theorem in analysis that we used when we claimed that the moment-generating function uniquely determines a distribution. This is illustrated in the next example.

Example 1. Let Z have a p.d.f. that is a member of the family $\{h(z; \theta) : 0 < \theta < \infty\}$, where

$$h(z; \theta) = \frac{1}{\theta} e^{-z/\theta}, \qquad 0 < z < \infty,$$
$$= 0 \qquad \text{elsewhere.}$$

Let us say that $E[u(Z)] = 0$ for every $\theta > 0$. That is,

$$\frac{1}{\theta} \int_0^\infty u(z) e^{-z/\theta}\, dz = 0, \qquad \text{for } \theta > 0.$$

Readers acquainted with the theory of transforms will recognize the integral in the left-hand member as being essentially the Laplace transform of $u(z)$. In that theory we learn that the only function $u(z)$ transforming to a function of θ which is identically equal to zero is $u(z) = 0$, except (in our terminology) on a set of points that has probability zero for each $h(z; \theta)$, $0 < \theta$. That is, the family $\{h(z; \theta) : 0 < \theta < \infty\}$ is complete.

Let the parameter θ in the p.d.f. $f(x; \theta)$, $\theta \in \Omega$, have a sufficient statistic $Y_1 = u_1(X_1, X_2, \ldots, X_n)$, where $X_1, X_2, \ldots, X_n$ is a random sample from this distribution. Let the p.d.f. of Y_1 be $g_1(y_1; \theta)$, $\theta \in \Omega$. It has been seen that, if there is any unbiased estimator Y_2 (not a function of Y_1 alone) of θ, then there is at least one function of Y_1 that is an unbiased estimator of θ, and our search for a best estimator of θ may be restricted to functions of Y_1. Suppose it has been verified that a certain function $\varphi(Y_1)$, not a function of θ, is such that $E[\varphi(Y_1)] = \theta$ for all values of θ, $\theta \in \Omega$. Let $\psi(Y_1)$ be another function of the sufficient statistic Y_1 alone, so that we also have $E[\psi(Y_1)] = \theta$ for all values of θ, $\theta \in \Omega$. Hence

$$E[\varphi(Y_1) - \psi(Y_1)] = 0, \qquad \theta \in \Omega.$$

If the family $\{g_1(y_1; \theta) : \theta \in \Omega\}$ is complete, the function $\varphi(y_1) - \psi(y_1) = 0$, except on a set of points that has probability zero. That is, for every other unbiased estimator $\psi(Y_1)$ of θ, we have

$$\varphi(y_1) = \psi(y_1)$$

except possibly at certain special points. Thus, in this sense [namely $\varphi(y_1) = \psi(y_1)$, except on a set of points with probability zero], $\varphi(Y_1)$ is the unique function of Y_1, which is an unbiased estimator of θ. In accordance with the Rao–Blackwell theorem, $\varphi(Y_1)$ has a smaller variance than every other unbiased estimator of θ. That is, the statistic $\varphi(Y_1)$ is the unbiased minimum variance estimator of θ. This fact is stated in the following theorem of Lehmann and Scheffé.

Theorem 4. *Let $X_1, X_2, \ldots, X_n$, n a fixed positive integer, denote a random sample from a distribution that has* p.d.f. $f(x; \theta)$, $\theta \in \Omega$, *let $Y_1 = u_1(X_1, X_2, \ldots, X_n)$ be a sufficient statistic for θ, and let the family $\{g_1(y_1; \theta) : \theta \in \Omega\}$ of probability density functions be complete. If there is a function of Y_1 that is an unbiased estimator of θ, then this function of Y_1 is the unique unbiased minimum variance estimator of θ. Here "unique" is used in the sense described in the preceding paragraph.*

The statement that Y_1 is a sufficient statistic for a parameter θ, $\theta \in \Omega$, and that the family $\{g_1(y_1; \theta) : \theta \in \Omega\}$ of probability density functions is complete is lengthy and somewhat awkward. We shall adopt the less descriptive, but more convenient, terminology that Y_1 is a *complete sufficient statistic* for θ. In the next section we study a fairly large class of probability density functions for which a complete sufficient statistic Y_1 for θ can be determined by inspection.

EXERCISES

7.23. If $az^2 + bz + c = 0$ for more than two values of z, then $a = b = c = 0$. Use this result to show that the family $\{b(2, \theta) : 0 < \theta < 1\}$ is complete.

7.24. Show that each of the following families is not complete by finding at least one nonzero function $u(x)$ such that $E[u(X)] = 0$, for all $\theta > 0$.

(a) $$f(x; \theta) = \frac{1}{2\theta}, \qquad -\theta < x < \theta, \quad \text{where } 0 < \theta < \infty,$$
$$= 0 \qquad \text{elsewhere.}$$

(b) $N(0, \theta)$, where $0 < \theta < \infty$.

7.25. Let $X_1, X_2, \ldots, X_n$ represent a random sample from the discrete distribution having the probability density function

$$f(x; \theta) = \theta^x(1 - \theta)^{1-x}, \qquad x = 0, 1, \quad 0 < \theta < 1,$$
$$= 0 \qquad \text{elsewhere.}$$

Show that $Y_1 = \sum_1^n X_i$ is a complete sufficient statistic for θ. Find the unique function of Y_1 that is the unbiased minimum variance estimator of θ.

Hint: Display $E[u(Y_1)] = 0$, show that the constant term $u(0)$ is equal to zero, divide both members of the equation by $\theta \neq 0$, and repeat the argument.

7.26. Consider the family of probability density functions $\{h(z; \theta) : \theta \in \Omega\}$, where $h(z; \theta) = 1/\theta$, $0 < z < \theta$, zero elsewhere.

(a) Show that the family is complete provided that $\Omega = \{\theta : 0 < \theta < \infty\}$. *Hint:* For convenience, assume that $u(z)$ is continuous and note that the derivative of $E[u(Z)]$ with respect to θ is equal to zero also.

(b) Show that this family is not complete if $\Omega = \{\theta : 1 < \theta < \infty\}$. *Hint:* Concentrate on the interval $0 < z < 1$ and find a nonzero function $u(z)$ on that interval such that $E[u(Z)] = 0$ for all $\theta > 1$.

7.27. Show that the first order statistic Y_1 of a random sample of size n from the distribution having p.d.f. $f(x; \theta) = e^{-(x-\theta)}$, $\theta < x < \infty$, $-\infty < \theta < \infty$, zero elsewhere, is a complete sufficient statistic for θ. Find the unique function of this statistic which is the unbiased minimum variance estimator of θ.

7.28. Let a random sample of size n be taken from a distribution of the discrete type with p.d.f. $f(x; \theta) = 1/\theta$, $x = 1, 2, \ldots, \theta$, zero elsewhere, where θ is an unknown positive integer.

(a) Show that the largest observation, say Y, of the sample is a complete sufficient statistic for θ.

(b) Prove that

$$[Y^{n+1} - (Y-1)^{n+1}]/[Y^n - (Y-1)^n]$$

is the unique unbiased minimum variance estimator of θ.

7.5 The Exponential Class of Probability Density Functions

Consider a family $\{f(x; \theta) : \theta \in \Omega\}$ of probability density functions, where Ω is the interval set $\Omega = \{\theta : \gamma < \theta < \delta\}$, where γ and δ are known constants, and where

$$\begin{aligned} f(x; \theta) &= \exp\,[p(\theta)K(x) + S(x) + q(\theta)], \qquad a < x < b, \\ &= 0 \qquad \text{elsewhere.} \end{aligned} \tag{1}$$

A p.d.f. of the form (1) is said to be a member of the *exponential class* of probability density functions of the continuous type. If, in addition,

1. neither a nor b depends upon θ, $\gamma < \theta < \delta$,
2. $p(\theta)$ is a nontrivial continuous function of θ, $\gamma < \theta < \delta$,
3. each of $K'(x) \not\equiv 0$ and $S(x)$ is a continuous function of x, $a < x < b$,

we say that we have a *regular case* of the exponential class. A p.d.f.

$$f(x; \theta) = \exp[p(\theta)K(x) + S(x) + q(\theta)], \qquad x = a_1, a_2, a_3, \ldots,$$
$$= 0 \qquad \text{elsewhere},$$

is said to represent a regular case of the exponential class of probability density functions of the discrete type if

1. The set $\{x : x = a_1, a_2, \ldots\}$ does not depend upon θ.
2. $p(\theta)$ is a nontrivial continuous function of θ, $\gamma < \theta < \delta$.
3. $K(x)$ is a nontrivial function of x on the set $\{x : x = a_1, a_2, \ldots\}$.

For example, each member of the family $\{f(x; \theta) : 0 < \theta < \infty\}$, where $f(x; \theta)$ is $N(0, \theta)$, represents a regular case of the exponential class of the continuous type because

$$f(x; \theta) = \frac{1}{\sqrt{2\pi\theta}} e^{-x^2/2\theta}$$
$$= \exp\left(-\frac{1}{2\theta}x^2 - \ln\sqrt{2\pi\theta}\right), \qquad -\infty < x < \infty.$$

Let $X_1, X_2, \ldots, X_n$ denote a random sample from a distribution that has a p.d.f. that represents a regular case of the exponential class of the continuous type. The joint p.d.f. of $X_1, X_2, \ldots, X_n$ is

$$\exp\left[p(\theta)\sum_1^n K(x_i) + \sum_1^n S(x_i) + nq(\theta)\right]$$

for $a < x_i < b$, $i = 1, 2, \ldots, n$, $\gamma < \theta < \delta$, and is zero elsewhere. At points of positive probability density, this joint p.d.f. may be written as the product of the two nonnegative functions

$$\exp\left[p(\theta)\sum_1^n K(x_i) + nq(\theta)\right]\exp\left[\sum_1^n S(x_i)\right].$$

In accordance with the factorization theorem (Theorem 1, Section 7.2) $Y_1 = \sum_1^n K(X_i)$ is a sufficient statistic for the parameter θ. To prove that $Y_1 = \sum_1^n K(X_i)$ is a sufficient statistic for θ in the discrete case, we take the joint p.d.f. of $X_1, X_2, \ldots, X_n$ to be positive on a discrete set of points, say, when $x_i \in \{x : x = a_1, a_2, \ldots\}$, $i = 1, 2, \ldots, n$. We then use the factorization theorem. It is left as an exercise to show that in either the continuous or the discrete case the p.d.f. of Y_1 is of the form

$$g_1(y_1; \theta) = R(y_1)\exp[p(\theta)y_1 + nq(\theta)]$$

at points of positive probability density. The points of positive probability density and the function $R(y_1)$ do not depend upon θ.

At this time we use a theorem in analysis to assert that the family $\{g_1(y_1; \theta) : \gamma < \theta < \delta\}$ of probability density functions is complete. This is the theorem we used when we asserted that a moment-generating function (when it exists) uniquely determines a distribution. In the present context it can be stated as follows.

Theorem 5. *Let $f(x; \theta)$, $\gamma < \theta < \delta$, be a* p.d.f. *which represents a regular case of the exponential class. Then if $X_1, X_2, \ldots, X_n$ (where n is a fixed positive integer) is a random sample from a distribution with* p.d.f. *$f(x; \theta)$, the statistic $Y_1 = \sum_1^n K(X_i)$ is a sufficient statistic for θ and the family $\{g_1(y_1; \theta) : \gamma < \theta < \delta\}$ of probability density functions of Y_1 is complete. That is, Y_1 is a complete sufficient statistic for θ.*

This theorem has useful implications. In a regular case of form (1), we can see by inspection that the sufficient statistic is $Y_1 = \sum_1^n K(X_i)$. If we can see how to form a function of Y_1, say $\varphi(Y_1)$, so that $E[\varphi(Y_1)] = \theta$, then the statistic $\varphi(Y_1)$ is unique and is the unbiased minimum variance estimator of θ.

Example 1. Let $X_1, X_2, \ldots, X_n$ denote a random sample from a normal distribution that has p.d.f.

$$f(x; \theta) = \frac{1}{\sigma\sqrt{2\pi}} \exp\left[-\frac{(x-\theta)^2}{2\sigma^2}\right], \qquad -\infty < x < \infty, \quad -\infty < \theta < \infty,$$

or

$$f(x; \theta) = \exp\left(\frac{\theta}{\sigma^2}x - \frac{x^2}{2\sigma^2} - \ln\sqrt{2\pi\sigma^2} - \frac{\theta^2}{2\sigma^2}\right).$$

Here σ^2 is any fixed positive number. This is a regular case of the exponential class with

$$p(\theta) = \frac{\theta}{\sigma^2}, \qquad K(x) = x,$$

$$S(x) = -\frac{x^2}{2\sigma^2} - \ln\sqrt{2\pi\sigma^2}, \qquad q(\theta) = -\frac{\theta^2}{2\sigma^2}.$$

Accordingly, $Y_1 = X_1 + X_2 + \cdots + X_n = n\bar{X}$ is a complete sufficient statistic for the mean θ of a normal distribution for every fixed value of the variance σ^2. Since $E(Y_1) = n\theta$, then $\varphi(Y_1) = Y_1/n = \bar{X}$ is the only function of Y_1 that is an unbiased estimator of θ; and being a function of the sufficient statistic

Y_1, it has a minimum variance. That is, $\bar{X}$ is the unique unbiased minimum variance estimator of θ. Incidentally, since Y_1 is a one-to-one function of $\bar{X}$, $\bar{X}$ itself is also a complete sufficient statistic for θ.

Example 2. Consider a Poisson distribution with parameter $\theta, 0 < \theta < \infty$. The p.d.f. of this distribution is

$$f(x; \theta) = \frac{\theta^x e^{-\theta}}{x!} = \exp[(\ln \theta)x - \ln(x!) - \theta], \qquad x = 0, 1, 2, \ldots,$$

$$= 0 \qquad \text{elsewhere.}$$

In accordance with Theorem 5, $Y_1 = \sum_1^n X_i$ is a complete sufficient statistic for θ. Since $E(Y_1) = n\theta$, the statistic $\varphi(Y_1) = Y_1/n = \bar{X}$, which is also a complete sufficient statistic for θ, is the unique unbiased minimum variance estimator of θ.

EXERCISES

7.29. Write the p.d.f.

$$f(x; \theta) = \frac{1}{6\theta^4} x^3 e^{-x/\theta}, \qquad 0 < x < \infty, \quad 0 < \theta < \infty,$$

zero elsewhere, in the exponential form. If $X_1, X_2, \ldots, X_n$ is a random sample from this distribution, find a complete sufficient statistic Y_1 for θ and the unique function $\varphi(Y_1)$ of this statistic that is the unbiased minimum variance estimator of θ. Is $\varphi(Y_1)$ itself a complete sufficient statistic?

7.30. Let $X_1, X_2, \ldots, X_n$ denote a random sample of size $n > 1$ from a distribution with p.d.f. $f(x; \theta) = \theta e^{-\theta x}$, $0 < x < \infty$, zero elsewhere, and $\theta > 0$. Then $Y = \sum_1^n X_i$ is a sufficient statistic for θ. Prove that $(n-1)/Y$ is the unbiased minimum variance estimator of θ.

7.31. Let $X_1, X_2, \ldots, X_n$ denote a random sample of size n from a distribution with p.d.f. $f(x; \theta) = \theta x^{\theta - 1}$, $0 < x < 1$, zero elsewhere, and $\theta > 0$.
(a) Show that the *geometric mean* $(X_1 X_2 \cdots X_n)^{1/n}$ of the sample is a complete sufficient statistic for θ.
(b) Find the maximum likelihood estimator of θ, and observe that it is a function of this geometric mean.

7.32. Let $\bar{X}$ denote the mean of the random sample $X_1, X_2, \ldots, X_n$ from a gamma-type distribution with parameters $\alpha > 0$ and $\beta = \theta > 0$. Compute $E[X_1|\bar{x}]$.

Hint: Can you find directly a function $\psi(\bar{X})$ of $\bar{X}$ such that $E[\psi(\bar{X})] = \theta$? Is $E(X_1|\bar{x}) = \psi(\bar{x})$? Why?

7.33. Let X be a random variable with a p.d.f. of a regular case of the exponential class. Show that $E[K(X)] = -q'(\theta)/p'(\theta)$, provided these derivatives exist, by differentiating both members of the equality

$$\int_a^b \exp[p(\theta)K(x) + S(x) + q(\theta)]\,dx = 1$$

with respect to θ. By a second differentiation, find the variance of $K(X)$.

7.34. Given that $f(x; \theta) = \exp[\theta K(x) + S(x) + q(\theta)]$, $a < x < b$, $\gamma < \theta < \delta$, represents a regular case of the exponential class, show that the moment-generating function $M(t)$ of $Y = K(X)$ is $M(t) = \exp[q(\theta) - q(\theta + t)]$, $\gamma < \theta + t < \delta$.

7.35. Given, in the preceding exercise, that $E(Y) = E[K(X)] = \theta$. Prove that Y is $N(\theta, 1)$.

Hint: Consider $M'(0) = \theta$ and solve the resulting differential equation.

7.36. If $X_1, X_2, \ldots, X_n$ is a random sample from a distribution that has a p.d.f. which is a regular case of the exponential class, show that the p.d.f. of $Y_1 = \sum_1^n K(X_i)$ is of the form $g_1(y_1; \theta) = R(y_1)\exp[p(\theta)y_1 + nq(\theta)]$.

Hint: Let $Y_2 = X_2, \ldots, Y_n = X_n$ be $n - 1$ auxiliary random variables. Find the joint p.d.f. of $Y_1, Y_2, \ldots, Y_n$ and then the marginal p.d.f. of Y_1.

7.37. Let Y denote the median and let $\bar{X}$ denote the mean of a random sample of the size $n = 2k + 1$ from a distribution that is $N(\mu, \sigma^2)$. Compute $E(Y|\bar{X} = \bar{x})$.

Hint: See Exercise 7.32.

7.38. Let $X_1, X_2, \ldots, X_n$ be a random sample from a distribution with p.d.f. $f(x; \theta) = \theta^2 x e^{-\theta x}$, $0 < x < \infty$, where $\theta > 0$.

(a) Argue that $Y = \sum_1^n X_i$ is a complete sufficient statistic for θ.

(b) Compute $E(1/Y)$ and find the function of Y which is the unique unbiased minimum variance estimator of θ.

7.39. Let $X_1, X_2, \ldots, X_n$, $n > 2$, be a random sample from the binomial distribution $b(1, \theta)$.

(a) Show that $Y_1 = X_1 + X_2 + \cdots + X_n$ is a complete sufficient statistic for θ.

(b) Find the function $\varphi(Y_1)$ which is the unbiased minimum variance estimator of θ.

(c) Let $Y_2 = (X_1 + X_2)/2$ and compute $E(Y_2)$.

(d) Determine $E(Y_2|Y_1 = y_1)$.

7.6 Functions of a Parameter

Up to this point we have sought an unbiased and minimum variance estimator of a parameter θ. Not always, however, are we interested in θ but rather in a function of θ. This will be illustrated in the following examples.

Example 1. Let $X_1, X_2, \ldots, X_n$ denote the observations of a random sample of size $n > 1$ from a distribution that is $b(1, \theta)$, $0 < \theta < 1$. We know that if $Y = \sum_1^n X_i$, then Y/n is the unique unbiased minimum variance estimator of θ. Now the variance of Y/n is $\theta(1 - \theta)/n$. Suppose that an unbiased and minimum variance estimator of this variance is sought. Because Y is a sufficient statistic for θ, it is known that we can restrict our search to functions of Y. Consider the statistic $(Y/n)(1 - Y/n)/n$. This statistic is suggested by the fact that Y/n is an estimator of θ. The expectation of this statistic is given by

$$\frac{1}{n} E\left[\frac{Y}{n}\left(1 - \frac{Y}{n}\right)\right] = \frac{1}{n^2} E(Y) - \frac{1}{n^3} E(Y^2).$$

Now $E(Y) = n\theta$ and $E(Y^2) = n\theta(1 - \theta) + n^2\theta^2$. Hence

$$\frac{1}{n} E\left[\frac{Y}{n}\left(1 - \frac{Y}{n}\right)\right] = \frac{n-1}{n}\frac{\theta(1-\theta)}{n}.$$

If we multiply both members of this equation by $n/(n - 1)$, we find that the statistic $(Y/n)(1 - Y/n)/(n - 1)$ is the unique unbiased minimum variance estimator of the variance of Y/n.

A somewhat different, but also very important problem in point estimation is considered in the next example. In the example the distribution of a random variable X is described by a p.d.f. $f(x; \theta)$ that depends upon $\theta \in \Omega$. The problem is to estimate the fractional part of the probability for this distribution which is at, or to the left of, a fixed point c. Thus we seek an unbiased minimum variance estimator of $F(c; \theta)$, where $F(x; \theta)$ is the distribution function of X.

Example 2. Let $X_1, X_2, \ldots, X_n$ be a random sample of size $n > 1$ from a distribution that is $N(\theta, 1)$. Suppose that we wish to find an unbiased minimum variance estimator of the function of θ defined by

$$\Pr(X \le c) = \int_{-\infty}^{c} \frac{1}{\sqrt{2\pi}} e^{-(x-\theta)^2/2}\, dx = \Phi(c - \theta),$$

where c is a fixed constant. There are many unbiased estimators of $\Phi(c - \theta)$. We first exhibit one of these, say $u(X_1)$, a function of X_1 alone. We shall then compute the conditional expectation, $E[u(X_1)|\overline{X} = \overline{x}] = \varphi(\overline{x})$, of this unbiased statistic, given the sufficient statistic $\overline{X}$, the mean of the sample. In accordance with the theorems of Rao–Blackwell and Lehmann–Scheffé, $\varphi(\overline{X})$ is the unique unbiased minimum variance estimator of $\Phi(c - \theta)$.

Consider the function $u(x_1)$, where

$$u(x_1) = 1, \qquad x_1 \le c,$$
$$= 0, \qquad x_1 > c.$$

The expected value of the random variable $u(X_1)$ is given by

$$E[u(X_1)] = \int_{-\infty}^{\infty} u(x_1)\frac{1}{\sqrt{2\pi}}\exp\left[-\frac{(x_1 - \theta)^2}{2}\right]dx_1$$

$$= \int_{-\infty}^{c} (1)\frac{1}{\sqrt{2\pi}}\exp\left[-\frac{(x_1 - \theta)^2}{2}\right]dx_1,$$

because $u(x_1) = 0$, $x_1 > c$. But the latter integral has the value $\Phi(c - \theta)$. That is, $u(X_1)$ is an unbiased estimator of $\Phi(c - \theta)$.

We shall next discuss the joint distribution of X_1 and $\overline{X}$ and the conditional distribution of X_1, given $\overline{X} = \overline{x}$. This conditional distribution will enable us to compute $E[u(X_1)|\overline{X} = \overline{x}] = \varphi(\overline{x})$. In accordance with Exercise 4.92, Section 4.7, the joint distribution of X_1 and $\overline{X}$ is bivariate normal with means θ and θ, variances $\sigma_1^2 = 1$ and $\sigma_2^2 = 1/n$, and correlation coefficient $\rho = 1/\sqrt{n}$. Thus the conditional p.d.f. of X_1, given $\overline{X} = \overline{x}$, is normal with linear conditional mean

$$\theta + \frac{\rho\sigma_1}{\sigma_2}(\overline{x} - \theta) = \overline{x}$$

and with variance

$$\sigma_1^2(1 - \rho^2) = \frac{n - 1}{n}.$$

The conditional expectation of $u(X_1)$, given $\overline{X} = \overline{x}$, is then

$$\varphi(\overline{x}) = \int_{-\infty}^{\infty} u(x_1)\sqrt{\frac{n}{n-1}}\frac{1}{\sqrt{2\pi}}\exp\left[-\frac{n(x_1 - \overline{x})^2}{2(n-1)}\right]dx_1$$

$$= \int_{-\infty}^{c} \sqrt{\frac{n}{n-1}}\frac{1}{\sqrt{2\pi}}\exp\left[-\frac{n(x_1 - \overline{x})^2}{2(n-1)}\right]dx_1.$$

The change of variable $z = \sqrt{n}(x_1 - \overline{x})/\sqrt{n-1}$ enables us to write, with $c' = \sqrt{n}(c - \overline{x})/\sqrt{n-1}$, this conditional expectation is

$$\varphi(\overline{x}) = \int_{-\infty}^{c'} \frac{1}{\sqrt{2\pi}}e^{-z^2/2}\,dz = \Phi(c') = \Phi\left[\frac{\sqrt{n}(c - \overline{x})}{\sqrt{n-1}}\right].$$

Thus the unique, unbiased, and minimum variance estimator of $\Phi(c - \theta)$ is, for every fixed constant c, given by $\varphi(\bar{X}) = \Phi[\sqrt{n}(c - \bar{X})/\sqrt{n-1}]$.

Remark. We should like to draw the attention of the reader to a rather important fact. This has to do with the adoption of a *principle*, such as the principle of unbiasedness and minimum variance. A principle is not a theorem; and seldom does a principle yield satisfactory results in all cases. So far, this principle has provided quite satisfactory results. To see that this is not always the case, let X have a Poisson distribution with parameter θ, $0 < \theta < \infty$. We may look upon X as a random sample of size 1 from this distribution. Thus X is a complete sufficient statistic for θ. We seek the estimator of $e^{-2\theta}$ that is unbiased and has minimum variance. Consider $Y = (-1)^X$. We have

$$E(Y) = E[(-1)^X] = \sum_{x=0}^{\infty} \frac{(-\theta)^x e^{-\theta}}{x!} = e^{-2\theta}.$$

Accordingly, $(-1)^X$ is the unbiased minimum variance estimator of $e^{-2\theta}$. Here this estimator leaves much to be desired. We are endeavoring to elicit some information about the number $e^{-2\theta}$, where $0 < e^{-2\theta} < 1$. Yet our point estimate is either -1 or $+1$, each of which is a very poor estimate of a number between zero and 1. We do not wish to leave the reader with the impression that an unbiased minimum variance estimator is *bad*. That is not the case at all. We merely wish to point out that if one tries hard enough, he can find instances where such a statistic is *not good*. Incidentally, the maximum likelihood estimator of $e^{-2\theta}$ is, in the case where the sample size equals 1, e^{-2X}, which is probably a much better estimator in practice than is the unbiased estimator $(-1)^X$.

EXERCISES

7.40. Let $X_1, X_2, \ldots, X_n$ denote a random sample from a distribution that is $N(\theta, 1)$, $-\infty < \theta < \infty$. Find the unbiased minimum variance estimator of θ^2.

Hint: First determine $E(\bar{X}^2)$.

7.41. Let $X_1, X_2, \ldots, X_n$ denote a random sample from a distribution that is $N(0, \theta)$. Then $Y = \sum X_i^2$ is a complete sufficient statistic for θ. Find the unbiased minimum variance estimator of θ^2.

7.42. In the notation of Example 2 of this section, is there an unbiased minimum variance estimator of $\Pr(-c \le X \le c)$? Here $c > 0$.

7.43. Let $X_1, X_2, \ldots, X_n$ be a random sample from a Poisson distribution with parameter $\theta > 0$. Find the unbiased minimum variance estimator of $\Pr(X \le 1) = (1 + \theta)e^{-\theta}$.

Hint: Let $u(x_1) = 1$, $x_1 \le 1$, zero elsewhere, and find $E[u(X_1)|Y = y]$, where $Y = \sum_1^n X_i$. Make use of Example 2, Section 4.2.

7.44. Let $X_1, X_2, \ldots, X_n$ denote a random sample from a Poisson distribution with parameter $\theta > 0$. From the Remark of this section, we know that $E[(-1)^{X_1}] = e^{-2\theta}$.

(a) Show that $E[(-1)^{X_1}|Y_1 = y_1] = (1 - 2/n)^{y_1}$, where $Y_1 = X_1 + X_2 + \cdots + X_n$.
Hint: First show that the conditional p.d.f. of $X_1, X_2, \ldots, X_{n-1}$, given $Y_1 = y_1$, is multinomial, and hence that of X_1 given $Y_1 = y_1$ is $b(y_1, 1/n)$.

(b) Show that the m.l.e. of $e^{-2\theta}$ is $e^{-2\bar{X}}$.

(c) Since $y_1 = n\bar{x}$, show that $(1 - 2/n)^{y_1}$ is approximately equal to $e^{-2\bar{x}}$ when n is large.

7.45. Let a random sample of size n be taken from a distribution that has the p.d.f. $f(x; \theta) = (1/\theta) \exp(-x/\theta) I_{(0, \infty)}(x)$. Find the m.l.e. and the unbiased minimum variance estimator of $\Pr(X \le 2)$.

7.7 The Case of Several Parameters

In many of the interesting problems we encounter, the p.d.f. may not depend upon a single parameter θ, but perhaps upon two (or more) parameters, say θ_1 and θ_2, where $(\theta_1, \theta_2) \in \Omega$, a two-dimensional parameter space. We now define joint sufficient statistics for the parameters. For the moment we shall restrict ourselves to the case of two parameters.

Definition 4. Let $X_1, X_2, \ldots, X_n$ denote a random sample from a distribution that has p.d.f. $f(x; \theta_1, \theta_2)$, where $(\theta_1, \theta_2) \in \Omega$. Let $Y_1 = u_1(X_1, X_2, \ldots, X_n)$ and $Y_2 = u_2(X_1, X_2, \ldots, X_n)$ be two statistics whose joint p.d.f. is $g_{12}(y_1, y_2; \theta_1, \theta_2)$. The statistics Y_1 and Y_2 are called *joint sufficient statistics* for θ_1 and θ_2 if and only if

$$\frac{f(x_1; \theta_1, \theta_2) f(x_2; \theta_1, \theta_2) \cdots f(x_n; \theta_1, \theta_2)}{g_{12}[u_1(x_1, \ldots, x_n), u_2(x_1, \ldots, x_n); \theta_1, \theta_2]} = H(x_1, x_2, \ldots, x_n),$$

where $H(x_1, x_2, \ldots, x_n)$ does not depend upon θ_1 or θ_2.

As may be anticipated, the factorization theorem can be extended. In our notation it can be stated in the following manner. The statistics $Y_1 = u_1(X_1, X_2, \ldots, X_n)$ and $Y_2 = u_2(X_1, X_2, \ldots, X_n)$ are joint sufficient statistics for the parameters θ_1 and θ_2 if and only if we can find two nonnegative functions k_1 and k_2 such that

$$f(x_1; \theta_1, \theta_2)f(x_2; \theta_1, \theta_2) \cdots f(x_n; \theta_1, \theta_2)$$
$$= k_1[u_1(x_1, x_2, \ldots, x_n), u_2(x_1, x_2, \ldots, x_n); \theta_1, \theta_2]k_2(x_1, x_2, \ldots, x_n),$$

where the function $k_2(x_1, x_2, \ldots, x_n)$ does not depend upon both or either of θ_1 and θ_2.

Example 1. Let $X_1, X_2, \ldots, X_n$ be a random sample from a distribution having p.d.f.

$$f(x; \theta_1, \theta_2) = \frac{1}{2\theta_2}, \qquad \theta_1 - \theta_2 < x < \theta_1 + \theta_2,$$
$$= 0 \qquad \text{elsewhere,}$$

where $-\infty < \theta_1 < \infty$, $0 < \theta_2 < \infty$. Let $Y_1 < Y_2 < \cdots < Y_n$ be the order statistics. The joint p.d.f. of Y_1 and Y_n is given by

$$g_{1n}(y_1, y_n; \theta_1, \theta_2) = \frac{n(n-1)}{(2\theta_2)^n}(y_n - y_1)^{n-2}, \qquad \theta_1 - \theta_2 < y_1 < y_n < \theta_1 + \theta_2,$$

and equals zero elsewhere. Accordingly, the joint p.d.f. of $X_1, X_2, \ldots, X_n$ can be written, for points of positive probability density,

$$\left(\frac{1}{2\theta_2}\right)^n = \frac{n(n-1)[\max(x_i) - \min(x_i)]^{n-2}}{(2\theta_2)^n}$$
$$\times \left(\frac{1}{n(n-1)[\max(x_i) - \min(x_i)]^{n-2}}\right).$$

Since $\min(x_i) \le x_j \le \max(x_i)$, $j = 1, 2, \ldots, n$, the last factor does not depend upon the parameters. Either the definition or the factorization theorem assures us that Y_1 and Y_n are joint sufficient statistics for θ_1 and θ_2.

The extension of the notion of joint sufficient statistics for more than two parameters is a natural one. Suppose that a certain p.d.f. depends upon m parameters. Let a random sample of size n be taken from the distribution that has this p.d.f. and define m statistics. These m statistics are called joint sufficient statistics for the m parameters if and only if the ratio of the joint p.d.f. of the observations of the random sample and the joint p.d.f. of these m statistics does not depend upon the m parameters, whatever the fixed values of the m statistics. Again the factorization theorem is readily extended.

There is an extension of the Rao–Blackwell theorem that can be adapted to joint sufficient statistics for several parameters, but that extension will not be included in this book. However, the concept of a complete family of probability density functions is generalized as follows: Let

$$\{f(v_1, v_2, \ldots, v_k; \theta_1, \theta_2, \ldots, \theta_m) : (\theta_1, \theta_2, \ldots, \theta_m) \in \Omega\}$$

denote a family of probability density functions of k random variables $V_1, V_2, \ldots, V_k$ that depends upon m parameters $(\theta_1, \theta_2, \ldots, \theta_m) \in \Omega$. Let $u(v_1, v_2, \ldots, v_k)$ be a function of $v_1, v_2, \ldots, v_k$ (but not a function of any or all of the parameters). If

$$E[u(V_1, V_2, \ldots, V_k)] = 0$$

for all $(\theta_1, \theta_2, \ldots, \theta_m) \in \Omega$ implies that $u(v_1, v_2, \ldots, v_k) = 0$ at all points $(v_1, v_2, \ldots, v_k)$, except on a set of points that has probability zero for all members of the family of probability density functions, we shall say that the family of probability density functions is a complete family.

The remainder of our treatment of the case of several parameters will be restricted to probability density functions that represent what we shall call regular cases of the exponential class. Let $X_1, X_2, \ldots, X_n$, $n > m$, denote a random sample from a distribution that depends on m parameters and has a p.d.f. of the form

$$f(x; \theta_1, \theta_2, \ldots, \theta_m) = \exp\left[\sum_{j=1}^{m} p_j(\theta_1, \theta_2, \ldots, \theta_m)K_j(x) + S(x) + q(\theta_1, \theta_2, \ldots, \theta_m)\right] \tag{1}$$

for $a < x < b$, and equals zero elsewhere.

A p.d.f. of the form (1) is said to be a member of the *exponential class* of probability density functions of the continuous type. If, in addition,

1. neither a nor b depends upon any or all of the parameters $\theta_1, \theta_2, \ldots, \theta_m$,
2. the $p_j(\theta_1, \theta_2, \ldots, \theta_m)$, $j = 1, 2, \ldots, m$, are nontrivial, functionally independent, continuous functions of θ_j, $\gamma_j < \theta_j < \delta_j$, $j = 1, 2, \ldots, m$,
3. the $K_j'(x)$, $j = 1, 2, \ldots, m$, are continuous for $a < x < b$ and no one is a linear homogeneous function of the others,

4. $S(x)$ is a continuous function of x, $a < x < b$, we say that we have a *regular case* of the exponential class.

The joint p.d.f. of $X_1, X_2, \ldots, X_n$ is given, at points of positive probability density, by

$$\exp\left[\sum_{j=1}^{m} p_j(\theta_1, \ldots, \theta_m) \sum_{i=1}^{n} K_j(x_i) + \sum_{i=1}^{n} S(x_i) + nq(\theta_1, \ldots, \theta_m)\right]$$

$$= \exp\left[\sum_{j=1}^{m} p_j(\theta_1, \ldots, \theta_m) \sum_{i=1}^{n} K_j(x_i) + nq(\theta_1, \ldots, \theta_m)\right]$$

$$\times \exp\left[\sum_{i=1}^{n} S(x_i)\right].$$

In accordance with the factorization theorem, the statistics

$$Y_1 = \sum_{i=1}^{n} K_1(X_i), \quad Y_2 = \sum_{i=1}^{n} K_2(X_i), \ldots, Y_m = \sum_{i=1}^{n} K_m(X_i)$$

are joint sufficient statistics for the m parameters $\theta_1, \theta_2, \ldots, \theta_m$. It is left as an exercise to prove that the joint p.d.f. of $Y_1, \ldots, Y_m$ is of the form

$$R(y_1, \ldots, y_m) \exp\left[\sum_{j=1}^{m} p_j(\theta_1, \ldots, \theta_m)y_j + nq(\theta_1, \ldots, \theta_m)\right] \quad (2)$$

at points of positive probability density. These points of positive probability density and the function $R(y_1, \ldots, y_m)$ do not depend upon any or all of the parameters $\theta_1, \theta_2, \ldots, \theta_m$. Moreover, in accordance with a theorem in analysis, it can be asserted that, in a regular case of the exponential class, the family of probability density functions of these joint sufficient statistics $Y_1, Y_2, \ldots, Y_m$ is complete when $n > m$. In accordance with a convention previously adopted, we shall refer to $Y_1, Y_2, \ldots, Y_m$ as *joint complete sufficient statistics for the parameters* $\theta_1, \theta_2, \ldots, \theta_m$.

Example 2. Let $X_1, X_2, \ldots, X_n$ denote a random sample from a distribution that is $N(\theta_1, \theta_2)$, $-\infty < \theta_1 < \infty, 0 < \theta_2 < \infty$. Thus the p.d.f. $f(x; \theta_1, \theta_2)$ of the distribution may be written as

$$f(x; \theta_1, \theta_2) = \exp\left(\frac{-1}{2\theta_2} x^2 + \frac{\theta_1}{\theta_2} x - \frac{\theta_1^2}{2\theta_2} - \ln \sqrt{2\pi\theta_2}\right).$$

Therefore, we can take $K_1(x) = x^2$ and $K_2(x) = x$. Consequently, the statistics

$$Y_1 = \sum_1^n X_i^2 \quad \text{and} \quad Y_2 = \sum_1^n X_i$$

are joint complete sufficient statistics for θ_1 and θ_2. Since the relations

$$Z_1 = \frac{Y_2}{n} = \bar{X}, \qquad Z_2 = \frac{Y_1 - Y_2^2/n}{n-1} = \frac{\sum (X_i - \bar{X})^2}{n-1}$$

define a one-to-one transformation, Z_1 and Z_2 are also joint complete sufficient statistics for θ_1 and θ_2. Moreover,

$$E(Z_1) = \theta_1 \quad \text{and} \quad E(Z_2) = \theta_2.$$

From completeness, we have that Z_1 and Z_2 are the only functions of Y_1 and Y_2 that are unbiased estimators of θ_1 and θ_2, respectively.

A p.d.f.

$$f(x; \theta_1, \theta_2, \dots, \theta_m) = \exp\left[\sum_{j=1}^m p_j(\theta_1, \theta_2, \dots, \theta_m)K_j(x) + S(x) + q(\theta_1, \theta_2, \dots, \theta_m)\right], \qquad x = a_1, a_2, a_3, \dots,$$

zero elsewhere, is said to represent a regular case of the exponential class of probability density functions of the discrete type if

1. the set $\{x : x = a_1, a_2, \dots\}$ does not depend upon any or all of the parameters $\theta_1, \theta_2, \dots, \theta_m$,
2. the $p_j(\theta_1, \theta_2, \dots, \theta_m)$, $j = 1, 2, \dots, m$, are nontrivial, functionally independent, and continuous functions of θ_j, $\gamma_j < \theta_j < \delta_j$, $j = 1, 2, \dots, m$,
3. the $K_j(x)$, $j = 1, 2, \dots, m$, are nontrivial functions of x on the set $\{x : x = a_1, a_2, \dots\}$ and no one is a linear function of the others.

Let $X_1, X_2, \dots, X_n$ denote a random sample from a discrete-type distribution that represents a regular case of the exponential class. Then the statements made above in connection with the random variable of the continuous type are also valid here.

Not always do we sample from a distribution of one random variable X. We could, for instance, sample from a distribution of two random variables V and W with joint p.d.f. $f(v, w; \theta_1, \theta_2, \dots, \theta_m)$. Recall that by a random sample $(V_1, W_1), (V_2, W_2), \dots, (V_n, W_n)$ from a distribution of this sort, we mean that the joint p.d.f. of these $2n$ random variables is given by

$$f(v_1, w_1; \theta_1, \dots, \theta_m)f(v_2, w_2; \theta_1, \dots, \theta_m) \cdots f(v_n, w_n; \theta_1, \dots, \theta_m).$$

In particular, suppose that the random sample is taken from a distribution that has the p.d.f. of V and W of the exponential class

$$f(v, w; \theta_1, \ldots, \theta_m) = \exp\left[\sum_{j=1}^{m} p_j(\theta_1, \ldots, \theta_m)K_j(v, w) + S(v, w) + q(\theta_1, \ldots, \theta_m)\right] \quad (3)$$

for $a < v < b$, $c < w < d$, and equals zero elsewhere, where a, b, c, d do not depend on the parameters and conditions similar to 1 to 4, p. 343, are imposed. Then the m statistics

$$Y_1 = \sum_{i=1}^{n} K_1(V_i, W_i), \ldots, \qquad Y_m = \sum_{i=1}^{n} K_m(V_i, W_i)$$

are joint complete sufficient statistics for the m parameters $\theta_1, \theta_2, \ldots, \theta_m$.

EXERCISES

7.46. Let $Y_1 < Y_2 < Y_3$ be the order statistics of a random sample of size 3 from the distribution with p.d.f.

$$f(x; \theta_1, \theta_2) = \frac{1}{\theta_2}\exp\left(-\frac{x - \theta_1}{\theta_2}\right),$$

$$\theta_1 < x < \infty, \quad -\infty < \theta_1 < \infty, \quad 0 < \theta_2 < \infty,$$

zero elsewhere. Find the joint p.d.f. of $Z_1 = Y_1$, $Z_2 = Y_2$, and $Z_3 = Y_1 + Y_2 + Y_3$. The corresponding transformation maps the space $\{(y_1, y_2, y_3): \theta_1 < y_1 < y_2 < y_3 < \infty\}$ onto the space

$$\{(z_1, z_2, z_3): \theta_1 < z_1 < z_2 < (z_3 - z_1)/2 < \infty\}$$

Show that Z_1 and Z_3 are joint sufficient statistics for θ_1 and θ_2.

7.47. Let $X_1, X_2, \ldots, X_n$ be a random sample from a distribution that has a p.d.f. of form (1) of this section. Show that $Y_1 = \sum_{i=1}^{n} K_1(X_i)$, $\ldots, Y_m = \sum_{i=1}^{n} K_m(X_i)$ have a joint p.d.f. of form (2) of this section.

7.48. Let $(X_1, Y_1), (X_2, Y_2), \ldots, (X_n, Y_n)$ denote a random sample of size n from a bivariate normal distribution with means μ_1 and μ_2, positive variances σ_1^2 and σ_2^2, and correlation coefficient ρ. Show that $\sum_1^n X_i$, $\sum_1^n Y_i$, $\sum_1^n X_i^2$, $\sum_1^n Y_i^2$, and $\sum_1^n X_i Y_i$ are joint complete sufficient statistics for the five parameters. Are $\bar{X} = \sum_1^n X_i/n$, $\bar{Y} = \sum_1^n Y_i/n$, $S_1^2 = \sum_1^n (X_i - \bar{X})^2/n$, $S_2^2 = \sum_1^n (Y_i - \bar{Y})^2/n$, and $\sum_1^n (X_i - \bar{X})(Y_i - \bar{Y})/nS_1S_2$ also joint complete sufficient statistics for these parameters?

7.49. Let the p.d.f. $f(x; \theta_1, \theta_2)$ be of the form

$$\exp [p_1(\theta_1, \theta_2)K_1(x) + p_2(\theta_1, \theta_2)K_2(x) + S(x) + q(\theta_1, \theta_2)], \quad a < x < b,$$

zero elsewhere. Let $K_1'(x) = cK_2'(x)$. Show that $f(x; \theta_1, \theta_2)$ can be written in the form

$$\exp [p(\theta_1, \theta_2)K(x) + S(x) + q_1(\theta_1, \theta_2)], \quad a < x < b,$$

zero elsewhere. This is the reason why it is required that no one $K_j'(x)$ be a linear homogeneous function of the others, that is, so that the number of sufficient statistics equals the number of parameters.

7.50. Let $Y_1 < Y_2 < \cdots < Y_n$ be the order statistics of a random sample $X_1, X_2, \ldots, X_n$ of size n from a distribution of the continuous type with p.d.f. $f(x)$. Show that the ratio of the joint p.d.f. of $X_1, X_2, \ldots, X_n$ and that of $Y_1 < Y_2 < \cdots < Y_n$ is equal to $1/n!$, which does not depend upon the underlying p.d.f. This suggests that $Y_1 < Y_2 < \cdots < Y_n$ are joint sufficient statistics for the unknown "parameter" f.

7.51. Let $X_1, X_2, \ldots, X_n$ be a random sample from the uniform distribution with p.d.f. $f(x; \theta_1, \theta_2) = 1/(2\theta_2)$, $\theta_1 - \theta_2 < x < \theta_1 + \theta_2$, where $-\infty < \theta_1 < \infty$ and $\theta_2 > 0$, and the p.d.f. is equal to zero elsewhere.

(a) Show that $Y_1 = \min (X_i)$ and $Y_n = \max (X_i)$, the joint sufficient statistics for θ_1 and θ_2, are complete.

(b) Find the unbiased minimum variance estimators of θ_1 and θ_2.

7.52. Let $X_1, X_2, \ldots, X_n$ be a random sample from $N(\theta_1, \theta_2)$.

(a) If the constant b is defined by the equation $\Pr (X \le b) = 0.90$, find the m.l.e. and the unbiased minimum variance estimator of b.

(b) If c is a given constant, find the m.l.e. and the unbiased minimum variance estimator of $\Pr (X \le c)$.

7.8 Minimal Sufficient and Ancillary Statistics

In the study of statistics, it is clear that we want to reduce the data contained in the entire sample as much as possible without losing relevant information about the important characteristics of the underlying distribution. That is, a large collection of numbers in the sample is not as meaningful as a few good summary statistics of those data. Sufficient statistics, if they exist, are valuable because we know that the statistician with those summary measures is as well off as the statistician with the entire sample. Sometimes, however, there are several sets of joint sufficient statistics, and thus we would like to find the simplest one of these sets. For illustration, in a sense, the observations $X_1, X_2, \ldots, X_n, n > 2$, of a random sample from $N(\theta_1, \theta_2)$ could be thought of as joint sufficient statistics for θ_1 and θ_2. We know, however, that we can use $\bar{X}$ and S^2 as joint sufficient statistics for those parameters, which is a great simplification over using $X_1, X_2, \ldots, X_n$, particularly if n is large.

In most instances in this chapter, we have been able to find a single sufficient statistic for one parameter or two joint sufficient statistics for two parameters. Possibly the most complicated case considered so far is given in Exercise 7.48, in which we find five joint sufficient statistics for five parameters. Exercise 7.50 suggests the possibility of using the order statistics of a random sample for some completely unknown distribution of the continuous type.

What we would like to do is to change from one set of joint sufficient statistics to another, always reducing the number of statistics involved until we cannot go any further without losing the sufficiency of the resulting statistics. Those statistics that are there at the end of this process are called minimal sufficient statistics for the parameters. That is, *minimal sufficient statistics* are those that are sufficient for the parameters and are functions of every other set of sufficient statistics for those same parameters. Often, if there are k parameters, we can find k joint sufficient statistics that are minimal. In particular, if there is one parameter, we can often find a single sufficient statistic which is minimal. Most of the earlier examples that we have considered illustrate this point, but this is not always the case as shown by the following example.

Example 1. Let $X_1, X_2, \ldots, X_n$ be a random sample from the uniform distribution over the interval $(\theta - 1, \theta + 1)$ having p.d.f.

$$f(x; \theta) = (\tfrac{1}{2})I_{(\theta - 1, \theta + 1)}(x), \qquad \text{where } -\infty < \theta < \infty.$$

The joint p.d.f. of $X_1, X_2, \ldots, X_n$ equals the product of $(\frac{1}{2})^n$ and certain indicator functions, namely

$$(\tfrac{1}{2})^n \prod_{i=1}^{n} I_{(\theta - 1, \theta + 1)}(x_i) = (\tfrac{1}{2})^n \{I_{(\theta - 1, \theta + 1)}[\min (x_i)]\}\{I_{(\theta - 1, \theta + 1)}[\max (x_i)]\},$$

because $\theta - 1 < \min (x_i) \le x_j \le \max (x_i) < \theta + 1, j = 1, 2, \ldots, n$. Thus the order statistics $Y_1 = \min (X_i)$ and $Y_n = \max (X_i)$ are the sufficient statistics for θ. These two statistics actually are minimal for this one parameter, as we cannot reduce the number of them to less than two and still have sufficiency.

There is an observation that helps us observe that almost all the sufficient statistics that we have studied thus far are minimal. We have noted that the m.l.e. $\hat{\theta}$ of θ is a function of one or more sufficient statistics, when the latter exist. Suppose that this m.l.e. $\hat{\theta}$ is also sufficient. Since this sufficient statistic $\hat{\theta}$ is a function of the other sufficient statistics, it must be minimal. For example, we have

1. The m.l.e. $\hat{\theta} = \bar{X}$ of θ in $N(\theta, \sigma^2)$, σ^2 known, is a minimal sufficient statistic for θ.
2. The m.l.e. $\hat{\theta} = \bar{X}$ of θ in a Poisson distribution with mean θ is a minimal sufficient statistic for θ.
3. The m.l.e. $\hat{\theta} = Y_n = \max(X_i)$ of θ in the uniform distribution over $(0, \theta)$ is a minimal sufficient statistic for θ.
4. The maximum likelihood estimators $\hat{\theta}_1 = \bar{X}$ and $\hat{\theta}_2 = S^2$ of θ_1 and θ_2 in $N(\theta_1, \theta_2)$ are joint minimal sufficient statistics for θ_1 and θ_2.

From these examples we see that the minimal sufficient statistics do not need to be unique, for any one-to-one transformation of them also provides minimal sufficient statistics. For illustration, in 4, the ΣX_i and ΣX_i^2 are also minimal sufficient statistics for θ_1 and θ_2.

Example 2. Consider the model given in Example 1. There we noted that $Y_1 = \min(X_i)$ and $Y_n = \max(X_i)$ are joint sufficient statistics. Also, we have

$$\theta - 1 < Y_1 < Y_n < \theta + 1$$

or, equivalently,

$$Y_n - 1 < \theta < Y_1 + 1.$$

Hence, to maximize the likelihood function so that it equals $(\frac{1}{2})^n$, θ can be any value between $Y_n - 1$ and $Y_1 + 1$. For example, many statisticians take the m.l.e. to be the mean of these two end points, namely

$$\hat{\theta} = \frac{Y_n - 1 + Y_1 + 1}{2} = \frac{Y_1 + Y_n}{2},$$

which is the midrange. We recognize, however, that this m.l.e. is not unique. Some might argue that since $\hat{\theta}$ is an m.l.e. of θ and since it is a function of the joint sufficient statistics, Y_1 and Y_n, for θ, it will be a minimal sufficient statistic. This is not the case at all, for $\hat{\theta}$ is not even sufficient. Note that the m.l.e. must itself be a sufficient statistic for the parameter before it can be considered the minimal sufficient statistic.

There is also a relationship between a minimal sufficient statistic and completeness that is explained more fully in the 1950 article by Lehmann and Scheffé. Let us say simply and without explanation that for the cases in this book, complete sufficient statistics are minimal sufficient statistics. The converse is not true, however, by noting that in Example 1 we have

$$E\left[\frac{Y_n - Y_1}{2} - \frac{n-1}{n+1}\right] = 0, \qquad \text{for all } \theta.$$

That is, there is a nonzero function of those minimal sufficient statistics, Y_1 and Y_n, whose expectation is zero for all θ.

There are other statistics that almost seem opposites of sufficient statistics. That is, while sufficient statistics contain all the information about the parameters, these other statistics, called *ancillary statistics*, have distributions free of the parameters and seemingly contain no information about those parameters. As an illustration, we know that the variance S^2 of a random sample from $N(\theta, 1)$ has a distribution that does not depend upon θ and hence is an ancillary statistic. Another example is the ratio $Z = X_1/(X_1 + X_2)$, where X_1, X_2 is a random sample from a gamma distribution with known parameter $\alpha > 0$ and unknown parameter $\beta = \theta$, because Z has a beta distribution that is free of θ. There are a great number of examples of ancillary statistics, and we provide some rules that make them rather easy to find with certain models.

First consider the situation in which there is a location parameter. That is, let $X_1, X_2, \ldots, X_n$ be a random sample from a distribution that has a p.d.f. of the form $f(x - \theta)$, for every real θ; that is, θ is a location parameter. Let $Z = u(X_1, X_2, \ldots, X_n)$ be a statistic such that

$$u(x_1 + d, x_2 + d, \ldots, x_n + d) = u(x_1, x_2, \ldots, x_n),$$

for all real d. The one-to-one transformation defined by $W_i = X_i - \theta$, $i = 1, 2, \ldots, n$, requires that the joint p.d.f. of $W_1, W_2, \ldots, W_n$ be

$$f(w_1)f(w_2)\cdots f(w_n),$$

which does not depend upon θ. In addition, we have, because of the special functional nature of $u(x_1, x_2, \ldots, x_n)$, that

$$Z = u(W_1 + \theta, W_2 + \theta, \ldots, W_n + \theta) = u(W_1, W_2, \ldots, W_n)$$

is a function of $W_1, W_2, \ldots, W_n$ alone (not of θ). Hence Z must have a distribution that does not depend upon θ because, for illustration, the m.g.f. of Z, namely

$$E(e^{tZ}) = \int_{-\infty}^{\infty} \cdots \int_{-\infty}^{\infty} e^{tu(x_1, \ldots, x_n)} f(x_1 - \theta) \cdots f(x_n - \theta)\, dx_1 \cdots dx_n$$

$$= \int_{-\infty}^{\infty} \cdots \int_{-\infty}^{\infty} e^{tu(w_1, \ldots, w_n)} f(w_1) \cdots f(w_n)\, dw_1 \cdots dw_n$$

is free of θ. We call $Z = u(X_1, X_2, \ldots, X_n)$ a *location-invariant statistic.* We immediately see that we can construct many examples of location-invariant statistics: the sample variance $= S^2$, the sample range $= Y_n - Y_1$, the mean deviation from the sample median $= (1/n)\,\Sigma|X_i - \text{median}\,(X_i)|$, $X_1 + X_2 - X_3 - X_4$, $X_1 + X_3 - 2X_2$, $(1/n)\,\Sigma\,[X_i - \min\,(X_i)]$, and so on.

We now consider a *scale-invariant statistic.* Let $X_1, X_2, \ldots, X_n$ be a random sample from a distribution that has a p.d.f. of the form $(1/\theta)f(x/\theta)$, for all $\theta > 0$; that is, θ is a scale parameter. Say that $Z = u(X_1, X_2, \ldots, X_n)$ is a statistic such that

$$u(cx_1, cx_2, \ldots, cx_n) = u(x_1, x_2, \ldots, x_n)$$

for all $c > 0$. The one-to-one transformation defined by $W_i = X_i/\theta$, $i = 1, 2, \ldots, n$, requires the following: (1) that the joint p.d.f. of $W_1, W_2, \ldots, W_n$ be equal to

$$f(w_1)f(w_2) \cdots f(w_n),$$

and (2) that the statistic Z be equal to

$$Z = u(\theta W_1, \theta W_2, \ldots, \theta W_n) = u(W_1, W_2, \ldots, W_n).$$

Since neither the joint p.d.f. of $W_1, W_2, \ldots, W_n$ nor Z contain θ, the distribution of Z must not depend upon θ. There are also many examples of scale-invariant statistics like this: $Z: X_1/(X_1 + X_2)$, $X_1^2/\sum_1^n X_i^2$, $\min\,(X_i)/\max\,(X_i)$, and so on.

Finally, the location and the scale parameters can be combined in a p.d.f. of the form $(1/\theta_2)f[(x - \theta_1)/\theta_2]$, $-\infty < \theta_1 < \infty, 0 < \theta_2 < \infty$. Through a one-to-one transformation defined by $W_i = (X_i - \theta_1)/\theta_2$, $i = 1, 2, \ldots, n$, it is easy to show that a statistic $Z = u(X_1, X_2, \ldots, X_n)$ such that

$$u(cx_1 + d, \ldots, cx_n + d) = u(x_1, \ldots, x_n)$$

for $-\infty < d < \infty$, $0 < c < \infty$, has a distribution that does not depend upon θ_1 and θ_2. Statistics like this $Z = u(X_1, X_2, \ldots, X_n)$ are *location-and-scale-invariant statistics.* Again there are many examples: $[\max(X_i) - \min(X_i)]/S$, $\sum_{i=1}^{n-1} (X_{i+1} - X_i)^2/S^2$, $(X_i - \bar{X})/S$, $|X_i - X_j|/S$, $i \neq j$, and so on.

Thus these location-invariant, scale-invariant, and location-and-scale-invariant statistics provide good illustrations, with the appropriate model for the p.d.f., of ancillary statistics. Since an ancillary statistic and a complete (minimal) sufficient statistic are such opposites, we might believe that there is, in some sense, no relationship between the two. This is true and in the next section we show that they are independent statistics.

EXERCISES

7.53. Let $X_1, X_2, \ldots, X_n$ be a random sample from each of the following distributions involving the parameter θ. In each case find the m.l.e. of θ and show that it is a sufficient statistic for θ and hence a minimal sufficient statistic.

(a) $b(1, \theta)$, where $0 \le \theta \le 1$.

(b) Poisson with mean $\theta > 0$.

(c) Gamma with $\alpha = 3$ and $\beta = \theta > 0$.

(d) $N(\theta, 1)$, where $-\infty < \theta < \infty$.

(e) $N(0, \theta)$, where $0 < \theta < \infty$.

7.54. Let $Y_1 < Y_2 < \cdots < Y_n$ be the order statistics of a random sample of size n from the uniform distribution over the closed interval $[-\theta, \theta]$ having p.d.f. $f(x; \theta) = (1/2\theta)I_{[-\theta,\theta]}(x)$.

(a) Show that Y_1 and Y_n are joint sufficient statistics for θ.

(b) Argue that the m.l.e. of θ equals $\hat{\theta} = \max(-Y_1, Y_n)$.

(c) Demonstrate that the m.l.e. $\hat{\theta}$ is a sufficient statistic for θ and thus is a minimal sufficient statistic for θ.

7.55. Let $Y_1 < Y_2 < \cdots < Y_n$ be the order statistics of a random sample of size n from a distribution with p.d.f.

$$f(x; \theta_1, \theta_2) = \left(\frac{1}{\theta_2}\right) e^{-(x-\theta_1)/\theta_2} I_{(\theta_1,\infty)}(x),$$

where $-\infty < \theta_1 < \infty$ and $0 < \theta_2 < \infty$. Find joint minimal sufficient statistics for θ_1 and θ_2.

7.56. With random samples from each of the distributions given in Exercises 7.53(d), 7.54, and 7.55, define at least two ancillary statistics that are different from the examples given in the text. These examples illustrate, respectively, location-invariant, scale-invariant, and location-and-scale-invariant statistics.

7.9 Sufficiency, Completeness, and Independence

We have noted that if we have a sufficient statistic Y_1 for a parameter θ, $\theta \in \Omega$, then $h(z|y_1)$, the conditional p.d.f. of another statistic Z, given $Y_1 = y_1$, does not depend upon θ. If, moreover, Y_1 and Z are independent, the p.d.f. $g_2(z)$ of Z is such that $g_2(z) = h(z|y_1)$, and hence $g_2(z)$ must not depend upon θ either. So the independence of a statistic Z and the sufficient statistic Y_1 for a parameter θ means that the distribution of Z does not depend upon $\theta \in \Omega$. That is, Z is an ancillary statistic.

It is interesting to investigate a converse of that property. Suppose that the distribution of an ancillary statistic Z does not depend upon θ; then, are Z and the sufficient statistic Y_1 for θ independent? To begin our search for the answer, we know that the joint p.d.f. of Y_1 and Z is $g_1(y_1; \theta)h(z|y_1)$, where $g_1(y_1; \theta)$ and $h(z|y_1)$ represent the marginal p.d.f. of Y_1 and the conditional p.d.f. of Z given $Y_1 = y_1$, respectively. Thus the marginal p.d.f. of Z is

$$\int_{-\infty}^{\infty} g_1(y_1; \theta)h(z|y_1)\, dy_1 = g_2(z),$$

which, by hypothesis, does not depend upon θ. Because

$$\int_{-\infty}^{\infty} g_2(z)g_1(y_1; \theta)\, dy_1 = g_2(z),$$

it follows, by taking the difference of the last two integrals, that

$$\int_{-\infty}^{\infty} [g_2(z) - h(z|y_1)]g_1(y_1; \theta)\, dy_1 = 0 \tag{1}$$

for all $\theta \in \Omega$. Since Y_1 is a sufficient statistic for θ, $h(z|y_1)$ does not depend upon θ. By assumption, $g_2(z)$ and hence $g_2(z) - h(z|y_1)$ do not depend upon θ. Now if the family $\{g_1(y_1; \theta) : \theta \in \Omega\}$ is complete, Equation (1) would require that

$$g_2(z) - h(z|y_1) = 0 \qquad \text{or} \qquad g_2(z) = h(z|y_1).$$

That is, the joint p.d.f. of Y_1 and Z must be equal to

$$g_1(y_1; \theta)h(z|y_1) = g_1(y_1; \theta)g_2(z).$$

Accordingly, Y_1 and Z are independent, and we have proved the following theorem, which was considered in special cases by Neyman and Hogg and proved in general by Basu.

Theorem 6. *Let $X_1, X_2, \ldots, X_n$ denote a random sample from a distribution having a* p.d.f. $f(x; \theta)$, $\theta \in \Omega$, *where Ω is an interval set. Let $Y_1 = u_1(X_1, X_2, \ldots, X_n)$ be a sufficient statistic for θ, and let the family $\{g_1(y_1; \theta) : \theta \in \Omega\}$ of probability density functions of Y_1 be complete. Let $Z = u(X_1, X_2, \ldots, X_n)$ be any other statistic (not a function of Y_1 alone). If the distribution of Z does not depend upon θ, then Z is independent of the sufficient statistic Y_1.*

In the discussion above, it is interesting to observe that if Y_1 is a sufficient statistic for θ, then the independence of Y_1 and Z implies that the distribution of Z does not depend upon θ whether $\{g_1(y_1; \theta) : \theta \in \Omega\}$ is or is not complete. However, in the converse, to prove the independence from the fact that $g_2(z)$ does not depend upon θ, we definitely need the completeness. Accordingly, if we are dealing with situations in which we know that the family $\{g(y_1; \theta) : \theta \in \Omega\}$ is complete (such as a regular case of the exponential class), we can say that the statistic Z is independent of the sufficient statistic Y_1 if, and only if, the distribution of Z does not depend upon θ (i.e., Z is an ancillary statistic).

It should be remarked that the theorem (including the special formulation of it for regular cases of the exponential class) extends immediately to probability density functions that involve m parameters for which there exist m joint sufficient statistics. For example, let $X_1, X_2, \ldots, X_n$ be a random sample from a distribution having the p.d.f. $f(x; \theta_1, \theta_2)$ that represents a regular case of the exponential class such that there are two joint complete sufficient statistics for θ_1 and θ_2. Then any other statistic $Z = u(X_1, X_2, \ldots, X_n)$ is independent of the joint complete sufficient statistics if and only if the distribution of Z does not depend upon θ_1 or θ_2.

We give an example of the theorem that provides an alternative proof of the independence of $\bar{X}$ and S^2, the mean and the variance of a random sample of size n from a distribution that is $N(\mu, \sigma^2)$. This proof is presented as if we did not know that nS^2/σ^2 is $\chi^2(n-1)$ because that fact and the independence were established in the same argument (see Section 4.8).

Example 1. Let $X_1, X_2, \ldots, X_n$ denote a random sample of size n from a distribution that is $N(\mu, \sigma^2)$. We know that the mean $\bar{X}$ of the sample is, for every known σ^2, a complete sufficient statistic for the parameter μ, $-\infty < \mu < \infty$. Consider the statistic

$$S^2 = \frac{1}{n} \sum_{i=1}^{n} (X_i - \bar{X})^2,$$

which is location-invariant. Thus S^2 must have a distribution that does not depend upon μ; and hence, by the theorem, S^2 and $\bar{X}$, the complete sufficient statistic for μ, are independent.

Example 2. Let $X_1, X_2, \ldots, X_n$ be a random sample of size n from the distribution having p.d.f.

$$\begin{aligned} f(x; \theta) &= e^{-(x-\theta)}, \qquad \theta < x < \infty, \quad -\infty < \theta < \infty. \\ &= 0 \qquad \text{elsewhere.} \end{aligned}$$

Here the p.d.f. is of the form $f(x - \theta)$, where $f(x) = e^{-x}$, $0 < x < \infty$, zero elsewhere. Moreover, we know (Exercise 7.27) that the first order statistic $Y_1 = \min (X_i)$ is a complete sufficient statistic for θ. Hence Y_1 must be independent of each location-invariant statistic $u(X_1, X_2, \ldots, X_n)$, enjoying the property that

$$u(x_1 + d, x_2 + d, \ldots, x_n + d) = u(x_1, x_2, \ldots, x_n)$$

for all real d. Illustrations of such statistics are S^2, the sample range, and

$$\frac{1}{n} \sum_{i=1}^{n} [X_i - \min (X_i)].$$

Example 3. Let X_1, X_2 denote a random sample of size $n = 2$ from a distribution with p.d.f.

$$\begin{aligned} f(x; \theta) &= \frac{1}{\theta} e^{-x/\theta}, \qquad 0 < x < \infty, \quad 0 < \theta < \infty, \\ &= 0 \qquad \text{elsewhere.} \end{aligned}$$

The p.d.f. is of the form $(1/\theta)f(x/\theta)$, where $f(x) = e^{-x}$, $0 < x < \infty$, zero elsewhere. We know (Section 7.5) that $Y_1 = X_1 + X_2$ is a complete sufficient statistic for θ. Hence Y_1 is independent of every scale-invariant statistic $u(X_1, X_2)$ with the property $u(cx_1, cx_2) = u(x_1, x_2)$. Illustrations of these are X_1/X_2 and $X_1/(X_1 + X_2)$, statistics that have F and beta distributions, respectively.

Example 4. Let $X_1, X_2, \ldots, X_n$ denote a random sample from a distribution that is $N(\theta_1, \theta_2)$, $-\infty < \theta_1 < \infty$, $0 < \theta_2 < \infty$. In Example 2,

Section 7.7, it was proved that the mean $\bar{X}$ and the variance S^2 of the sample are joint complete sufficient statistics for θ_1 and θ_2. Consider the statistic

$$Z = \frac{\sum_{1}^{n-1} (X_{i+1} - X_i)^2}{\sum_{1}^{n} (X_i - \bar{X})^2} = u(X_1, X_2, \ldots, X_n),$$

which satisfies the property that $u(cx_1 + d, \ldots, cx_n + d) = u(x_1, \ldots, x_n)$. That is, the ancillary statistic Z is independent of both $\bar{X}$ and S^2.

Let $N(\theta_1, \theta_3)$ and $N(\theta_2, \theta_4)$ denote two normal distributions. Recall that in Example 2, Section 6.5, a statistic, which was denoted by T, was used to test the hypothesis that $\theta_1 = \theta_2$, provided that the unknown variances θ_3 and θ_4 were equal. The hypothesis that $\theta_1 = \theta_2$ is rejected if the computed $|T| \geq c$, where the constant c is selected so that $\alpha_2 = \Pr(|T| \geq c; \theta_1 = \theta_2, \theta_3 = \theta_4)$ is the assigned significance level of the test. We shall show that, if $\theta_3 = \theta_4$, F of Exercise 6.52 and T are independent. Among other things, this means that if these two tests based on F and T, respectively, are performed sequentially, with significance levels α_1 and α_2, the probability of accepting both these hypotheses, when they are true, is $(1 - \alpha_1)(1 - \alpha_2)$. Thus the significance level of this joint test is $\alpha = 1 - (1 - \alpha_1)(1 - \alpha_2)$.

The independence of F and T, when $\theta_3 = \theta_4$, can be established by an appeal to sufficiency and completeness. The three statistics $\bar{X}$, $\bar{Y}$, and $\sum_{1}^{n} (X_i - \bar{X})^2 + \sum_{1}^{m} (Y_i - \bar{Y})^2$ are joint complete sufficient statistics for the three parameters θ_1, θ_2, and $\theta_3 = \theta_4$. Obviously, the distribution of F does not depend upon θ_1, θ_2, or $\theta_3 = \theta_4$, and hence F is independent of the three joint complete sufficient statistics. However, T is a function of these three joint complete sufficient statistics alone, and, accordingly, T is independent of F. It is important to note that these two statistics are independent whether $\theta_1 = \theta_2$ or $\theta_1 \neq \theta_2$. This permits us to calculate probabilities other than the significance level of the test. For example, if $\theta_3 = \theta_4$ and $\theta_1 \neq \theta_2$, then

$$\Pr(c_1 < F < c_2, |T| \geq c) = \Pr(c_1 < F < c_2) \Pr(|T| \geq c).$$

The second factor in the right-hand member is evaluated by using the probabilities for what is called a noncentral t-distribution. Of course, if $\theta_3 = \theta_4$ and the difference $\theta_1 - \theta_2$ is large, we would want the preceding probability to be close to 1 because the event $\{c_1 < F < c_2, |T| \geq c\}$ leads to a correct decision, namely accept $\theta_3 = \theta_4$ and reject $\theta_1 = \theta_2$.

In this section we have given several examples in which the complete sufficient statistics are independent of ancillary statistics. Thus, in those cases, the ancillary statistics provide no information about the parameters. However, if the sufficient statistics are not complete, the ancillary statistics could provide some information as the following example demonstrates.

Example 5. We refer back to Examples 1 and 2 of Section 7.8. There the first and nth order statistics, Y_1 and Y_n, were minimal sufficient statistics for θ, where the sample arose from an underlying distribution having p.d.f. $(\frac{1}{2})I_{(\theta - 1, \theta + 1)}(x)$. Often $T_1 = (Y_1 + Y_n)/2$ is used as an estimator of θ as it is a function of those sufficient statistics which is unbiased. Let us find a relationship between T_1 and the ancillary statistic $T_2 = Y_n - Y_1$.

The joint p.d.f. of Y_1 and Y_n is

$$g(y_1, y_n; \theta) = n(n-1)(y_n - y_1)^{n-2}/2^n, \qquad \theta - 1 < y_1 < y_n < \theta + 1,$$

zero elsewhere. Accordingly, the joint p.d.f. of T_1 and T_2 is, since the absolute value of the Jacobian equals 1,

$$h(t_1, t_2; \theta) = n(n-1)t_2^{n-2}/2^n, \qquad \theta - 1 + \frac{t_2}{2} < t_1 < \theta + 1 - \frac{t_2}{2}, \quad 0 < t_2 < 2,$$

zero elsewhere. Thus the p.d.f. of T_2 is

$$h_2(t_2; \theta) = n(n-1)t_2^{n-2}(2 - t_2)/2^n, \qquad 0 < t_2 < 2,$$

zero elsewhere, which of course is free of θ as T_2 is an ancillary statistic. Thus the conditional p.d.f. of T_1, given $T_2 = t_2$, is

$$h_{1|2}(t_1|t_2; \theta) = \frac{1}{2 - t_2}, \qquad \theta - 1 + \frac{t_2}{2} < t_1 < \theta + 1 - \frac{t_2}{2}, \quad 0 < t_2 < 2,$$

zero elsewhere. Note that this is uniform on the interval $(\theta - 1 + t_2/2, \theta + 1 - t_2/2)$; so the conditional mean and variance of T_1 are, respectively,

$$E(T_1|t_2) = \theta \qquad \text{and} \qquad \text{var}\,(T_1|t_2) = \frac{(2 - t_2)^2}{12}.$$

That is, given $T_2 = t_2$, we know something about the conditional variance of T_1. In particular, if that observed value of T_2 is large (close to 2), that variance is small and we can place more reliance on the estimator T_1. On the other hand, a small value of t_2 means that we have less confidence in T_1 as an estimator of θ. It is extremely interesting to note that this conditional variance does not depend upon the sample size n but only on the given value of $T_2 = t_2$. Of course, as the sample size increases, T_2 tends to become larger and, in those cases, T_1 has smaller conditional variance.

While Example 5 is a special one demonstrating mathematically that an ancillary statistic can provide some help in point estimation, this does actually happen in practice too. For illustration, we know that if the sample size is large enough, then

$$T = \frac{\bar{X} - \mu}{S/\sqrt{n-1}}$$

has an approximate standard normal distribution. Of course, if the sample arises from a normal distribution, $\bar{X}$ and S are independent and T has a t-distribution with $n - 1$ degrees of freedom. Even if the sample arises from a symmetric distribution, $\bar{X}$ and S are uncorrelated and T has an approximate t-distribution and certainly an approximate standard normal distribution with sample sizes around 30 or 40. On the other hand, if the sample arises from a highly skewed distribution (say to the right), then $\bar{X}$ and S are highly correlated and the probability $\Pr(-1.96 < T < 1.96)$ is not necessarily close to 0.95 unless the sample size is extremely large (certainly much greater than 30). Intuitively, one can understand why this correlation exists if the underlying distribution is highly skewed to the right. While S has a distribution free of μ (and hence is an ancillary), a large value of S implies a large value of $\bar{X}$, since the underlying p.d.f. is like the one depicted in Figure 7.1. Of course, a small value of $\bar{X}$ (say less than the mode) requires a relatively small value of S. This means that unless n is extremely large, it is risky to say that

$$\bar{x} - \frac{1.96s}{\sqrt{n-1}}, \qquad \bar{x} + \frac{1.96s}{\sqrt{n-1}}$$

provides an approximate 95 percent confidence interval with data from a very skewed distribution. As a matter of fact, the authors have seen situations in which this confidence coefficient is closer to 70 percent, rather than 95 percent, with sample sizes of 30 to 40.

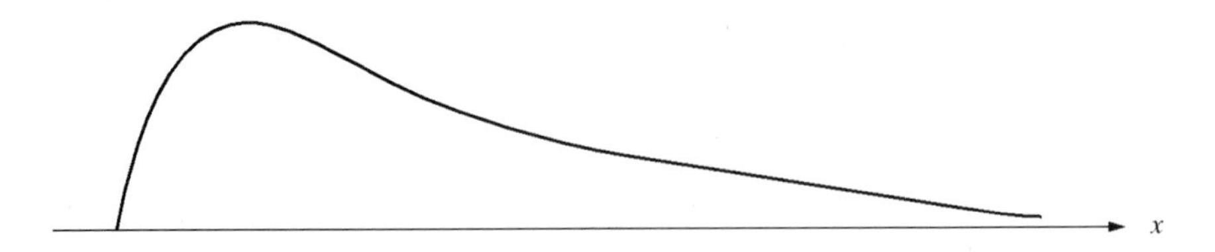

FIGURE 7.1

EXERCISES

7.57. Let $Y_1 < Y_2 < Y_3 < Y_4$ denote the order statistics of a random sample of size $n = 4$ from a distribution having p.d.f. $f(x; \theta) = 1/\theta, 0 < x < \theta$, zero elsewhere, where $0 < \theta < \infty$. Argue that the complete sufficient statistic Y_4 for θ is independent of each of the statistics Y_1/Y_4 and $(Y_1 + Y_2)/(Y_3 + Y_4)$.

Hint: Show that the p.d.f. is of the form $(1/\theta)f(x/\theta)$, where $f(x) = 1$, $0 < x < 1$, zero elsewhere.

7.58. Let $Y_1 < Y_2 < \cdots < Y_n$ be the order statistics of a random sample from the normal distribution $N(\theta, \sigma^2)$, $-\infty < \theta < \infty$. Show that the distribution of $Z = Y_n - \bar{Y}$ does not depend upon θ. Thus $\bar{Y} = \sum_1^n Y_i/n$, a complete sufficient statistic for θ, is independent of Z.

7.59. Let $X_1, X_2, \ldots, X_n$ be a random sample from the normal distribution $N(\theta, \sigma^2)$, $-\infty < \theta < \infty$. Prove that a necessary and sufficient condition that the statistics $Z = \sum_1^n a_i X_i$ and $Y = \sum_1^n X_i$, a complete sufficient statistic for θ, be independent is that $\sum_1^n a_i = 0$.

7.60. Let X and Y be random variables such that $E(X^k)$ and $E(Y^k) \neq 0$ exist for $k = 1, 2, 3, \ldots$. If the ratio X/Y and its denominator Y are independent, prove that $E[(X/Y)^k] = E(X^k)/E(Y^k)$, $k = 1, 2, 3, \ldots$.

Hint: Write $E(X^k) = E[Y^k(X/Y)^k]$.

7.61. Let $Y_1 < Y_2 < \cdots < Y_n$ be the order statistics of a random sample of size n from a distribution that has p.d.f. $f(x; \theta) = (1/\theta)e^{-x/\theta}$, $0 < x < \infty$, $0 < \theta < \infty$, zero elsewhere. Show that the ratio $R = nY_1 \Big/ \sum_1^n Y_i$ and its denominator (a complete sufficient statistic for θ) are independent. Use the result of the preceding exercise to determine $E(R^k)$, $k = 1, 2, 3, \ldots$.

7.62. Let $X_1, X_2, \ldots, X_5$ be a random sample of size 5 from the distribution that has p.d.f. $f(x) = e^{-x}$, $0 < x < \infty$, zero elsewhere. Show that $(X_1 + X_2)/(X_1 + X_2 + \cdots + X_5)$ and its denominator are independent.

Hint: The p.d.f. $f(x)$ is a member of $\{f(x; \theta) : 0 < \theta < \infty\}$, where $f(x; \theta) = (1/\theta)e^{-x/\theta}$, $0 < x < \infty$, zero elsewhere.

7.63. Let $Y_1 < Y_2 < \cdots < Y_n$ be the order statistics of a random sample from the normal distribution $N(\theta_1, \theta_2)$, $-\infty < \theta_1 < \infty, 0 < \theta_2 < \infty$. Show that the joint complete sufficient statistics $\bar{X} = \bar{Y}$ and S^2 for θ_1 and θ_2 are independent of each of $(Y_n - \bar{Y})/S$ and $(Y_n - Y_1)/S$.

7.64. Let $Y_1 < Y_2 < \cdots < Y_n$ be the order statistics of a random sample from a distribution with the p.d.f.

$$f(x; \theta_1, \theta_2) = \frac{1}{\theta_2} \exp\left(-\frac{x - \theta_1}{\theta_2}\right),$$

$\theta_1 < x < \infty$, zero elsewhere, where $-\infty < \theta_1 < \infty, 0 < \theta_2 < \infty$. Show that the joint complete sufficient statistics Y_1 and $\bar{X} = \bar{Y}$ for θ_1 and θ_2 are independent of $(Y_2 - Y_1) \Big/ \sum_1^n (Y_i - Y_1)$.

7.65. Let $X_1, X_2, \ldots, X_5$ be a random sample of size $n = 5$ from the normal distribution $N(0, \theta)$.

(a) Argue that the ratio $R = (X_1^2 + X_2^2)/(X_1^2 + \cdots + X_5^2)$ and its denominator $(X_1^2 + \cdots + X_5^2)$ are independent.

(b) Does $5R/2$ have an F-distribution with 2 and 5 degrees of freedom? Explain your answer.

(c) Compute $E(R)$ using Exercise 7.60.

7.66. Let $Y_1 < Y_2 < \cdots < Y_n$ be the order statistics of a random sample of size n from a distribution having p.d.f.

$$f(x; \theta) = (1/\theta) \exp\left(\frac{-x}{\theta}\right), \qquad 0 < x < \infty,$$

and equal zero elsewhere, where $0 < \theta < \infty$. Show that $W = \sum_1^n Y_i$ and $Z = nY_1 \Big/ \sum_1^n Y_i$ are independent. Find $E(Z^k)$, $k = 1, 2, 3, \ldots$ using the result of Exercise 7.60. What is the distribution of Z?

7.67. Referring to Example 5 of this section, determine c so that

$$\Pr(-c < T_1 - \theta < c | T_2 = t_2) = 0.95.$$

Use this result to find a 95 percent confidence interval for θ, given $T_2 = t_2$; and note how its length is smaller when the range t_2 is larger.

ADDITIONAL EXERCISES

7.68. Let $X_1, X_2, \ldots, X_n$ be a random sample from a distribution with p.d.f. $f(x; \theta) = \theta e^{-\theta x}$, $0 < x < \infty$, zero elsewhere where $0 < \theta$.

(a) What is the complete sufficient statistic, say Y, for θ?

(b) What function of Y is an unbiased estimator of θ?

7.69. Let $Y_1 < Y_2 < \cdots < Y_n$ be the order statistics of a random sample of size n from a distribution with p.d.f. $f(x; \theta) = 1/\theta$, $0 < x < \theta$, zero elsewhere. The statistic Y_n is a complete sufficient statistic for θ and it has p.d.f.

$$g(y_n; \theta) = \frac{ny_n^{n-1}}{\theta^n}, \qquad 0 < y_n < \theta,$$

and zero elsewhere.

(a) Find the distribution function $H_n(z; \theta)$ of $Z = n(\theta - Y_n)$.
(b) Find the $\lim_{n\to\infty} H_n(z; \theta)$ and thus the limiting distribution of Z.

7.70. Let $X_1, \ldots, X_n$; $Y_1, \ldots, Y_n$; $Z_1, \ldots, Z_n$ be respective independent random samples from three normal distributions $N(\mu_1 = \alpha + \beta, \sigma^2)$ $N(\mu_2 = \beta + \gamma, \sigma^2)$, $N(\mu_3 = \alpha + \gamma, \sigma^2)$. Find a point estimator for β that is based on $\bar{X}$, $\bar{Y}$, $\bar{Z}$. Is this estimator unique? Why? If σ^2 is unknown, explain how to find a confidence interval for β.

7.71. Let $X_1, X_2, \ldots, X_n$ be a random sample from a Poisson distribution with mean θ. Find the conditional expectation $E(X_1 + 2X_2 + 3X_3 | \sum_1^n X_i)$.

7.72. Let $X_1, X_2, \ldots, X_n$ be a random sample of size n from the normal distribution $N(\theta, 1)$. Find the unbiased minimum variance estimator of θ^2.

7.73. Let $X_1, X_2, \ldots, X_n$ be a random sample from a Poisson distribution with mean θ. Find the unbiased minimum variance estimator of θ^2.

7.74. We consider a random sample $X_1, X_2, \ldots, X_n$ from a distribution with p.d.f. $f(x; \theta) = (1/\theta) \exp(-x/\theta)$, $0 < x < \infty$, zero elsewhere, where $0 < \theta$. Possibly, in a life testing situation, however, we only observe the first r order statistics, $Y_1 < Y_2 < \cdots < Y_r$.

(a) Record the joint p.d.f. of these order statistics and denote it by $L(\theta)$.
(b) Under these conditions, find the m.l.e., $\hat{\theta}$, by maximizing $L(\theta)$.
(c) Find the m.g.f. and p.d.f. of $\hat{\theta}$.
(d) With a slight extension of the definition of sufficiency, is $\hat{\theta}$ a sufficient statistic?
(e) Find the unbiased minimum variance estimator for θ.
(f) Show that $Y_1/\hat{\theta}$ and $\hat{\theta}$ are independent.

7.75. Let us repeat Bernoulli trials with parameter θ until k successes occur. If Y is the number of trials needed:

(a) Show that the p.d.f. of Y is $g(y; \theta) = \binom{y-1}{k-1}\theta^k(1 - \theta)^{y-k}$, $y = k$, $k + 1, \ldots$, zero elsewhere, where $0 \le \theta \le 1$.
(b) Prove that this family of probability density functions is complete.
(c) Demonstrate that $E[(k - 1)/(Y - 1)] = \theta$.
(d) Is it possible to find another statistic, which is a function of Y alone, that is unbiased? Why?

7.76. Let $X_1, X_2, \ldots, X_n$ be a random sample from a distribution with p.d.f. $f(x; \theta) = \theta^x(1 - \theta)$, $x = 0, 1, 2, \ldots$, zero elsewhere, where $0 \le \theta \le 1$.

(a) Find the m.l.e., $\hat{\theta}$, of θ.

(b) Show that $\sum_{i=1}^{n} X_i$ is a complete sufficient statistic for θ.

(c) Determine the unbiased minimum variance estimator of θ.

7.77. If $X_1, X_2, \ldots, X_n$ is a random sample from a distribution with p.d.f. $f(x; \theta) = \frac{1}{2}\theta^3 x^2 e^{-\theta x}$, $0 < x < \infty$, zero elsewhere, where $0 < \theta < \infty$:

(a) Find the m.l.e., $\hat{\theta}$, of θ. Is $\hat{\theta}$ unbiased?

Hint: First find the p.d.f. of $Y = \sum_{i=1}^{n} X_i$ and then compute $E(\hat{\theta})$.

(b) Argue that Y is a complete sufficient statistic for θ.

(c) Find the unbiased minimum variance estimator of θ.

(d) Show that X_1/Y and Y are independent.

(e) What is the distribution of X_1/Y?

CHAPTER 8

More About Estimation

8.1 Bayesian Estimation

In Chapter 6 we introduced point and interval estimation for various parameters. In Chapter 7 we observed how such inferences should be based upon sufficient statistics for the parameters if they exist. In this chapter we introduce other concepts related to estimation and begin this by considering *Bayesian estimates*, which are also based upon sufficient statistics if the latter exist.

In introducing the interesting and sometimes controversial Bayesian method of estimation, the student should constantly keep in mind that making statistical inferences from the data does not strictly follow a mathematical approach. Clearly, up to now, we have had to construct models before we have been able to make such inferences. These models are subjective, and the resulting inference depends greatly on the model selected. For illustration, two statisticians could very well select different models for exactly the same situation and make different inferences with exactly the same data. Most statisticians would use some type of model diagnostics to see if the models seem to be reasonable ones, but we must still recognize that there can be differences among statisticians' inferences.

We shall now describe the Bayesian approach to the problem of estimation. This approach takes into account any prior knowledge of the experiment that the statistician has and it is one application of a principle of statistical inference that may be called *Bayesian statistics*. Consider a random variable X that has a distribution of probability that depends upon the symbol θ, where θ is an element of a well-defined set Ω. For example, if the symbol θ is the mean of a normal distribution, Ω may be the real line. We have previously looked upon θ as being some constant, although an unknown constant. Let us now introduce a random variable Θ that has a distribution of probability over the set Ω; and, just as we look upon x as a possible value of the random variable X, we now look upon θ as a possible value of the random variable Θ. Thus the distribution of X depends upon θ, an experimental value of the random variable Θ. We shall denote the p.d.f. of Θ by $h(\theta)$ and we take $h(\theta) = 0$ when θ is not an element of Ω. Moreover, we now denote the p.d.f. of X by $f(x|\theta)$ since we think of it as a conditional p.d.f. of X, given $\Theta = \theta$.

Say $X_1, X_2, \ldots, X_n$ is a random sample from this conditional distribution of X. Thus we can write the joint conditional p.d.f. of $X_1, X_2, \ldots, X_n$, given $\Theta = \theta$, as

$$f(x_1|\theta)f(x_2|\theta)\cdots f(x_n|\theta).$$

Thus the joint p.d.f. of $X_1, X_2, \ldots, X_n$ and Θ is

$$g(x_1, x_2, \ldots, x_n, \theta) = f(x_1|\theta)f(x_2|\theta)\cdots f(x_n|\theta)h(\theta).$$

If Θ is a random variable of the continuous type, the joint marginal p.d.f. of $X_1, X_2, \ldots, X_n$ is given by

$$g_1(x_1, x_2, \ldots, x_n) = \int_{-\infty}^{\infty} g(x_1, x_2, \ldots, x_n, \theta)\, d\theta.$$

If Θ is a random variable of the discrete type, integration would be replaced by summation. In either case the conditional p.d.f. of Θ, given $X_1 = x_1, \ldots, X_n = x_n$, is

$$k(\theta|x_1, x_2, \ldots, x_n) = \frac{g(x_1, x_2, \ldots, x_n, \theta)}{g_1(x_1, x_2, \ldots, x_n)} = \frac{f(x_1|\theta)f(x_2|\theta)\cdots f(x_n|\theta)h(\theta)}{g_1(x_1, x_2, \ldots, x_n)}.$$

This relationship is another form of Bayes' formula.

Example 1. Let $X_1, X_2, \ldots, X_n$ be a random sample from a Poisson distribution with mean θ, where θ is the observed value of a random variable Θ having a gamma distribution with known parameters α and β. Thus

$$g(x_1, \ldots, x_n, \theta) = \left[\frac{\theta^{x_1}e^{-\theta}}{x_1!} \cdots \frac{\theta^{x_n}e^{-\theta}}{x_n!}\right]\left[\frac{\theta^{\alpha-1}e^{-\theta/\beta}}{\Gamma(\alpha)\beta^\alpha}\right],$$

provided that $x_i = 0, 1, 2, 3, \ldots, i = 1, 2, \ldots, n$ and $0 < \theta < \infty$, and is equal to zero elsewhere. Then

$$g_1(x_1, \ldots, x_n) = \int_0^\infty \frac{\theta^{\Sigma x_i + \alpha - 1}e^{-(n+1/\beta)\theta}}{x_1! \cdots x_n!\,\Gamma(\alpha)\beta^\alpha}\,d\theta$$

$$= \frac{\Gamma\left(\sum_1^n x_i + \alpha\right)}{x_1! \cdots x_n!\,\Gamma(\alpha)\beta^\alpha(n + 1/\beta)^{\Sigma x_i + \alpha}}.$$

Finally, the conditional p.d.f. of Θ, given $X_1 = x_1, \ldots, X_n = x_n$, is

$$k(\theta|x_1, \ldots, x_n) = \frac{g(x_1, \ldots, x_n, \theta)}{g_1(x_1, \ldots, x_n)}$$

$$= \frac{\theta^{\Sigma x_i + \alpha - 1}e^{-\theta/[\beta/(n\beta+1)]}}{\Gamma\left(\sum x_i + \alpha\right)[\beta/(n\beta + 1)]^{\Sigma x_i + \alpha}},$$

provided that $0 < \theta < \infty$, and is equal to zero elsewhere. This conditional p.d.f. is one of the gamma type with parameters $\alpha^* = \Sigma\, x_i + \alpha$ and $\beta^* = \beta/(n\beta + 1)$.

In Example 1 it is extremely convenient to notice that it is not really necessary to determine $g_1(x_1, \ldots, x_n)$ to find $k(\theta|x_1, \ldots, x_n)$. If we divide

$$f(x_1|\theta)f(x_2|\theta) \cdots f(x_n|\theta)h(\theta)$$

by $g_1(x_1, \ldots, x_n)$, we must get the product of a factor, which depends upon $x_1, \ldots, x_n$ but does *not* depend upon θ, say $c(x_1, \ldots, x_n)$, and

$$\theta^{\Sigma\, x_i + \alpha - 1}e^{-\theta/[\beta/(n\beta+1)]}.$$

That is,

$$k(\theta|x_1, \ldots, x_n) = c(x_1, \ldots, x_n)\theta^{\Sigma\, x_i + \alpha - 1}e^{-\theta/[\beta/(n\beta+1)]},$$

provided that $0 < \theta < \infty$ and $x_i = 0, 1, 2, \ldots,$ $i = 1, 2, \ldots, n$. However, $c(x_1, \ldots, x_n)$ must be that "constant" needed to make $k(\theta|x_1, \ldots, x_n)$ a p.d.f., namely

$$c(x_1, \ldots, x_n) = \frac{1}{\Gamma\left(\sum x_i + \alpha\right)[\beta/(n\beta + 1)]^{\Sigma\, x_i + \alpha}}.$$

Accordingly, Bayesian statisticians frequently write that $k(\theta|x_1, \ldots, x_n)$ is proportional to

$$g(x_1, x_2, \ldots, x_n, \theta);$$

that is,

$$k(\theta|x_1, \ldots, x_n) \propto f(x_1|\theta) \cdots f(x_n|\theta)h(\theta).$$

Note that in the right-hand member of this expression all factors involving constants and $x_1, \ldots, x_n$ alone (not θ) can be dropped. For illustration, in solving the problem presented in Example 1, the Bayesian statistician would simply write

$$k(\theta|x_1, \ldots, x_n) \propto \theta^{\Sigma x_i}e^{-n\theta}\theta^{\alpha - 1}e^{-\theta/\beta}$$

or, equivalently,

$$k(\theta|x_1, \ldots, x_n) \propto \theta^{\Sigma x_i + \alpha - 1}e^{-\theta/[\beta/(n\beta + 1)]},$$

$0 < \theta < \infty$ and is equal to zero elsewhere. Clearly, $k(\theta|x_1, \ldots, x_n)$ must be a gamma p.d.f. with parameters $\alpha^* = \Sigma\, x_i + \alpha$ and $\beta^* = \beta/(n\beta + 1)$.

There is another observation that can be made at this point. Suppose that there exists a sufficient statistic $Y = u(X_1, \ldots, X_n)$ for the parameter so that

$$f(x_1|\theta) \cdots f(x_n|\theta) = g[u(x_1, \ldots, x_n)|\theta]H(x_1, \ldots, x_n),$$

where now $g(y|\theta)$ is the p.d.f. of Y, given $\Theta = \theta$. Then we note that

$$k(\theta|x_1, \ldots, x_n) \propto g[u(x_1, \ldots, x_n)|\theta]h(\theta)$$

because the factor $H(x_1, \ldots, x_n)$ that does not depend upon θ can be dropped. Thus, if a sufficient statistic Y for the parameter exists, we can begin with the p.d.f. of Y if we wish and write

$$k(\theta|y) \propto g(y|\theta)h(\theta),$$

where now $k(\theta|y)$ is the conditional p.d.f. of Θ, given the sufficient statistic $Y = y$. The following discussion assumes that a sufficient statistic Y does exist; but more generally, we could replace Y by $X_1, X_2, \ldots, X_n$ in what follows. Also, we now use $g_1(y)$ to be the marginal p.d.f. of Y; that is, in the continuous case,

$$g_1(y) = \int_{-\infty}^{\infty} g(y|\theta)h(\theta)\, d\theta.$$

In Bayesian statistics, the p.d.f. $h(\theta)$ is called the *prior p.d.f.* of Θ, and the conditional p.d.f. $k(\theta|y)$ is called the *posterior p.d.f.* of Θ. This is because $h(\theta)$ is the p.d.f. of Θ prior to the observation of Y, whereas $k(\theta|y)$ is the p.d.f. of Θ after the observation of Y has been made. In many instances, $h(\theta)$ is not known; yet the choice of $h(\theta)$ affects the p.d.f. $k(\theta|y)$. In these instances the statistician takes into account all prior knowledge of the experiment and *assigns* the prior p.d.f. $h(\theta)$. This, of course, injects the problem of *personal* or *subjective probability* (see the Remark, Section 1.1).

Suppose that we want a point estimate of θ. From the Bayesian viewpoint, this really amounts to selecting a decision function δ, so that $\delta(y)$ is a predicted value of θ (an experimental value of the random variable Θ) when both the computed value y and the conditional p.d.f. $k(\theta|y)$ are known. Now, in general, how would we predict an experimental value of any random variable, say W, if we want our prediction to be "reasonably close" to the value to be observed? Many statisticians would predict the mean, $E(W)$, of the distribution of W; others would predict a median (perhaps unique) of the distribution of W; some would predict a mode (perhaps unique) of the distribution of W; and some would have other predictions. However, it seems desirable that the choice of the decision function should depend upon the loss function $\mathscr{L}[\theta, \delta(y)]$. One way in which this dependence upon the loss function can be reflected is to select the decision function δ in such a way that the conditional expectation of the loss is a minimum. A *Bayes' solution* is a decision function δ that minimizes

$$E\{\mathscr{L}[\Theta, \delta(y)]|Y = y\} = \int_{-\infty}^{\infty} \mathscr{L}[\theta, \delta(y)]k(\theta|y)\, d\theta,$$

if Θ is a random variable of the continuous type. The usual modification of the right-hand member of this equation is made for random variables of the discrete type. If, for example, the loss function is given by $\mathscr{L}[\theta, \delta(y)] = [\theta - \delta(y)]^2$, the Bayes' solution is given by $\delta(y) = E(\Theta|y)$, the mean of the conditional distribution of Θ, given $Y = y$. This follows from the fact that $E[(W - b)^2]$, if it exists, is a minimum when $b = E(W)$. If the loss function is given by $\mathscr{L}[\theta, \delta(y)] = |\theta - \delta(y)|$, then a median of the conditional distribution of Θ, given $Y = y$, is the Bayes' solution. This follows from the fact that $E(|W - b|)$, if it exists, is a minimum when b is equal to any median of the distribution of W.

The conditional expectation of the loss, given $Y = y$, defines a random variable that is a function of the statistic Y. The expected value of that function of Y, in the notation of this section, is given by

$$\int_{-\infty}^{\infty} \left\{ \int_{-\infty}^{\infty} \mathscr{L}[\theta, \delta(y)] k(\theta|y)\, d\theta \right\} g_1(y)\, dy$$

$$= \int_{-\infty}^{\infty} \left\{ \int_{-\infty}^{\infty} \mathscr{L}[\theta, \delta(y)] g(y|\theta)\, dy \right\} h(\theta)\, d\theta,$$

in the continuous case. The integral within the braces in the latter expression is, for every given $\theta \in \Omega$, the risk function $R(\theta, \delta)$; accordingly, the latter expression is the mean value of the risk, or the expected risk. Because a Bayes' solution minimizes

$$\int_{-\infty}^{\infty} \mathscr{L}[\theta, \delta(y)] k(\theta|y)\, d\theta$$

for every y for which $g_1(y) > 0$, it is evident that a Bayes' solution $\delta(y)$ minimizes this mean value of the risk. We now give an illustrative example.

Example 2. Let $X_1, X_2, \ldots, X_n$ denote a random sample from a distribution that is $b(1, \theta)$, $0 < \theta < 1$. We seek a decision function δ that is a Bayes' solution. The sufficient statistic $Y = \sum_1^n X_i$, and Y is $b(n, \theta)$. That is, the conditional p.d.f. of Y, given $\Theta = \theta$, is

$$g(y|\theta) = \binom{n}{y} \theta^y (1 - \theta)^{n-y}, \qquad y = 0, 1, \ldots, n,$$

$$= 0 \qquad \text{elsewhere.}$$

We take the prior p.d.f. of the random variable Θ to be

$$h(\theta) = \frac{\Gamma(\alpha + \beta)}{\Gamma(\alpha)\Gamma(\beta)} \theta^{\alpha - 1} (1 - \theta)^{\beta - 1}, \qquad 0 < \theta < 1,$$

$$= 0 \qquad \text{elsewhere.}$$

where α and β are assigned positive constants. Thus the conditional p.d.f. of Θ, given $Y = y$, is, at points of positive probability density,

$$k(\theta|y) \propto \theta^y (1 - \theta)^{n-y} \theta^{\alpha - 1} (1 - \theta)^{\beta - 1}, \qquad 0 < \theta < 1.$$

That is,

$$k(\theta|y) = \frac{\Gamma(n+\alpha+\beta)}{\Gamma(\alpha+y)\Gamma(n+\beta-y)}\,\theta^{\alpha+y-1}(1-\theta)^{\beta+n-y-1}, \qquad 0<\theta<1,$$

and $y = 0, 1, \ldots, n$. We take the loss function to be $\mathscr{L}[\theta, \delta(y)] = [\theta - \delta(y)]^2$. Because Y is a random variable of the discrete type, whereas Θ is of the continuous type, we have for the expected risk,

$$\int_0^1 \left\{ \sum_{y=0}^{n} [\theta - \delta(y)]^2 \binom{n}{y} \theta^y (1-\theta)^{n-y} \right\} h(\theta)\, d\theta$$

$$= \sum_{y=0}^{n} \left\{ \int_0^1 [\theta - \delta(y)]^2 k(\theta|y)\, d\theta \right\} g_1(y).$$

The Bayes' solution $\delta(y)$ is the mean of the conditional distribution of Θ, given $Y = y$. Thus

$$\begin{aligned}\delta(y) &= \int_0^1 \theta k(\theta|y)\, d\theta \\ &= \frac{\Gamma(n+\alpha+\beta)}{\Gamma(\alpha+y)\Gamma(n+\beta-y)} \int_0^1 \theta^{\alpha+y}(1-\theta)^{\beta+n-y-1}\, d\theta \\ &= \frac{\alpha+y}{\alpha+\beta+n}.\end{aligned}$$

This decision function $\delta(y)$ minimizes

$$\int_0^1 [\theta - \delta(y)]^2 k(\theta|y)\, d\theta$$

for $y = 0, 1, \ldots, n$ and, accordingly, it minimizes the expected risk. It is very instructive to note that this Bayes' solution can be written as

$$\delta(y) = \left(\frac{n}{\alpha+\beta+n}\right)\frac{y}{n} + \left(\frac{\alpha+\beta}{\alpha+\beta+n}\right)\frac{\alpha}{\alpha+\beta}$$

which is a weighted average of the maximum likelihood estimate y/n of θ and the mean $\alpha/(\alpha+\beta)$ of the prior p.d.f. of the parameter. Moreover, the respective weights are $n/(\alpha+\beta+n)$ and $(\alpha+\beta)/(\alpha+\beta+n)$. Thus we see that α and β should be selected so that not only is $\alpha/(\alpha+\beta)$ the desired prior mean, but the sum $\alpha+\beta$ indicates the worth of the prior opinion, relative to a sample of size n. That is, if we want our prior opinion to have as much weight as a sample size of 20, we would take $\alpha+\beta = 20$. So if our prior mean is $\frac{3}{4}$; we have that α and β are selected so that $\alpha = 15$ and $\beta = 5$.

Example 3. Suppose that $Y = \bar{X}$, the sufficient statistic, is the mean of a random sample of size n that arises from the normal distribution $N(\theta, \sigma^2)$, where σ^2 is known. Then $g(y|\theta)$ is $N(\theta, \sigma^2/n)$. Further suppose that we are able to assign prior knowledge to θ through a prior p.d.f. $h(\theta)$ that is $N(\theta_0, \sigma_0^2)$. Then we have that

$$k(\theta|y) \propto \frac{1}{\sqrt{2\pi}\sigma/\sqrt{n}} \frac{1}{\sqrt{2\pi}\sigma_0} \exp\left[-\frac{(y-\theta)^2}{2(\sigma^2/n)} - \frac{(\theta-\theta_0)^2}{2\sigma_0^2}\right].$$

If we eliminate all constant factors (including factors involving y only), we have

$$k(\theta|y) \propto \exp\left[-\frac{(\sigma_0^2 + \sigma^2/n)\theta^2 - 2(y\sigma_0^2 + \theta_0\sigma^2/n)\theta}{2(\sigma^2/n)\sigma_0^2}\right].$$

This can be simplified, by completing the square, to read (after eliminating factors not involving θ)

$$k(\theta|y) \propto \exp\left[-\frac{\left(\theta - \dfrac{y\sigma_0^2 + \theta_0\sigma^2/n}{\sigma_0^2 + \sigma^2/n}\right)^2}{\dfrac{2(\sigma^2/n)\sigma_0^2}{(\sigma_0^2 + \sigma^2/n)}}\right].$$

That is, the posterior p.d.f. of the parameter is obviously normal with mean

$$\frac{y\sigma_0^2 + \theta_0\sigma^2/n}{\sigma_0^2 + \sigma^2/n} = \left(\frac{\sigma_0^2}{\sigma_0^2 + \sigma^2/n}\right)y + \left(\frac{\sigma^2/n}{\sigma_0^2 + \sigma^2/n}\right)\theta_0$$

and variance $(\sigma^2/n)\sigma_0^2/(\sigma_0^2 + \sigma^2/n)$. If the square-error loss function is used, this posterior mean is the Bayes' solution. Again, note that it is a weighted average of the maximum likelihood estimate $y = \bar{x}$ and the prior mean θ_0. Observe here and in Example 2 that the Bayes' solution gets closer to the maximum likelihood estimate as n increases. Thus the Bayesian procedures permit the decision maker to enter his or her prior opinions into the solution in a very formal way such that the influences of these prior notions will be less and less as n increases.

In Bayesian statistics all the information is contained in the posterior p.d.f. $k(\theta|y)$. In Examples 2 and 3 we found Bayesian point estimates using the square-error loss function. It should be noted that if $\mathscr{L}[\delta(y), \theta] = |\delta(y) - \theta|$, the absolute value of the error, then the Bayes' solution would be the median of the posterior distribution of the parameter, which is given by $k(\theta|y)$. Hence the Bayes' solution changes, *as it should*, with different loss functions.

If an interval estimate of θ is desired, we can now find two functions $u(y)$ and $v(y)$ so that the conditional probability

$$\Pr[u(y) < \Theta < v(y)|Y = y] = \int_{u(y)}^{v(y)} k(\theta|y)\, d\theta,$$

is large, say 0.95. The experimental values of $X_1, X_2, \ldots, X_n$, say

$x_1, x_2, \ldots, x_n$, provide us with an experimental value of Y, say y. Then the interval $u(y)$ to $v(y)$ is an interval estimate of θ in the sense that the conditional probability of Θ belonging to that interval is equal to 0.95. For illustration, in Example 3 where the posterior p.d.f. of the parameter was normal, the interval, whose end points are found by taking the mean of that distribution and adding and subtracting 1.96 of its standard deviation,

$$\frac{y\sigma_0^2 + \theta_0\sigma^2/n}{\sigma_0^2 + \sigma^2/n} \pm 1.96\sqrt{\frac{(\sigma^2/n)\sigma_0^2}{\sigma_0^2 + \sigma^2/n}}$$

serves as an interval estimate for θ with posterior probability of 0.95.

EXERCISES

8.1. Let $X_1, X_2, \ldots, X_n$ be a random sample from a distribution that is $b(1, \theta)$. Let the prior p.d.f. of Θ be a beta one with parameters α and β. Show that the posterior p.d.f. $k(\theta|x_1, x_2, \ldots, x_n)$ is exactly the same as $k(\theta|y)$ given in Example 2.

8.2. Let $X_1, X_2, \ldots, X_n$ denote a random sample from a distribution that is $N(\theta, \sigma^2)$, $-\infty < \theta < \infty$, where σ^2 is a given positive number. Let $Y = \bar{X}$, the mean of the random sample. Take the loss function to be $\mathscr{L}[\theta, \delta(y)] = |\theta - \delta(y)|$. If θ is an observed value of the random variable Θ that is $N(\mu, \tau^2)$, where $\tau^2 > 0$ and μ are known numbers, find the Bayes' solution $\delta(y)$ for a point estimate of θ.

8.3. Let $X_1, X_2, \ldots, X_n$ denote a random sample from a Poisson distribution with mean θ, $0 < \theta < \infty$. Let $Y = \sum_1^n X_i$ and take the loss function to be $\mathscr{L}[\theta, \delta(y)] = [\theta - \delta(y)]^2$. Let θ be an observed value of the random variable Θ. If Θ has the p.d.f. $h(\theta) = \theta^{\alpha-1}e^{-\theta/\beta}/\Gamma(\alpha)\beta^\alpha$, $0 < \theta < \infty$, zero elsewhere, where $\alpha > 0$, $\beta > 0$ are known numbers, find the Bayes' solution $\delta(y)$ for a point estimate of θ.

8.4. Let Y_n be the nth order statistic of a random sample of size n from a distribution with p.d.f. $f(x|\theta) = 1/\theta$, $0 < x < \theta$, zero elsewhere. Take the loss function to be $\mathscr{L}[\theta, \delta(y_n)] = [\theta - \delta(y_n)]^2$. Let θ be an observed value of the random variable Θ, which has p.d.f. $h(\theta) = \beta\alpha^\beta/\theta^{\beta+1}$, $\alpha < \theta < \infty$, zero elsewhere, with $\alpha > 0$, $\beta > 0$. Find the Bayes' solution $\delta(y_n)$ for a point estimate of θ.

8.5. Let Y_1 and Y_2 be statistics that have a trinomial distribution with parameters n, θ_1, and θ_2. Here θ_1 and θ_2 are observed values of the random variables Θ_1 and Θ_2, which have a Dirichlet distribution with known parameters α_1, α_2, and α_3 (see Example 1, Section 4.5). Show that the conditional distribution of Θ_1 and Θ_2 is Dirichlet and determine the conditional means $E(\Theta_1|y_1, y_2)$ and $E(\Theta_2|y_1, y_2)$.

8.6. Let X be $N(0, 1/\theta)$. Assume that the unknown θ is a value of a random variable Θ which has a gamma distribution with parameters $\alpha = r/2$ and $\beta = 2/r$, where r is a positive integer. Show that X has a marginal t-distribution with r degrees of freedom. This procedure is called *compounding*, and it may be used by a Bayesian statistician as a way of first presenting the t-distribution, as well as other distributions.

8.7. Let X have a Poisson distribution with parameter θ. Assume that the unknown θ is a value of a random variable Θ that has a gamma distribution with parameters $\alpha = r$ and $\beta = (1 - p)/p$, where r is a positive integer and $0 < p < 1$. Show, by the procedure of compounding, that X has a marginal distribution which is negative binomial, a distribution that was introduced earlier (Section 3.1) under very different assumptions.

8.8. In Example 2 let $n = 30$, $\alpha = 10$, and $\beta = 5$ so that $\delta(y) = (10 + y)/45$ is the Bayes' estimate of θ.
(a) If Y has the binomial distribution $b(30, \theta)$, compute the risk $E\{[\theta - \delta(Y)]^2\}$.
(b) Determine those values of θ for which the risk of part (a) is less than $\theta(1 - \theta)/30$, the risk associated with the maximum likelihood estimator Y/n of θ.

8.9. Let Y_4 be the largest order statistic of a sample of size $n = 4$ from a distribution with uniform p.d.f. $f(x; \theta) = 1/\theta$, $0 < x < \theta$, zero elsewhere. If the prior p.d.f. of the parameter is $g(\theta) = 2/\theta^3$, $1 < \theta < \infty$, zero elsewhere, find the Bayesian estimator $\delta(Y_4)$ of θ, based upon the sufficient statistic Y_4, using the loss function $|\delta(y_4) - \theta|$.

8.10. Consider a random sample $X_1, X_2, \ldots, X_n$ from the *Weibull distribution* with p.d.f. $f(x; \theta, \tau) = \theta\tau x^{\tau-1}e^{-\theta x^\tau}$, $0 < x < \infty$, where $0 < \theta$, $0 < \tau$, zero elsewhere.
(a) If τ is known, find the m.l.e. of θ.
(b) If the parameter θ has a prior gamma p.d.f. $g(\theta)$ with parameters α and $\beta^* = 1/\beta$, show that the compound distribution is a *Burr type* with p.d.f. $h(x) = \alpha\tau\beta^\alpha x^{\tau-1}/(x^\tau + \beta)^{\alpha+1}$, $0 < x < \infty$, zero elsewhere.
(c) If, in the Burr distribution, τ and β are known, find the m.l.e. of α based on a random sample of size n.

8.2 Fisher Information and the Rao–Cramér Inequality

Let X be a random variable with p.d.f. $f(x; \theta)$, $\theta \in \Omega$, where the parameter space Ω is an interval. We consider only special cases, sometimes called *regular cases*, of probability density functions as we wish to differentiate under an integral (summation) sign. In particular, this means that the parameter θ does not appear in endpoints of the interval in which $f(x; \theta) > 0$.

With these assumptions, we have (in the continuous case, but the discrete case can be handled in a similar manner) that

$$\int_{-\infty}^{\infty} f(x; \theta)\, dx = 1$$

and, by taking the derivative with respect to θ,

$$\int_{-\infty}^{\infty} \frac{\partial f(x; \theta)}{\partial \theta}\, dx = 0. \tag{1}$$

The latter expression can be rewritten as

$$\int_{-\infty}^{\infty} \frac{\dfrac{\partial f(x; \theta)}{\partial \theta}}{f(x; \theta)} f(x; \theta)\, dx = 0$$

or, equivalently,

$$\int_{-\infty}^{\infty} \frac{\partial \ln f(x; \theta)}{\partial \theta} f(x; \theta)\, dx = 0.$$

If we differentiate again, it follows that

$$\int_{-\infty}^{\infty} \left[\frac{\partial^2 \ln f(x; \theta)}{\partial \theta^2} f(x; \theta) + \frac{\partial \ln f(x; \theta)}{\partial \theta} \frac{\partial f(x; \theta)}{\partial \theta} \right] dx = 0. \tag{2}$$

We rewrite the second term of the left-hand member of this equation as

$$\int_{-\infty}^{\infty} \frac{\partial \ln f(x; \theta)}{\partial \theta} \frac{\dfrac{\partial f(x; \theta)}{\partial \theta}}{f(x; \theta)} f(x; \theta)\, dx = \int_{-\infty}^{\infty} \left[\frac{\partial \ln f(x; \theta)}{\partial \theta} \right]^2 f(x; \theta)\, dx.$$

This is called *Fisher information* and is denoted by $I(\theta)$. That is,

$$I(\theta) = \int_{-\infty}^{\infty} \left[\frac{\partial \ln f(x; \theta)}{\partial \theta} \right]^2 f(x; \theta)\, dx;$$

but, from Equation (2), we see that $I(\theta)$ can be computed from

$$I(\theta) = -\int_{-\infty}^{\infty} \frac{\partial^2 \ln f(x; \theta)}{\partial \theta^2} f(x; \theta)\, dx.$$

Sometimes, one expression is easier to compute than the other, but often we prefer the second expression.

Remark. Note that the information is the weighted mean of either

$$\left[\frac{\partial \ln f(x;\theta)}{\partial\theta}\right]^2 \quad \text{or} \quad -\frac{\partial^2 \ln f(x;\theta)}{\partial\theta^2},$$

where the weights are given by the p.d.f. $f(x;\theta)$. That is, the greater these derivatives on the average, the more information that we get about θ. Clearly, if they were equal to zero [so that θ would not be in $\ln f(x;\theta)$], there would be zero information about θ. As we study more and more statistics, we learn to recognize that the function

$$\frac{\partial \ln f(x;\theta)}{\partial\theta}$$

is a very important one. For example, it played a major role in finding the m.l.e. $\hat{\theta}$ by solving

$$\sum_{i=1}^{n} \frac{\partial \ln f(x_i;\theta)}{\partial\theta} = 0$$

for θ.

Example 1. Let X be $N(\theta, \sigma^2)$, where $-\infty < \theta < \infty$ and σ^2 is known. Then

$$f(x;\theta) = \frac{1}{\sqrt{2\pi\sigma^2}} \exp\left[-\frac{(x-\theta)^2}{2\sigma^2}\right], \qquad -\infty < x < \infty,$$

where $-\infty < \theta < \infty$, and

$$\ln f(x;\theta) = -\frac{1}{2}\ln(2\pi\sigma^2) - \frac{(x-\theta)^2}{2\sigma^2}.$$

Thus

$$\frac{\partial \ln f(x;\theta)}{\partial\theta} = \frac{x-\theta}{\sigma^2}$$

and

$$\frac{\partial^2 \ln f(x;\theta)}{\partial\theta^2} = \frac{-1}{\sigma^2}.$$

Clearly, $E[(X-\theta)^2/\sigma^4] = -E[-1/\sigma^2] = 1/\sigma^2$. That is, in this case, it does not matter much which way we compute $I(\theta)$, as

$$I(\theta) = E\left\{\left[\frac{\partial \ln f(X;\theta)}{\partial\theta}\right]^2\right\} \quad \text{or} \quad I(\theta) = -E\left[\frac{\partial^2 \ln f(X;\theta)}{\partial\theta^2}\right],$$

because each is very easy. Of course, the information is greater with smaller values of σ^2.

Example 2. Let X be binomial $b(1, \theta)$. Thus

$$\ln f(x; \theta) = x \ln \theta + (1 - x) \ln (1 - \theta),$$

$$\frac{\partial \ln f(x; \theta)}{\partial \theta} = \frac{x}{\theta} - \frac{1 - x}{1 - \theta},$$

and

$$\frac{\partial^2 \ln f(x; \theta)}{\partial \theta^2} = -\frac{x}{\theta^2} - \frac{1 - x}{(1 - \theta)^2}.$$

Clearly,

$$I(\theta) = -E\left[\frac{-X}{\theta^2} - \frac{1 - X}{(1 - \theta)^2}\right]$$

$$= \frac{\theta}{\theta^2} + \frac{1 - \theta}{(1 - \theta)^2} = \frac{1}{\theta} + \frac{1}{1 - \theta} = \frac{1}{\theta(1 - \theta)},$$

which is larger for θ values close to zero or 1.

Suppose that $X_1, X_2, \ldots, X_n$ is a random sample from a distribution having p.d.f. $f(x; \theta)$. Thus the likelihood function (the joint p.d.f. of $X_1, X_2, \ldots, X_n$) is

$$L(\theta) = f(x_1; \theta) f(x_2; \theta) \cdots f(x_n; \theta).$$

Of course,

$$\ln L(\theta) = \ln f(x_1; \theta) + \ln f(x_2; \theta) + \cdots + \ln f(x_n; \theta)$$

and

$$\frac{\partial \ln L(\theta)}{\partial \theta} = \frac{\partial \ln f(x_1; \theta)}{\partial \theta} + \frac{\partial \ln f(x_2; \theta)}{\partial \theta} + \cdots + \frac{\partial \ln f(x_n; \theta)}{\partial \theta}. \quad (3)$$

It seems reasonable to define the *Fisher information* in the random sample as

$$I_n(\theta) = E\left\{\left[\frac{\partial \ln L(\theta)}{\partial \theta}\right]^2\right\}.$$

Note if we square Equation (3), we obtain cross-product terms like

$$2E\left[\frac{\partial \ln f(X_i; \theta)}{\partial \theta} \frac{\partial \ln f(X_j; \theta)}{\partial \theta}\right], \qquad i \neq j,$$

which from the independence of X_i and X_j equals

$$2E\left[\frac{\partial \ln f(X_i;\theta)}{\partial \theta}\right]E\left[\frac{\partial \ln f(X_j;\theta)}{\partial \theta}\right] = 0.$$

The fact that this product equals zero follows immediately from Equation (1). Hence we have the result that

$$I_n(\theta) = \sum_{i=1}^{n} E\left\{\left[\frac{\partial \ln f(X_i;\theta)}{\partial \theta}\right]^2\right\}.$$

However, each term of this summation equals $I(\theta)$, and hence

$$I_n(\theta) = nI(\theta).$$

That is, the Fisher information in a random sample of size n is n times the Fisher information in one observation. So, in the two examples of this section, the Fisher information in a random sample of size n is n/σ^2 in Example 1 and $n/[\theta(1-\theta)]$ in Example 2.

We can now prove a very important inequality involving the variance of an estimator, say $Y = u(X_1, X_2, \ldots, X_n)$, of θ, which can be biased. Suppose that

$$E(Y) = E[u(X_1, X_2, \ldots, X_n)] = k(\theta).$$

That is, in the continuous case,

$$k(\theta) = \int_{-\infty}^{\infty} \cdots \int_{-\infty}^{\infty} u(x_1, \ldots, x_n) f(x_1;\theta) \cdots f(x_n;\theta)\, dx_1 \cdots dx_n;$$

$$k'(\theta) = \int_{-\infty}^{\infty} \cdots \int_{-\infty}^{\infty} u(x_1, x_2, \ldots, x_n)\left[\sum_{1}^{n} \frac{1}{f(x_i;\theta)} \frac{\partial f(x_i;\theta)}{\partial \theta}\right]$$

$$\times f(x_1;\theta) \cdots f(x_n;\theta)\, dx_1 \cdots dx_n$$

$$= \int_{-\infty}^{\infty} \cdots \int_{-\infty}^{\infty} u(x_1, x_2, \ldots, x_n)\left[\sum_{1}^{n} \frac{\partial \ln f(x_i;\theta)}{\partial \theta}\right]$$

$$\times f(x_1;\theta) \cdots f(x_n;\theta)\, dx_1 \cdots dx_n. \tag{4}$$

Define the random variable Z by $Z = \sum_{1}^{n} [\partial \ln f(X_i;\theta)/\partial\theta]$. In accordance with Equation (1) we have $E(Z) = \sum_{1}^{n} E[\partial \ln f(X_i;\theta)/\partial\theta] = 0$. Moreover, Z is the sum of n independent random variables each with mean zero and consequently with variance $E\{[\partial \ln f(X;\theta)/\partial\theta]^2\}$. Hence the variance of Z is the sum of the n variances,

$$\sigma_Z^2 = nE\left[\left(\frac{\partial \ln f(X;\theta)}{\partial \theta}\right)^2\right] = I_n(\theta) = nI(\theta).$$

Because $Y = u(X_1, \ldots, X_n)$ and $Z = \sum_1^n [\partial \ln f(X_i; \theta)/\partial\theta]$, Equation (4) shows that $E(YZ) = k'(\theta)$. Recall that

$$E(YZ) = E(Y)E(Z) + \rho\sigma_Y\sigma_Z,$$

where ρ is the correlation coefficient of Y and Z. Since $E(Y) = k(\theta)$ and $E(Z) = 0$, we have

$$k'(\theta) = k(\theta)\cdot 0 + \rho\sigma_Y\sigma_Z \qquad \text{or} \qquad \rho = \frac{k'(\theta)}{\sigma_Y\sigma_Z}.$$

Now $\rho^2 \le 1$. Hence

$$\frac{[k'(\theta)]^2}{\sigma_Y^2\sigma_Z^2} \le 1 \qquad \text{or} \qquad \frac{[k'(\theta)]^2}{\sigma_Z^2} \le \sigma_Y^2.$$

If we replace σ_Z^2 by its value, we have

$$\sigma_Y^2 \ge \frac{[k'(\theta)]^2}{nE\left[\left(\frac{\partial \ln f(X;\theta)}{\partial\theta}\right)^2\right]} = \frac{[k'(\theta)]^2}{nI(\theta)}.$$

This inequality is known as the *Rao–Cramér inequality*.

If $Y = u(X_1, X_2, \ldots, X_n)$ is an unbiased estimator of θ, so that $k(\theta) = \theta$, then the Rao–Cramér inequality becomes, since $k'(\theta) = 1$,

$$\sigma_Y^2 \ge \frac{1}{nI(\theta)}.$$

Note that in Examples 1 and 2 of this section $1/nI(\theta)$ equals σ^2/n and $\theta(1-\theta)/n$, respectively. In each case, the unbiased estimator, $\bar{X}$, of θ, which is based upon the sufficient statistic for θ, has a variance that is equal to this Rao–Cramér lower bound of $1/nI(\theta)$.

We now make the following definitions.

Definition 1. Let Y be an unbiased estimator of a parameter θ in such a case of point estimation. The statistic Y is called an *efficient estimator* of θ if and only if the variance of Y attains the Rao–Cramér lower bound.

Definition 2. In cases in which we can differentiate with respect to a parameter under an integral or summation symbol, the ratio of the Rao–Cramér lower bound to the actual variance of any unbiased estimation of a parameter is called the *efficiency* of that statistic.

Example 3. Let $X_1, X_2, \ldots, X_n$ denote a random sample from a Poisson distribution that has the mean $\theta > 0$. It is known that $\overline{X}$ is an m.l.e. of θ; we shall show that it is also an efficient estimator of θ. We have

$$\frac{\partial \ln f(x;\theta)}{\partial \theta} = \frac{\partial}{\partial \theta}(x \ln \theta - \theta - \ln x!)$$

$$= \frac{x}{\theta} - 1 = \frac{x-\theta}{\theta}.$$

Accordingly,

$$E\left[\left(\frac{\partial \ln f(X;\theta)}{\partial \theta}\right)^2\right] = \frac{E(X-\theta)^2}{\theta^2} = \frac{\sigma^2}{\theta^2} = \frac{\theta}{\theta^2} = \frac{1}{\theta}.$$

The Rao–Cramér lower bound in this case is $1/[n(1/\theta)] = \theta/n$. But θ/n is the variance of $\overline{X}$. Hence $\overline{X}$ is an efficient estimator of θ.

Example 4. Let S^2 denote the variance of a random sample of size $n > 1$ from a distribution that is $N(\mu, \theta)$, $0 < \theta < \infty$, where μ is known. We know that $E[nS^2/(n-1)] = \theta$. What is the efficiency of the estimator $nS^2/(n-1)$? We have

$$\ln f(x;\theta) = -\frac{(x-\mu)^2}{2\theta} - \frac{\ln(2\pi\theta)}{2},$$

$$\frac{\partial \ln f(x;\theta)}{\partial \theta} = \frac{(x-\mu)^2}{2\theta^2} - \frac{1}{2\theta},$$

and

$$\frac{\partial^2 \ln f(x;\theta)}{\partial \theta^2} = -\frac{(x-\mu)^2}{\theta^3} + \frac{1}{2\theta^2}.$$

Accordingly,

$$-E\left[\frac{\partial^2 \ln f(X;\theta)}{\partial \theta^2}\right] = \frac{\theta}{\theta^3} - \frac{1}{2\theta^2} = \frac{1}{2\theta^2}.$$

Thus the Rao–Cramér lower bound is $2\theta^2/n$. Now nS^2/θ is $\chi^2(n-1)$, so the variance of nS^2/θ is $2(n-1)$. Accordingly, the variance of $nS^2/(n-1)$ is $2(n-1)[\theta^2/(n-1)^2] = 2\theta^2/(n-1)$. Thus the efficiency of the estimator $nS^2/(n-1)$ is $(n-1)/n$. With μ known, what is the efficient estimator of the variance?

Example 5. Let $X_1, X_2, \ldots, X_n$ denote a random sample of size $n > 2$ from a distribution with p.d.f.

$$f(x;\theta) = \theta x^{\theta-1} = \exp(\theta \ln x - \ln x + \ln \theta), \qquad 0 < x < 1,$$
$$= 0 \qquad \text{elsewhere.}$$

It is easy to verify that the Rao–Cramér lower bound is θ^2/n. Let

$Y_i = -\ln X_i$. We shall indicate that each Y_i has a gamma distribution. The associated transform $y_i = -\ln x_i$, with inverse $x_i = e^{-y_i}$, is one-to-one and the transformation maps the space $\{x_i: 0 < x_i < 1\}$ onto the space $\{y_i: 0 < y_i < \infty\}$. We have $|J| = e^{-y_i}$. Thus Y_i has a gamma distribution with $\alpha = 1$ and $\beta = 1/\theta$. Let $Z = -\sum_1^n \ln X_i$. Then Z has a gamma distribution with $\alpha = n$ and $\beta = 1/\theta$. Accordingly, we have $E(Z) = \alpha\beta = n/\theta$. This suggests that we compute the expectation of $1/Z$ to see if we can find an unbiased estimator of θ. A simple integration shows that $E(1/Z) = \theta/(n-1)$. Hence $(n-1)/Z$ is an unbiased estimator of θ. With $n > 2$, the variance of $(n-1)/Z$ exists and is found to be $\theta^2/(n-2)$, so that the efficiency of $(n-1)/Z$ is $(n-2)/n$. This efficiency tends to 1 as n increases. In such an instance, the estimator is said to be *asymptotically efficient*.

The concept of joint efficient estimators of several parameters has been developed along with the associated concept of joint efficiency of several estimators. But limitations of space prevent their inclusion in this book.

EXERCISES

8.11. Prove that $\bar{X}$, the mean of a random sample of size n from a distribution that is $N(\theta, \sigma^2)$, $-\infty < \theta < \infty$, is, for every known $\sigma^2 > 0$, an efficient estimator of θ.

8.12. Show that the mean $\bar{X}$ of a random sample of size n from a distribution which is $b(1, \theta)$, $0 < \theta < 1$, is an efficient estimator of θ.

8.13. Given $f(x; \theta) = 1/\theta$, $0 < x < \theta$, zero elsewhere, with $\theta > 0$, formally compute the reciprocal of

$$nE\left\{\left[\frac{\partial \ln f(X; \theta)}{\partial \theta}\right]^2\right\}.$$

Compare this with the variance of $(n+1)Y_n/n$, where Y_n is the largest item of a random sample of size n from this distribution. Comment.

8.14. Given the p.d.f.

$$f(x; \theta) = \frac{1}{\pi[1 + (x-\theta)^2]}, \qquad -\infty < x < \infty, \quad -\infty < \theta < \infty.$$

Show that the Rao–Cramér lower bound is $2/n$, where n is the size of a random sample from this Cauchy distribution.

8.15. Let X have a gamma distribution with $\alpha = 4$ and $\beta = \theta > 0$.
(a) Find the Fisher information $I(\theta)$.

(b) If $X_1, X_2, \ldots, X_n$ is a random sample from this distribution, show that the m.l.e. of θ is an efficient estimator of θ.

8.16. Let X be $N(0, \theta)$, $0 < \theta < \infty$.

(a) Find the Fisher information $I(\theta)$.

(b) If $X_1, X_2, \ldots, X_n$ is a random sample from this distribution, show that the m.l.e. of θ is an efficient estimator of θ.

8.3 Limiting Distributions of Maximum Likelihood Estimators

We use the notation and assumptions of Section 8.2 as much as possible here. In particular, $f(x; \theta)$ is the p.d.f., $I(\theta)$ is the Fisher information, and the likelihood function is

$$L(\theta) = f(x_1; \theta)f(x_2; \theta) \cdots f(x_n; \theta).$$

Also, we can differentiate under the integral (summation) sign, so that

$$Z = \frac{\partial \ln L(\theta)}{\partial \theta} = \sum_{i=1}^{n} \frac{\partial \ln f(X_i; \theta)}{\partial \theta}$$

has mean zero and variance $nI(\theta)$. In addition, we want to be able to find the maximum likelihood estimator $\hat{\theta}$ by solving

$$\frac{\partial[\ln L(\theta)]}{\partial \theta} = 0.$$

That is,

$$\frac{\partial[\ln L(\hat{\theta})]}{\partial \theta} = 0,$$

where now, with $\hat{\theta}$ in this expression, $L(\hat{\theta}) = f(X_1; \hat{\theta}) \cdots f(X_n; \hat{\theta})$. We can approximate the left-hand member of this latter equation by a linear function found from the first two terms of a Taylor's series expanded about θ, namely

$$\frac{\partial[\ln L(\theta)]}{\partial \theta} + (\hat{\theta} - \theta)\frac{\partial^2[\ln L(\theta)]}{\partial \theta^2} \approx 0,$$

when $L(\theta) = f(X_1; \theta)f(X_2; \theta) \cdots f(X_n; \theta)$.

Obviously, this approximation is good enough only if $\hat{\theta}$ is close to θ, and an adequate mathematical proof involves certain regularity conditions, all of which we have not given here. But a heuristic argument can be made by solving for $\hat{\theta} - \theta$ to obtain

$$\hat{\theta} - \theta = \frac{\dfrac{\partial[\ln L(\theta)]}{\partial\theta}}{-\dfrac{\partial^2[\ln L(\theta)]}{\partial\theta^2}} = \frac{Z}{-\dfrac{\partial^2[\ln L(\theta)]}{\partial\theta^2}}.$$

Let us rewrite this equation as

$$\frac{\hat{\theta} - \theta}{\sqrt{\dfrac{1}{nI(\theta)}}} = \frac{Z/\sqrt{nI(\theta)}}{-\dfrac{1}{n}\dfrac{\partial^2[\ln L(\theta)]}{\partial\theta^2}\Big/ I(\theta)}. \tag{1}$$

Since Z is the sum of the i.i.d. random variables

$$\frac{\partial \ln f(X_i; \theta)}{\partial\theta}, \qquad i = 1, 2, \ldots, n,$$

each with mean zero and variance $I(\theta)$, the numerator of the right-hand member of Equation (1) is limiting $N(0, 1)$ by the central limit theorem. Moreover, the mean

$$\frac{1}{n}\sum_{i=1}^{n} \frac{-\partial^2 \ln f(X_i; \theta)}{\partial\theta^2}$$

converges in probability to its expected value, namely $I(\theta)$. So the denominator of the right-hand member of Equation (1) converges in probability 1. Thus, by Slutsky's theorem given in Section 5.5, the right-hand member of Equation (1) is limiting $N(0, 1)$. Hence the left-hand member also has this limiting standard normal distribution. That means that we can say that $\hat{\theta}$ has an approximate normal distribution with mean θ and variance $1/nI(\theta)$.

The preceding result means that in a regular case of estimation and in some limiting sense, the m.l.e. $\hat{\theta}$ is unbiased and its variance achieves the Rao–Cramér lower bound. That is, the m.l.e. $\hat{\theta}$ is asymptotically efficient.

Example 1. In Exercise 8.14 we examined the Rao–Cramér lower bound of the variance of an unbiased estimator of θ, the median of a certain Cauchy distribution. We now know that the m.l.e. $\hat{\theta}$ of θ has an approximate normal distribution with mean θ and variance equal to the lower bound of $2/n$. Hence, once we compute $\hat{\theta}$, we can say, for illustration, that $\hat{\theta} \pm 1.96\sqrt{2/n}$ provides an approximate 95 percent confidence interval for θ.

To determine $\hat{\theta}$, there are many numerical methods that can be used. In the Cauchy case, one of the easiest is given by the following:

$$0 = \frac{\partial \ln L(\theta)}{\partial \theta} = \sum_{i=1}^{n} \frac{2(x_i - \theta)}{1 + (x_i - \theta)^2}.$$

In the denominator of the right-hand member, we use a preliminary estimate of θ that is not influenced too much by extreme observations. For illustration, the sample median, say $\hat{\theta}_0$, is very good one while the sample mean $\bar{x}$ would be a poor choice. This provides weights

$$w_{i1} = \frac{2}{1 + (x_i - \hat{\theta}_0)^2}, \qquad i = 1, 2, \ldots, n,$$

so that we can solve

$$0 = \sum_{i=1}^{n} (w_{i1})(x_i - \theta) \quad \text{to get} \quad \hat{\theta}_1 = \frac{\sum w_{i1} x_i}{\sum w_{i1}}.$$

Now $\hat{\theta}_1$ can be used to obtain new weights and $\hat{\theta}_2$:

$$w_{i2} = \frac{2}{1 + (x_i - \hat{\theta}_1)^2}, \qquad \hat{\theta}_2 = \frac{\sum w_{i2} x_i}{\sum w_{i2}}.$$

This iterative process can be continued until adequate convergence is obtained; that is, at some step k, $\hat{\theta}_k$ will be close enough to $\hat{\theta}$ to be used as the m.l.e.

Example 2. Suppose that the random sample arises from a distribution with p.d.f.

$$f(x; \theta) = \theta x^{\theta - 1}, \qquad 0 < x < 1, \quad \theta \in \Omega = \{\theta : 0 < \theta < \infty\},$$

zero elsewhere. We have

$$\ln f(x; \theta) = \ln \theta + (\theta - 1) \ln x,$$

$$\frac{\partial \ln f(x; \theta)}{\partial \theta} = \frac{1}{\theta} + \ln x,$$

and

$$\frac{\partial^2 \ln f(x; \theta)}{\partial \theta^2} = -\frac{1}{\theta^2}.$$

Since $E(-1/\theta^2) = -1/\theta^2$, the lower bound of the variance of every unbiased estimator of θ is θ^2/n. Moreover, the maximum likelihood estimator $\hat{\theta} = -n/\ln \prod_{i=1}^{n} X_i$ has an approximate normal distribution with mean θ and variance θ^2/n. Thus, in a limiting sense, $\hat{\theta}$ is the unbiased minimum variance estimator of θ; that is, $\hat{\theta}$ is asymptotically efficient.

Example 3. The m.l.e. for θ in

$$f(x; \theta) = \frac{\theta^x e^{-\theta}}{x!}, \qquad x = 0, 1, 2, \ldots, \quad \theta \in \Omega = \{\theta : 0 < \theta < \infty\},$$

is $\hat{\theta} = \bar{X}$, the mean of a random sample. Now

$$\ln f(x; \theta) = x \ln \theta - \theta - \ln x!$$

and

$$\frac{\partial[\ln f(x; \theta)]}{\partial \theta} = \frac{x}{\theta} - 1 \quad \text{and} \quad \frac{\partial^2[\ln f(x; \theta)]}{\partial \theta^2} = -\frac{x}{\theta^2}.$$

Thus

$$-E\left(-\frac{X}{\theta^2}\right) = \frac{\theta}{\theta^2} = \frac{1}{\theta}$$

and $\hat{\theta} = \bar{X}$ has an approximate normal distribution with mean θ and standard deviation $\sqrt{\theta/n}$. That is, $Y = (\bar{X} - \theta)/\sqrt{\theta/n}$ has a limiting standard normal distribution. The problem in practice is how best to estimate the standard deviation in the denominator of Y. Clearly, we might use $\bar{X}$ for θ there, but does that create too much dependence between the numerator and denominator? If so, this requires a very large sample size for $(\bar{X} - \theta)/\sqrt{\bar{X}/n}$ to have an approximate normal distribution. It might be better to approximate $I(\theta)$ by

$$\frac{1}{n} \sum_{i=1}^{n} \left\{\frac{\partial[\ln f(x_i; \hat{\theta})]}{\partial \hat{\theta}}\right\}^2 = \frac{1}{n} \sum_{i=1}^{n} \left(\frac{x_i}{\bar{x}} - 1\right)^2 = \frac{s^2}{\bar{x}^2}.$$

Thus $nI(\theta)$ is approximated by $ns^2/\bar{x}^2$ and we can say that

$$\frac{\sqrt{n}(\bar{X} - \theta)}{\bar{X}/S}$$

is approximately $N(0, 1)$. We do not know exactly which of these two solutions, or others like simply using $s/\sqrt{n}$ in the denominator, is best. Fortunately, however, if the Poisson model is correct, usually

$$\sqrt{\frac{\bar{x}}{n}} \approx \frac{\bar{x}}{\sqrt{n}\, s} \approx \frac{s}{\sqrt{n}}.$$

If this is not true, we should check the Poisson assumption, which requires, among other things, that $\mu = \sigma^2$. Hence, for illustration, either

$$\bar{x} \pm 1.96 \sqrt{\frac{\bar{x}}{n}} \quad \text{or} \quad \bar{x} \pm \frac{1.96\bar{x}}{\sqrt{n}\, s} \quad \text{or} \quad \bar{x} \pm \frac{1.96 s}{\sqrt{n}}$$

serves as an approximate 95 percent confidence interval for θ. In situations like this, we recommend that a person try all three because they should be in substantial agreement. If not, check the Poisson assumption.

The fact that the m.l.e. $\hat{\theta}$ has an approximate normal distribution with mean θ and variance $1/nI(\theta)$ suggests that $\hat{\theta}$ (really a sequence $\hat{\theta}_1, \hat{\theta}_2, \hat{\theta}_3, \ldots, \hat{\theta}_n, \ldots$) converges in probability to θ. Of course, $\hat{\theta}_n$ can be biased; say $E(\hat{\theta}_n - \theta) = b_n(\theta)$, where $b_n(\theta)$ is the bias. However, $b_n(\theta)$ equals zero in the limit. Moreover, if we assume that the variances exist and

$$\lim_{n\to\infty} [\text{var}(\hat{\theta}_n)] = \lim_{n\to\infty} \left[\frac{1}{nI(\theta)}\right],$$

then the limit of the variances is obviously zero. Hence, from Chebyshev's inequality, we have

$$\Pr[|\hat{\theta}_n - \theta| \geq \epsilon] \leq \frac{E[(\hat{\theta}_n - \theta)^2]}{\epsilon^2}.$$

However,

$$\lim_{n\to\infty} E[(\hat{\theta}_n - \theta)^2] = \lim_{n\to\infty} [b_n^2(\theta) + \text{var}(\hat{\theta}_n)] = 0$$

and thus

$$\lim_{n\to\infty} \Pr[|\hat{\theta}_n - \theta| \geq \epsilon] = 0$$

for each fixed $\epsilon > 0$. Any estimator, not just maximum likelihood estimators, that enjoys this property is said to be a *consistent estimator* of θ. As illustrations, we note that all the unbiased estimators based upon the complete sufficient statistics in Chapter 7 and all the estimators in Sections 8.1 and 8.2 are consistent ones.

We close this section by considering the extension of these limiting distributions to maximum likelihood estimators of two or more parameters. For convenience, we restrict ourselves to the regular case involving two parameters, but the extension to more than two is obvious once the reader understands multivariate normal distributions (Section 4.10).

Suppose that the random sample $X_1, X_2, \ldots, X_n$ arises from a distribution with p.d.f. $f(x; \theta_1, \theta_2)$, $(\theta_1, \theta_2) \in \Omega$, in which regularity conditions exist. Without describing these conditions in any detail, let us simply say that the space of X where $f(x; \theta_1, \theta_2) > 0$ does not involve θ_1 and θ_2, and we are able to differentiate under the integral (summation) signs. The information matrix of the sample is equal to

$$\mathbf{I}_n = n \times$$

$$\begin{bmatrix} E\left\{\left[\frac{\partial \ln f(X; \theta_1, \theta_2)}{\partial \theta_1}\right]^2\right\}, & E\left\{\frac{\partial \ln f(X; \theta_1, \theta_2)}{\partial \theta_1}\frac{\partial \ln f(X; \theta_1, \theta_2)}{\partial \theta_2}\right\}, \\ E\left\{\frac{\partial \ln f(X; \theta_1, \theta_2)}{\partial \theta_1}\frac{\partial \ln f(X; \theta_1, \theta_2)}{\partial \theta_2}\right\} & E\left\{\left[\frac{\partial \ln f(X; \theta_1, \theta_2)}{\partial \theta_2}\right]^2\right\} \end{bmatrix}$$

$$= -n \begin{pmatrix} E\left[\frac{\partial^2 \ln f(X; \theta_1, \theta_2)}{\partial \theta_1^2}\right] & E\left[\frac{\partial^2 \ln f(X; \theta_1, \theta_2)}{\partial \theta_1\, \partial \theta_2}\right] \\ E\left[\frac{\partial^2 \ln f(X; \theta_1, \theta_2)}{\partial \theta_1\, \partial \theta_2}\right] & E\left[\frac{\partial^2 \ln f(X; \theta_1, \theta_2)}{\partial \theta_2^2}\right] \end{pmatrix}.$$

One can immediately see the similarity of this to the one-parameter case.

If $\hat{\theta}_1$ and $\hat{\theta}_2$ are maximum likelihood estimators of θ_1 and θ_2, then $\hat{\theta}_1$ and $\hat{\theta}_2$ have an approximate bivariate normal distribution with means θ_1 and θ_2 and variance–covariance matrix $\mathbf{I}_n^{-1}$. That is, the approximate variances and covariances are found, respectively, in the matrix

$$\mathbf{I}_n^{-1} \approx \begin{pmatrix} \text{var}\,(\hat{\theta}_1) & \text{cov}\,(\hat{\theta}_1, \hat{\theta}_2) \\ \text{cov}\,(\hat{\theta}_1, \hat{\theta}_2) & \text{var}\,(\hat{\theta}_2) \end{pmatrix}.$$

An illustration will help us understand this result that has simply been given to the reader to accept without any mathematical derivation.

Example 4. Let the random sample $X_1, X_2, \ldots, X_n$ arise from $N(\theta_1, \theta_2)$. Then

$$\ln f(x; \theta_1, \theta_2) = -\frac{1}{2}\ln (2\pi\theta_2) - \frac{(x - \theta_1)^2}{2\theta_2},$$

$$\frac{\partial \ln f(x; \theta_1, \theta_2)}{\partial \theta_1} = \frac{x - \theta_1}{\theta_2},$$

$$\frac{\partial \ln f(x; \theta_1, \theta_2)}{\partial \theta_2} = -\frac{1}{2\theta_2} + \frac{(x - \theta_1)^2}{2\theta_2^2},$$

$$\frac{\partial^2 \ln f(x; \theta_1, \theta_2)}{\partial \theta_1^2} = \frac{-1}{\theta_2},$$

$$\frac{\partial^2 \ln f(x; \theta_1, \theta_2)}{\partial \theta_1 \, \partial \theta_2} = \frac{-(x - \theta_1)}{\theta_2^2},$$

$$\frac{\partial^2 \ln f(x; \theta_1, \theta_2)}{\partial \theta_2^2} = \frac{1}{2\theta_2^2} - \frac{(x - \theta_1)^2}{\theta_2^3}.$$

If we take the expected value of these three second partial derivatives and multiply by $-n$, we obtain the information matrix of the sample, namely,

$$\mathbf{I}_n = \begin{bmatrix} \dfrac{n}{\theta_2} & 0 \\ 0 & \dfrac{n}{2\theta_2^2} \end{bmatrix}.$$

Hence the approximate variance–covariance matrix of the maximum likelihood estimators $\hat{\theta}_1 = \bar{X}$ and $\hat{\theta}_2 = S^2$ is

$$\mathbf{I}_n^{-1} = \begin{bmatrix} \dfrac{\theta_2}{n} & 0 \\ 0 & \dfrac{2\theta_2^2}{n} \end{bmatrix}.$$

It is not surprising that the covariance equals zero as we know that $\bar{X}$ and S^2 are independent. In addition, we know that

$$\operatorname{var}(\bar{X}) = \frac{\theta_2}{n}$$

and

$$\operatorname{var}(S^2) = \operatorname{var}\left[\left(\frac{\theta_2}{n}\right)\left(\frac{nS^2}{\theta_2}\right)\right] = \frac{\theta_2^2}{n^2} \operatorname{var}\left(\frac{nS^2}{\theta_2}\right) = \frac{2(n-1)\theta_2^2}{n^2}$$

since nS^2/θ_2 is $\chi^2(n-1)$. While $\operatorname{var}(S^2) \neq 2\theta_2^2/n$, it is true that

$$\frac{2\theta_2^2}{n} \approx \frac{2(n-1)\theta_2^2}{n^2}$$

for large n.

EXERCISES

8.17. Let $X_1, X_2, \ldots, X_n$ be a random sample from each of the following distributions. In each case, find the m.l.e. $\hat{\theta}$, $\operatorname{var}(\hat{\theta})$, $1/nI(\theta)$, where $I(\theta)$ is the Fisher information of a single observation X, and compare $\operatorname{var}(\hat{\theta})$ and $1/nI(\theta)$.

(a) $b(1, \theta)$, $0 \le \theta \le 1$.

(b) $N(\theta, 1)$, $-\infty < \theta < \infty$.
(c) $N(0, \theta)$, $0 < \theta < \infty$.
(d) Gamma ($\alpha = 5, \beta = \theta$), $0 < \theta < \infty$.

8.18. Referring to Exercise 8.17 and using the fact that $\hat{\theta}$ has an approximate $N[\theta, 1/nI(\theta)]$, in each case construct an approximate 95 percent confidence interval for θ.

8.19. Let $(X_1, Y_1), (X_2, Y_2), \ldots, (X_n, Y_n)$ be a random sample from a bivariate normal distribution with unknown means θ_1 and θ_2 and with known variances and correlation coefficient, σ_1^2, σ_2^2, and ρ, respectively. Find the maximum likelihood estimators $\hat{\theta}_1$ and $\hat{\theta}_2$ of θ_1 and θ_2 and their approximate variance–covariance matrix. In this case, does the latter provide the exact variances and covariance?

8.20. Let $(X_1, Y_1), (X_2, Y_2), \ldots, (X_n, Y_n)$ be a random sample from a bivariate normal distribution with means equal to zero and variances θ_1 and θ_2, respectively, and known correlation coefficient ρ. Find the maximum likelihood estimators $\hat{\theta}_1$ and $\hat{\theta}_2$ of θ_1 and θ_2 and their approximate variance–covariance matrix.

8.4 Robust *M*-Estimation

In Example 1 of Section 8.3 we found the m.l.e. of the center θ of the Cauchy distribution with p.d.f.

$$f(x; \theta) = \frac{1}{\pi[1 + (x - \theta)^2]}, \qquad -\infty < x < \infty,$$

where $-\infty < \theta < \infty$. The logarithm of the likelihood function of a random sample $X_1, X_2, \ldots, X_n$ from this distribution is

$$\ln L(\theta) = -n \ln \pi - \sum_{i=1}^{n} \ln [1 + (x_i - \theta)^2].$$

To maximize, we differentiated $\ln L(\theta)$ to obtain

$$\frac{d \ln L(\theta)}{d\theta} = \sum_{i=1}^{n} \frac{2(x_i - \theta)}{1 + (x_i - \theta)^2} = 0.$$

The solution of this equation cannot be found in closed form, but the equation can be solved by some iterative process. There, to do this, we used the weight function

$$w(x - \hat{\theta}_0) = \frac{2}{1 + (x - \hat{\theta}_0)^2},$$

where $\hat{\theta}_0$ is some preliminary estimator of θ, like the sample median. Note that values of x for which $|x - \hat{\theta}_0|$ is relatively large do not have much weight. That is, in finding the maximum likelihood estimator of θ, the outlying values are downweighted greatly.

The generalization of this special case is described as follows. Let $X_1, X_2, \ldots, X_n$ be a random sample from a distribution with a p.d.f. of the form $f(x - \theta)$, where θ is a location parameter such that $-\infty < \theta < \infty$. Thus

$$\ln L(\theta) = \sum_{i=1}^{n} \ln f(x_i - \theta) = -\sum_{i=1}^{n} \rho(x_i - \theta),$$

where $\rho(x) = -\ln f(x)$, and

$$\frac{d \ln L(\theta)}{d\theta} = -\sum_{i=1}^{n} \frac{f'(x_i - \theta)}{f(x_i - \theta)} = \sum_{i=1}^{n} \Psi(x_i - \theta),$$

where $\rho'(x) = \Psi(x)$. For the Cauchy distribution, we have that these functions are

$$\rho(x) = \ln \pi + \ln (1 + x^2),$$

and

$$\Psi(x) = \frac{2x}{1 + x^2}.$$

In addition, we define a weight function as

$$w(x) = \frac{\Psi(x)}{x},$$

which equals $2/(1 + x^2)$ in the Cauchy case.

To appreciate how outlying observations are handled in estimating a center θ of different models progressing from a fairly light-tailed distribution like the normal to a very heavy-tailed distribution like the Cauchy, it is an easy exercise (Exercise 8.21) to show that standard normal distribution, with p.d.f. $\varphi(x)$, has

$$\rho(x) = \frac{1}{2} \ln 2\pi + \frac{x^2}{2}, \qquad \Psi(x) = x, \quad w(x) = 1.$$

That is, in estimating the center θ in $\varphi(x - \theta)$ each value of x has the weight 1 to yield the estimator $\hat{\theta} = \overline{X}$.

Also, the double exponential distribution, with p.d.f.

$$f(x) = \frac{1}{2} e^{-|x|}, \qquad -\infty < x < \infty,$$

has, provided that $x \neq 0$,

$$\rho(x) = \ln 2 + |x|, \qquad \Psi(x) = \text{sign}\,(x), \quad w(x) = \frac{\text{sign}\,(x)}{x} = \frac{1}{|x|}.$$

Here $\hat{\theta} = \text{median}\,(X_i)$ because in solving

$$\sum_{i=1}^{n} \Psi(x_i - \theta) = \sum_{i=1}^{n} \text{sign}\,(x_i - \theta) = 0$$

we need as many positive values of $x_i - \theta$ as negative values. The weights in the double exponential case are of the order $1/|x - \theta|$, while those in the Cauchy case are $2/[1 + (x - \theta)^2]$. That is, in estimating the center, outliers are downweighted more severely in a Cauchy situation, as the tails of the distribution are heavier than those of the double exponential distribution. On the other hand, extreme values from the double exponential distribution are downweighted more than those under normal assumptions in arriving at an estimate of the center θ.

Thus we suspect that the m.l.e. associated with one of these three distributions would not necessarily be a good estimator in another situation. This is true; for example, $\bar{X}$ is a very poor estimator of the median of a Cauchy distribution, as the variance of $\bar{X}$ does not even exist if the sample arises from a Cauchy distribution. Intuitively, $\bar{X}$ is not a good estimator with the Cauchy distribution, because the very small or very large values (outliers) that can arise from that distribution influence the mean $\bar{X}$ of the sample too much.

An estimator that is fairly good (small variance, say) for a wide variety of distributions (not necessarily the best for any one of them) is called a *robust estimator*. Also estimators associated with the solution of the equation

$$\sum_{i=1}^{n} \Psi(x_i - \theta) = 0$$

are frequently called *robust M-estimators* (denoted by $\hat{\theta}$) because they can be thought of as *m*aximum likelihood estimators. So in finding a robust M-estimator we must select a Ψ function which will provide an estimator that is good for each distribution in the collection under consideration. For certain theoretical reasons that we cannot explain at this level, Huber suggested a Ψ function that is a combination of those associated with the normal and double exponential distributions,

$$\begin{aligned}\Psi(x) &= -k, && x < -k\\ &= x, && -k \le x \le k,\\ &= k, && k < x,\end{aligned}$$

with weight $w(x) = 1$, $|x| \le k$, and $k/|x|$, provided that $k < |x|$. In Exercise 8.23 the reader is asked to find the p.d.f. $f(x)$ so that the M-estimator associated with this Ψ function is the m.l.e. of the location parameter θ in the p.d.f. $f(x - \theta)$.

With Huber's Ψ function, another problem arises. Note that if we double (for illustration) each $X_1, X_2, \ldots, X_n$, estimators such as $\bar{X}$ and median (X_i) also double. This is not at all true with the solution of the equation

$$\sum_{i=1}^{n} \Psi(x_i - \theta) = 0,$$

where the Ψ function is that of Huber. One way to avoid this difficulty is to solve another, but similar, equation instead,

$$\sum_{i=1}^{n} \Psi\left(\frac{x_i - \theta}{d}\right) = 0, \tag{1}$$

where d is a robust estimate of the scale. A popular d to use is

$$d = \frac{\text{median } |x_i - \text{median } (x_i)|}{0.6745}.$$

The divisor 0.6745 is inserted in the definition of d because then d is a consistent estimate of σ and thus is about equal to σ, if the sample arises from a normal distribution. That is, σ can be approximated by d under normal assumptions.

That scheme of selecting d also provides us with a clue for selecting k. For if the sample actually arises from a normal distribution, we would want most of the values $x_1, x_2, \ldots, x_n$ to satisfy the inequality

$$\left|\frac{x_i - \theta}{d}\right| \le k$$

because then

$$\Psi\left(\frac{x_i - \theta}{d}\right) = \frac{x_i - \theta}{d}.$$

That is, for illustration, if *all* the values satisfy this inequality, then Equation (1) becomes

$$\sum_{i=1}^{n} \Psi\left(\frac{x_i - \theta}{d}\right) = \sum_{i=1}^{n} \frac{x_i - \theta}{d} = 0.$$

This has the solution $\overline{x}$, which of course is most desirable with normal distributions. Since d approximates σ, popular values of k to use are 1.5 and 2.0, because with those selections most normal variables would satisfy the desired inequality.

Again an iterative process must usually be used to solve Equation (1). One such scheme, Newton's method, is described. Let $\hat{\theta}_0$ be a first estimate of θ, such as $\hat{\theta}_0 = \text{median}\,(x_i)$. Approximate the left-hand member of Equation (1) by the first two terms of Taylor's expansion about $\hat{\theta}_0$ to obtain

$$\sum_{i=1}^{n} \Psi\left(\frac{x_i - \hat{\theta}_0}{d}\right) + (\theta - \hat{\theta}_0) \sum_{i=1}^{n} \Psi'\left(\frac{x_i - \hat{\theta}_0}{d}\right)\left(-\frac{1}{d}\right) = 0,$$

approximately. The solution of this provides a second estimate of θ,

$$\hat{\theta}_1 = \hat{\theta}_0 + \frac{d \displaystyle\sum_{i=1}^{n} \Psi\left(\frac{x_i - \hat{\theta}_0}{d}\right)}{\displaystyle\sum_{i=1}^{n} \Psi'\left(\frac{x_i - \hat{\theta}_0}{d}\right)},$$

which is called the one-step M-estimate of θ. If we use $\hat{\theta}_1$ in place of $\hat{\theta}_0$, we obtain $\hat{\theta}_2$, the two-step M-estimate of θ. This process can continue to obtain any desired degree of accuracy. With Huber's Ψ function, the denominator of the second term,

$$\sum_{i=1}^{n} \Psi'\left(\frac{x_i - \hat{\theta}_0}{d}\right),$$

is particularly easy to compute because $\Psi'(x) = 1$, $-k \le x \le k$, and zero elsewhere. Thus that denominator simply counts the number of $x_1, x_2, \ldots, x_n$ such that $|x_i - \hat{\theta}_0|/d \le k$.

Say that the scale parameter σ is known (here σ is not necessarily the standard deviation for it does not exist for a distribution like the Cauchy). Two terms of Taylor's expansion of

$$\sum_{i=1}^{n} \Psi\left(\frac{X_i - \hat{\theta}}{\sigma}\right) = 0$$

about θ provides the approximation

$$\sum_{i=1}^{n} \Psi\left(\frac{X_i - \theta}{\sigma}\right) + (\hat{\theta} - \theta) \sum_{i=1}^{n} \Psi'\left(\frac{X_i - \theta}{\sigma}\right)\left(-\frac{1}{\sigma}\right) = 0.$$

This can be rewritten

$$\hat{\theta} - \theta = \frac{\sigma \sum \Psi\left(\frac{X_i - \theta}{\sigma}\right)}{\sum \Psi'\left(\frac{X_i - \theta}{\sigma}\right)}. \tag{2}$$

For the asymmetric Ψ functions that we have considered

$$E\left[\Psi\left(\frac{X - \theta}{\sigma}\right)\right] = 0,$$

provided that X has a symmetric distribution about θ. Clearly,

$$\operatorname{var}\left[\Psi\left(\frac{X - \theta}{\sigma}\right)\right] = E\left[\Psi^2\left(\frac{X - \theta}{\sigma}\right)\right].$$

Thus Equation (2) can be rewritten as

$$\frac{\sqrt{n}(\hat{\theta} - \theta)}{\sigma\sqrt{E\left[\Psi^2\left(\frac{X - \theta}{\sigma}\right)\right] \Big/ \left\{E\left[\Psi'\left(\frac{X - \theta}{\sigma}\right)\right]\right\}^2}}$$

$$= \frac{\sum \Psi\left(\frac{X_i - \theta}{\sigma}\right) \Big/ \sqrt{nE\left[\Psi^2\left(\frac{X - \theta}{\sigma}\right)\right]}}{\sum \Psi'\left(\frac{X_i - \theta}{\sigma}\right) \Big/ nE\left[\Psi'\left(\frac{X_i - \theta}{\sigma}\right)\right]}. \tag{3}$$

Clearly, by the central limit theorem, the numerator of the right-hand member of Equation (3) has a limiting standardized normal distribution, while the denominator converges in probability to 1. Thus the left-hand member has a limiting distribution that is $N(0, 1)$. In application we must approximate the denominator of the left-hand member. So we say that the robust M-estimator $\hat{\theta}$ has an approximate normal distribution with mean θ and variance

$$v = \frac{\left(\frac{d^2}{n}\right)\frac{1}{n}\sum_{i=1}^{n} \Psi^2\left(\frac{x_i - \hat{\theta}_k}{d}\right)}{\left\{\frac{1}{n}\sum_{i=1}^{n} \Psi'\left(\frac{x_i - \hat{\theta}_k}{d}\right)\right\}^2},$$

where $\hat{\theta}_k$ is the (last) k-step estimator of θ. Of course, $\hat{\theta}$ is approximated by $\hat{\theta}_k$; and an approximate 95 percent confidence interval for θ is given by $\hat{\theta}_k - 1.96\sqrt{v}$ to $\hat{\theta}_k + 1.96\sqrt{v}$.

EXERCISES

8.21. Verify that the functions $\rho(x)$, $\Psi(x)$, and $w(x)$ given in the text for the normal and double exponential distributions are correct.

8.22. Compute the one-step M-estimate $\hat{\theta}_1$ using Huber's Ψ with $k = 1.5$ if $n = 7$ and the seven observations are 2.1, 5.2, 2.3, 1.4, 2.2, 2.3, and 1.6. Here take $\hat{\theta}_0 = 2.2$, the median of the sample. Compare $\hat{\theta}_1$ with $\bar{x}$.

8.23. Let the p.d.f. $f(x)$ be such that the M-estimator associated with Huber's Ψ function is a maximum likelihood estimator of the location parameter in $f(x - \theta)$. Show that $f(x)$ is of the form $ce^{-\rho_1(x)}$, where $\rho_1(x) = x^2/2$, $|x| \le k$ and $\rho_1(x) = k|x| - k^2/2$, $k < |x|$.

8.24. Plot the Ψ functions associated with the normal, double exponential, and Cauchy distributions in addition to that of Huber. Why is the M-estimator associated with the Ψ function of the Cauchy distribution called a *redescending* M-estimator?

8.25. Use the data in Exercise 8.22 to find the one-step redescending M-estimator $\hat{\theta}_1$ associated with $\Psi(x) = \sin(x/1.5)$, $|x| \le 1.5\pi$, zero elsewhere. This was first proposed by D. F. Andrews. Compare this to $\bar{x}$ and the one-step M-estimator of Exercise 8.22. [It should be noted that there is no p.d.f. $f(x)$ that could be associated with this $\Psi(x)$ because $\Psi(x) = 0$ if $|x| > 1.5\pi$.]

ADDITIONAL EXERCISES

8.26. Let $X_1, X_2, \ldots, X_n$ be a random sample from a gamma distribution with $\alpha = 2$ and $\beta = 1/\theta$, $0 < \theta < \infty$.
(a) Find the m.l.e., $\hat{\theta}$, of θ. Is $\hat{\theta}$ unbiased?
(b) What is the approximating distribution of $\hat{\theta}$?
(c) If the prior distribution of the parameter is exponential with mean 2, determine the Bayes' estimator associated with a square-error loss function.

8.27. If $X_1, X_2, \ldots, X_n$ is a random sample from a distribution with p.d.f. $f(x; \theta) = 3\theta^3(x + \theta)^{-4}$, $0 < x < \infty$, zero elsewhere, where $0 < \theta$, show that $Y = 2\bar{X}$ is an unbiased estimator of θ and determine its efficiency.

8.28. Let $X_1, X_2, \ldots, X_n$ be a random sample from a distribution with p.d.f.

$$f(x; \theta) = \frac{\theta}{(1 + x)^{\theta + 1}}, \quad 0 < x < \infty, \text{ zero elsewhere, where } 0 < \theta.$$

(a) Find the m.l.e., $\hat{\theta}$, of θ and argue that it is a complete sufficient statistic for θ. Is $\hat{\theta}$ unbiased?

(b) If $\hat{\theta}$ is adjusted so that it is an unbiased estimator of θ, what is a lower bound for the variance of this unbiased estimator?

8.29. If $X_1, X_2, \ldots, X_n$ is a random sample from $N(\theta, 1)$, find a lower bound for the variance of an estimator of $k(\theta) = \theta^2$. Determine an unbiased minimum variance estimator of θ^2 and then compute its efficiency.

8.30. Suppose that we want to estimate the middle, θ, of a symmetric distribution using a robust estimator because we believe that the tails of this distribution are much thicker than those of a normal distribution. A t-distribution with 3 degrees of freedom with center at θ (not at zero) is such a distribution, so we decide to use the m.l.e., $\hat{\theta}$, associated with that distribution as our robust estimator. Evaluate $\hat{\theta}$ for the five observations: 10.1, 20.7, 11.3, 12.5, 6.0. Here we assume that the spread parameter is equal to 1.

8.31. Consider the normal distribution $N(0, \theta)$. With a random sample $X_1, X_2, \ldots, X_n$ we want to estimate the standard deviation $\sqrt{\theta}$. Find the constant c so that $Y = c \sum_{i=1}^{n} |X_i|$ is an unbiased estimator of $\sqrt{\theta}$ and determine its efficiency.

CHAPTER 9

Theory of Statistical Tests

9.1 Certain Best Tests

In Chapter 6 we introduced many concepts associated with tests of statistical hypotheses. In this chapter we consider some methods of constructing good statistical tests, beginning with testing a simple hypothesis H_0 against a simple alternative hypothesis H_1. Thus, in all instances, the parameter space is a set that consists of exactly two points. Under this restriction, we shall do three things:

1. Define a best test for testing H_0 against H_1.
2. Prove a theorem that provides a method of determining a best test.
3. Give two examples.

Before we define a best test, one important observation should be made. Certainly, a test specifies a critical region; but it can also be said that a choice of a critical region defines a test. For instance, if one is given the critical region $C = \{(x_1, x_2, x_3) : x_1^2 + x_2^2 + x_3^2 \geq 1\}$, the test is determined: Three random variables X_1, X_2, X_3 are to be considered; if the observed values are x_1, x_2, x_3, accept H_0 if

$x_1^2 + x_2^2 + x_3^2 < 1$; otherwise, reject H_0. That is, the terms "test" and "critical region" can, in this sense, be used interchangeably. Thus, if we define a best critical region, we have defined a best test.

Let $f(x; \theta)$ denote the p.d.f. of a random variable X. Let $X_1, X_2, \ldots, X_n$ denote a random sample from this distribution, and consider the two simple hypotheses $H_0 : \theta = \theta'$ and $H_1 : \theta = \theta''$. Thus $\Omega = \{\theta : \theta = \theta', \theta''\}$. We now define a best critical region (and hence a best test) for testing the simple hypothesis H_0 against the alternative simple hypothesis H_1. In this definition the symbols $\Pr[(X_1, X_2, \ldots, X_n) \in C; H_0]$ and $\Pr[(X_1, X_2, \ldots, X_n) \in C; H_1]$ mean $\Pr[(X_1, X_2, \ldots, X_n) \in C]$ when, respectively, H_0 and H_1 are true.

Definition 1. Let C denote a subset of the sample space. Then C is called a *best critical region* of size α for testing the simple hypothesis $H_0 : \theta = \theta'$ against the alternative simple hypothesis $H_1 : \theta = \theta''$ if, for every subset A of the sample space for which $\Pr[(X_1, \ldots, X_n) \in A; H_0] = \alpha$:

(a) $\Pr[(X_1, X_2, \ldots, X_n) \in C; H_0] = \alpha$.
(b) $\Pr[(X_1, X_2, \ldots, X_n) \in C; H_1] \geq \Pr[(X_1, X_2, \ldots, X_n) \in A; H_1]$.

This definition states, in effect, the following: First assume H_0 to be true. In general, there will be a multiplicity of subsets A of the sample space such that $\Pr[(X_1, X_2, \ldots, X_n) \in A] = \alpha$. Suppose that there is one of these subsets, say C, such that when H_1 is true, the power of the test associated with C is at least as great as the power of the test associated with each other A. Then C is defined as a best critical region of size α for testing H_0 against H_1.

In the following example we shall examine this definition in some detail and in a very simple case.

Example 1. Consider the one random variable X that has a binomial distribution with $n = 5$ and $p = \theta$. Let $f(x; \theta)$ denote the p.d.f. of X and let $H_0 : \theta = \frac{1}{2}$ and $H_1 : \theta = \frac{3}{4}$. The following tabulation gives, at points of positive probability density, the values of $f(x; \frac{1}{2})$, $f(x; \frac{3}{4})$, and the ratio $f(x; \frac{1}{2})/f(x; \frac{3}{4})$.

x	0	1	2	3	4	5
$f(x; \frac{1}{2})$	$\frac{1}{32}$	$\frac{5}{32}$	$\frac{10}{32}$	$\frac{10}{32}$	$\frac{5}{32}$	$\frac{1}{32}$
$f(x; \frac{3}{4})$	$\frac{1}{1024}$	$\frac{15}{1024}$	$\frac{90}{1024}$	$\frac{270}{1024}$	$\frac{405}{1024}$	$\frac{243}{1024}$
$f(x; \frac{1}{2})/f(x; \frac{3}{4})$	32	$\frac{32}{3}$	$\frac{32}{9}$	$\frac{32}{27}$	$\frac{32}{81}$	$\frac{32}{243}$

We shall use one random value of X to test the simple hypothesis $H_0 : \theta = \frac{1}{2}$ against the alternative simple hypothesis $H_1 : \theta = \frac{3}{4}$, and we shall first assign the significance level of the test to be $\alpha = \frac{1}{32}$. We seek a best critical region of size $\alpha = \frac{1}{32}$. If $A_1 = \{x : x = 0\}$ and $A_2 = \{x : x = 5\}$, then $\Pr(X \in A_1; H_0) = \Pr(X \in A_2; H_0) = \frac{1}{32}$ and there is no other subset A_3 of the space $\{x : x = 0, 1, 2, 3, 4, 5\}$ such that $\Pr(X \in A_3; H_0) = \frac{1}{32}$. Then either A_1 or A_2 is the best critical region C of size $\alpha = \frac{1}{32}$ for testing H_0 against H_1. We note that $\Pr(X \in A_1; H_0) = \frac{1}{32}$ and that $\Pr(X \in A_1; H_1) = \frac{1}{1024}$. Thus, if the set A_1 is used as a critical region of size $\alpha = \frac{1}{32}$, we have the intolerable situation that the probability of rejecting H_0 when H_1 is true (H_0 is false) is much less than the probability of rejecting H_0 when H_0 is true.

On the other hand, if the set A_2 is used as a critical region, then $\Pr(X \in A_2; H_0) = \frac{1}{32}$ and $\Pr(X \in A_2; H_1) = \frac{243}{1024}$. That is, the probability of rejecting H_0 when H_1 is true is much greater than the probability of rejecting H_0 when H_0 is true. Certainly, this is a more desirable state of affairs, and actually A_2 is the best critical region of size $\alpha = \frac{1}{32}$. The latter statement follows from the fact that, when H_0 is true, there are but two subsets, A_1 and A_2, of the sample space, each of whose probability measure is $\frac{1}{32}$ and the fact that

$$\frac{243}{1024} = \Pr(X \in A_2; H_1) > \Pr(X \in A_1; H_1) = \frac{1}{1024}.$$

It should be noted, in this problem, that the best critical region $C = A_2$ of size $\alpha = \frac{1}{32}$ is found by including in C the point (or points) at which $f(x; \frac{1}{2})$ is *small* in comparison with $f(x; \frac{3}{4})$. This is seen to be true once it is observed that the ratio $f(x; \frac{1}{2})/f(x; \frac{3}{4})$ is a minimum at $x = 5$. Accordingly, the ratio $f(x; \frac{1}{2})/f(x; \frac{3}{4})$, which is given in the last line of the above tabulation, provides us with a precise tool by which to find a best critical region C for certain given values of α. To illustrate this, take $\alpha = \frac{6}{32}$. When H_0 is true, each of the subsets $\{x : x = 0, 1\}$, $\{x : x = 0, 4\}$, $\{x : x = 1, 5\}$, $\{x : x = 4, 5\}$ has probability measure $\frac{6}{32}$. By direct computation it is found that the best critical region of this size is $\{x : x = 4, 5\}$. This reflects the fact that the ratio $f(x; \frac{1}{2})/f(x; \frac{3}{4})$ has its two smallest values for $x = 4$ and $x = 5$. The power of this test, which has $\alpha = \frac{6}{32}$, is

$$\Pr(X = 4, 5; H_1) = \frac{405}{1024} + \frac{243}{1024} = \frac{648}{1024}.$$

The preceding example should make the following theorem, due to Neyman and Pearson, easier to understand. It is an important theorem because it provides a systematic method of determining a best critical region.

Neyman–Pearson Theorem. *Let $X_1, X_2, \ldots, X_n$, where n is a fixed positive integer, denote a random sample from a distribution that has* p.d.f. $f(x; \theta)$. *Then the joint* p.d.f. *of $X_1, X_2, \ldots, X_n$ is*

$$L(\theta; x_1, x_2, \ldots, x_n) = f(x_1; \theta)f(x_2; \theta) \cdots f(x_n; \theta).$$

Let θ' and θ'' be distinct fixed values of θ so that $\Omega = \{\theta : \theta = \theta', \theta''\}$, and let k be a positive number. Let C be a subset of the sample space such that:

(a) $\dfrac{L(\theta'; x_1, x_2, \ldots, x_n)}{L(\theta''; x_1, x_2, \ldots, x_n)} \le k,$ *for each point* $(x_1, x_2, \ldots, x_n) \in C$.

(b) $\dfrac{L(\theta'; x_1, x_2, \ldots, x_n)}{L(\theta''; x_1, x_2, \ldots, x_n)} \ge k,$ *for each point* $(x_1, x_2, \ldots, x_n) \in C^*$.

(c) $\alpha = \Pr[(X_1, X_2, \ldots, X_n) \in C; H_0]$.

Then C is a best critical region of size α for testing the simple hypothesis $H_0 : \theta = \theta'$ against the alternative simple hypothesis $H_1 : \theta = \theta''$.

Proof. We shall give the proof when the random variables are of the continuous type. If C is the only critical region of size α, the theorem is proved. If there is another critical region of size α, denote it by A. For convenience, we shall let $\int \cdot \overset{\cdot}{\underset{R}{\cdot}} \cdot \int L(\theta; x_1, \ldots, x_n)\, dx_1 \cdots dx_n$ be denoted by $\int_R L(\theta)$. In this notation we wish to show that

$$\int_C L(\theta'') - \int_A L(\theta'') \ge 0.$$

Since C is the union of the disjoint sets $C \cap A$ and $C \cap A^*$ and A is the union of the disjoint sets $A \cap C$ and $A \cap C^*$, we have

$$\int_C L(\theta'') - \int_A L(\theta'')$$

$$= \int_{C \cap A} L(\theta'') + \int_{C \cap A^*} L(\theta'') - \int_{A \cap C} L(\theta'') - \int_{A \cap C^*} L(\theta'')$$

$$= \int_{C \cap A^*} L(\theta'') - \int_{A \cap C^*} L(\theta''). \tag{1}$$

However, by the hypothesis of the theorem, $L(\theta'') \ge (1/k)L(\theta')$ at each point of C, and hence at each point of $C \cap A^*$; thus

$$\int_{C \cap A^*} L(\theta'') \ge \frac{1}{k} \int_{C \cap A^*} L(\theta').$$

But $L(\theta'') \le (1/k)L(\theta')$ at each point of C^*, and hence at each point of $A \cap C^*$; accordingly,

$$\int_{A \cap C^*} L(\theta'') \le \frac{1}{k} \int_{A \cap C^*} L(\theta').$$

These inequalities imply that

$$\int_{C \cap A^*} L(\theta'') - \int_{A \cap C^*} L(\theta'') \ge \frac{1}{k} \int_{C \cap A^*} L(\theta') - \frac{1}{k} \int_{A \cap C^*} L(\theta');$$

and, from Equation (1), we obtain

$$\int_C L(\theta'') - \int_A L(\theta'') \ge \frac{1}{k} \left[\int_{C \cap A^*} L(\theta') - \int_{A \cap C^*} L(\theta') \right]. \tag{2}$$

However,

$$\begin{aligned} &\int_{C \cap A^*} L(\theta') - \int_{A \cap C^*} L(\theta') \\ &= \int_{C \cap A^*} L(\theta') + \int_{C \cap A} L(\theta') - \int_{A \cap C} L(\theta') - \int_{A \cap C^*} L(\theta') \\ &= \int_C L(\theta') - \int_A L(\theta') \\ &= \alpha - \alpha = 0. \end{aligned}$$

If this result is substituted in inequality (2), we obtain the desired result,

$$\int_C L(\theta'') - \int_A L(\theta'') \ge 0.$$

If the random variables are of the discrete type, the proof is the same, with integration replaced by summation.

Remark. As stated in the theorem, conditions (a), (b), and (c) are sufficient ones for region C to be a best critical region of size α. However, they are also necessary. We discuss this briefly. Suppose there is a region A of size α that does not satisfy (a) and (b) and that is as powerful at $\theta = \theta''$ as C, which satisfies (a), (b), and (c). Then expression (1) would be zero, since the power at θ'' using A is equal to that using C. It can be proved that to have expression (1) equal zero A must be of the same form as C. As a matter of fact, in the continuous case, A and C would essentially be the same region; that is, they could differ only by a set having probability zero. However, in the discrete case, if $\Pr[L(\theta') = kL(\theta''); H_0]$ is positive, A and C could be different sets, but each would necessarily enjoy conditions (a), (b), and (c) to be a best critical region of size α.

One aspect of the theorem to be emphasized is that if we take C to be the set of all points $(x_1, x_2, \ldots, x_n)$ which satisfy

$$\frac{L(\theta'; x_1, x_2, \ldots, x_n)}{L(\theta''; x_1, x_2, \ldots, x_n)} \le k, \qquad k > 0,$$

then, in accordance with the theorem, C will be a best critical region. This inequality can frequently be expressed in one of the forms (where c_1 and c_2 are constants)

$$u_1(x_1, x_2, \ldots, x_n; \theta', \theta'') \le c_1,$$

or

$$u_2(x_1, x_2, \ldots, x_n; \theta', \theta'') \ge c_2.$$

Suppose that it is the first form, $u_1 \le c_1$. Since θ' and θ'' are given constants, $u_1(X_1, X_2, \ldots, X_n; \theta', \theta'')$ is a statistic; and if the p.d.f. of this statistic can be found when H_0 is true, then the significance level of the test of H_0 against H_1 can be determined from this distribution. That is,

$$\alpha = \Pr [u_1(X_1, X_2, \ldots, X_n; \theta', \theta'') \le c_1; H_0].$$

Moreover, the test may be based on this statistic; for, if the observed values of $X_1, X_2, \ldots, X_n$ are $x_1, x_2, \ldots, x_n$, we reject H_0 (accept H_1) if $u_1(x_1, x_2, \ldots, x_n) \le c_1$.

A positive number k determines a best critical region C whose size is $\alpha = \Pr [(X_1, X_2, \ldots, X_n) \in C; H_0]$ for that particular k. It may be that this value of α is unsuitable for the purpose at hand; that is, it is too large or too small. However, if there is a statistic $u_1(X_1, X_2, \ldots, X_n)$, as in the preceding paragraph, whose p.d.f. can be determined when H_0 is true, we need not experiment with various values of k to obtain a desirable significance level. For if the distribution of the statistic is known, or can be found, we may determine c_1 such that $\Pr [u_1(X_1, X_2, \ldots, X_n) \le c_1; H_0]$ is a desirable significance level.

An illustrative example follows.

Example 2. Let $X_1, X_2, \ldots, X_n$ denote a random sample from the distribution that has the p.d.f.

$$f(x; \theta) = \frac{1}{\sqrt{2\pi}} \exp\left(-\frac{(x-\theta)^2}{2}\right), \qquad -\infty < x < \infty.$$

It is desired to test the simple hypothesis $H_0: \theta = \theta' = 0$ against the alternative simple hypothesis $H_1: \theta = \theta'' = 1$. Now

$$\frac{L(\theta'; x_1, \ldots, x_n)}{L(\theta''; x_1, \ldots, x_n)} = \frac{(1/\sqrt{2\pi})^n \exp\left[-\left(\sum_1^n x_i^2\right)\Big/2\right]}{(1/\sqrt{2\pi})^n \exp\left[-\left(\sum_1^n (x_i - 1)^2\right)\Big/2\right]}$$

$$= \exp\left(-\sum_1^n x_i + \frac{n}{2}\right).$$

If $k > 0$, the set of all points $(x_1, x_2, \ldots, x_n)$ such that

$$\exp\left(-\sum_1^n x_i + \frac{n}{2}\right) \le k$$

is a best critical region. This inequality holds if and only if

$$-\sum_1^n x_i + \frac{n}{2} \le \ln k$$

or, equivalently,

$$\sum_1^n x_i \ge \frac{n}{2} - \ln k = c.$$

In this case, a best critical region is the set $C = \left\{(x_1, x_2, \ldots, x_n): \sum_1^n x_i \ge c\right\}$, where c is a constant that can be determined so that the size of the critical region is a desired number α. The event $\sum_1^n X_i \ge c$ is equivalent to the event $\bar{X} \ge c/n = c_1$, say, so the test may be based upon the statistic $\bar{X}$. If H_0 is true, that is, $\theta = \theta' = 0$, then $\bar{X}$ has a distribution that is $N(0, 1/n)$. For a given positive integer n, the size of the sample, and a given significance level α, the number c_1 can be found from Table III in Appendix B, so that $\Pr(\bar{X} \ge c_1; H_0) = \alpha$. Hence, if the experimental values of $X_1, X_2, \ldots, X_n$ were, respectively, $x_1, x_2, \ldots, x_n$, we would compute $\bar{x} = \sum_1^n x_i/n$. If $\bar{x} \ge c_1$, the simple hypothesis $H_0: \theta = \theta' = 0$ would be rejected at the significance level α; if $\bar{x} < c_1$, the hypothesis H_0 would be accepted. The probability of rejecting H_0, when H_0 is true, is α; the probability of rejecting H_0, when H_0 is false, is the value of the power of the test at $\theta = \theta'' = 1$. That is,

$$\Pr(\bar{X} \ge c_1; H_1) = \int_{c_1}^{\infty} \frac{1}{\sqrt{2\pi}\sqrt{1/n}} \exp\left[-\frac{(\bar{x} - 1)^2}{2(1/n)}\right] d\bar{x}.$$

For example, if $n = 25$ and if α is selected to be 0.05, then from Table III we find that $c_1 = 1.645/\sqrt{25} = 0.329$. Thus the power of this best test of H_0 against H_1 is 0.05, when H_0 is true, and is

$$\int_{0.329}^{\infty} \frac{1}{\sqrt{2\pi}\sqrt{\frac{1}{25}}} \exp\left[-\frac{(\bar{x}-1)^2}{2(\frac{1}{25})}\right] d\bar{x} = \int_{-3.355}^{\infty} \frac{1}{\sqrt{2\pi}} e^{-w^2/2}\, dw = 0.999+,$$

when H_1 is true.

There is another aspect of this theorem that warrants special mention. It has to do with the number of parameters that appear in the p.d.f. Our notation suggests that there is but one parameter. However, a careful review of the proof will reveal that nowhere was this needed or assumed. The p.d.f. may depend upon any finite number of parameters. What is essential is that the hypothesis H_0 and the alternative hypothesis H_1 be simple, namely that they completely specify the distributions. With this in mind, we see that the simple hypotheses H_0 and H_1 do not need to be hypotheses about the parameters of a distribution, nor, as a matter of fact, do the random variables $X_1, X_2, \ldots, X_n$ need to be independent. That is, if H_0 is the simple hypothesis that the joint p.d.f. is $g(x_1, x_2, \ldots, x_n)$, and if H_1 is the alternative simple hypothesis that the joint p.d.f. is $h(x_1, x_2, \ldots, x_n)$, then C is a best critical region of size α for testing H_0 against H_1 if, for $k > 0$:

1′. $\dfrac{g(x_1, x_2, \ldots, x_n)}{h(x_1, x_2, \ldots, x_n)} \le k$ for $(x_1, x_2, \ldots, x_n) \in C$.

2′. $\dfrac{g(x_1, x_2, \ldots, x_n)}{h(x_1, x_2, \ldots, x_n)} \ge k$ for $(x_1, x_2, \ldots, x_n) \in C^*$.

3′. $\alpha = \Pr[(X_1, X_2, \ldots, X_n) \in C; H_0]$.

An illustrative example follows.

Example 3. Let $X_1, \ldots, X_n$ denote a random sample from a distribution which has a p.d.f. $f(x)$ that is positive on and only on the nonnegative integers. It is desired to test the simple hypothesis

$$H_0 : f(x) = \frac{e^{-1}}{x!}, \qquad x = 0, 1, 2, \ldots,$$
$$= 0 \qquad \text{elsewhere},$$

against the alternative simple hypothesis

$$H_1 : f(x) = (\tfrac{1}{2})^{x+1}, \qquad x = 0, 1, 2, \ldots,$$
$$= 0 \qquad \text{elsewhere}.$$

Here

$$\frac{g(x_1, \ldots, x_n)}{h(x_1, \ldots, x_n)} = \frac{e^{-n}/(x_1!\, x_2! \cdots x_n!)}{(\frac{1}{2})^n(\frac{1}{2})^{x_1 + x_2 + \cdots + x_n}}$$

$$= \frac{(2e^{-1})^n 2^{\Sigma x_i}}{\prod_1^n (x_i!)}.$$

If $k > 0$, the set of points $(x_1, x_2, \ldots, x_n)$ such that

$$\left(\sum_1^n x_i\right) \ln 2 - \ln\left[\prod_1^n (x_i!)\right] \le \ln k - n \ln (2e^{-1}) = c$$

is a best critical region C. Consider the case of $k = 1$ and $n = 1$. The preceding inequality may be written $2^{x_1}/x_1! \le e/2$. This inequality is satisfied by all points in the set $C = \{x_1 : x_1 = 0, 3, 4, 5, \ldots\}$. Thus the power of the test when H_0 is true is

$$\Pr(X_1 \in C; H_0) = 1 - \Pr(X_1 = 1, 2; H_0) = 0.448,$$

approximately, in accordance with Table I of Appendix B. The power of the test when H_1 is true is given by

$$\Pr(X_1 \in C; H_1) = 1 - \Pr(X_1 = 1, 2; H_1)$$

$$= 1 - (\tfrac{1}{4} + \tfrac{1}{8}) = 0.625.$$

Remark. In the notation of this section, say C is a critical region such that

$$\alpha = \int_C L(\theta') \qquad \text{and} \qquad \beta = \int_{C^*} L(\theta''),$$

so that here α and β equal the respective probabilities of the type I and type II errors associated with C. Let d_1 and d_2 be two given positive constants. Consider a certain linear function of α and β, namely

$$d_1 \int_C L(\theta') + d_2 \int_{C^*} L(\theta'') = d_1 \int_C L(\theta') + d_2\left[1 - \int_C L(\theta'')\right]$$

$$= d_2 + \int_C [d_1 L(\theta') - d_2 L(\theta'')].$$

If we wished to minimize this expression, we would select C to be the set of all $(x_1, x_2, \ldots, x_n)$ such that

$$d_1 L(\theta') - d_2 L(\theta'') < 0$$

or, equivalently,

$$\frac{L(\theta')}{L(\theta'')} < \frac{d_2}{d_1}, \qquad \text{for all } (x_1, x_2, \ldots, x_n) \in C,$$

which according to the Neyman–Pearson theorem provides a best critical region with $k = d_2/d_1$. That is, this critical region C is one that minimizes $d_1\alpha + d_2\beta$. There could be others, for example, including points on which $L(\theta')/L(\theta'') = d_2/d_1$, but these would still be best critical regions according to the Neyman–Pearson theorem.

EXERCISES

9.1. In Example 2 of this section, let the simple hypotheses read $H_0 : \theta = \theta' = 0$ and $H_1 : \theta = \theta'' = -1$. Show that the best test of H_0 against H_1 may be carried out by use of the statistic $\bar{X}$, and that if $n = 25$ and $\alpha = 0.05$, the power of the test is 0.999+ when H_1 is true.

9.2. Let the random variable X have the p.d.f. $f(x; \theta) = (1/\theta)e^{-x/\theta}$, $0 < x < \infty$, zero elsewhere. Consider the simple hypothesis $H_0 : \theta = \theta' = 2$ and the alternative hypothesis $H_1 : \theta = \theta'' = 4$. Let X_1, X_2 denote a random sample of size 2 from this distribution. Show that the best test of H_0 against H_1 may be carried out by use of the statistic $X_1 + X_2$ and that the assertion in Example 2 of Section 6.4 is correct.

9.3. Repeat Exercise 9.2 when $H_1 : \theta = \theta'' = 6$. Generalize this for every $\theta'' > 2$.

9.4. Let $X_1, X_2, \ldots, X_{10}$ be a random sample of size 10 from a normal distribution $N(0, \sigma^2)$. Find a best critical region of size $\alpha = 0.05$ for testing $H_0 : \sigma^2 = 1$ against $H_1 : \sigma^2 = 2$. Is this a best critical region of size 0.05 for testing $H_0 : \sigma^2 = 1$ against $H_1 : \sigma^2 = 4$? Against $H_1 : \sigma^2 = \sigma_1^2 > 1$?

9.5. If $X_1, X_2, \ldots, X_n$ is a random sample from a distribution having p.d.f. of the form $f(x; \theta) = \theta x^{\theta - 1}$, $0 < x < 1$, zero elsewhere, show that a best critical region for testing $H_0 : \theta = 1$ against $H_1 : \theta = 2$ is

$$C = \left\{(x_1, x_2, \ldots, x_n) : c \le \prod_{i=1}^{n} x_i\right\}.$$

9.6. Let $X_1, X_2, \ldots, X_{10}$ be a random sample from a distribution that is $N(\theta_1, \theta_2)$. Find a best test of the simple hypothesis $H_0 : \theta_1 = \theta_1' = 0$, $\theta_2 = \theta_2' = 1$ against the alternative simple hypothesis $H_1 : \theta_1 = \theta_1'' = 1$, $\theta_2 = \theta_2'' = 4$.

9.7. Let $X_1, X_2, \ldots, X_n$ denote a random sample from a normal distribution $N(\theta, 100)$. Show that $C = \left\{(x_1, x_2, \ldots, x_n) : c \le \bar{x} = \sum_{1}^{n} x_i/n\right\}$ is a best critical region for testing $H_0 : \theta = 75$ against $H_1 : \theta = 78$. Find n and c so that

$$\Pr[(X_1, X_2, \ldots, X_n) \in C; H_0] = \Pr(\bar{X} \ge c; H_0) = 0.05$$

and

$$\Pr[(X_1, X_2, \ldots, X_n) \in C; H_1] = \Pr(\bar{X} \ge c; H_1) = 0.90, \text{ approximately.}$$

9.8. If $X_1, X_2, \ldots, X_n$ is a random sample from a beta distribution with parameters $\alpha = \beta = \theta > 0$, find a best critical region for testing $H_0: \theta = 1$ against $H_1: \theta = 2$.

9.9. Let $X_1, X_2, \ldots, X_n$ denote a random sample from a distribution having the p.d.f. $f(x; p) = p^x(1-p)^{1-x}$, $x = 0, 1$, zero elsewhere. Show that $C = \left\{(x_1, \ldots, x_n) : \sum_1^n x_i \le c\right\}$ is a best critical region for testing $H_0: p = \frac{1}{2}$ against $H_1: p = \frac{1}{3}$. Use the central limit theorem to find n and c so that approximately $\Pr\left(\sum_1^n X_i \le c; H_0\right) = 0.10$ and $\Pr\left(\sum_1^n X_i \le c; H_1\right) = 0.80$.

9.10. Let $X_1, X_2, \ldots, X_{10}$ denote a random sample of size 10 from a Poisson distribution with mean θ. Show that the critical region C defined by $\sum_1^{10} x_i \ge 3$ is a best critical region for testing $H_0: \theta = 0.1$ against $H_1: \theta = 0.5$. Determine, for this test, the significance level α and the power at $\theta = 0.5$.

9.2 Uniformly Most Powerful Tests

This section will take up the problem of a test of a simple hypothesis H_0 against an alternative composite hypothesis H_1. We begin with an example.

Example 1. Consider the p.d.f.

$$f(x; \theta) = \frac{1}{\theta} e^{-x/\theta}, \qquad 0 < x < \infty,$$
$$= 0 \qquad \text{elsewhere,}$$

of Example 2, Section 6.4, and later of Exercise 9.3. It is desired to test the simple hypothesis $H_0: \theta = 2$ against the alternative composite hypothesis $H_1: \theta > 2$. Thus $\Omega = \{\theta : \theta \ge 2\}$. A random sample, X_1, X_2, of size $n = 2$ will be used, and the critical region is $C = \{(x_1, x_2) : 9.5 \le x_1 + x_2 < \infty\}$. It was shown in the example cited that the significance level of the test is approximately 0.05 and that the power of the test when $\theta = 4$ is approximately 0.31. The power function $K(\theta)$ of the test for all $\theta \ge 2$ will now be obtained. We have

$$K(\theta) = 1 - \int_0^{9.5} \int_0^{9.5 - x_2} \frac{1}{\theta^2} \exp\left(-\frac{x_1 + x_2}{\theta}\right) dx_1\, dx_2$$
$$= \left(\frac{\theta + 9.5}{\theta}\right) e^{-9.5/\theta}, \qquad 2 \le \theta.$$

For example, $K(2) = 0.05$, $K(4) = 0.31$, and $K(9.5) = 2/e$. It is known (Exercise 9.3) that $C = \{(x_1, x_2) : 9.5 \le x_1 + x_2 < \infty\}$ is a best critical region of size 0.05 for testing the simple hypothesis $H_0 : \theta = 2$ against each simple hypothesis in the composite hypothesis $H_1 : \theta > 2$.

The preceding example affords an illustration of a test of a simple hypothesis H_0 that is a best test of H_0 against every simple hypothesis in the alternative composite hypothesis H_1. We now define a critical region, when it exists, which is a best critical region for testing a simple hypothesis H_0 against an alternative composite hypothesis H_1. It seems desirable that this critical region should be a best critical region for testing H_0 against each simple hypothesis in H_1. That is, the power function of the test that corresponds to this critical region should be at least as great as the power function of any other test with the same significance level for every simple hypothesis in H_1.

Definition 2. The critical region C is a *uniformly most powerful critical region* of size α for testing the simple hypothesis H_0 against an alternative composite hypothesis H_1 if the set C is a best critical region of size α for testing H_0 against each simple hypothesis in H_1. A test defined by this critical region C is called a *uniformly most powerful test*, with significance level α, for testing the simple hypothesis H_0 against the alternative composite hypothesis H_1.

As will be seen presently, uniformly most powerful tests do not always exist. However, when they do exist, the Neyman–Pearson theorem provides a technique for finding them. Some illustrative examples are given here.

Example 2. Let $X_1, X_2, \ldots, X_n$ denote a random sample from a distribution that is $N(0, \theta)$, where the variance θ is an unknown positive number. It will be shown that there exists a uniformly most powerful test with significance level α for testing the simple hypothesis $H_0 : \theta = \theta'$, where θ' is a fixed positive number, against the alternative composite hypothesis $H_1 : \theta > \theta'$. Thus $\Omega = \{\theta : \theta \geq \theta'\}$. The joint p.d.f. of $X_1, X_2, \ldots, X_n$ is

$$L(\theta; x_1, x_2, \ldots, x_n) = \left(\frac{1}{2\pi\theta}\right)^{n/2} \exp\left(-\frac{\sum_1^n x_i^2}{2\theta}\right).$$

Let θ'' represent a number greater than θ', and let k denote a positive number. Let C be the set of points where

$$\frac{L(\theta'; x_1, x_2, \ldots, x_n)}{L(\theta''; x_1, x_2, \ldots, x_n)} \leq k,$$

that is, the set of points where

$$\left(\frac{\theta''}{\theta'}\right)^{n/2} \exp\left[-\left(\frac{\theta''-\theta'}{2\theta'\theta''}\right)\sum_1^n x_i^2\right] \le k$$

or, equivalently,

$$\sum_1^n x_i^2 \ge \frac{2\theta'\theta''}{\theta''-\theta'}\left[\frac{n}{2}\ln\left(\frac{\theta''}{\theta'}\right) - \ln k\right] = c.$$

The set $C = \left\{(x_1, x_2, \ldots, x_n) : \sum_1^n x_i^2 \ge c\right\}$ is then a best critical region for testing the simple hypothesis $H_0 : \theta = \theta'$ against the simple hypothesis $\theta = \theta''$. It remains to determine c, so that this critical region has the desired size α. If H_0 is true, the random variable $\sum_1^n X_i^2/\theta'$ has a chi-square distribution with n degrees of freedom. Since $\alpha = \Pr\left(\sum_1^n X_i^2/\theta' \ge c/\theta'; H_0\right)$, c/θ' may be read from Table II in Appendix B and c determined. Then $C = \left\{(x_1, x_2, \ldots, x_n) : \sum_1^n x_i^2 \ge c\right\}$ is a best critical region of size α for testing $H_0 : \theta = \theta'$ against the hypothesis $\theta = \theta''$. Moreover, for each number θ'' greater than θ', the foregoing argument holds. That is, if θ''' is another number greater than θ', then $C = \left\{(x_1, \ldots, x_n) : \sum_1^n x_i^2 \ge c\right\}$ is a best critical region of size α for testing $H_0 : \theta = \theta'$ against the hypothesis $\theta = \theta'''$. Accordingly, $C = \left\{(x_1, \ldots, x_n) : \sum_1^n x_i^2 \ge c\right\}$ is a uniformly most powerful critical region of size α for testing $H_0 : \theta = \theta'$ against $H_1 : \theta > \theta'$. If $x_1, x_2, \ldots, x_n$ denote the experimental values of $X_1, X_2, \ldots, X_n$, then $H_0 : \theta = \theta'$ is rejected at the significance level α, and $H_1 : \theta > \theta'$ is accepted, if $\sum_1^n x_i^2 \ge c$; otherwise, $H_0 : \theta = \theta'$ is accepted.

If, in the preceding discussion, we take $n = 15$, $\alpha = 0.05$, and $\theta' = 3$, then here the two hypotheses will be $H_0 : \theta = 3$ and $H_1 : \theta > 3$. From Table II, $c/3 = 25$ and hence $c = 75$.

Example 3. Let $X_1, X_2, \ldots, X_n$ denote a random sample from a distribution that is $N(\theta, 1)$, where the mean θ is unknown. It will be shown that there is no uniformly most powerful test of the simple hypothesis $H_0 : \theta = \theta'$, where θ' is a fixed number, against the alternative composite hypothesis $H_1 : \theta \ne \theta'$. Thus $\Omega = \{\theta : -\infty < \theta < \infty\}$. Let θ'' be a number not equal to θ'. Let k be a positive number and consider

$$\frac{(1/2\pi)^{n/2}\exp\left[-\sum_{1}^{n}(x_i-\theta')^2/2\right]}{(1/2\pi)^{n/2}\exp\left[-\sum_{1}^{n}(x_i-\theta'')^2/2\right]}\le k.$$

The preceding inequality may be written as

$$\exp\left\{-(\theta''-\theta')\sum_{1}^{n}x_i+\frac{n}{2}[(\theta'')^2-(\theta')^2]\right\}\le k$$

or

$$(\theta''-\theta')\sum_{1}^{n}x_i\ge\frac{n}{2}[(\theta'')^2-(\theta')^2]-\ln k.$$

This last inequality is equivalent to

$$\sum_{1}^{n}x_i\ge\frac{n}{2}(\theta''+\theta')-\frac{\ln k}{\theta''-\theta'},$$

provided that $\theta''>\theta'$, and it is equivalent to

$$\sum_{1}^{n}x_i\le\frac{n}{2}(\theta''+\theta')-\frac{\ln k}{\theta''-\theta'}$$

if $\theta''<\theta'$. The first of these two expressions defines a best critical region for testing $H_0:\theta=\theta'$ against the hypothesis $\theta=\theta''$ provided that $\theta''>\theta'$, while the second expression defines a best critical region for testing $H_0:\theta=\theta'$ against the hypothesis $\theta=\theta''$ provided that $\theta''<\theta'$. That is, a best critical region for testing the simple hypothesis against an alternative simple hypothesis, say $\theta=\theta'+1$, will not serve as a best critical region for testing $H_0:\theta=\theta'$ against the alternative simple hypothesis $\theta=\theta'-1$, say. By definition, then, there is no uniformly most powerful test in the case under consideration.

It should be noted that had the alternative composite hypothesis been either $H_1:\theta>\theta'$ or $H_1:\theta<\theta'$, a uniformly most powerful test would exist in each instance.

Example 4. In Exercise 9.10 the reader was asked to show that if a random sample of size $n=10$ is taken from a Poisson distribution with mean θ, the critical region defined by $\sum_{1}^{10}x_i\ge 3$ is a best critical region for testing $H_0:\theta=0.1$ against $H_1:\theta=0.5$. This critical region is also a uniformly most powerful one for testing $H_0:\theta=0.1$ against $H_1:\theta>0.1$ because, with $\theta''>0.1$,

$$\frac{(0.1)^{\Sigma x_i}e^{-10(0.1)}/(x_1!\,x_2!\cdots x_n!)}{(\theta'')^{\Sigma x_i}e^{-10(\theta'')}/(x_1!\,x_2!\cdots x_n!)}\le k$$

is equivalent to

$$\left(\frac{0.1}{\theta''}\right)^{\Sigma x_i} e^{-10(0.1-\theta'')} \le k.$$

The preceding inequality may be written as

$$\left(\sum_1^n x_i\right)(\ln 0.1 - \ln \theta'') \le \ln k + 10(0.1 - \theta'')$$

or, since $\theta'' > 0.1$, equivalently as

$$\sum_1^n x_i \ge \frac{\ln k + 1 - 10\theta''}{\ln 0.1 - \ln \theta''}.$$

Of course, $\sum_1^{10} x_i \ge 3$ is of the latter form.

Let us make an observation, although obvious when pointed out, that is important. Let $X_1, X_2, \ldots, X_n$ denote a random sample from a distribution that has p.d.f. $f(x; \theta)$, $\theta \in \Omega$. Suppose that $Y = u(X_1, X_2, \ldots, X_n)$ is a sufficient statistic for θ. In accordance with the factorization theorem, the joint p.d.f. of $X_1, X_2, \ldots, X_n$ may be written

$$L(\theta; x_1, x_2, \ldots, x_n) = k_1[u(x_1, x_2, \ldots, x_n); \theta]k_2(x_1, x_2, \ldots, x_n),$$

where $k_2(x_1, x_2, \ldots, x_n)$ does not depend upon θ. Consequently, the ratio

$$\frac{L(\theta'; x_1, x_2, \ldots, x_n)}{L(\theta''; x_1, x_2, \ldots, x_n)} = \frac{k_1[u(x_1, x_2, \ldots, x_n); \theta']}{k_1[u(x_1, x_2, \ldots, x_n); \theta'']}$$

depends upon $x_1, x_2, \ldots, x_n$ only through $u(x_1, x_2, \ldots, x_n)$. Accordingly, if there is a sufficient statistic $Y = u(X_1, X_2, \ldots, X_n)$ for θ and if a best test or a uniformly most powerful test is desired, there is no need to consider tests which are based upon any statistic other than the sufficient statistic. This result supports the importance of sufficiency.

Often, when $\theta'' < \theta'$ the ratio

$$\frac{L(\theta'; x_1, x_2, \ldots, x_n)}{L(\theta''; x_1, x_2, \ldots, x_n)},$$

which depends upon $x_1, x_2, \ldots, x_n$ only through $y = u(x_1, x_2, \ldots, x_n)$, is an increasing function of $y = u(x_1, x_2, \ldots, x_n)$. In such a case we say that we have a *monotone likelihood ratio* in the statistic $Y = u(X_1, X_2, \ldots, X_n)$.

Example 5. Let $X_1, X_2, \ldots, X_n$ be a random sample from a Bernoulli distribution with parameter $p = \theta$, where $0 < \theta < 1$. Let $\theta'' < \theta'$. Then the ratio

$$\frac{L(\theta'; x_1, x_2, \ldots, x_n)}{L(\theta''; x_1, x_2, \ldots, x_n)} = \frac{(\theta')^{\Sigma x_i}(1 - \theta')^{n - \Sigma x_i}}{(\theta'')^{\Sigma x_i}(1 - \theta'')^{n - \Sigma x_i}} = \left[\frac{\theta'(1 - \theta'')}{\theta''(1 - \theta')}\right]^{\Sigma x_i}\left(\frac{1 - \theta'}{1 - \theta''}\right)^n.$$

Since $\theta'/\theta'' > 1$ and $(1 - \theta'')/(1 - \theta') > 1$, so that $\theta'(1 - \theta'')/\theta''(1 - \theta') > 1$, the ratio is an increasing function of $y = \Sigma x_i$. Thus we have a monotone likelihood ratio in the statistic $Y = \Sigma X_i$.

We can generalize Example 5 by noting the following. Suppose that the random sample $X_1, X_2, \ldots, X_n$ arises from a p.d.f. representing a regular case of the exponential class, namely

$$\begin{aligned} f(x; \theta) &= \exp[p(\theta)K(x) + S(x) + q(\theta)], \qquad x \in \mathscr{A}, \\ &= 0 \qquad \text{elsewhere,} \end{aligned}$$

where the space $\mathscr{A}$ of X is free of θ. Further assume that $p(\theta)$ is an increasing function of θ. Then

$$\begin{aligned} \frac{L(\theta')}{L(\theta'')} &= \frac{\exp\left[p(\theta')\sum_{i=1}^{n} K(x_i) + \sum_{i=1}^{n} S(x_i) + nq(\theta')\right]}{\exp\left[p(\theta'')\sum_{i=1}^{n} K(x_i) + \sum_{i=1}^{n} S(x_i) + nq(\theta'')\right]} \\ &= \exp\left\{[p(\theta') - p(\theta'')]\sum_{i=1}^{n} K(x_i) + n[q(\theta') - q(\theta'')]\right\}. \end{aligned}$$

If $\theta'' < \theta'$, $p(\theta)$ being an increasing function requires this ratio to be an increasing function of $y = \sum_{i=1}^{n} K(x_i)$. Thus we have a monotone likelihood ratio in the statistic $Y = \sum_{i=1}^{n} K(X_i)$. Moreover, if we test $H_0 : \theta = \theta'$ against $H_1 : \theta < \theta'$, then, with $\theta'' < \theta'$, we see that

$$\frac{L(\theta')}{L(\theta'')} \le k$$

is equivalent to $\Sigma K(x_i) \le c$ for every $\theta'' < \theta'$. That is, this provides a uniformly most powerful critical region.

If, in the preceding situation with monotone likelihood ratio, we test $H_0 : \theta = \theta'$ against $H_1 : \theta > \theta'$, then $\Sigma K(x_i) \ge c$ would be a uniformly most powerful critical region. From the likelihood ratios displayed in Examples 2, 3, 4, and 5 we see immediately that the respective critical regions

$$\sum_{i=1}^{n} x_i^2 \geq c, \qquad \sum_{i=1}^{n} x_i \geq c, \qquad \sum_{i=1}^{n} x_i \geq c, \qquad \sum_{i=1}^{n} x_i \geq c$$

are uniformly most powerful for testing $H_0 : \theta = \theta'$ against $H_1 : \theta > \theta'$.

There is a final remark that should be made about uniformly most powerful tests. Of course, in Definition 2, the word *uniformly* is associated with θ; that is, C is a best critical region of size α for testing $H_0 : \theta = \theta_0$ against all θ values given by the composite alternative H_1. However, suppose that the form of such a region is

$$u(x_1, x_2, \ldots, x_n) \leq c.$$

Then this form provides uniformly most powerful critical regions for all attainable α values by, of course, appropriately changing the value of c. That is, there is a certain uniformity property, also associated with α, that is not always noted in statistics texts.

EXERCISES

9.11. Let X have the p.d.f. $f(x; \theta) = \theta^x(1 - \theta)^{1-x}$, $x = 0, 1$, zero elsewhere. We test the simple hypothesis $H_0 : \theta = \frac{1}{4}$ against the alternative composite hypothesis $H_1 : \theta < \frac{1}{4}$ by taking a random sample of size 10 and rejecting $H_0 : \theta = \frac{1}{4}$ if and only if the observed values $x_1, x_2, \ldots, x_{10}$ of the sample observations are such that $\sum_{1}^{10} x_i \leq 1$. Find the power function $K(\theta)$, $0 < \theta \leq \frac{1}{4}$, of this test.

9.12. Let X have a p.d.f. of the form $f(x; \theta) = 1/\theta, 0 < x < \theta$, zero elsewhere. Let $Y_1 < Y_2 < Y_3 < Y_4$ denote the order statistics of a random sample of size 4 from this distribution. Let the observed value of Y_4 be y_4. We reject $H_0 : \theta = 1$ and accept $H_1 : \theta \neq 1$ if either $y_4 \leq \frac{1}{2}$ or $y_4 \geq 1$. Find the power function $K(\theta)$, $0 < \theta$, of the test.

9.13. Consider a normal distribution of the form $N(\theta, 4)$. The simple hypothesis $H_0 : \theta = 0$ is rejected, and the alternative composite hypothesis $H_1 : \theta > 0$ is accepted if and only if the observed mean $\bar{x}$ of a random sample of size 25 is greater than or equal to $\frac{3}{5}$. Find the power function $K(\theta)$, $0 \leq \theta$, of this test.

9.14. Consider the two normal distributions $N(\mu_1, 400)$ and $N(\mu_2, 225)$. Let $\theta = \mu_1 - \mu_2$. Let $\bar{x}$ and $\bar{y}$ denote the observed means of two independent random samples, each of size n, from these two distributions. We reject $H_0 : \theta = 0$ and accept $H_1 : \theta > 0$ if and only if $\bar{x} - \bar{y} \geq c$. If $K(\theta)$ is the power function of this test, find n and c so that $K(0) = 0.05$ and $K(10) = 0.90$, approximately.

9.15. If, in Example 2 of this section, $H_0 : \theta = \theta'$, where θ' is a fixed positive number, and $H_1 : \theta < \theta'$, show that the set $\left\{(x_1, x_2, \ldots, x_n) : \sum_1^n x_i^2 \le c\right\}$ is a uniformly most powerful critical region for testing H_0 against H_1.

9.16. If, in Example 2 of this section, $H_0 : \theta = \theta'$, where θ' is a fixed positive number, and $H_1 : \theta \ne \theta'$, show that there is no uniformly most powerful test for testing H_0 against H_1.

9.17. Let $X_1, X_2, \ldots, X_{25}$ denote a random sample of size 25 from a normal distribution $N(\theta, 100)$. Find a uniformly most powerful critical region of size $\alpha = 0.10$ for testing $H_0 : \theta = 75$ against $H_1 : \theta > 75$.

9.18. Let $X_1, X_2, \ldots, X_n$ denote a random sample from a normal distribution $N(\theta, 16)$. Find the sample size n and a uniformly most powerful test of $H_0 : \theta = 25$ against $H_1 : \theta < 25$ with power function $K(\theta)$ so that approximately $K(25) = 0.10$ and $K(23) = 0.90$.

9.19. Consider a distribution having a p.d.f. of the form $f(x; \theta) = \theta^x(1 - \theta)^{1-x}$, $x = 0, 1$, zero elsewhere. Let $H_0 : \theta = \frac{1}{20}$ and $H_1 : \theta > \frac{1}{20}$. Use the central limit theorem to determine the sample size n of a random sample so that a uniformly most powerful test of H_0 against H_1 has a power function $K(\theta)$, with approximately $K(\frac{1}{20}) = 0.05$ and $K(\frac{1}{10}) = 0.90$.

9.20. Illustrative Example 1 of this section dealt with a random sample of size $n = 2$ from a gamma distribution with $\alpha = 1$, $\beta = \theta$. Thus the m.g.f. of the distribution is $(1 - \theta t)^{-1}$, $t < 1/\theta$, $\theta \ge 2$. Let $Z = X_1 + X_2$. Show that Z has a gamma distribution with $\alpha = 2$, $\beta = \theta$. Express the power function $K(\theta)$ of Example 1 in terms of a single integral. Generalize this for a random sample of size n.

9.21. Let $X_1, X_2, \ldots, X_n$ be a random sample from a distribution with p.d.f. $f(x; \theta) = \theta x^{\theta - 1}$, $0 < x < \infty$, zero elsewhere, where $\theta > 0$. Find a sufficient statistic for θ and show that a uniformly most powerful test of $H_0 : \theta = 6$ against $H_1 : \theta < 6$ is based on this statistic.

9.22. Let X have the p.d.f. $f(x; \theta) = \theta^x(1 - \theta)^{1-x}$, $x = 0, 1$, zero elsewhere. We test $H_0 : \theta = \frac{1}{2}$ against $H_1 : \theta < \frac{1}{2}$ by taking a random sample $X_1, X_2, \ldots, X_5$ of size $n = 5$ and rejecting H_0 if $Y = \sum_1^5 X_i$ is observed to be less than or equal to a constant c.

(a) Show that this is a uniformly most powerful test.
(b) Find the significance level when $c = 1$.
(c) Find the significance level when $c = 0$.
(d) By using a *randomized test*, modify the tests given in parts (b) and (c) to find a test with significance level $\alpha = \frac{2}{32}$.

9.3 Likelihood Ratio Tests

The notion of using the magnitude of the ratio of two probability density functions as the basis of a best test or of a uniformly most powerful test can be modified, and made intuitively appealing, to provide a method of constructing a test of a composite hypothesis against an alternative composite hypothesis or of constructing a test of a simple hypothesis against an alternative composite hypothesis when a uniformly most powerful test does not exist. This method leads to tests called *likelihood ratio tests.* A likelihood ratio test, as just remarked, is not necessarily a uniformly most powerful test, but it has been proved in the literature that such a test often has desirable properties.

A certain terminology and notation will be introduced by means of an example.

Example 1. Let the random variable X be $N(\theta_1, \theta_2)$ and let the parameter space be $\Omega = \{(\theta_1, \theta_2) : -\infty < \theta_1 < \infty, 0 < \theta_2 < \infty\}$. Let the composite hypothesis be $H_0 : \theta_1 = 0, \theta_2 > 0$, and let the alternative composite hypothesis be $H_1 : \theta_1 \neq 0, \theta_2 > 0$. The set $\omega = \{(\theta_1, \theta_2) : \theta_1 = 0, 0 < \theta_2 < \infty\}$ is a subset of Ω and will be called the *subspace* specified by the hypothesis H_0. Then, for instance, the hypothesis H_0 may be described as $H_0 : (\theta_1, \theta_2) \in \omega$. It is proposed that we test H_0 against all alternatives in H_1.

Let $X_1, X_2, \ldots, X_n$ denote a random sample of size $n > 1$ from the distribution of this example. The joint p.d.f. of $X_1, X_2, \ldots, X_n$ is, at each point in Ω,

$$L(\theta_1, \theta_2; x_1, \ldots, x_n) = \left(\frac{1}{2\pi\theta_2}\right)^{n/2} \exp\left[-\frac{\sum_1^n (x_i - \theta_1)^2}{2\theta_2}\right] = L(\Omega).$$

At each point $(\theta_1, \theta_2) \in \omega$, the joint p.d.f. of $X_1, X_2, \ldots, X_n$ is

$$L(0, \theta_2; x_1, \ldots, x_n) = \left(\frac{1}{2\pi\theta_2}\right)^{n/2} \exp\left[-\frac{\sum_1^n x_i^2}{2\theta_2}\right] = L(\omega).$$

The joint p.d.f., now denoted by $L(\omega)$, is not completely specified, since θ_2 may be any positive number; nor is the joint p.d.f., now denoted by $L(\Omega)$, completely specified, since θ_1 may be any real number and θ_2 any positive number. Thus the ratio of $L(\omega)$ to $L(\Omega)$ could not provide a basis for a test of H_0 against H_1. Suppose, however, that we modify this ratio in the following manner. We shall find the maximum of $L(\omega)$ in ω, that is, the maximum of $L(\omega)$ with respect to θ_2. And we shall find the maximum of

$L(\Omega)$ in Ω, that is, the maximum of $L(\Omega)$ with respect to θ_1 and θ_2. The ratio of these maxima will be taken as the criterion for a test of H_0 against H_1. Let the maximum of $L(\omega)$ in ω be denoted by $L(\hat{\omega})$ and let the maximum of $L(\Omega)$ in Ω be denoted by $L(\hat{\Omega})$. Then the criterion for the test of H_0 against H_1 is the likelihood ratio

$$\lambda(x_1, x_2, \ldots, x_n) = \lambda = \frac{L(\hat{\omega})}{L(\hat{\Omega})}.$$

Since $L(\omega)$ and $L(\Omega)$ are probability density functions, $\lambda \geq 0$; and since ω is a subset of Ω, $\lambda \leq 1$.

In our example the maximum, $L(\hat{\omega})$, of $L(\omega)$ is obtained by first setting

$$\frac{d \ln L(\omega)}{d\theta_2} = -\frac{n}{2\theta_2} + \frac{\sum_1^n x_i^2}{2\theta_2^2}$$

equal to zero and solving for θ_2. The solution of θ_2 is $\sum_1^n x_i^2/n$, and this number maximizes $L(\omega)$. Thus the maximum is

$$L(\hat{\omega}) = \left(\frac{1}{2\pi \sum_1^n x_i^2/n}\right)^{n/2} \exp\left[-\frac{\sum_1^n x_i^2}{2\sum_1^n x_i^2/n}\right]$$

$$= \left(\frac{ne^{-1}}{2\pi \sum_1^n x_i^2}\right)^{n/2}.$$

On the other hand, by using Example 4, Section 6.1, the maximum, $L(\hat{\Omega})$, of $L(\Omega)$ is obtained by replacing θ_1 and θ_2 by $\sum_1^n x_i/n = \bar{x}$ and $\sum_1^n (x_i - \bar{x})^2/n$, respectively. That is

$$L(\hat{\Omega}) = \left[\frac{1}{2\pi \sum_1^n (x_i - \bar{x})^2/n}\right]^{n/2} \exp\left[-\frac{\sum_1^n (x_i - \bar{x})^2}{2\sum_1^n (x_i - \bar{x})^2/n}\right]$$

$$= \left[\frac{ne^{-1}}{2\pi \sum_1^n (x_i - \bar{x})^2}\right]^{n/2}.$$

Thus here

$$\lambda = \left[\frac{\sum_1^n (x_i - \bar{x})^2}{\sum_1^n x_i^2}\right]^{n/2}.$$

Because $\sum_1^n x_i^2 = \sum_1^n (x_i - \bar{x})^2 + n\bar{x}^2$, λ may be written

$$\lambda = \frac{1}{\left\{1 + \left[n\bar{x}^2 \Big/ \sum_1^n (x_i - \bar{x})^2\right]\right\}^{n/2}}.$$

Now the hypothesis H_0 is $\theta_1 = 0$, $\theta_2 > 0$. If the observed number $\bar{x}$ were zero, the experiment tends to confirm H_0. But if $\bar{x} = 0$ and $\sum_1^n x_i^2 > 0$, then $\lambda = 1$. On the other hand, if $\bar{x}$ and $n\bar{x}^2 \Big/ \sum_1^n (x_i - \bar{x})^2$ deviate considerably from zero, the experiment tends to negate H_0. Now the greater the deviation of $n\bar{x}^2 \Big/ \sum_1^n (x_i - \bar{x})^2$ from zero, the smaller λ becomes. That is, if λ is used as a test criterion, then an intuitively appealing critical region for testing H_0 is a set defined by $0 \le \lambda \le \lambda_0$, where λ_0 is a positive proper fraction. Thus we reject H_0 if $\lambda \le \lambda_0$. A test that has the critical region $\lambda \le \lambda_0$ is a *likelihood ratio test*. In this example $\lambda \le \lambda_0$ when and only when

$$\frac{\sqrt{n}|\bar{x}|}{\sqrt{\sum_1^n (x_i - \bar{x})^2/(n-1)}} \ge \sqrt{(n-1)(\lambda_0^{-2/n} - 1)} = c.$$

If $H_0 : \theta_1 = 0$ is true, the results in Section 4.8 show that the statistic

$$t(X_1, X_2, \ldots, X_n) = \frac{\sqrt{n}(\bar{X} - 0)}{\sqrt{\sum_1^n (X_i - \bar{X})^2/(n-1)}}$$

has a t-distribution with $n - 1$ degrees of freedom. Accordingly, in this example the likelihood ratio test of H_0 against H_1 may be based on a T-statistic. For a given positive integer n, Table IV in Appendix B may be used (with $n - 1$ degrees of freedom) to determine the number c such that $\alpha = \Pr [|t(X_1, X_2, \ldots, X_n)| \ge c; H_0]$ is the desired significance level of the test. If the experimental values of $X_1, X_2, \ldots, X_n$ are, respectively, $x_1, x_2, \ldots, x_n$, then we reject H_0 if and only if $|t(x_1, x_2, \ldots, x_n)| \ge c$. If, for instance, $n = 6$ and $\alpha = 0.05$, then from Table IV, $c = 2.571$.

The preceding example should make the following generalization easier to read: Let $X_1, X_2, \ldots, X_n$ denote n independent random variables having, respectively, the probability density functions $f_i(x_i; \theta_1, \theta_2, \ldots, \theta_m)$, $i = 1, 2, \ldots, n$. The set that consists of all parameter points $(\theta_1, \theta_2, \ldots, \theta_m)$ is denoted by Ω, which we have called the parameter space. Let ω be a subset of the parameter space Ω. We wish to test the (simple or composite) hypothesis

$H_0 : (\theta_1, \theta_2, \ldots, \theta_m) \in \omega$ against all alternative hypotheses. Define the likelihood functions

$$L(\omega) = \prod_{i=1}^{n} f_i(x_i; \theta_1, \theta_2, \ldots, \theta_m), \qquad (\theta_1, \theta_2, \ldots, \theta_m) \in \omega,$$

and

$$L(\Omega) = \prod_{i=1}^{n} f_i(x_i; \theta_1, \theta_2, \ldots, \theta_m), \qquad (\theta_1, \theta_2, \ldots, \theta_m) \in \Omega.$$

Let $L(\hat{\omega})$ and $L(\hat{\Omega})$ be the maxima, which we assume to exist, of these two likelihood functions. The ratio of $L(\hat{\omega})$ to $L(\hat{\Omega})$ is called the *likelihood ratio* and is denoted by

$$\lambda(x_1, x_2, \ldots, x_n) = \lambda = \frac{L(\hat{\omega})}{L(\hat{\Omega})}.$$

Let λ_0 be a positive proper function. The *likelihood ratio test principle* states that the hypothesis $H_0 : (\theta_1, \theta_2, \ldots, \theta_m) \in \omega$ is rejected if and only if

$$\lambda(x_1, x_2, \ldots, x_n) = \lambda \le \lambda_0.$$

The function λ defines a random variable $\lambda(X_1, X_2, \ldots, X_n)$, and the significance level of the test is given by

$$\alpha = \Pr[\lambda(X_1, X_2, \ldots, X_n) \le \lambda_0; H_0].$$

The likelihood ratio test principle is an intuitive one. However, the principle does lead to the same test, when testing a simple hypothesis H_0 against an alternative simple hypothesis H_1, as that given by the Neyman–Pearson theorem (Exercise 9.25). Thus it might be expected that a test based on this principle has some desirable properties.

An example of the preceding generalization will be given.

Example 2. Let the independent random variables X and Y have distributions that are $N(\theta_1, \theta_3)$ and $N(\theta_2, \theta_3)$, where the means θ_1 and θ_2 and common variance θ_3 are unknown. Then $\Omega = \{(\theta_1, \theta_2, \theta_3) : -\infty < \theta_1 < \infty, -\infty < \theta_2 < \infty, 0 < \theta_3 < \infty\}$. Let $X_1, X_2, \ldots, X_n$ and $Y_1, Y_2, \ldots, Y_m$ denote independent random samples from these distributions. The hypothesis $H_0 : \theta_1 = \theta_2$, unspecified, and θ_3 unspecified, is to be tested against all alternatives. Then $\omega = \{(\theta_1, \theta_2, \theta_3) : -\infty < \theta_1 = \theta_2 < \infty, 0 < \theta_3 < \infty\}$. Here

$X_1, X_2, \ldots, X_n, Y_1, Y_2, \ldots, Y_m$ are $n + m > 2$ mutually independent random variables having the likelihood functions

$$L(\omega) = \left(\frac{1}{2\pi\theta_3}\right)^{(n+m)/2} \exp\left[-\frac{\sum_1^n (x_i - \theta_1)^2 + \sum_1^m (y_i - \theta_1)^2}{2\theta_3}\right]$$

and

$$L(\Omega) = \left(\frac{1}{2\pi\theta_3}\right)^{(n+m)/2} \exp\left[-\frac{\sum_1^n (x_i - \theta_1)^2 + \sum_1^m (y_i - \theta_2)^2}{2\theta_3}\right].$$

If

$$\frac{\partial \ln L(\omega)}{\partial \theta_1} \quad \text{and} \quad \frac{\partial \ln L(\omega)}{\partial \theta_3}$$

are equated to zero, then (Exercise 9.26)

$$\sum_1^n (x_i - \theta_1) + \sum_1^m (y_i - \theta_1) = 0, \tag{1}$$

$$-(n + m) + \frac{1}{\theta_3}\left[\sum_1^n (x_i - \theta_1)^2 + \sum_1^m (y_i - \theta_1)^2\right] = 0.$$

The solutions for θ_1 and θ_3 are, respectively,

$$u = \frac{\sum_1^n x_i + \sum_1^m y_i}{n + m}$$

and

$$w = \frac{\sum_1^n (x_i - u)^2 + \sum_1^m (y_i - u)^2}{n + m},$$

and u and w maximize $L(\omega)$. The maximum is

$$L(\hat{\omega}) = \left(\frac{e^{-1}}{2\pi w}\right)^{(n+m)/2}.$$

In like manner, if

$$\frac{\partial \ln L(\Omega)}{\partial \theta_1}, \quad \frac{\partial \ln L(\Omega)}{\partial \theta_2}, \quad \frac{\partial \ln L(\Omega)}{\partial \theta_3}$$

are equated to zero, then (Exercise 9.27)

$$\sum_1^n (x_i - \theta_1) = 0,$$

$$\sum_1^m (y_i - \theta_2) = 0, \tag{2}$$

$$-(n+m) + \frac{1}{\theta_3}\left[\sum_1^n (x_i - \theta_1)^2 + \sum_1^m (y_i - \theta_2)^2\right] = 0.$$

The solutions for θ_1, θ_2, and θ_3 are, respectively,

$$u_1 = \frac{\sum_1^n x_i}{n},$$

$$u_2 = \frac{\sum_1^m y_i}{m},$$

$$w' = \frac{\sum_1^n (x_i - u_1)^2 + \sum_1^m (y_i - u_2)^2}{n+m},$$

and u_1, u_2, and w' maximize $L(\Omega)$. The maximum is

$$L(\hat{\Omega}) = \left(\frac{e^{-1}}{2\pi w'}\right)^{(n+m)/2},$$

so that

$$\lambda(x_1, \ldots, x_n, y_1, \ldots, y_m) = \lambda = \frac{L(\hat{\omega})}{L(\hat{\Omega})} = \left(\frac{w'}{w}\right)^{(n+m)/2}.$$

The random variable defined by $\lambda^{2/(n+m)}$ is

$$\frac{\sum_1^n (X_i - \bar{X})^2 + \sum_1^m (Y_i - \bar{Y})^2}{\sum_1^n \{X_i - [(n\bar{X} + m\bar{Y})/(n+m)]\}^2 + \sum_1^m \{Y_i - [(n\bar{X} + m\bar{Y})/(n+m)]\}^2}.$$

Now

$$\sum_1^n \left(X_i - \frac{n\bar{X} + m\bar{Y}}{n+m}\right)^2 = \sum_1^n \left[(X_i - \bar{X}) + \left(\bar{X} - \frac{n\bar{X} + m\bar{Y}}{n+m}\right)\right]^2$$

$$= \sum_1^n (X_i - \bar{X})^2 + n\left(\bar{X} - \frac{n\bar{X} + m\bar{Y}}{n+m}\right)^2$$

and

$$\sum_1^m \left(Y_i - \frac{n\bar{X} + m\bar{Y}}{n+m}\right)^2 = \sum_1^m \left[(Y_i - \bar{Y}) + \left(\bar{Y} - \frac{n\bar{X} + m\bar{Y}}{n+m}\right)\right]^2$$

$$= \sum_1^m (Y_i - \bar{Y})^2 + m\left(\bar{Y} - \frac{n\bar{X} + m\bar{Y}}{n+m}\right)^2.$$

But

$$n\left(\bar{X} - \frac{n\bar{X} + m\bar{Y}}{n+m}\right)^2 = \frac{m^2 n}{(n+m)^2}(\bar{X} - \bar{Y})^2$$

and

$$m\left(\bar{Y} - \frac{n\bar{X} + m\bar{Y}}{n+m}\right)^2 = \frac{n^2 m}{(n+m)^2}(\bar{X} - \bar{Y})^2.$$

Hence the random variable defined by $\lambda^{2/(n+m)}$ may be written

$$\frac{\sum_1^n (X_i - \bar{X})^2 + \sum_1^m (Y_i - \bar{Y})^2}{\sum_1^n (X_i - \bar{X})^2 + \sum_1^m (Y_i - \bar{Y})^2 + [nm/(n+m)](\bar{X} - \bar{Y})^2}$$

$$= \frac{1}{1 + \dfrac{[nm/(n+m)](\bar{X} - \bar{Y})^2}{\sum_1^n (X_i - \bar{X})^2 + \sum_1^m (Y_i - \bar{Y})^2}}.$$

If the hypothesis $H_0 : \theta_1 = \theta_2$ is true, the random variable

$$T = \frac{\sqrt{\dfrac{nm}{n+m}}(\bar{X} - \bar{Y})}{\sqrt{\dfrac{\sum_1^n (X_i - \bar{X})^2 + \sum_1^m (Y_i - \bar{Y})^2}{n+m-2}}}$$

has, in accordance with Section 6.3, a t-distribution with $n + m - 2$ degrees of freedom. Thus the random variable defined by $\lambda^{2/(n+m)}$ is

$$\frac{n+m-2}{(n+m-2) + T^2}.$$

The test of H_0 against all alternatives may then be based on a t-distribution with $n + m - 2$ degrees of freedom.

The likelihood ratio principle calls for the rejection of H_0 if and only if $\lambda \le \lambda_0 < 1$. Thus the significance level of the test is

$$\alpha = \Pr\,[\lambda(X_1, \ldots, X_n, Y_1, \ldots, Y_m) \le \lambda_0; H_0].$$

However, $\lambda(X_1, \ldots, X_n, Y_1, \ldots, Y_m) \le \lambda_0$ is equivalent to $|T| \ge c$, and so

$$\alpha = \Pr(|T| \ge c; H_0).$$

For given values of n and m, the number c is determined from Table IV in Appendix B (with $n + m - 2$ degrees of freedom) in such a manner as to yield a desired α. Then H_0 is rejected at a significance level α if and only if $|t| \ge c$, where t is the experimental value of T. If, for instance, $n = 10$, $m = 6$, and $\alpha = 0.05$, then $c = 2.145$.

In each of the two examples of this section it was found that the likelihood ratio test could be based on a statistic which, when the hypothesis H_0 is true, has a t-distribution. To help us compute the powers of these tests at parameter points other than those described by the hypothesis H_0, we turn to the following definition.

Definition 3. Let the random variable W be $N(\delta, 1)$; let the random variable V be $\chi^2(r)$, and W and V be independent. The quotient

$$T = \frac{W}{\sqrt{V/r}}$$

is said to have a *noncentral t-distribution* with r degrees of freedom and noncentrality parameter δ. If $\delta = 0$, we say that T has a central t-distribution.

In the light of this definition, let us reexamine the statistics of the examples of this section. In Example 1 we had

$$t(X_1, \ldots, X_n) = \frac{\sqrt{n}\,\bar{X}}{\sqrt{\sum_1^n (X_i - \bar{X})^2/(n-1)}}$$

$$= \frac{\sqrt{n}\,\bar{X}/\sigma}{\sqrt{\sum_1^n (X_i - \bar{X})^2/[\sigma^2(n-1)]}}.$$

Here $W_1 = \sqrt{n}\,\bar{X}/\sigma$ is $N(\sqrt{n}\,\theta_1/\sigma, 1)$, $V_1 = \sum_1^n (X_i - \bar{X})^2/\sigma^2$ is $\chi^2(n-1)$, and W_1 and V_1 are independent. Thus, if $\theta_1 \ne 0$, we see, in accordance with the definition, that $t(X_1, \ldots, X_n)$ has a noncentral t-distribution with $n - 1$ degrees of freedom and noncentrality parameter $\delta_1 = \sqrt{n}\,\theta_1/\sigma$. In Example 2 we had

$$T = \frac{W_2}{\sqrt{V_2/(n+m-2)}},$$

where

$$W_2 = \sqrt{\frac{nm}{n+m}}\,(\bar{X} - \bar{Y}) \Big/ \sigma$$

and

$$V_2 = \frac{\sum_1^n (X_i - \bar{X})^2 + \sum_1^m (Y_i - \bar{Y})^2}{\sigma^2}.$$

Here W_2 is $N[\sqrt{nm/(n+m)}(\theta_1 - \theta_2)/\sigma, 1]$, V_2 is $\chi^2(n+m-2)$, and W_2 and V_2 are independent. Accordingly, if $\theta_1 \neq \theta_2$, T has a noncentral t-distribution with $n+m-2$ degrees of freedom and noncentrality parameter $\delta_2 = \sqrt{nm/(n+m)}(\theta_1 - \theta_2)/\sigma$. It is interesting to note that $\delta_1 = \sqrt{n}\,\theta_1/\sigma$ measures the deviation of θ_1 from $\theta_1 = 0$ in units of the standard deviation $\sigma/\sqrt{n}$ of $\bar{X}$. The noncentrality parameter $\delta_2 = \sqrt{nm/(n+m)}(\theta_1 - \theta_2)/\sigma$ is equal to the deviation of $\theta_1 - \theta_2$ from $\theta_1 - \theta_2 = 0$ in units of the standard deviation $\sigma\sqrt{(n+m)/nm}$ of $\bar{X} - \bar{Y}$.

There are various tables of the noncentral t-distribution, but they are much too cumbersome to be included in this book. However, with the aid of such tables, we can determine the power functions of these tests as functions of the noncentrality parameters.

In Example 2, in testing the equality of the means of two normal distributions, it was assumed that the unknown variances of the distributions were equal. Let us now consider the problem of testing the equality of these two unknown variances.

Example 3. We are given the independent random samples $X_1, \ldots, X_n$ and $Y_1, \ldots, Y_m$ from the distributions, which are $N(\theta_1, \theta_3)$ and $N(\theta_2, \theta_4)$, respectively. We have

$$\Omega = \{(\theta_1, \theta_2, \theta_3, \theta_4) : -\infty < \theta_1, \theta_2 < \infty, 0 < \theta_3, \theta_4 < \infty\}.$$

The hypothesis $H_0 : \theta_3 = \theta_4$, unspecified, with θ_1 and θ_2 also unspecified, is to be tested against all alternatives. Then

$$\omega = \{(\theta_1, \theta_2, \theta_3, \theta_4) : -\infty < \theta_1, \theta_2 < \infty, 0 < \theta_3 = \theta_4 < \infty\}.$$

It is easy to show (see Exercise 9.30) that the statistic defined by $\lambda = L(\hat{\omega})/L(\hat{\Omega})$ is a function of the statistic

$$F = \frac{\sum_1^n (X_i - \bar{X})^2/(n-1)}{\sum_1^m (Y_i - \bar{Y})^2/(m-1)}.$$

If $\theta_3 = \theta_4$, this statistic F has an F-distribution with $n - 1$ and $m - 1$ degrees of freedom. The hypothesis that $(\theta_1, \theta_2, \theta_3, \theta_4) \in \omega$ is rejected if the computed $F \le c_1$ or if the computed $F \ge c_2$. The constants c_1 and c_2 are usually selected so that, if $\theta_3 = \theta_4$,

$$\Pr(F \le c_1) = \Pr(F \ge c_2) = \frac{\alpha_1}{2},$$

where α_1 is the desired significance level of this test.

Often, under H_0, it is difficult to determine the distribution of $\lambda = \lambda(X_1, X_2, \ldots, X_n)$ or the distribution of an equivalent statistic upon which to base the likelihood ratio test. Hence it is impossible to find λ_0 such that $\Pr[\lambda \le \lambda_0; H_0]$ equals an appropriate value of α. The fact that the maximum likelihood estimators in a regular case have a joint normal distribution does, however, provide a solution. Using this fact, in a more advanced course, we can show that $-2 \ln \lambda$ has, given H_0 is true, an approximate chi-square distribution with r degrees of freedom, where $r =$ the dimension of $\Omega -$ the dimension of ω. For illustration, in Example 1, the dimension of $\Omega = 2$ and the dimension of $\omega = 1$ and $r = 2 - 1 = 1$.

Also, in that example, note that

$$-2 \ln \lambda = n \ln \left\{1 + \frac{n\bar{x}^2}{\sum (x_i - \bar{x})^2}\right\} = n \ln \left\{1 + \frac{\bar{x}^2}{s^2}\right\}.$$

Hence, with n large so that $\bar{x}^2/s^2$ is close to zero under $H_0 : \theta_1 = 0$, let us approximate the right-hand member by two terms of a Taylor's series expanded about zero:

$$-2 \ln \lambda \approx 0 + \frac{n\bar{x}^2}{s^2}.$$

Since n is large, we can replace n by $n - 1$ to get the approximation

$$-2 \ln \lambda \approx \left(\frac{\bar{x}}{s/\sqrt{n-1}}\right)^2 = t^2.$$

But $T = \bar{X}/(S/\sqrt{n-1})$ under $H_0 : \theta_1 = 0$ has a t-distribution with $n - 1$ degrees of freedom. Moreover, with large $n - 1$, the distribution of T is approximately $N(0, 1)$ and the square of a standardized normal variable is $\chi^2(1)$, which is in agreement with the stated result. Exercise 9.31 provides another illustration of the fact that $-2 \ln \lambda$ has an approximate chi-square distribution.

EXERCISES

9.23. In Example 1 let $n = 10$, and let the experimental values of the random variables yield $\bar{x} = 0.6$ and $\sum_{1}^{10} (x_i - \bar{x})^2 = 3.6$. If the test derived in that example is used, do we accept or reject $H_0: \theta_1 = 0$ at the 5 percent significance level?

9.24. In Example 2 let $n = m = 8$, $\bar{x} = 75.2$, $\bar{y} = 78.6$, $\sum_{1}^{8} (x_i - \bar{x})^2 = 71.2$, $\sum_{1}^{8} (y_i - \bar{y})^2 = 54.8$. If we use the test derived in that example, do we accept or reject $H_0: \theta_1 = \theta_2$ at the 5 percent significance level?

9.25. Show that the likelihood ratio principle leads to the same test, when testing a simple hypothesis H_0 against an alternative simple hypothesis H_1, as that given by the Neyman–Pearson theorem. Note that there are only two points in Ω.

9.26. Verify Equations (1) of Example 2 of this section.

9.27. Verify Equations (2) of Example 2 of this section.

9.28. Let $X_1, X_2, \ldots, X_n$ be a random sample from the normal distribution $N(\theta, 1)$. Show that the likelihood ratio principle for testing $H_0: \theta = \theta'$, where θ' is specified, against $H_1: \theta \neq \theta'$ leads to the inequality $|\bar{x} - \theta'| \geq c$. Is this a uniformly most powerful test of H_0 against H_1?

9.29. Let $X_1, X_2, \ldots, X_n$ be a random sample from the normal distribution $N(\theta_1, \theta_2)$. Show that the likelihood ratio principle for testing $H_0: \theta_2 = \theta_2'$ specified, and θ_1 unspecified, against $H_1: \theta_2 \neq \theta_2'$, θ_1 unspecified, leads to a test that rejects when $\sum_{1}^{n} (x_i - \bar{x})^2 \leq c_1$ or $\sum_{1}^{n} (x_i - \bar{x})^2 \geq c_2$, where $c_1 < c_2$ are selected appropriately.

9.30. Let $X_1, \ldots, X_n$ and $Y_1, \ldots, Y_m$ be independent random samples from the distributions $N(\theta_1, \theta_3)$ and $N(\theta_2, \theta_4)$, respectively.

(a) Show that the likelihood ratio for testing $H_0: \theta_1 = \theta_2, \theta_3 = \theta_4$ against all alternatives is given by

$$\frac{\left[\sum_{1}^{n} (x_i - \bar{x})^2/n\right]^{n/2} \left[\sum_{1}^{m} (y_i - \bar{y})^2/m\right]^{m/2}}{\left\{\left[\sum_{1}^{n} (x_i - u)^2 + \sum_{1}^{m} (y_i - u)^2\right] \Big/ (m + n)\right\}^{(n+m)/2}},$$

where $u = (n\bar{x} + m\bar{y})/(n + m)$.

(b) Show that the likelihood ratio test for testing $H_0: \theta_3 = \theta_4$, θ_1 and θ_2 unspecified, against $H_1: \theta_3 \neq \theta_4$, θ_1 and θ_2 unspecified, can be based on the random variable

$$F = \frac{\sum_1^n (X_i - \bar{X})^2/(n-1)}{\sum_1^m (Y_i - \bar{Y})^2/(m-1)}.$$

(c) If $\theta_3 = \theta_4$, argue that the F-statistic in part (b) is independent of the T-statistic of Example 2 of this section.

9.31. Let n independent trials of an experiment be such that $x_1, x_2, \ldots, x_k$ are the respective numbers of times that the experiment ends in the mutually exclusive and exhaustive events $A_1, A_2, \ldots, A_k$. If $p_i = P(A_i)$ is constant throughout the n trials, then the probability of that particular sequence of trials is $L = p_1^{x_1} p_2^{x_2} \cdots p_k^{x_k}$.

(a) Recalling that $p_1 + p_2 + \cdots + p_k = 1$, show that the likelihood ratio for testing $H_0: p_i = p_{i0} > 0$, $i = 1, 2, \ldots, k$, against all alternatives is given by

$$\lambda = \prod_{i=1}^k \left(\frac{(p_{i0})^{x_i}}{(x_i/n)^{x_i}} \right).$$

(b) Show that

$$-2 \ln \lambda = \sum_{i=1}^k \frac{x_i(x_i - np_{0i})^2}{(np_i')^2},$$

where p_i' is between p_{0i} and x_i/n.

Hint: Expand $\ln p_{i0}$ in a Taylor's series with the remainder in the term involving $(p_{i0} - x_i/n)^2$.

(c) For large n, argue that $x_i/(np_i')^2$ is approximated by $1/(np_{i0})$ and hence

$$-2 \ln \lambda \approx \sum_{i=1}^k \frac{(x_i - np_{0i})^2}{np_{0i}}, \qquad \text{when } H_0 \text{ is true.}$$

In Section 6.6 we said the right-hand member of this last equation defines a statistic that has an approximate chi-square distribution with $k - 1$ degrees of freedom. Note that

$$\text{dimension of } \Omega - \text{dimension of } \omega = (k-1) - 0 = k - 1.$$

9.32. Let $Y_1 < Y_2 < \cdots < Y_5$ be the order statistics of a random sample of size $n = 5$ from a distribution with p.d.f. $f(x; \theta) = \frac{1}{2}e^{-|x-\theta|}$, $-\infty < x < \infty$, for all real θ. Find the likelihood ratio test λ for testing $H_0: \theta = \theta_0$ against $H_1: \theta \neq \theta_0$.

9.33. Let $X_1, X_2, \ldots, X_n$ and $Y_1, Y_2, \ldots, Y_m$ be independent random samples from the two normal distributions $N(0, \theta_1)$ and $N(0, \theta_2)$.

(a) Find the likelihood ratio λ for testing the composite hypothesis $H_0 : \theta_1 = \theta_2$ against the composite alternative $H_1 : \theta_1 \neq \theta_2$.
(b) This λ is a function of what F-statistic that would actually be used in this test?

9.34. A random sample $X_1, X_2, \ldots, X_n$ arises from a distribution given by

$$H_0 : f(x; \theta) = \frac{1}{\theta}, \qquad 0 < x < \theta, \quad \text{zero elsewhere,}$$

or

$$H_1 : f(x; \theta) = \frac{1}{\theta} e^{-x/\theta}, \qquad 0 < x < \infty, \quad \text{zero elsewhere.}$$

Determine the likelihood ratio (λ) test associated with the test of H_0 against H_1.

9.35. Let X and Y be two independent random variables with respective probability density functions

$$f(x; \theta_i) = \left(\frac{1}{\theta_i}\right) e^{-x/\theta_i}, \qquad 0 < x < \infty,$$

zero elsewhere, $i = 1, 2$. To test $H_0 : \theta_1 = \theta_2$ against $H_1 : \theta_1 \neq \theta_2$, two independent random samples of sizes n_1 and n_2, respectively, were taken from these distributions. Find the likelihood ratio λ and show that λ can be written as a function of a statistic having an F-distribution, under H_0.

9.36. Consider the two uniform distributions with respective probability density functions

$$f(x; \theta_i) = \frac{1}{2\theta_i}, \qquad -\theta_i < x < \theta_i,$$

zero elsewhere, $i = 1, 2$. The null hypothesis is $H_0 : \theta_1 = \theta_2$ while the alternative is $H_1 : \theta_1 \neq \theta_2$. Let $X_1 < X_2 < \cdots < X_{n_1}$ and $Y_1 < Y_2 < \cdots < Y_{n_2}$ be the order statistics of two independent random samples from the two distributions, respectively. Using the likelihood ratio λ, find the statistic used to test H_0 against H_1. Find the distribution of $-2 \ln \lambda$ when H_0 is true. Note that in this nonregular case the number of degrees of freedom is two times the difference of the dimensions of Ω and ω.

9.4 The Sequential Probability Ratio Test

In Section 9.1 we proved a theorem that provided us with a method for determining a best critical region for testing a simple hypothesis against an alternative simple hypothesis. The theorem was as follows. Let $X_1, X_2, \ldots, X_n$ be a random sample with fixed sample size n from a distribution that has p.d.f. $f(x; \theta)$, where $\theta \in \{\theta : \theta = \theta', \theta''\}$ and θ' and θ'' are known numbers. Let the joint p.d.f. of $X_1, X_2, \ldots, X_n$ be denoted by

$$L(\theta, n) = f(x_1; \theta)f(x_2; \theta) \cdots f(x_n; \theta),$$

a notation that reveals both the parameter θ and the sample size n. If we reject $H_0 : \theta = \theta'$ and accept $H_1 : \theta = \theta''$ when and only when

$$\frac{L(\theta', n)}{L(\theta'', n)} \leq k,$$

where $k > 0$, then this is a best test of H_0 against H_1.

Let us now suppose that the sample size n is *not* fixed in advance. In fact, let the sample size be a random variable N with sample space $\{n : n = 1, 2, 3, \ldots\}$. An interesting procedure for testing the simple hypothesis $H_0 : \theta = \theta'$ against the simple hypothesis $H_1 : \theta = \theta''$ is the following. Let k_0 and k_1 be two positive constants with $k_0 < k_1$. Observe the independent outcomes $X_1, X_2, X_3, \ldots$ in sequence, say $x_1, x_2, x_3, \ldots$, and compute

$$\frac{L(\theta', 1)}{L(\theta'', 1)}, \frac{L(\theta', 2)}{L(\theta'', 2)}, \frac{L(\theta', 3)}{L(\theta'', 3)}, \ldots.$$

The hypothesis $H_0 : \theta = \theta'$ is rejected (and $H_1 : \theta = \theta''$ is accepted) if and only if there exists a positive integer n so that $(x_1, x_2, \ldots, x_n)$ belongs to the set

$$C_n = \left\{(x_1, \ldots, x_n) : k_0 < \frac{L(\theta', j)}{L(\theta'', j)} < k_1, j = 1, \ldots, n-1, \right.$$
$$\left. \text{and} \quad \frac{L(\theta', n)}{L(\theta'', n)} \leq k_0 \right\}.$$

On the other hand, the hypothesis $H_0 : \theta = \theta'$ is accepted (and $H_1 : \theta = \theta''$ is rejected) if and only if there exists a positive integer n so that $(x_1, x_2, \ldots, x_n)$ belongs to the set

$$B_n = \left\{(x_1, \ldots, x_n) : k_0 < \frac{L(\theta', j)}{L(\theta'', j)} < k_1, j = 1, 2, \ldots, n-1, \right.$$
$$\left. \text{and} \quad \frac{L(\theta', n)}{L(\theta'', n)} \geq k_1 \right\}.$$

That is, we continue to observe sample observations as long as

$$k_0 < \frac{L(\theta', n)}{L(\theta'', n)} < k_1. \tag{1}$$

We stop these observations in one of two ways:

1. With rejection of $H_0 : \theta = \theta'$ as soon as

$$\frac{L(\theta', n)}{L(\theta'', n)} \le k_0,$$

or

2. with acceptance of $H_0 : \theta = \theta'$ as soon as

$$\frac{L(\theta', n)}{L(\theta'', n)} \ge k_1.$$

A test of this kind is called Wald's *sequential probability ratio test*. Now, frequently inequality (1) can be conveniently expressed in an equivalent form

$$c_0(n) < u(x_1, x_2, \ldots, x_n) < c_1(n),$$

where $u(X_1, X_2, \ldots, X_n)$ is a statistic and $c_0(n)$ and $c_1(n)$ depend on the constants $k_0, k_1, \theta', \theta''$, and on n. Then the observations are stopped and a decision is reached as soon as

$$u(x_1, x_2, \ldots, x_n) \le c_0(n) \qquad \text{or} \qquad u(x_1, x_2, \ldots, x_n) \ge c_1(n).$$

We now give an illustrative example.

Example 1. Let X have a p.d.f.

$$f(x; \theta) = \theta^x(1 - \theta)^{1-x}, \qquad x = 0, 1,$$
$$= 0 \qquad \text{elsewhere.}$$

In the preceding discussion of a sequential probability ratio test, let $H_0 : \theta = \frac{1}{3}$ and $H_1 : \theta = \frac{2}{3}$; then, with $\sum x_i = \sum_{1}^{n} x_i$,

$$\frac{L(\frac{1}{3}, n)}{L(\frac{2}{3}, n)} = \frac{(\frac{1}{3})^{\Sigma x_i}(\frac{2}{3})^{n - \Sigma x_i}}{(\frac{2}{3})^{\Sigma x_i}(\frac{1}{3})^{n - \Sigma x_i}} = 2^{n - 2\Sigma x_i}.$$

If we take logarithms to the base 2, the inequality

$$k_0 < \frac{L(\frac{1}{3}, n)}{L(\frac{2}{3}, n)} < k_1,$$

with $0 < k_0 < k_1$, becomes

$$\log_2 k_0 < n - 2\sum_1^n x_i < \log_2 k_1,$$

or, equivalently,

$$c_0(n) = \frac{n}{2} - \frac{1}{2}\log_2 k_1 < \sum_1^n x_i < \frac{n}{2} - \frac{1}{2}\log_2 k_0 = c_1(n).$$

Note that $L(\frac{1}{3}, n)/L(\frac{2}{3}, n) \leq k_0$ if and only if $c_1(n) \leq \sum_1^n x_i$; and $L(\frac{1}{3}, n)/L(\frac{2}{3}, n) \geq k_1$ if and only if $c_0(n) \geq \sum_1^n x_i$. Thus we continue to observe outcomes as long as $c_0(n) < \sum_1^n x_i < c_1(n)$. The observation of outcomes is discontinued with the first value n of N for which either $c_1(n) \leq \sum_1^n x_i$ or $c_0(n) \geq \sum_1^n x_i$. The inequality $c_1(n) \leq \sum_1^n x_i$ leads to the rejection of $H_0 : \theta = \frac{1}{3}$ (the acceptance of H_1), and the inequality $c_0(n) \geq \sum_1^n x_i$ leads to the acceptance of $H_0 : \theta = \frac{1}{3}$ (the rejection of H_1).

Remarks. At this point, the reader undoubtedly sees that there are many questions that should be raised in connection with the sequential probability ratio test. Some of these questions are possibly among the following:

1. What is the probability of the procedure continuing indefinitely?
2. What is the value of the power function of this test at each of the points $\theta = \theta'$ and $\theta = \theta''$?
3. If θ'' is one of several values of θ specified by an alternative composite hypothesis, say $H_1 : \theta > \theta'$, what is the power function at each point $\theta \geq \theta'$?
4. Since the sample size N is a random variable, what are some of the properties of the distribution of N? In particular, what is the expected value $E(N)$ of N?
5. How does this test compare with tests that have a fixed sample size n?

A course in sequential analysis would investigate these and many other problems. However, in this book our objective is largely that of acquainting the reader with this kind of test procedure. Accordingly, we assert that the answer to question 1 is zero. Moreover, it can be proved that if $\theta = \theta'$ or if $\theta = \theta''$, $E(N)$ is smaller, for this sequential procedure, than the sample size of a fixed-sample-size test which has the same values of the power function at those points. We now consider question 2 in some detail.

In this section we shall denote the power of the test when H_0 is true by the symbol α and the power of the test when H_1 is true by the symbol $1 - \beta$. Thus α is the probability of committing a type I error (the rejection of H_0 when H_0 is true), and β is the probability of committing a type II error (the acceptance of H_0 when H_0 is false). With the sets C_n and B_n as previously defined, and with random variables of the continuous type, we then have

$$\alpha = \sum_{n=1}^{\infty} \int_{C_n} L(\theta', n), \qquad 1 - \beta = \sum_{n=1}^{\infty} \int_{C_n} L(\theta'', n).$$

Since the probability is 1 that the procedure will terminate, we also have

$$1 - \alpha = \sum_{n=1}^{\infty} \int_{B_n} L(\theta', n), \qquad \beta = \sum_{n=1}^{\infty} \int_{B_n} L(\theta'', n).$$

If $(x_1, x_2, \ldots, x_n) \in C_n$, we have $L(\theta', n) \le k_0 L(\theta'', n)$; hence it is clear that

$$\alpha = \sum_{n=1}^{\infty} \int_{C_n} L(\theta', n) \le \sum_{n=1}^{\infty} \int_{C_n} k_0 L(\theta'', n) = k_0(1 - \beta).$$

Because $L(\theta', n) \ge k_1 L(\theta'', n)$ at each point of the set B_n, we have

$$1 - \alpha = \sum_{n=1}^{\infty} \int_{B_n} L(\theta', n) \ge \sum_{n=1}^{\infty} \int_{B_n} k_1 L(\theta'', n) = k_1 \beta.$$

Accordingly, it follows that

$$\frac{\alpha}{1 - \beta} \le k_0, \qquad k_1 \le \frac{1 - \alpha}{\beta}, \tag{2}$$

provided that β is not equal to zero or 1.

Now let α_a and β_a be preassigned proper fractions; some typical values in the applications are 0.01, 0.05, and 0.10. If we take

$$k_0 = \frac{\alpha_a}{1 - \beta_a}, \qquad k_1 = \frac{1 - \alpha_a}{\beta_a},$$

then inequalities (2) become

$$\frac{\alpha}{1 - \beta} \le \frac{\alpha_a}{1 - \beta_a}, \qquad \frac{1 - \alpha_a}{\beta_a} \le \frac{1 - \alpha}{\beta}; \tag{3}$$

or, equivalently,

$$\alpha(1 - \beta_a) \le (1 - \beta)\alpha_a, \qquad \beta(1 - \alpha_a) \le (1 - \alpha)\beta_a.$$

If we add corresponding members of the immediately preceding inequalities, we find that

$$\alpha + \beta - \alpha\beta_a - \beta\alpha_a \leq \alpha_a + \beta_a - \beta\alpha_a - \alpha\beta_a$$

and hence

$$\alpha + \beta \leq \alpha_a + \beta_a.$$

That is, the sum $\alpha + \beta$ of the probabilities of the two kinds of errors is bounded above by the sum $\alpha_a + \beta_a$ of the preassigned numbers. Moreover, since α and β are positive proper fractions, inequalities (3) imply that

$$\alpha \leq \frac{\alpha_a}{1 - \beta_a}, \qquad \beta \leq \frac{\beta_a}{1 - \alpha_a};$$

consequently, we have an upper bound on each of α and β. Various investigations of the sequential probability ratio test seem to indicate that in most practical cases, the values of α and β are quite close to α_a and β_a. This prompts us to approximate the power function at the points $\theta = \theta'$ and $\theta = \theta''$ by α_a and $1 - \beta_a$, respectively.

Example 2. Let X be $N(\theta, 100)$. To find the sequential probability ratio test for testing $H_0 : \theta = 75$ against $H_1 : \theta = 78$ such that each of α and β is approximately equal to 0.10, take

$$k_0 = \frac{0.10}{1 - 0.10} = \frac{1}{9}, \qquad k_1 = \frac{1 - 0.10}{0.10} = 9.$$

Since

$$\frac{L(75, n)}{L(78, n)} = \frac{\exp\left[-\sum (x_i - 75)^2/2(100)\right]}{\exp\left[-\sum (x_i - 78)^2/2(100)\right]} = \exp\left(-\frac{6\sum x_i - 459n}{200}\right),$$

the inequality

$$k_0 = \frac{1}{9} < \frac{L(75, n)}{L(78, n)} < 9 = k_1$$

can be rewritten, by taking logarithms, as

$$-\ln 9 < \frac{6\sum x_i - 459\text{n}}{200} < \ln 9.$$

This inequality is equivalent to the inequality

$$c_0(n) = \tfrac{153}{2}n - \tfrac{100}{3}\ln 9 < \sum_1^n x_i < \tfrac{153}{2}n + \tfrac{100}{3}\ln 9 = c_1(n).$$

Moreover, $L(75, n)/L(78, n) \leq k_0$ and $L(75, n)/L(78, n) \geq k_1$ are equivalent to the inequalities $\sum_1^n x_i \geq c_1(n)$ and $\sum_1^n x_i \leq c_0(n)$, respectively. Thus the observation of outcomes is discontinued with the first value n of N for which either $\sum_1^n x_i \geq c_1(n)$ or $\sum_1^n x_i \leq c_0(n)$. The inequality $\sum_1^n x_i \geq c_1(n)$ leads to the rejection of $H_0: \theta = 75$, and the inequality $\sum_1^n x_i \leq c_0(n)$ leads to the acceptance of $H_0: \theta = 75$. The power of the test is approximately 0.10 when H_0 is true, and approximately 0.90 when H_1 is true.

Remark. It is interesting to note that a sequential probability ratio test can be thought of as a *random-walk procedure*. For illustrations, the final inequalities of Examples 1 and 2 can be rewritten as

$$-\log_2 k_1 < \sum_1^n 2(x_i - 0.5) < -\log_2 k_0$$

and

$$-\frac{100}{3} \ln 9 < \sum_1^n (x_i - 76.5) < \frac{100}{3} \ln 9,$$

respectively. In each instance, we can think of starting at the point zero and taking random steps until one of the boundaries is reached. In the first situation the random steps are $2(X_1 - 0.5)$, $2(X_2 - 0.5)$, $2(X_3 - 0.5)$, . . . and hence are of the same length, 1, but with random directions. In the second instance, both the length and the direction of the steps are random variables, $X_1 - 76.5$, $X_2 - 76.5$, $X_3 - 76.5$, ...

In recent years, there has been much attention to improving quality of products using statistical methods. One such simple method was developed by Walter Shewhart in which a sample of size n of the items being produced is taken and they are measured, resulting in n values. The mean $\bar{x}$ of these n measurements has an approximate normal distribution with mean μ and variance σ^2/n. In practice, μ and σ^2 must be estimated, but in this discussion, we assume that they are known. From theory we know that the probability is 0.997 that $\bar{x}$ is between

$$\text{LCL} = \mu - \frac{3\sigma}{\sqrt{n}} \quad \text{and} \quad \text{UCL} = \mu + \frac{3\sigma}{\sqrt{n}}.$$

These two values are called the lower (LCL) and upper (UCL) control limits, respectively. Samples like this are taken periodically, resulting in a sequence of means, say $\bar{x}_1, \bar{x}_2, \bar{x}_3, \ldots$. These are usually plotted; and if they are between the LCL and UCL, we say that the process is *in control*. If one falls outside the limits, this would suggest that the mean μ has shifted, and the process would be investigated.

It was recognized by some that there could be a shift in the mean, say from μ to $\mu + (\sigma/\sqrt{n})$; and it would still be difficult to detect that shift with a single sample mean as now the probability of a single $\bar{x}$ exceeding UCL is only about 0.023. This means that we would need about $1/0.023 \approx 43$ samples, each of size n, on the average before detecting such a shift. This seems too long; so statisticians recognized that they should be cumulating experience as the sequence $\bar{x}_1, \bar{x}_2, \bar{x}_3, \ldots$ is observed in order to help them detect the shift sooner. It is the practice to compute the standardized variable $Z = (\bar{X} - \mu)/(\sigma/\sqrt{n})$; thus we state the problem in these terms and provide the solution given by a sequential probability ratio test.

Here Z is $N(\theta, 1)$, and we wish to test $H_0 : \theta = 0$ against $H_1 : \theta = 1$ using the sequence of i.i.d. random variables $Z_1, Z_2, \ldots, Z_m, \ldots$. We use m rather than n, as the latter is the size of the samples taken periodically. We have

$$\frac{L(0, m)}{L(1, m)} = \frac{\exp\left[-\sum z_i^2/2\right]}{\exp\left[-\sum (z_i - 1)^2/2\right]} = \exp\left[-\sum_{i=1}^{m} (z_i - 0.5)\right].$$

Thus

$$k_0 < \exp\left[-\sum_{i=1}^{m} (z_i - 0.5)\right] < k_1$$

can be rewritten as

$$h = -\ln k_0 > \sum_{i=1}^{m} (z_i - 0.5) > -\ln k_1 = -h.$$

It is true that $-\ln k_0 = \ln k_1$ when $\alpha_a = \beta_a$. Often, $h = -\ln k_0$ is taken to be about 4 or 5, suggesting that $\alpha_a = \beta_a$ is small, like 0.01. As $\Sigma\,(z_i - 0.5)$ is cumulating the sum of $z_i - 0.5$, $i = 1, 2, 3, \ldots$, these procedures are often called CUSUMS. If the CUSUM $= \Sigma\,(z_i - 0.5)$ exceeds h, we would investigate the process, as it seems that the mean has shifted upward. If this shift is to $\theta = 1$, the theory associated with these procedures shows that we need only 8 or 9 samples on the average, rather than 43, to detect this shift. For more information about these methods, the reader is referred to one of the many books on quality improvement through statistical methods. What we would like to emphasize here is that, through sequential methods (not only the sequential probability ratio test), we should take advantage of all past experience that we can gather in making inferences.

EXERCISES

9.37. Let X be $N(0, \theta)$ and, in the notation of this section, let $\theta' = 4$, $\theta'' = 9$, $\alpha_a = 0.05$, and $\beta_a = 0.10$. Show that the sequential probability ratio test can be based upon the statistic $\sum_1^n X_i^2$. Determine $c_0(n)$ and $c_1(n)$.

9.38. Let X have a Poisson distribution with mean θ. Find the sequential probability ratio test for testing $H_0: \theta = 0.02$ against $H_1: \theta = 0.07$. Show that this test can be based upon the statistic $\sum_1^n X_i$. If $\alpha_a = 0.20$ and $\beta_a = 0.10$, find $c_0(n)$ and $c_1(n)$.

9.39. Let the independent random variables Y and Z be $N(\mu_1, 1)$ and $N(\mu_2, 1)$, respectively. Let $\theta = \mu_1 - \mu_2$. Let us observe independent observations from each distribution, say $Y_1, Y_2, \ldots$ and $Z_1, Z_2, \ldots$. To test sequentially the hypothesis $H_0: \theta = 0$ against $H_1: \theta = \frac{1}{2}$, use the sequence $X_i = Y_i - Z_i$, $i = 1, 2, \ldots$. If $\alpha_a = \beta_a = 0.05$, show that the test can be based upon $\bar{X} = \bar{Y} - \bar{Z}$. Find $c_0(n)$ and $c_1(n)$.

9.40. Say that a manufacturing process makes about 3 percent defective items, which is considered satisfactory for this particular product. The managers would like to decrease this to about 1 percent and clearly want to guard against a substantial increase, say to 5 percent. To monitor the process, periodically $n = 100$ items are taken and the number X of defectives counted. Assume that X is $b(n = 100, p = \theta)$. Based on a sequence $X_1, X_2, \ldots, X_n, \ldots$, determine a sequential probability ratio test that tests $H_0: \theta = 0.01$ against $H_1: \theta = 0.05$. (Note that $\theta = 0.03$, the present level, is in between these two values.) Write this test in the form

$$h_0 > \sum_{i=1}^{m} (x_i - nd) > h_1$$

and determine d, h_0, and h_1 if $\alpha_a = \beta_a = 0.02$.

9.5 Minimax, Bayesian, and Classification Procedures

In Chapters 7 and 8 we considered several procedures which may be used in problems of point estimation. Among these were decision function procedures (in particular, minimax decisions) and Bayesian procedures. In this section, we apply these same principles to the problem of testing a simple hypothesis H_0 against an alternative simple hypothesis H_1. It is important to observe that each of these procedures yields, in accordance with the Neyman–Pearson theorem, a best test of H_0 against H_1.

We first investigate the decision function approach to the problem of testing a simple hypothesis against a simple alternative hypothesis. Let the joint p.d.f. of n random variables $X_1, X_2, \ldots, X_n$ depend upon the parameter θ. Here n is a fixed positive integer. This p.d.f. is denoted by $L(\theta; x_1, x_2, \ldots, x_n)$ or, for brevity, by $L(\theta)$. Let θ' and θ'' be distinct and fixed values of θ. We wish to test the simple hypothesis $H_0 : \theta = \theta'$ against the simple hypothesis $H_1 : \theta = \theta''$. Thus the parameter space is $\Omega = \{\theta : \theta = \theta', \theta''\}$. In accordance with the decision function procedure, we need a function δ of the observed values of $X_1, \ldots, X_n$ (or, of the observed value of a statistic Y) that decides which of the two values of θ, θ' or θ'', to accept. That is, the function δ selects either $H_0 : \theta = \theta'$ or $H_1 : \theta = \theta''$. We denote these decisions by $\delta = \theta'$ and $\delta = \theta''$, respectively. Let $\mathscr{L}(\theta, \delta)$ represent the loss function associated with this decision problem. Because the pairs $(\theta = \theta', \delta = \theta')$ and $(\theta = \theta'', \delta = \theta'')$ represent correct decisions, we shall always take $\mathscr{L}(\theta', \theta') = \mathscr{L}(\theta'', \theta'') = 0$. On the other hand, if either $\delta = \theta''$ when $\theta = \theta'$ or $\delta = \theta'$ when $\theta = \theta''$, then a positive value should be assigned to the loss function; that is, $\mathscr{L}(\theta', \theta'') > 0$ and $\mathscr{L}(\theta'', \theta') > 0$.

It has previously been emphasized that a test of $H_0 : \theta = \theta'$ against $H_1 : \theta = \theta''$ can be described in terms of a critical region in the sample space. We can do the same kind of thing with the decision function. That is, we can choose a subset C of the sample space and if $(x_1, x_2, \ldots, x_n) \in C$, we can make the decision $\delta = \theta''$; whereas, if $(x_1, x_2, \ldots, x_n) \in C^*$, the complement of C, we make the decision $\delta = \theta'$. Thus a given critical region C determines the decision function. In this sense, we may denote the risk function by $R(\theta, C)$ instead of $R(\theta, \delta)$. That is, in a notation used in Section 9.1,

$$R(\theta, C) = R(\theta, \delta) = \int_{C \cup C^*} \mathscr{L}(\theta, \delta) L(\theta).$$

Since $\delta = \theta''$ if $(x_1, \ldots, x_n) \in C$ and $\delta = \theta'$ if $(x_1, \ldots, x_n) \in C^*$, we have

$$R(\theta, C) = \int_C \mathscr{L}(\theta, \theta'') L(\theta) + \int_{C^*} \mathscr{L}(\theta, \theta') L(\theta). \tag{1}$$

If, in Equation (1), we take $\theta = \theta'$, then $\mathscr{L}(\theta', \theta') = 0$ and hence

$$R(\theta', C) = \int_C \mathscr{L}(\theta', \theta'') L(\theta') = \mathscr{L}(\theta', \theta'') \int_C L(\theta').$$

On the other hand, if in Equation (1) we let $\theta = \theta''$, then $\mathscr{L}(\theta'', \theta'') = 0$ and, accordingly,

$$R(\theta'', C) = \int_{C^*} \mathscr{L}(\theta'', \theta')L(\theta'') = \mathscr{L}(\theta'', \theta') \int_{C^*} L(\theta'').$$

It is enlightening to note that, if $K(\theta)$ is the power function of the test associated with the critical region C, then

$$R(\theta', C) = \mathscr{L}(\theta', \theta'')K(\theta') = \mathscr{L}(\theta', \theta'')\alpha,$$

where $\alpha = K(\theta')$ is the significance level; and

$$R(\theta'', C) = \mathscr{L}(\theta'', \theta')[1 - K(\theta'')] = \mathscr{L}(\theta'', \theta')\beta,$$

where $\beta = 1 - K(\theta'')$ is the probability of the type II error.

Let us now see if we can find a minimax solution to our problem. That is, we want to find a critical region C so that

$$\max [R(\theta', C), R(\theta'', C)]$$

is minimized. We shall show that the solution is the region

$$C = \left\{(x_1, \ldots, x_n) : \frac{L(\theta'; x_1, \ldots, x_n)}{L(\theta''; x_1, \ldots, x_n)} \le k\right\},$$

provided the positive constant k is selected so that $R(\theta', C) = R(\theta'', C)$. That is, if k is chosen so that

$$\mathscr{L}(\theta', \theta'') \int_C L(\theta') = \mathscr{L}(\theta'', \theta') \int_{C^*} L(\theta''),$$

then the critical region C provides a minimax solution. In the case of random variables of the continuous type, k can always be selected so that $R(\theta', C) = R(\theta'', C)$. However, with random variables of the discrete type, we may need to consider an auxiliary random experiment when $L(\theta')/L(\theta'') = k$ in order to achieve the exact equality $R(\theta', C) = R(\theta'', C)$.

To see that this region C is the minimax solution, consider every other region A for which $R(\theta', C) \ge R(\theta', A)$. Obviously, a region A for which $R(\theta', C) < R(\theta', A)$ is not a candidate for a minimax solution, for then $R(\theta', C) = R(\theta'', C) < \max [R(\theta', A), R(\theta'', A)]$. Since $R(\theta', C) \ge R(\theta', A)$ means that

$$\mathscr{L}(\theta', \theta'') \int_C L(\theta') \ge \mathscr{L}(\theta', \theta'') \int_A L(\theta'),$$

we have

$$\alpha = \int_C L(\theta') \geq \int_A L(\theta').$$

That is, the significance level of the test associated with the critical region A is less than or equal to α. But C, in accordance with the Neyman–Pearson theorem, is a best critical region of size α. Thus

$$\int_C L(\theta'') \geq \int_A L(\theta'')$$

and

$$\int_{C^*} L(\theta'') \leq \int_{A^*} L(\theta'').$$

Accordingly,

$$\mathscr{L}(\theta'', \theta') \int_{C^*} L(\theta'') \leq \mathscr{L}(\theta'', \theta') \int_{A^*} L(\theta''),$$

or, equivalently,

$$R(\theta'', C) \leq R(\theta'', A).$$

That is,

$$R(\theta', C) = R(\theta'', C) \leq R(\theta'', A).$$

This means that

$$\max\,[R(\theta', C), R(\theta'', C)] \leq R(\theta'', A).$$

Then certainly,

$$\max\,[R(\theta', C), R(\theta'', C)] \leq \max\,[R(\theta', A), R(\theta'', A)],$$

and the critical region C provides a minimax solution, as we wanted to show.

Example 1. Let $X_1, X_2, \ldots, X_{100}$ denote a random sample of size 100 from a distribution that is $N(\theta, 100)$. We again consider the problem of testing $H_0: \theta = 75$ against $H_1: \theta = 78$. We seek a minimax solution with $\mathscr{L}(75, 78) = 3$ and $\mathscr{L}(78, 75) = 1$. Since $L(75)/L(78) \leq k$ is equivalent to $\bar{x} \geq c$, we want to determine c, and thus k, so that

$$3 \Pr(\bar{X} \geq c; \theta = 75) = \Pr(\bar{X} < c; \theta = 78).$$

Because $\bar{X}$ is $N(\theta, 1)$, the preceding equation can be rewritten as

$$3[1 - \Phi(c - 75)] = \Phi(c - 78).$$

If we use Table III of the appendix, we see, by trial and error, that the solution is $c = 76.8$, approximately. The significance level of the test is $1 - \Phi(1.8) = 0.036$, approximately, and the power of the test when H_1 is true is $1 - \Phi(-1.2) = 0.885$, approximately.

Next, let us consider the Bayesian approach to the problem of testing the simple hypothesis $H_0 : \theta = \theta'$ against the simple hypothesis $H_1 : \theta = \theta''$. We continue to use the notation already presented in this section. In addition, we recall that we need the p.d.f. $h(\theta)$ of the random variable Θ. Since the parameter space consists of but two points, θ' and θ'', Θ is a random variable of the discrete type; and we have $h(\theta') + h(\theta'') = 1$. Since $L(\theta; x_1, x_2, \ldots, x_n) = L(\theta)$ is the conditional p.d.f. of $X_1, X_2, \ldots, X_n$, given $\Theta = \theta$, the joint p.d.f. of $X_1, X_2, \ldots, X_n$ and Θ is

$$h(\theta)L(\theta; x_1, x_2, \ldots, x_n) = h(\theta)L(\theta).$$

Because

$$\sum_{\Omega} h(\theta)L(\theta) = h(\theta')L(\theta') + h(\theta'')L(\theta'')$$

is the marginal p.d.f. of $X_1, X_2, \ldots, X_n$, the conditional p.d.f. of Θ, given $X_1 = x_1, \ldots, X_n = x_n$, is

$$k(\theta|x_1, \ldots, x_n) = \frac{h(\theta)L(\theta)}{h(\theta')L(\theta') + h(\theta'')L(\theta'')}.$$

Now a Bayes' solution to a decision problem is defined in Section 8.1 as a $\delta(y)$ such that $E\{\mathscr{L}[\theta, \delta(y)]|Y = y\}$ is a minimum. In this problem if $\delta = \theta'$, the conditional expectation of $\mathscr{L}(\theta, \delta)$, given $X_1 = x_1, \ldots, X_n = x_n$, is

$$\sum_{\Omega} \mathscr{L}(\theta, \theta')k(\theta|x_1, \ldots, x_n) = \frac{\mathscr{L}(\theta'', \theta')h(\theta'')L(\theta'')}{h(\theta')L(\theta') + h(\theta'')L(\theta'')},$$

because $\mathscr{L}(\theta', \theta') = 0$; and if $\delta = \theta''$, this expectation is

$$\sum_{\Omega} \mathscr{L}(\theta, \theta'')k(\theta|x_1, \ldots, x_n) = \frac{\mathscr{L}(\theta', \theta'')h(\theta')L(\theta')}{h(\theta')L(\theta') + h(\theta'')L(\theta'')},$$

because $\mathscr{L}(\theta'', \theta'') = 0$. Accordingly, the Bayes' solution requires that the decision $\delta = \theta''$ be made if

$$\frac{\mathscr{L}(\theta', \theta'')h(\theta')L(\theta')}{h(\theta')L(\theta') + h(\theta'')L(\theta'')} < \frac{\mathscr{L}(\theta'', \theta')h(\theta'')L(\theta'')}{h(\theta')L(\theta') + h(\theta'')L(\theta'')},$$

or, equivalently, if

$$\frac{L(\theta')}{L(\theta'')} < \frac{\mathscr{L}(\theta'', \theta')h(\theta'')}{\mathscr{L}(\theta', \theta'')h(\theta')}. \tag{2}$$

If the sign of inequality in expression (2) is reversed, we make the decision $\delta = \theta'$; and if the two members of expression (2) are equal, we can use some auxiliary random experiment to make the decision. It is important to note that expression (2) describes, in accordance with the Neyman–Pearson theorem, a best test.

Example 2. In addition to the information given in Example 1, suppose that we know the prior probabilities for $\theta = \theta' = 75$ and for $\theta = \theta'' = 78$ to be given, respectively, by $h(75) = \frac{1}{7}$ and $h(78) = \frac{6}{7}$. Then the Bayes' solution is, in this case,

$$\frac{L(75)}{L(78)} < \frac{(1)(\frac{6}{7})}{(3)(\frac{1}{7})} = 2,$$

which is equivalent to $\bar{x} > 76.3$, approximately. The power of the test when H_0 is true is $1 - \Phi(1.3) = 0.097$, approximately, and the power of the test when H_1 is true is $1 - \Phi(-1.7) = \Phi(1.7) = 0.955$, approximately.

In summary, we make the following comments. In testing the simple hypothesis $H_0 : \theta = \theta'$ against the simple hypothesis $H_1 : \theta = \theta''$, it is emphasized that each principle leads to critical regions of the form

$$\left\{(x_1, x_2, \ldots, x_n) : \frac{L(\theta'; x_1, \ldots, x_n)}{L(\theta''; x_1, \ldots, x_n)} \le k\right\},$$

where k is a positive constant. In the classical approach, we determine k by requiring that the power function of the test have a certain value at the point $\theta = \theta'$ or at the point $\theta = \theta''$ (usually, the value α at the point $\theta = \theta'$). The minimax decision requires k to be selected so that

$$\mathscr{L}(\theta', \theta'') \int_C L(\theta') = \mathscr{L}(\theta'', \theta') \int_{C^*} L(\theta'').$$

Finally, the Bayes' procedure requires that

$$k = \frac{\mathscr{L}(\theta'', \theta')h(\theta'')}{\mathscr{L}(\theta', \theta'')h(\theta')}.$$

Each of these tests is a best test for testing a simple hypothesis $H_0 : \theta = \theta'$ against a simple alternative hypothesis $H_1 : \theta = \theta''$.

The summary above has an interesting application to the problem of *classification*, which can be described as follows. An investigator makes a number of measurements on an item and wants to place it into one of several categories (or classify it). For convenience in our discussion, we assume that only two measurements, say X and Y, are made on the item to be classified. Moreover, let X and Y have a joint p.d.f. $f(x, y; \theta)$, where the parameter θ represents one or more parameters. In our simplification, suppose that there are only two possible joint distributions (categories) for X and Y, which are indexed by the parameter values θ' and θ'', respectively. In this case, the problem then reduces to one of observing $X = x$ and $Y = y$ and then testing the hypothesis $\theta = \theta'$ against the hypothesis $\theta = \theta''$, with the classification of X and Y being in accord with which hypothesis is accepted. From the Neyman–Pearson theorem, we know that a best decision of this sort is of the form: If

$$\frac{f(x, y; \theta')}{f(x, y; \theta'')} \le k,$$

choose the distribution indexed by θ''; that is, we classify (x, y) as coming from the distribution indexed by θ''. Otherwise, choose the distribution indexed by θ'; that is, we classify (x, y) as coming from the distribution indexed by θ'. Here k can be selected by considering the power function, a minimax decision, or a Bayes' procedure. We favor the latter if the losses and prior probabilities are known.

Example 3. Let (x, y) be an observation of the random pair (X, Y), which has a bivariate normal distribution with parameters $\mu_1, \mu_2, \sigma_1^2, \sigma_2^2$, and ρ. In Section 3.5 that joint p.d.f. is given by

$$f(x, y; \mu_1, \mu_2, \sigma_1^2, \sigma_2^2, \rho) = \frac{1}{2\pi\sigma_1\sigma_2\sqrt{1-\rho^2}} e^{-q(x,y;\mu_1,\mu_2)/2},$$

$$-\infty < x < \infty, \quad -\infty < y < \infty,$$

where $\sigma_1 > 0$, $\sigma_2 > 0$, $-1 < \rho < 1$, and

$$q(x, y; \mu_1, \mu_2) = \frac{1}{1-\rho^2}\left[\left(\frac{x-\mu_1}{\sigma_1}\right)^2 - 2\rho\left(\frac{x-\mu_1}{\sigma_1}\right)\left(\frac{y-\mu_2}{\sigma_2}\right) + \left(\frac{y-\mu_2}{\sigma_2}\right)^2\right].$$

Assume that σ_1^2, σ_2^2, and ρ are known but that we do not know whether the respective means of (X, Y) are (μ_1', μ_2') or (μ_1'', μ_2''). The inequality

$$\frac{f(x, y; \mu_1', \mu_2', \sigma_1^2, \sigma_2^2, \rho)}{f(x, y; \mu_1'', \mu_2'', \sigma_1^2, \sigma_2^2, \rho)} \le k$$

is equivalent to

$$\tfrac{1}{2}[q(x, y; \mu_1'', \mu_2'') - q(x, y; \mu_1', \mu_2')] \le \ln k.$$

Moreover, it is clear that the difference in the left-hand member of this inequality does not contain terms involving x^2, xy, and y^2. In particular, this inequality is the same as

$$\frac{1}{1-\rho^2}\left\{\left[\frac{\mu_1' - \mu_1''}{\sigma_1^2} - \frac{\rho(\mu_2' - \mu_2'')}{\sigma_1\sigma_2}\right]x + \left[\frac{\mu_2' - \mu_2''}{\sigma_2^2} - \frac{\rho(\mu_1' - \mu_1'')}{\sigma_1\sigma_2}\right]y\right\}$$
$$\le \ln k + \tfrac{1}{2}[q(0, 0; \mu_1', \mu_2') - q(0, 0; \mu_1'', \mu_2'')], \quad (3)$$

or, for brevity,

$$ax + by \le c.$$

That is, if this linear function of x and y in the left-hand member of inequality (3) is less than or equal to a certain constant, we would classify that (x, y) as coming from the bivariate normal distribution with means μ_1'' and μ_2''. Otherwise, we would classify (x, y) as arising from the bivariate normal distribution with means μ_1' and μ_2'. Of course, if the prior probabilities and losses are given, k and thus c can be found easily; this will be illustrated in Exercise 9.43.

Once the rule for classification is established, the statistician might be interested in the two probabilities of misclassifications using that rule. The first of these two is associated with the classification of (x, y) as arising from the distribution indexed by θ'' if, in fact, it comes from that index by θ'. The second misclassification is similar, but with the interchange of θ' and θ''. In the preceding example, the probabilities of these respective misclassifications are

$$\Pr(aX + bY \le c; \mu_1', \mu_2') \qquad \text{and} \qquad \Pr(aX + bY > c; \mu_1'', \mu_2'').$$

Fortunately, the distribution of $Z = aX + bY$ is easy to determine, so each of these probabilities is easy to calculate. The m.g.f. of Z is

$$E(e^{tZ}) = E[e^{t(aX + bY)}] = E(e^{atX + btY}).$$

Hence in the joint m.g.f. of X and Y found in Section 3.5, simply replace t_1 by at and t_2 by bt to obtain

$$E(e^{tZ}) = \exp\left[\mu_1 at + \mu_2 bt + \frac{\sigma_1^2(at)^2 + 2\rho\sigma_1\sigma_2(at)(bt) + \sigma_2^2(bt)^2}{2}\right]$$

$$= \exp\left[(a\mu_1 + b\mu_2)t + \frac{(a^2\sigma_1^2 + 2ab\rho\sigma_1\sigma_2 + b^2\sigma_2^2)t^2}{2}\right].$$

However, this is the m.g.f. of the normal distribution

$$N(a\mu_1 + b\mu_2, a^2\sigma_1^2 + 2ab\rho\sigma_1\sigma_2 + b^2\sigma_2^2).$$

With this information, it is easy to compute the probabilities of misclassifications, and this will also be demonstrated in Exercise 9.43.

One final remark must be made with respect to the use of the important classification rule established in Example 3. In most instances the parameter values μ_1', μ_2' and μ_1'', μ_2'' as well as σ_1^2, σ_2^2, and ρ are unknown. In such cases the statistician has usually observed a random sample (frequently called a *training sample*) from each of the two distributions. Let us say the samples have sizes n' and n'', respectively, with sample characteristics

$$\bar{x}', \bar{y}', (s_x')^2, (s_y')^2, r' \qquad \text{and} \qquad \bar{x}'', \bar{y}'', (s_x'')^2, (s_y'')^2, r''.$$

Accordingly, if in inequality (3) the parameters $\mu_1', \mu_2', \mu_1'', \mu_2'', \sigma_1^2, \sigma_2^2$, and $\rho\sigma_1\sigma_2$ are replaced by the unbiased estimates

$$\bar{x}', \bar{y}', \bar{x}'', \bar{y}'', \frac{n'(s_x')^2 + n''(s_x'')^2}{n' + n'' - 2}, \frac{n'(s_y')^2 + n''(s_y'')^2}{n' + n'' - 2},$$

$$\frac{n'r's_x's_y' + n''r''s_x''s_y''}{n' + n'' - 2},$$

the resulting expression in the left-hand member is frequently called Fisher's *linear discriminant function*. Since those parameters have been estimated, the distribution theory associated with $aX + bY$ is not appropriate for Fisher's function. However, if n' and n'' are large, the distribution of $aX + bY$ does provide an approximation.

Although we have considered only bivariate distributions in this section, the results can easily be extended to multivariate normal distributions after a study of Sections 4.10, 10.8, and 10.9.

EXERCISES

9.41. Let $X_1, X_2, \ldots, X_{20}$ be a random sample of size 20 from a distribution which is $N(\theta, 5)$. Let $L(\theta)$ represent the joint p.d.f. of $X_1, X_2, \ldots, X_{20}$. The problem is to test $H_0: \theta = 1$ against $H_1: \theta = 0$. Thus $\Omega = \{\theta : \theta = 0, 1\}$.
(a) Show that $L(1)/L(0) \le k$ is equivalent to $\bar{x} \le c$.
(b) Find c so that the significance level is $\alpha = 0.05$. Compute the power of this test if H_1 is true.
(c) If the loss function is such that $\mathscr{L}(1, 1) = \mathscr{L}(0, 0) = 0$ and $\mathscr{L}(1, 0) = \mathscr{L}(0, 1) > 0$, find the minimax test. Evaluate the power function of this test at the points $\theta = 1$ and $\theta = 0$.

(d) If, in addition, the prior probabilities of $\theta = 1$ and $\theta = 0$ are, respectively, $h(1) = \frac{3}{4}$ and $h(0) = \frac{1}{4}$, find the Bayes' test. Evaluate the power function of this test at the points $\theta = 1$ and $\theta = 0$.

9.42. Let $X_1, X_2, \ldots, X_{10}$ be a random sample of size 10 from a Poisson distribution with parameter θ. Let $L(\theta)$ be the joint p.d.f. of $X_1, X_2, \ldots, X_{10}$. The problem is to test $H_0 : \theta = \frac{1}{2}$ against $H_1 : \theta = 1$.

(a) Show that $L(\frac{1}{2})/L(1) \le k$ is equivalent to $y = \sum_1^{10} x_i \ge c$.

(b) In order to make $\alpha = 0.05$, show that H_0 is rejected if $y > 9$ and, if $y = 9$, reject H_0 with probability $\frac{1}{2}$ (using some auxiliary random experiment).

(c) If the loss function is such that $\mathscr{L}(\frac{1}{2}, \frac{1}{2}) = \mathscr{L}(1, 1) = 0$ and $\mathscr{L}(\frac{1}{2}, 1) = 1$ and $\mathscr{L}(1, \frac{1}{2}) = 2$ show that the minimax procedure is to reject H_0 if $y > 6$ and, if $y = 6$, reject H_0 with probability 0.08 (using some auxiliary random experiment).

(d) If, in addition, we are given that the prior probabilities of $\theta = \frac{1}{2}$ and $\theta = 1$ are $h(\frac{1}{2}) = \frac{1}{3}$ and $h(1) = \frac{2}{3}$, respectively, show that the Bayes' solution is to reject H_0 if $y > 5.2$, that is, reject H_0 if $y \ge 6$.

9.43. In Example 3 let $\mu_1' = \mu_2' = 0$, $\mu_1'' = \mu_2'' = 1$, $\sigma_1^2 = 1$, $\sigma_2^2 = 1$, and $\rho = \frac{1}{2}$.

(a) Evaluate inequality (3) when the prior probabilities are $h(\mu_1', \mu_2') = \frac{1}{3}$ and $h(\mu_1'', \mu_2'') = \frac{2}{3}$ and the losses are $\mathscr{L}[\theta = (\mu_1', \mu_2'), \delta = (\mu_1'', \mu_2'')] = 4$ and $\mathscr{L}[\theta = (\mu_1'', \mu_2''), \delta = (\mu_1', \mu_2')] = 1$.

(b) Find the distribution of the linear function $aX + bY$ that results from part (a).

(c) Compute $\Pr(aX + bY \le c; \mu_1' = \mu_2' = 0)$ and $\Pr(aX + bY > c; \mu_1'' = \mu_2'' = 1)$.

9.44. Let X and Y have the joint p.d.f.

$$f(x, y; \theta_1, \theta_2) = \frac{1}{\theta_1\theta_2} \exp\left(-\frac{x}{\theta_1} - \frac{y}{\theta_2}\right), \qquad 0 < x < \infty, \quad 0 < y < \infty,$$

zero elsewhere, where $0 < \theta_1$, $0 < \theta_2$. An observation (x, y) arises from the joint distribution with parameters equal to either $(\theta_1' = 1, \theta_2' = 5)$ or $(\theta_1'' = 3, \theta_2'' = 2)$. Determine the form of the classification rule.

9.45. Let X and Y have a joint bivariate normal distribution. An observation (x, y) arises from the joint distribution with parameters equal to either

$$\mu_1' = \mu_2' = 0, \quad (\sigma_1^2)' = (\sigma_2^2)' = 1, \quad \rho' = \tfrac{1}{2}$$

or

$$\mu_1'' = \mu_2'' = 1, \quad (\sigma_1^2)'' = 4, \quad (\sigma_2^2)'' = 9, \quad \rho'' = \tfrac{1}{2}.$$

Show that the classification rule involves a second degree polynomial in x and y.

9.46. Let $X_1, X_2, \ldots, X_n$ be a random sample from a distribution with one of the two probability density functions $(1/\rho)f_i[(x-\theta)/\rho]$, $-\infty < \theta < \infty$, $\rho > 0$, $i = 1, 2$. We wish to decide from which of these distributions the sample arose. We assign the respective prior probabilities p_1 and p_2 to f_1 and f_2, where $p_1 + p_2 = 1$. If the prior p.d.f. assigned to the nuisance parameters θ and ρ is $g(\theta, \rho)$, the posterior probability of f_i is proportional to $p_i I(f_i|x_1, \ldots, x_n)$, where

$$I(f_i|x_1, \ldots, x_n) = \int_0^\infty \int_{-\infty}^\infty \left(\frac{1}{\rho}\right)^n f_i\left(\frac{x_1-\theta}{\rho}\right) \cdots f_i\left(\frac{x_n-\theta}{\rho}\right) g(\theta, \rho)\, d\theta\, d\rho,$$

$$i = 1, 2.$$

If the losses associated with the two wrong decisions are equal, we would select the p.d.f. with the largest posterior probability.

(a) If $g(\theta; \rho)$ is a vague noninformative prior proportional to $1/\rho$, show that

$$I(f_i|x_1, \ldots, x_n) = \int_0^\infty \int_{-\infty}^\infty \left(\frac{1}{\rho}\right)^{n+1} f_i\left(\frac{x_1-\theta}{\rho}\right) \cdots f_i\left(\frac{x_n-\theta}{\rho}\right) d\theta\, d\rho$$

$$= \int_0^\infty \int_{-\infty}^\infty \lambda^{n-2} f_i(\lambda x_1 - u) \cdots f_i(\lambda x_n - u)\, du\, d\lambda$$

by changing variables through $\theta = \mu/\lambda$, $\rho = 1/\lambda$. Hájek and Šidák show that using this last expression, the Bayesian procedure of selecting f_2 over f_1 if

$$p_2 I(f_2|x_1, \ldots, x_n) \ge p_1 I(f_1|x_1, \ldots, x_n)$$

provides a most powerful location and scale invariant test of one model against another.

(b) Evaluate $I(f_i|x_1, \ldots, x_n)$, $i = 1, 2$, given in (a) for $f_1(x) = \frac{1}{2}$, $-1 < x < 1$, zero elsewhere, and $f_2(x)$ is the p.d.f. of $N(0, 1)$. Show that the most powerful location and scale invariant test for selecting the normal distribution over the uniform is of the form $(Y_n - Y_1)/S \le k$, where $Y_1 < Y_2 < \cdots < Y_n$ are the order statistics and S is the sample standard deviation.

ADDITIONAL EXERCISES

9.47. Consider a random sample $X_1, X_2, \ldots, X_n$ from a distribution with p.d.f. $f(x; \theta) = \theta(1-x)^{\theta-1}$, $0 < x < 1$, zero elsewhere, where $\theta > 0$.

(a) Find the form of the uniformly most powerful test of $H_0: \theta = 1$ against $H_1: \theta > 1$.

(b) What is the likelihood ratio λ for testing $H_0: \theta = 1$ against $H_1: \theta \ne 1$?

9.48. Let $X_1, X_2, \ldots, X_n$ be a random sample from a distribution with p.d.f.

$f(x; \theta) = \theta x^{\theta-1}$, $0 < x < 1$, zero elsewhere.

(a) Find a complete sufficient statistic for θ.

(b) If $\alpha = \beta = \frac{1}{10}$, find the sequential probability ratio test of $H_0: \theta = 2$ against $H_1: \theta = 3$.

9.49. Let X have a Poisson p.d.f. with parameter θ. We shall use a random sample of size n to test $H_0: \theta = 1$ against $H_1: \theta \neq 1$.

(a) Find the likelihood ratio λ for making this test.

(b) Show that λ can be expressed in terms of $\bar{X}$, the mean of the sample, so that the test can be based upon $\bar{X}$.

9.50. Let $X_1, X_2, \ldots, X_n$ and $Y_1, Y_2, \ldots, Y_n$ be independent random samples from two normal distributions $N(\mu_1, \sigma^2)$ and $N(\mu_2, \sigma^2)$, respectively, where σ^2 is the common but unknown variance.

(a) Find the likelihood ratio λ for testing $H_0: \mu_1 = \mu_2 = 0$ against all alternatives.

(b) Rewrite λ so that it is a function of a statistic Z which has a well-known distribution.

(c) Give the distribution of Z under both null and alternative hypotheses.

9.51. Let $X_1, \ldots, X_n$ denote a random sample from a gamma-type distribution with alpha equal to 2 and beta equal to θ. Let $H_0: \theta = 1$ and $H_1: \theta > 1$.

(a) Show that there exists a uniformly most powerful test for H_0 against H_1, determine the statistic Y upon which the test may be based, and indicate the nature of the best critical region.

(b) Find the p.d.f. of the statistic Y in part (a). If we want a significance level of 0.05, write an equation which can be used to determine the critical region. Let $K(\theta)$, $\theta \geq 1$, be the power function of the test. Express the power function as an integral.

9.52. Let $(X_1, Y_1), (X_2, Y_2), \ldots, (X_n, Y_n)$ be a random sample from a bivariate normal distribution with $\mu_1, \mu_2, \sigma_1^2 = \sigma_2^2 = \sigma^2, \rho = \frac{1}{2}$, where μ_1, μ_2, and $\sigma^2 > 0$ are unknown real numbers. Find the likelihood ratio λ for testing $H_0: \mu_1 = \mu_2 = 0$, σ^2 unknown against all alternatives. The likelihood ratio λ is a function of what statistic that has a well-known distribution?

9.53. Let $\mathbf{W}' = (W_1, W_2)$ be an observation from one of two bivariate normal distributions, I and II, each with $\mu_1 = \mu_2 = 0$ but with the respective variance–covariance matrices

$$\mathbf{V}_1 = \begin{pmatrix} 1 & 0 \\ 0 & 4 \end{pmatrix} \quad \text{and} \quad \mathbf{V}_2 = \begin{pmatrix} 3 & 0 \\ 0 & 12 \end{pmatrix}.$$

How would you classify $\mathbf{W}$ into I or II?

9.54. Let X be Poisson θ. Find the sequential probability ratio test for testing $H_0: \theta = 0.05$ against $H_1: \theta = 0.03$. Write this in the form

$c_0(n) < \sum_{i=1}^{n} X_i < c_1(n)$, determining $c_0(n)$ and $c_1(n)$ when $\alpha_a = 0.10$ and $\beta_a = 0.05$.

9.55. Let X and Y have the joint p.d.f.

$$f(x, y; \theta_1, \theta_2) = \frac{1}{\theta_1\theta_2} \exp\left(-\frac{x}{\theta_1} - \frac{y}{\theta_2}\right), \qquad 0 < x < \infty, \quad 0 < y < \infty,$$

zero elsewhere, where $0 < \theta_1, 0 < \theta_2$. An observation (x, y) arises from the joint distribution with $\theta_1' = 10$, $\theta_2' = 5$ or $\theta_1'' = 3$, $\theta_2'' = 2$. Determine the form of the classification rule.

CHAPTER 10

Inferences About Normal Models

10.1 The Distributions of Certain Quadratic Forms

A homogeneous polynomial of degree 2 in n variables is called a *quadratic* form in those variables. If both the variables and the coefficients are real, the form is called a *real quadratic* form. Only real quadratic forms will be considered in this book. To illustrate, the form $X_1^2 + X_1X_2 + X_2^2$ is a quadratic form in the two variables X_1 and X_2; the form $X_1^2 + X_2^2 + X_3^2 - 2X_1X_2$ is a quadratic form in the three variables X_1, X_2, and X_3; but the form $(X_1 - 1)^2 + (X_2 - 2)^2 = X_1^2 + X_2^2 - 2X_1 - 4X_2 + 5$ is not a quadratic form in X_1 and X_2, although it is a quadratic form in the variables $X_1 - 1$ and $X_2 - 2$.

Let $\bar{X}$ and S^2 denote, respectively, the mean and the variance of a random sample $X_1, X_2, \ldots, X_n$ from an arbitrary distribution. Thus

$$nS^2 = \sum_1^n (X_i - \bar{X})^2 = \sum_1^n \left(X_i - \frac{X_1 + X_2 + \cdots + X_n}{n}\right)^2$$

$$= \frac{n-1}{n}(X_1^2 + X_2^2 + \cdots + X_n^2)$$

$$-\frac{2}{n}(X_1X_2 + \cdots + X_1X_n + \cdots + X_{n-1}X_n)$$

is a quadratic form in the n variables $X_1, X_2, \ldots, X_n$. If the sample arises from a distribution that is $N(\mu, \sigma^2)$, we know that the random variable nS^2/σ^2 is $\chi^2(n-1)$ regardless of the value of μ. This fact proved useful in our search for a confidence interval for σ^2 when μ is unknown.

It has been seen that tests of certain statistical hypotheses require a statistic that is a quadratic form. For instance, Example 2, Section 9.2, made use of the statistic $\sum_1^n X_i^2$, which is a quadratic form in the variables $X_1, X_2, \ldots, X_n$. Later in this chapter, tests of other statistical hypotheses will be investigated, and it will be seen that functions of statistics that are quadratic forms will be needed to carry out the tests in an expeditious manner. But first we shall make a study of the distribution of certain quadratic forms in normal and independent random variables.

The following theorem will be proved in Section 10.9.

Theorem 1. *Let $Q = Q_1 + Q_2 + \cdots + Q_{k-1} + Q_k$, where $Q, Q_1, \ldots, Q_k$ are $k+1$ random variables that are real quadratic forms in n independent random variables which are normally distributed with the means $\mu_1, \mu_2, \ldots, \mu_n$ and the same variance σ^2. Let Q/σ^2, $Q_1/\sigma^2, \ldots, Q_{k-1}/\sigma^2$ have chi-square distributions with degrees of freedom $r, r_1, \ldots, r_{k-1}$, respectively. Let Q_k be nonnegative. Then:*

(a) *$Q_1, \ldots, Q_k$ are independent, and hence*
(b) *Q_k/σ^2 has a chi-square distribution with $r - (r_1 + \cdots + r_{k-1}) = r_k$ degrees of freedom.*

Three examples illustrative of the theorem will follow. Each of these examples will deal with a distribution problem that is based on the remarks made in the subsequent paragraph.

Let the random variable X have a distribution that is $N(\mu, \sigma^2)$. Let a and b denote positive integers greater than 1 and let $n = ab$. Consider a random sample of size $n = ab$ from this normal distribution. The observations of the random sample will be denoted by the symbols

$$\begin{array}{cccccc} X_{11}, & X_{12}, & \ldots, & X_{1j}, & \ldots, & X_{1b} \\ X_{21}, & X_{22}, & \ldots, & X_{2j}, & \ldots, & X_{2b} \\ \vdots & & & & & \\ X_{i1}, & X_{i2}, & \ldots, & X_{ij}, & \ldots, & X_{ib} \\ \vdots & & & & & \\ X_{a1}, & X_{a2}, & \ldots, & X_{aj}, & \ldots, & X_{ab}. \end{array}$$

In this notation, the first subscript indicates the row, and the second subscript indicates the column in which the observation appears. Thus X_{ij} is in row i and column j, $i = 1, 2, \ldots, a$ and $j = 1, 2, \ldots, b$. By assumption these $n = ab$ random variables are independent, and each has the same normal distribution with mean μ and variance σ^2. Thus, if we wish, we may consider each row as being a random sample of size b from the given distribution; and we may consider each column as being a random sample of size a from the given distribution. We now define $a + b + 1$ statistics. They are

$$\bar{X}_{..} = \frac{X_{11} + \cdots + X_{1b} + \cdots + X_{a1} + \cdots + X_{ab}}{ab} = \frac{\sum_{i=1}^{a} \sum_{j=1}^{b} X_{ij}}{ab},$$

$$\bar{X}_{i.} = \frac{X_{i1} + X_{i2} + \cdots + X_{ib}}{b} = \frac{\sum_{j=1}^{b} X_{ij}}{b}, \qquad i = 1, 2, \ldots, a,$$

and

$$\bar{X}_{.j} = \frac{X_{1j} + X_{2j} + \cdots + X_{aj}}{a} = \frac{\sum_{i=1}^{a} X_{ij}}{a}, \qquad j = 1, 2, \ldots, b.$$

The statistic $\bar{X}_{..}$ is the mean of the random sample of size $n = ab$; the statistics $\bar{X}_{1.}, \bar{X}_{2.}, \ldots, \bar{X}_{a.}$ are, respectively, the means of the rows; and the statistics $\bar{X}_{.1}, \bar{X}_{.2}, \ldots, \bar{X}_{.b}$ are, respectively, the means of the columns. Three examples illustrative of the theorem will follow.

Example 1. Consider the variance S^2 of the random sample of size $n = ab$. We have the algebraic identity

$$
\begin{aligned}
abS^2 &= \sum_{i=1}^{a}\sum_{j=1}^{b}(X_{ij} - \bar{X}_{..})^2 \\
&= \sum_{i=1}^{a}\sum_{j=1}^{b}[(X_{ij} - \bar{X}_{i.}) + (\bar{X}_{i.} - \bar{X}_{..})]^2 \\
&= \sum_{i=1}^{a}\sum_{j=1}^{b}(X_{ij} - \bar{X}_{i.})^2 + \sum_{i=1}^{a}\sum_{j=1}^{b}(\bar{X}_{i.} - \bar{X}_{..})^2 \\
&\quad + 2\sum_{i=1}^{a}\sum_{j=1}^{b}(X_{ij} - \bar{X}_{i.})(\bar{X}_{i.} - \bar{X}_{..}).
\end{aligned}
$$

The last term of the right-hand member of this identity may be written

$$2\sum_{i=1}^{a}\left[(\bar{X}_{i.} - \bar{X}_{..})\sum_{j=1}^{b}(X_{ij} - \bar{X}_{i.})\right] = 2\sum_{i=1}^{a}[(\bar{X}_{i.} - \bar{X}_{..})(b\bar{X}_{i.} - b\bar{X}_{i.})] = 0,$$

and the term

$$\sum_{i=1}^{a}\sum_{j=1}^{b}(\bar{X}_{i.} - \bar{X}_{..})^2$$

may be written

$$b\sum_{i=1}^{a}(\bar{X}_{i.} - \bar{X}_{..})^2.$$

Thus

$$abS^2 = \sum_{i=1}^{a}\sum_{j=1}^{b}(X_{ij} - \bar{X}_{i.})^2 + b\sum_{i=1}^{a}(\bar{X}_{i.} - \bar{X}_{..})^2,$$

or, for brevity,

$$Q = Q_1 + Q_2.$$

Clearly, Q, Q_1, and Q_2 are quadratic forms in the $n = ab$ variables X_{ij}. We shall use the theorem with $k = 2$ to show that Q_1 and Q_2 are independent. Since S^2 is the variance of a random sample of size $n = ab$ from the given normal distribution, then abS^2/σ^2 has a chi-square distribution with $ab - 1$ degrees of freedom. Now

$$\frac{Q_1}{\sigma^2} = \sum_{i=1}^{a}\left[\frac{\sum_{j=1}^{b}(X_{ij} - \bar{X}_{i.})^2}{\sigma^2}\right].$$

For each fixed value of i, $\sum_{j=1}^{b}(X_{ij} - \bar{X}_{i.})^2/b$ is the variance of a random sample of size b from the given normal distribution, and, accordingly, $\sum_{j=1}^{b} (X_{ij} - \bar{X}_{i.})^2/\sigma^2$ has a chi-square distribution with $b - 1$ degrees of freedom. Because the X_{ij} are independent, Q_1/σ^2 is the sum of a independent random variables, each having a chi-square distribution with $b - 1$ degrees of freedom. Hence Q_1/σ^2 has a chi-square distribution with $a(b - 1)$ degrees of freedom. Now $Q_2 = b \sum_{i=1}^{a} (\bar{X}_{i.} - \bar{X}_{..})^2 \geq 0$. In accordance with the theorem, Q_1 and Q_2 are independent, and Q_2/σ^2 has a chi-square distribution with $ab - 1 - a(b - 1) = a - 1$ degrees of freedom.

Example 2. In abS^2 replace $X_{ij} - \bar{X}_{..}$ by $(X_{ij} - \bar{X}_{.j}) + (\bar{X}_{.j} - \bar{X}_{..})$ to obtain

$$abS^2 = \sum_{j=1}^{b} \sum_{i=1}^{a} [(X_{ij} - \bar{X}_{.j}) + (\bar{X}_{.j} - \bar{X}_{..})]^2,$$

or

$$abS^2 = \sum_{j=1}^{b} \sum_{i=1}^{a} (X_{ij} - \bar{X}_{.j})^2 + a \sum_{j=1}^{b} (\bar{X}_{.j} - \bar{X}_{..})^2,$$

or, for brevity,

$$Q = Q_3 + Q_4.$$

It is easy to show (Exercise 10.1) that Q_3/σ^2 has a chi-square distribution with $b(a - 1)$ degrees of freedom. Since $Q_4 = a \sum_{j=1}^{b} (\bar{X}_{.j} - \bar{X}_{..})^2 \geq 0$, the theorem enables us to assert that Q_3 and Q_4 are independent and that Q_4/σ^2 has a chi-square distribution with $ab - 1 - b(a - 1) = b - 1$ degrees of freedom.

Example 3. In abS^2 replace $X_{ij} - \bar{X}_{..}$ by $(\bar{X}_{i.} - \bar{X}_{..}) + (\bar{X}_{.j} - \bar{X}_{..}) + (X_{ij} - \bar{X}_{i.} - \bar{X}_{.j} + \bar{X}_{..})$ to obtain (Exercise 10.2)

$$abS^2 = b \sum_{i=1}^{a} (\bar{X}_{i.} - \bar{X}_{..})^2 + a \sum_{j=1}^{b} (\bar{X}_{.j} - \bar{X}_{..})^2$$

$$+ \sum_{j=1}^{b} \sum_{i=1}^{a} (X_{ij} - \bar{X}_{i.} - \bar{X}_{.j} + \bar{X}_{..})^2,$$

or, for brevity,

$$Q = Q_2 + Q_4 + Q_5,$$

where Q_2 and Q_4 are as defined in Examples 1 and 2. From Examples 1 and 2, Q/σ^2, Q_2/σ^2, and Q_4/σ^2 have chi-square distributions with $ab - 1$, $a - 1$, and $b - 1$ degrees of freedom, respectively. Since $Q_5 \geq 0$, the theorem asserts that Q_2, Q_4, and Q_5 are independent and that Q_5/σ^2 has a chi-square distribution with $ab - 1 - (a - 1) - (b - 1) = (a - 1)(b - 1)$ degrees of freedom.

Once these quadratic form statistics have been shown to be independent, a multiplicity of F-statistics can be defined. For instance,

$$\frac{Q_4/[\sigma^2(b-1)]}{Q_3/[\sigma^2 b(a-1)]} = \frac{Q_4/(b-1)}{Q_3/[b(a-1)]}$$

has an F-distribution with $b - 1$ and $b(a - 1)$ degrees of freedom; and

$$\frac{Q_4/[\sigma^2(b-1)]}{Q_5/[\sigma^2(a-1)(b-1)]} = \frac{Q_4/(b-1)}{Q_5/(a-1)(b-1)}$$

has an F-distribution with $b - 1$ and $(a - 1)(b - 1)$ degrees of freedom. In the subsequent sections it will be seen that some likelihood ratio tests of certain statistical hypotheses can be based on these F-statistics.

EXERCISES

10.1. In Example 2 verify that $Q = Q_3 + Q_4$ and that Q_3/σ^2 has a chi-square distribution with $b(a - 1)$ degrees of freedom.

10.2. In Example 3 verify that $Q = Q_2 + Q_4 + Q_5$.

10.3. Let $X_1, X_2, \ldots, X_n$ be a random sample from a normal distribution $N(\mu, \sigma^2)$. Show that

$$\sum_{i=1}^{n} (X_i - \bar{X})^2 = \sum_{i=2}^{n} (X_i - \bar{X}')^2 + \frac{n-1}{n}(X_1 - \bar{X}')^2,$$

where $\bar{X} = \sum_{i=1}^{n} X_i/n$ and $\bar{X}' = \sum_{i=2}^{n} X_i/(n-1)$.

Hint: Replace $X_i - \bar{X}$ by $(X_i - \bar{X}') - (X_1 - \bar{X}')/n$. Show that $\sum_{i=2}^{n} (X_i - \bar{X}')^2/\sigma^2$ has a chi-square distribution with $n - 2$ degrees of freedom. Prove that the two terms in the right-hand member are independent. What then is the distribution of

$$\frac{[(n-1)/n](X_1 - \bar{X}')^2}{\sigma^2}?$$

10.4. Let X_{ijk}, $i = 1, \ldots, a$; $j = 1, \ldots, b$; $k = 1, \ldots, c$, be a random sample of size $n = abc$ from a normal distribution $N(\mu, \sigma^2)$. Let $\bar{X}_{\ldots} = \sum_{k=1}^{c} \sum_{j=1}^{b} \sum_{i=1}^{a} X_{ijk}/n$ and $\bar{X}_{i..} = \sum_{k=1}^{c} \sum_{j=1}^{b} X_{ijk}/bc$. Show that

$$\sum_{i=1}^{a} \sum_{j=1}^{b} \sum_{k=1}^{c} (X_{ijk} - \bar{X}_{\ldots})^2 = \sum_{i=1}^{a} \sum_{j=1}^{b} \sum_{k=1}^{c} (X_{ijk} - \bar{X}_{i..})^2 + bc \sum_{i=1}^{a} (\bar{X}_{i..} - \bar{X}_{\ldots})^2.$$

Show that $\sum_{i=1}^{a}\sum_{j=1}^{b}\sum_{k=1}^{c}(X_{ijk}-\bar{X}_{i..})^2/\sigma^2$ has a chi-square distribution with $a(bc-1)$ degrees of freedom. Prove that the two terms in the right-hand member are independent. What, then, is the distribution of $bc\sum_{i=1}^{a}(\bar{X}_{i..}-\bar{X}_{...})^2/\sigma^2$? Furthermore, let $\bar{X}_{.j.}=\sum_{k=1}^{c}\sum_{i=1}^{a}X_{ijk}/ac$ and $\bar{X}_{ij.}=\sum_{k=1}^{c}X_{ijk}/c$. Show that

$$\begin{aligned}\sum_{i=1}^{a}\sum_{j=1}^{b}\sum_{k=1}^{c}(X_{ijk}-\bar{X}_{...})^2 &= \sum_{i=1}^{a}\sum_{j=1}^{b}\sum_{k=1}^{c}(X_{ijk}-\bar{X}_{ij.})^2\\ &\quad + bc\sum_{i=1}^{a}(\bar{X}_{i..}-\bar{X}_{...})^2 + ac\sum_{j=1}^{b}(\bar{X}_{.j.}-\bar{X}_{...})^2\\ &\quad + c\sum_{i=1}^{a}\sum_{j=1}^{b}(\bar{X}_{ij.}-\bar{X}_{i..}-\bar{X}_{.j.}+\bar{X}_{...})^2.\end{aligned}$$

Show that the four terms in the right-hand member, when divided by σ^2, are independent chi-square variables with $ab(c-1)$, $a-1$, $b-1$, and $(a-1)(b-1)$ degrees of freedom, respectively.

10.5. Let X_1, X_2, X_3, X_4 be a random sample of size $n=4$ from the normal distribution $N(0, 1)$. Show that $\sum_{i=1}^{4}(X_i-\bar{X})^2$ equals

$$\frac{(X_1-X_2)^2}{2}+\frac{[X_3-(X_1+X_2)/2]^2}{3/2}+\frac{[X_4-(X_1+X_2+X_3)/3]^2}{4/3}$$

and argue that these three terms are independent, each with a chi-square distribution with 1 degree of freedom.

10.2 A Test of the Equality of Several Means

Consider b independent random variables that have normal distributions with unknown means $\mu_1, \mu_2, \ldots, \mu_b$, respectively, and unknown but common variance σ^2. Let $X_{1j}, X_{2j}, \ldots, X_{aj}$ represent a random sample of size a from the normal distribution with mean μ_j and variance σ^2, $j=1, 2, \ldots, b$. It is desired to test the composite hypothesis $H_0: \mu_1=\mu_2=\cdots=\mu_b=\mu$, μ unspecified, against all possible alternative hypotheses H_1. A likelihood ratio test will be used. Here the total parameter space is

$$\Omega=\{(\mu_1, \mu_2, \ldots, \mu_b, \sigma^2): \quad -\infty<\mu_j<\infty, \quad 0<\sigma^2<\infty\}$$

and

$$\omega = \{(\mu_1, \mu_2, \ldots, \mu_b, \sigma^2): \quad -\infty < \mu_1 = \mu_2 = \cdots = \mu_b = \mu < \infty, \quad 0 < \sigma^2 < \infty\}.$$

The likelihood functions, denoted by $L(\omega)$ and $L(\Omega)$ are, respectively,

$$L(\omega) = \left(\frac{1}{2\pi\sigma^2}\right)^{ab/2} \exp\left[-\frac{1}{2\sigma^2}\sum_{j=1}^{b}\sum_{i=1}^{a}(x_{ij} - \mu)^2\right]$$

and

$$L(\Omega) = \left(\frac{1}{2\pi\sigma^2}\right)^{ab/2} \exp\left[-\frac{1}{2\sigma^2}\sum_{j=1}^{b}\sum_{i=1}^{a}(x_{ij} - \mu_j)^2\right].$$

Now

$$\frac{\partial \ln L(\omega)}{\partial \mu} = \frac{\sum_{j=1}^{b}\sum_{i=1}^{a}(x_{ij} - \mu)}{\sigma^2}$$

and

$$\frac{\partial \ln L(\omega)}{\partial(\sigma^2)} = -\frac{ab}{2\sigma^2} + \frac{1}{2\sigma^4}\sum_{j=1}^{b}\sum_{i=1}^{a}(x_{ij} - \mu)^2.$$

If we equate these partial derivatives to zero, the solutions for μ and σ^2 are, respectively, in ω,

$$\frac{\sum_{j=1}^{b}\sum_{i=1}^{a} x_{ij}}{ab} = \bar{x}_{..},$$

$$\frac{\sum_{j=1}^{b}\sum_{i=1}^{a}(x_{ij} - \bar{x}_{..})^2}{ab} = v, \tag{1}$$

and these values maximize $L(\omega)$. Furthermore,

$$\frac{\partial \ln L(\Omega)}{\partial \mu_j} = \frac{\sum_{i=1}^{a}(x_{ij} - \mu_j)}{\sigma^2}, \qquad j = 1, 2, \ldots, b,$$

and

$$\frac{\partial \ln L(\Omega)}{\partial(\sigma^2)} = -\frac{ab}{2\sigma^2} + \frac{1}{2\sigma^4}\sum_{j=1}^{b}\sum_{i=1}^{a}(x_{ij} - \mu_j)^2.$$

If we equate these partial derivatives to zero, the solutions for $\mu_1, \mu_2, \ldots, \mu_b$, and σ^2 are, respectively, in Ω,

$$\frac{\sum_{i=1}^{a} x_{ij}}{a} = \bar{x}_{.j}, \qquad j = 1, 2, \ldots, b,$$

$$\frac{\sum_{j=1}^{b}\sum_{i=1}^{a} (x_{ij} - \bar{x}_{.j})^2}{ab} = w, \tag{2}$$

and these values maximize $L(\Omega)$. These maxima are, respectively,

$$L(\hat{\omega}) = \left[\frac{ab}{2\pi \sum_{j=1}^{b}\sum_{i=1}^{a} (x_{ij} - \bar{x}_{..})^2}\right]^{ab/2} \exp\left[-\frac{ab \sum_{j=1}^{b}\sum_{i=1}^{a} (x_{ij} - \bar{x}_{..})^2}{2 \sum_{j=1}^{b}\sum_{i=1}^{a} (x_{ij} - \bar{x}_{..})^2}\right]$$

$$= \left[\frac{ab}{2\pi \sum_{j=1}^{b}\sum_{i=1}^{a} (x_{ij} - \bar{x}_{..})^2}\right]^{ab/2} e^{-ab/2}$$

and

$$L(\hat{\Omega}) = \left[\frac{ab}{2\pi \sum_{j=1}^{b}\sum_{i=1}^{a} (x_{ij} - \bar{x}_{.j})^2}\right]^{ab/2} e^{-ab/2}.$$

Finally,

$$\lambda = \frac{L(\hat{\omega})}{L(\hat{\Omega})} = \left[\frac{\sum_{j=1}^{b}\sum_{i=1}^{a} (x_{ij} - \bar{x}_{.j})^2}{\sum_{j=1}^{b}\sum_{i=1}^{a} (x_{ij} - \bar{x}_{..})^2}\right]^{ab/2}.$$

In the notation of Section 10.1, the statistics defined by the functions $\bar{x}_{..}$ and v given by Equations (1) of this section are

$$\bar{X}_{..} = \sum_{j=1}^{b}\sum_{i=1}^{a} \frac{X_{ij}}{ab} \qquad \text{and} \qquad S^2 = \sum_{j=1}^{b}\sum_{i=1}^{a} \frac{(X_{ij} - \bar{X}_{..})^2}{ab} = \frac{Q}{ab};$$

while the statistics defined by the functions $\bar{x}_{.1}, \bar{x}_{.2}, \ldots, \bar{x}_{.b}$ and w given by Equations (2) in this section are, respectively, $\bar{X}_{.j} = \sum_{i=1}^{a} X_{ij}/a$,

$j = 1, 2, \ldots, b$, and $Q_3/ab = \sum_{j=1}^{b} \sum_{i=1}^{a} (X_{ij} - \bar{X}_{.j})^2/ab$. Thus, in the notation of Section 10.1, $\lambda^{2/ab}$ defines the statistic Q_3/Q.

We reject the hypothesis H_0 if $\lambda \le \lambda_0$. To find λ_0 so that we have a desired significance level α, we must assume that the hypothesis H_0 is true. If the hypothesis H_0 is true, the random variables X_{ij} constitute a random sample of size $n = ab$ from a distribution that is normal with mean μ and variance σ^2. This being the case, it was shown in Example 2, Section 10.1, that $Q = Q_3 + Q_4$, where $Q_4 = a \sum_{j=1}^{b} (\bar{X}_{.j} - \bar{X}_{..})^2$; that Q_3 and Q_4 are independent; and that Q_3/σ^2 and Q_4/σ^2 have chi-square distributions with $b(a - 1)$ and $b - 1$ degrees of freedom, respectively. Thus the statistic defined by $\lambda^{2/ab}$ may be written

$$\frac{Q_3}{Q_3 + Q_4} = \frac{1}{1 + Q_4/Q_3}.$$

The significance level of the test of H_0 is

$$\alpha = \Pr\left[\frac{1}{1 + Q_4/Q_3} \le \lambda_0^{2/ab}; H_0\right]$$

$$= \Pr\left[\frac{Q_4/(b - 1)}{Q_3/[b(a - 1)]} \ge c; H_0\right],$$

where

$$c = \frac{b(a - 1)}{b - 1}(\lambda_0^{-2/ab} - 1).$$

But

$$F = \frac{Q_4/[\sigma^2(b - 1)]}{Q_3/[\sigma^2 b(a - 1)]} = \frac{Q_4/(b - 1)}{Q_3/[b(a - 1)]}$$

has an F-distribution with $b - 1$ and $b(a - 1)$ degrees of freedom. Hence the test of the composite hypothesis $H_0 : \mu_1 = \mu_2 = \cdots = \mu_b = \mu$, μ unspecified, against all possible alternatives may be based on an F-statistic. The constant c is so selected as to yield the desired value of α.

Remark. It should be pointed out that a test of the equality of the b means μ_j, $j = 1, 2, \ldots, b$, does not require that we take a random sample of size a from each of the b normal distributions. That is, the samples may be of different sizes, say $a_1, a_2, \ldots, a_b$. A consideration of this procedure is left to Exercise 10.6.

Suppose now that we wish to compute the power of the test of H_0 against H_1 when H_0 is false, that is, when we do not have $\mu_1 = \mu_2 = \cdots = \mu_b = \mu$. It will be seen in Section 10.3 that when H_1 is true, no longer is Q_4/σ^2 a random variable that is $\chi^2(b-1)$. Thus we cannot use an F-statistic to compute the power of the test when H_1 is true. This problem is discussed in Section 10.3.

An observation should be made in connection with maximizing a likelihood function with respect to certain parameters. Sometimes it is easier to avoid the use of the calculus. For example, $L(\Omega)$ of this section can be maximized with respect to μ_j, for every fixed positive σ^2, by minimizing

$$z = \sum_{j=1}^{b} \sum_{i=1}^{a} (x_{ij} - \mu_j)^2$$

with respect to μ_j, $j = 1, 2, \ldots, b$. Now z can be written as

$$z = \sum_{j=1}^{b} \sum_{i=1}^{a} [(x_{ij} - \bar{x}_{.j}) + (\bar{x}_{.j} - \mu_j)]^2$$

$$= \sum_{j=1}^{b} \sum_{i=1}^{a} (x_{ij} - \bar{x}_{.j})^2 + a \sum_{j=1}^{b} (\bar{x}_{.j} - \mu_j)^2.$$

Since each term in the right-hand member of the preceding equation is nonnegative, clearly z is a minimum, with respect to μ_j, if we take $\mu_j = \bar{x}_{.j}$, $j = 1, 2, \ldots, b$.

EXERCISES

10.6. Let $X_{1j}, X_{2j}, \ldots, X_{a_j j}$ represent independent random samples of sizes a_j from normal distributions with means μ_j and variances σ^2, $j = 1, 2, \ldots, b$. Show that

$$\sum_{j=1}^{b} \sum_{i=1}^{a_j} (X_{ij} - \bar{X}_{..})^2 = \sum_{j=1}^{b} \sum_{i=1}^{a_j} (X_{ij} - \bar{X}_{.j})^2 + \sum_{j=1}^{b} a_j(\bar{X}_{.j} - \bar{X}_{..})^2,$$

or $Q' = Q_3' + Q_4'$. Here $\bar{X}_{..} = \sum_{j=1}^{b} \sum_{i=1}^{a_j} X_{ij} / \sum_{j=1}^{b} a_j$ and $\bar{X}_{.j} = \sum_{i=1}^{a_j} X_{ij}/a_j$. If $\mu_1 = \mu_2 = \cdots = \mu_b$, show that Q'/σ^2 and Q_3'/σ^2 have chi-square distributions. Prove that Q_3' and Q_4' are independent, and hence Q_4'/σ^2 also has a chi-square distribution. If the likelihood ratio λ is used to test $H_0: \mu_1 = \mu_2 = \cdots = \mu_b = \mu$, μ unspecified and σ^2 unknown, against all possible alternatives, show that $\lambda \le \lambda_0$ is equivalent to the computed $F \ge c$, where

$$F = \frac{\left(\sum_{j=1}^{b} a_j - b\right)Q_4'}{(b-1)Q_3'}.$$

What is the distribution of F when H_0 is true?

10.7. Consider the T-statistic that was derived through a likelihood ratio for testing the equality of the means of two normal distributions having common variance in Example 2 in Section 9.3. Show that T^2 is exactly the F-statistic of Exercise 10.6 with $a_1 = n$, $a_2 = m$, and $b = 2$. Of course, $X_1, \ldots, X_n, \bar{X}$ are replaced with $X_{11}, \ldots, X_{1n}, \bar{X}_{1.}$ and $Y_1, \ldots, Y_m, \bar{Y}$ by $X_{21}, \ldots, X_{2m}, \bar{X}_{2.}$.

10.8. In Exercise 10.6, show that the linear functions $X_{ij} - \bar{X}_{.j}$ and $\bar{X}_{.j} - \bar{X}_{..}$ are uncorrelated.

Hint: Recall the definitions of $\bar{X}_{.j}$ and $\bar{X}_{..}$ and, without loss of generality, we can let $E(X_{ij}) = 0$ for all i, j.

10.9. The following are observations associated with independent random samples from three normal distributions having equal variances and respective means μ_1, μ_2, μ_3.

I	II	III
0.5	2.1	3.0
1.3	3.3	5.1
−1.0	0.0	1.9
1.8	2.3	2.4
	2.5	4.2
		4.1

Compute the F-statistic that is used to test $H_0: \mu_1 = \mu_2 = \mu_3$.

10.10. Using the notation of this section, assume that the means satisfy the condition that $\mu = \mu_1 + (b-1)d = \mu_2 - d = \mu_3 - d = \cdots = \mu_b - d$. That is, the last $b - 1$ means are equal but differ from the first mean μ_1, provided that $d \ne 0$. Let independent random samples of size a be taken from the b normal distributions with common unknown variance σ^2.

(a) Show that the maximum likelihood estimators of μ and d are $\hat{\mu} = \bar{X}_{..}$ and

$$\hat{d} = \frac{\sum_{j=2}^{b} \bar{X}_{.j}/(b-1) - \bar{X}_{.1}}{b}.$$

(b) Using Exercise 10.3, find Q_6 and $Q_7 = c\hat{d}^2$ so that, when $d = 0$, Q_7/σ^2 is $\chi^2(1)$ and

$$\sum_{i=1}^{a} \sum_{j=1}^{b} (X_{ij} - \bar{X}_{..})^2 = Q_3 + Q_6 + Q_7.$$

(c) Argue that the three terms in the right-hand member of part (b), once divided by σ^2, are independent random variables with chi-square distributions, provided that $d = 0$.

(d) The ratio $Q_7/(Q_3 + Q_6)$ times what constant has an F-distribution, provided that $d = 0$? Note that this F is really the square of the two-sample T used to test the equality of the mean of the first distribution and the common mean of the other distributions, in which the last $b - 1$ samples are combined into one.

10.3 Noncentral χ^2 and Noncentral F

Let $X_1, X_2, \ldots, X_n$ denote independent random variables that are $N(\mu_i, \sigma^2)$, $i = 1, 2, \ldots, n$, and let $Y = \sum_1^n X_i^2/\sigma^2$. If each μ_i is zero, we know that Y is $\chi^2(n)$. We shall now investigate the distribution of Y when each μ_i is not zero. The m.g.f. of Y is given by

$$M(t) = E\left[\exp\left(t \sum_{i=1}^{n} \frac{X_i^2}{\sigma^2}\right)\right]$$

$$= \prod_{i=1}^{n} E\left[\exp\left(t \frac{X_i^2}{\sigma^2}\right)\right].$$

Consider

$$E\left[\exp\left(\frac{tX_i^2}{\sigma^2}\right)\right] = \int_{-\infty}^{\infty} \frac{1}{\sigma\sqrt{2\pi}} \exp\left[\frac{tx_i^2}{\sigma^2} - \frac{(x_i - \mu_i)^2}{2\sigma^2}\right] dx_i.$$

The integral exists if $t < \frac{1}{2}$. To evaluate the integral, note that

$$\frac{tx_i^2}{\sigma^2} - \frac{(x_i - \mu_i)^2}{2\sigma^2} = -\frac{x_i^2(1 - 2t)}{2\sigma^2} + \frac{2\mu_i x_i}{2\sigma^2} - \frac{\mu_i^2}{2\sigma^2}$$

$$= \frac{t\mu_i^2}{\sigma^2(1 - 2t)} - \frac{1 - 2t}{2\sigma^2}\left(x_i - \frac{\mu_i}{1 - 2t}\right)^2.$$

Accordingly, with $t < \frac{1}{2}$, we have

$$E\left[\exp\left(\frac{tX_i^2}{\sigma^2}\right)\right] = \exp\left[\frac{t\mu_i^2}{\sigma^2(1-2t)}\right]\int_{-\infty}^{\infty}\frac{1}{\sigma\sqrt{2\pi}}$$
$$\times \exp\left[-\frac{1-2t}{2\sigma^2}\left(x_i - \frac{\mu_i}{1-2t}\right)^2\right] dx_i.$$

If we multiply the integrand by $\sqrt{1-2t}$, $t < \frac{1}{2}$, we have the integral of a normal p.d.f. with mean $\mu_i/(1-2t)$ and variance $\sigma^2/(1-2t)$. Thus

$$E\left[\exp\left(\frac{tX_i^2}{\sigma^2}\right)\right] = \frac{1}{\sqrt{1-2t}}\exp\left[\frac{t\mu_i^2}{\sigma^2(1-2t)}\right],$$

and the m.g.f. of $Y = \sum_1^n X_i^2/\sigma^2$ is given by

$$M(t) = \frac{1}{(1-2t)^{n/2}}\exp\left[\frac{t\sum_1^n \mu_i^2}{\sigma^2(1-2t)}\right], \qquad t < \frac{1}{2}.$$

A random variable that has an m.g.f. of the functional form

$$M(t) = \frac{1}{(1-2t)^{r/2}}e^{t\theta/(1-2t)},$$

where $t < \frac{1}{2}$, $0 < \theta$, and r is a positive integer, is said to have a *noncentral chi-square distribution* with r degrees of freedom and noncentrality parameter θ. If one sets the noncentrality parameter $\theta = 0$, one has $M(t) = (1-2t)^{-r/2}$, which is the m.g.f. of a random variable that is $\chi^2(r)$. Such a random variable can appropriately be called a *central chi-square variable*. We shall use the symbol $\chi^2(r, \theta)$ to denote a noncentral chi-square distribution that has the parameters r and θ; and we shall say that a random variable is $\chi^2(r, \theta)$ when that random variable has this kind of distribution. The symbol $\chi^2(r, 0)$ is equivalent to $\chi^2(r)$. Thus our random variable $Y = \sum_1^n X_i^2/\sigma^2$ of this section is $\chi^2\left(n, \sum_1^n \mu_i^2/\sigma^2\right)$. If each μ_i is equal to zero, then Y is $\chi^2(n, 0)$ or, more simply, Y is $\chi^2(n)$.

The noncentral chi-square variables in which we have interest are certain quadratic forms, in normally distributed variables, divided by a variance σ^2. In our example it is worth noting that the noncentrality parameter of $\sum_1^n X_i^2/\sigma^2$, which is $\sum_1^n \mu_i^2/\sigma^2$, may be computed by replacing each X_i in the quadratic form by its mean μ_i, $i = 1, 2, \ldots, n$. This is no fortuitous circumstance; any quadratic form $Q = Q(X_1, \ldots, X_n)$ in normally distributed variables, which is such that Q/σ^2 is $\chi^2(r, \theta)$, has $\theta = Q(\mu_1, \mu_2, \ldots, \mu_n)/\sigma^2$; and if Q/σ^2 is a chi-square variable (central or noncentral) for certain real values of $\mu_1, \mu_2, \ldots, \mu_n$, it is chi-square (central or noncentral) for *all* real values of these means.

It should be pointed out that Theorem 1, Section 10.1, is valid whether the random variables are central or noncentral chi-square variables.

We next discuss a noncentral F-variable. If U and V are independent and are, respectively, $\chi^2(r_1)$ and $\chi^2(r_2)$, the random variable F has been defined by $F = r_2 U/r_1 V$. Now suppose, in particular, that U is $\chi^2(r_1, \theta)$, V is $\chi^2(r_2)$, and that U and V are independent. The random variable $r_2 U/r_1 V$ is called a *noncentral F-variable* with r_1 and r_2 degrees of freedom and with noncentrality parameter θ. Note that the noncentrality parameter of F is precisely the noncentrality parameter of the random variable U, which is $\chi^2(r_1, \theta)$.

Tables of noncentral chi-square and noncentral F are available in the literature. However, like those of noncentral t, they are too bulky to be put in this book.

EXERCISES

10.11. Let Y_i, $i = 1, 2, \ldots, n$, denote independent random variables that are, respectively, $\chi^2(r_i, \theta_i)$, $i = 1, 2, \ldots, n$. Prove that $Z = \sum_1^n Y_i$ is $\chi^2\left(\sum_1^n r_i, \sum_1^n \theta_i\right)$.

10.12. Compute the mean and the variance of a random variable that is $\chi^2(r, \theta)$.

10.13. Compute the mean of a random variable that has a noncentral F-distribution with degrees of freedom r_1 and $r_2 > 2$ and noncentrality parameter θ.

10.14. Show that the square of a noncentral T random variable is a noncentral F random variable.

10.15. Let X_1 and X_2 be two independent random variables. Let X_1 and $Y = X_1 + X_2$ be $\chi^2(r_1, \theta_1)$ and $\chi^2(r, \theta)$, respectively. Here $r_1 < r$ and $\theta_1 \le \theta$. Show that X_2 is $\chi^2(r - r_1, \theta - \theta_1)$.

10.16. In Exercise 10.6, if $\mu_1, \mu_2, \ldots, \mu_b$ are not equal, what are the distributions of Q_3'/σ^2, Q_4'/σ^2, and F?

10.4 Multiple Comparisons

Consider b independent random variables that have normal distributions with unknown means $\mu_1, \mu_2, \ldots, \mu_b$, respectively, and with unknown but common variance σ^2. Let $k_1, k_2, \ldots, k_b$ represent b known real constants that are not all zero. We want to find a confidence interval for $\sum_1^b k_j\mu_j$, a linear function of the means $\mu_1, \mu_2, \ldots, \mu_b$. To do this, we take a random sample $X_{1j}, X_{2j}, \ldots, X_{aj}$ of size a from the distribution $N(\mu_j, \sigma^2)$, $j = 1, 2, \ldots, b$. If we denote $\sum_{i=1}^a X_{ij}/a$ by $\bar{X}_{.j}$, then we know that $\bar{X}_{.j}$ is $N(\mu_j, \sigma^2/a)$, that $\sum_{i=1}^a (X_{ij} - \bar{X}_{.j})^2/\sigma^2$ is $\chi^2(a-1)$, and that the two random variables are independent. Since the independent random samples are taken from the b distributions, the $2b$ random variables $\bar{X}_{.j}$, $\sum_{i=1}^a (X_{ij} - \bar{X}_{.j})^2/\sigma^2$, $j = 1, 2, \ldots, b$, are independent. Moreover, $\bar{X}_{.1}, \bar{X}_{.2}, \ldots, \bar{X}_{.b}$ and

$$\sum_{j=1}^b \sum_{i=1}^a \frac{(X_{ij} - \bar{X}_{.j})^2}{\sigma^2}$$

are independent and the latter is $\chi^2[b(a-1)]$. Let $Z = \sum_1^b k_j \bar{X}_{.j}$. Then Z is normal with mean $\sum_1^b k_j \mu_j$ and variance $\left(\sum_1^b k_j^2\right)\sigma^2/a$, and Z is independent of

$$V = \frac{1}{b(a-1)} \sum_{j=1}^b \sum_{i=1}^a (X_{ij} - \bar{X}_{.j})^2.$$

Hence the random variable

$$T = \frac{\dfrac{\sum_1^b k_j \bar{X}_{.j} - \sum_1^b k_j \mu_j}{\sqrt{\left(\sum_1^b k_i^2\right)\sigma^2/a}}}{\sqrt{V/\sigma^2}} = \frac{\sum_1^b k_j \bar{X}_{.j} - \sum_1^b k_j \mu_j}{\sqrt{\left(\sum_1^b k_j^2\right)V/a}}$$

has a t-distribution with $b(a - 1)$ degrees of freedom. A positive number c can be found in Table IV in Appendix B, for certain values of α, $0 < \alpha < 1$, such that $\Pr(-c \le T \le c) = 1 - \alpha$. It follows that the probability is $1 - \alpha$ that

$$\sum_1^b k_j \bar{X}_{.j} - c\sqrt{\left(\sum_1^b k_j^2\right)\frac{V}{a}} \le \sum_1^b k_j \mu_j \le \sum_1^b k_j \bar{X}_{.j} + c\sqrt{\left(\sum_1^b k_j^2\right)\frac{V}{a}}.$$

The experimental values of $\bar{X}_{.j}$, $j = 1, 2, \ldots, b$, and V will provide a $100(1 - \alpha)$ percent confidence interval for $\sum_1^b k_j \mu_j$.

It should be observed that the confidence interval for $\sum_1^b k_j \mu_j$ depends upon the particular choice of $k_1, k_2, \ldots, k_b$. It is conceivable that we may be interested in more than one linear function of $\mu_1, \mu_2, \ldots, \mu_b$, such as $\mu_2 - \mu_1$, $\mu_3 - (\mu_1 + \mu_2)/2$, or $\mu_1 + \cdots + \mu_b$. We can, of course, find for each $\sum_1^b k_j \mu_j$ a random interval that has a preassigned probability of including that particular $\sum_1^b k_j \mu_j$. But how can we compute the probability that *simultaneously* these random intervals include their respective linear functions of $\mu_1, \mu_2, \ldots, \mu_b$? The following procedure of multiple comparisons, due to Scheffé, is one solution to this problem.

The random variable

$$\frac{\sum_{j=1}^b (\bar{X}_{.j} - \mu_j)^2}{\sigma^2/a}$$

is $\chi^2(b)$ and, because it is a function of $\bar{X}_{.1}, \ldots, \bar{X}_{.b}$ alone, it is independent of the random variable

$$V = \frac{1}{b(a-1)} \sum_{j=1}^b \sum_{i=1}^a (X_{ij} - \bar{X}_{.j})^2.$$

Hence the random variable

$$F = \frac{a \sum_{j=1}^{b} (\bar{X}_{.j} - \mu_j)^2/b}{V}$$

has an F-distribution with b and $b(a - 1)$ degrees of freedom. From Table V in Appendix B, for certain values of α, we can find a constant d such that $\Pr(F \le d) = 1 - \alpha$ or

$$\Pr\left[\sum_{j=1}^{b} (\bar{X}_{.j} - \mu_j)^2 \le bd\frac{V}{a}\right] = 1 - \alpha.$$

Note that $\sum_{j=1}^{b} (\bar{X}_{.j} - \mu_j)^2$ is the square of the distance, in b-dimensional space, from the point $(\mu_1, \mu_2, \ldots, \mu_b)$ to the random point $(\bar{X}_{.1}, \bar{X}_{.2}, \ldots, \bar{X}_{.b})$. Consider a space of dimension b and let $(t_1, t_2, \ldots, t_b)$ denote the coordinates of a point in that space. An equation of a hyperplane that passes through the point $(\mu_1, \mu_2, \ldots, \mu_b)$ is given by

$$k_1(t_1 - \mu_1) + k_2(t_2 - \mu_2) + \cdots + k_b(t_b - \mu_b) = 0, \tag{1}$$

where not all the real numbers k_j, $j = 1, 2, \ldots, b$, are equal to zero. The square of the distance from this hyperplane to the point $(t_1 = \bar{X}_{.1}, t_2 = \bar{X}_{.2}, \ldots, t_b = \bar{X}_{.b})$ is

$$\frac{[k_1(\bar{X}_{.1} - \mu_1) + k_2(\bar{X}_{.2} - \mu_2) + \cdots + k_b(\bar{X}_{.b} - \mu_b)]^2}{k_1^2 + k_2^2 + \cdots + k_b^2}. \tag{2}$$

From the geometry of the situation it follows that $\sum_{1}^{b} (\bar{X}_{.j} - \mu_j)^2$ is equal to the maximum of expression (2) with respect to $k_1, k_2, \ldots, k_b$. Thus the inequality $\sum_{1}^{b} (\bar{X}_{.j} - \mu_j)^2 \le (bd)(V/a)$ holds if and only if

$$\frac{\left[\sum_{j=1}^{b} k_j(\bar{X}_{.j} - \mu_j)\right]^2}{\sum_{j=1}^{b} k_j^2} \le bd\frac{V}{a}, \tag{3}$$

for every real $k_1, k_2, \ldots, k_b$, not all zero. Accordingly, these two equivalent events have the same probability, $1 - \alpha$. However, inequality (3) may be written in the form

$$\left|\sum_{1}^{b} k_j\bar{X}_{.j} - \sum_{1}^{b} k_j\mu_j\right| \le \sqrt{bd\left(\sum_{1}^{b} k_j^2\right)\frac{V}{a}}.$$

Thus the probability is $1 - \alpha$ that simultaneously, for *all* real $k_1, k_2, \ldots, k_b$, not all zero,

$$\sum_1^b k_j\bar{X}_{.j} - \sqrt{bd\left(\sum_1^b k_j^2\right)\frac{V}{a}} \le \sum_1^b k_j\mu_j \le \sum_1^b k_j\bar{X}_{.j} + \sqrt{bd\left(\sum_1^b k_j^2\right)\frac{V}{a}}. \quad (4)$$

Denote by A the event where inequality (4) is true for all real $k_1, \ldots, k_b$, and denote by B the event where that inequality is true for a finite number of b-tuples $(k_1, \ldots, k_b)$. If the event A occurs, certainly the event B occurs. Hence $P(A) \le P(B)$. In the applications, one is often interested only in a finite number of linear functions $\sum_1^b k_j\mu_j$. Once the experimental values are available, we obtain from (4) a confidence interval for each of these linear functions. Since $P(B) \ge P(A) = 1 - \alpha$, we have a confidence coefficient of at least $100(1 - \alpha)$ percent that the linear functions are in these respective confidence intervals.

Remarks. If the sample sizes, say $a_1, a_2, \ldots, a_b$, are unequal, inequality (4) becomes

$$\sum_1^b k_j\bar{X}_{.j} - \sqrt{bd\sum_1^b \frac{k_j^2}{a_j} V} \le \sum_1^b k_j\mu_j \le \sum_1^b k_j\bar{X}_{.j} + \sqrt{bd\sum_1^b \frac{k_j^2}{a_j} V}, \quad (4')$$

where

$$\bar{X}_{.j} = \frac{\sum_{i=1}^{a_j} X_{ij}}{a_j}, \qquad V = \frac{\sum_{j=1}^{b}\sum_{i=1}^{a_j} (X_{ij} - \bar{X}_{.j})^2}{\sum_1^b (a_j - 1)},$$

and d is selected from Table V with b and $\sum_1^b (a_j - 1)$ degrees of freedom. Inequality (4′) reduces to inequality (4) when $a_1 = a_2 = \cdots = a_b$.

Moreover, if we restrict our attention to linear functions of the form $\sum_1^b k_j\mu_j$ with $\sum_1^b k_j = 0$ (such linear functions are called *contrasts*), the radical in inequality (4′) is replaced by

$$\sqrt{d(b-1)\sum_1^b \frac{k_j^2}{a_j} V},$$

where d is now found in Table V with $b - 1$ and $\sum_1^b (a_j - 1)$ degrees of freedom.

In these multiple comparisons, one often finds that the length of a confidence interval is much greater than the length of a $100(1-\alpha)$ percent confidence interval for a particular linear function $\sum_1^b k_j\mu_j$. But this is to be expected because in one case the probability $1-\alpha$ applies to just one event, and in the other it applies to the simultaneous occurrence of many events. One reasonable way to reduce the length of these intervals is to take a larger value of α, say 0.25, instead of 0.05. After all, it is still a very strong statement to say that the probability is 0.75 that *all* these events occur.

EXERCISES

10.17. If $A_1, A_2, \ldots, A_k$ are events, prove, by induction, Boole's inequality $P(A_1 \cup A_2 \cup \cdots \cup A_k) \le \sum_1^k P(A_i)$. Then show that

$$P(A_1^* \cap A_2^* \cap \cdots \cap A_k^*) \ge 1 - \sum_1^k P(A_i).$$

10.18. In the notation of this section, let $(k_{i1}, k_{i2}, \ldots, k_{ib})$, $i = 1, 2, \ldots, m$, represent a finite number of b-tuples. The problem is to find simultaneous confidence intervals for $\sum_{j=1}^b k_{ij}\mu_j$, $i = 1, 2, \ldots, m$, by a method different from that of Scheffé. Define the random variable T_i by

$$\left(\sum_{j=1}^b k_{ij}\bar{X}_{.j} - \sum_{j=1}^b k_{ij}\mu_j\right)\Bigg/\sqrt{\left(\sum_{j=1}^b k_{ij}^2\right)V/a}, \qquad i = 1, 2, \ldots, m.$$

(a) Let the event A_i^* be given by $-c_i \le T_i \le c_i$, $i = 1, 2, \ldots, m$. Find the random variables U_i and W_i such that $U_i \le \sum_{j=1}^b k_{ij}\mu_j \le W_i$ is equivalent to A_i^*.

(b) Select c_i such that $P(A_i^*) = 1 - \alpha/m$; that is, $P(A_i) = \alpha/m$. Use the results of Exercise 10.17 to determine a lower bound on the probability that simultaneously the random intervals $(U_1, W_1), \ldots, (U_m, W_m)$ include $\sum_{j=1}^b k_{1j}\mu_j, \ldots, \sum_{j=1}^b k_{mj}\mu_j$, respectively.

(c) Let $a = 3$, $b = 6$, and $\alpha = 0.05$. Consider the linear functions $\mu_1 - \mu_2$, $\mu_2 - \mu_3$, $\mu_3 - \mu_4$, $\mu_4 - (\mu_5 + \mu_6)/2$, and $(\mu_1 + \mu_2 + \cdots + \mu_6)/6$. Here $m = 5$. Show that the lengths of the confidence intervals given by the results of part (b) are shorter than the corresponding ones given by the method of Scheffé, as described in the text. If m becomes sufficiently large, however, this is not the case.

10.5 The Analysis of Variance

The problem considered in Section 10.2 is an example of a method of statistical inference called the *analysis of variance*. This method derives its name from the fact that the quadratic form abS^2, which is a total sum of squares, is resolved into several component parts. In this section other problems in the analysis of variance will be investigated.

Let X_{ij}, $i = 1, 2, \ldots, a$ and $j = 1, 2, \ldots, b$, denote $n = ab$ random variables that are independent and have normal distributions with common variance σ^2. The means of these normal distributions are $\mu_{ij} = \mu + \alpha_i + \beta_j$, where $\sum_1^a \alpha_i = 0$ and $\sum_1^b \beta_j = 0$. For example, take $a = 2$, $b = 3$, $\mu = 5$, $\alpha_1 = 1$, $\alpha_2 = -1$, $\beta_1 = 1$, $\beta_2 = 0$, and $\beta_3 = -1$. Then the $ab =$ six random variables have means

$$\begin{array}{lll} \mu_{11} = 7, & \mu_{12} = 6, & \mu_{13} = 5, \\ \mu_{21} = 5, & \mu_{22} = 4, & \mu_{23} = 3. \end{array}$$

Had we taken $\beta_1 = \beta_2 = \beta_3 = 0$, the six random variables would have had means

$$\begin{array}{lll} \mu_{11} = 6, & \mu_{12} = 6, & \mu_{13} = 6, \\ \mu_{21} = 4, & \mu_{22} = 4, & \mu_{23} = 4. \end{array}$$

Thus, if we wish to test the composite hypothesis that

$$\begin{gathered} \mu_{11} = \mu_{12} = \cdots = \mu_{1b}, \\ \mu_{21} = \mu_{22} = \cdots = \mu_{2b}, \\ \vdots \\ \mu_{a1} = \mu_{a2} = \cdots = \mu_{ab}, \end{gathered}$$

we could say that we are testing the composite hypothesis that $\beta_1 = \beta_2 = \cdots = \beta_b$ (and hence each $\beta_j = 0$, since their sum is zero). On the other hand, the composite hypothesis

$$\begin{gathered} \mu_{11} = \mu_{21} = \cdots = \mu_{a1}, \\ \mu_{12} = \mu_{22} = \cdots = \mu_{a2}, \\ \vdots \\ \mu_{1b} = \mu_{2b} = \cdots = \mu_{ab}, \end{gathered}$$

is the same as the composite hypothesis that $\alpha_1 = \alpha_2 = \cdots = \alpha_a = 0$.

Remarks. The model just described, and others similar to it, are widely used in statistical applications. Consider a situation in which it is desirable to investigate the effects of two factors that influence an outcome. Thus the variety of a grain and the type of fertilizer used influence the yield; or the teacher and the size of a class may influence the score on a standard test. Let X_{ij} denote the yield from the use of variety i of a grain and type j of fertilizer. A test of the hypothesis that $\beta_1 = \beta_2 = \cdots = \beta_b = 0$ would then be a test of the hypothesis that the mean yield of each variety of grain is the same regardless of the type of fertilizer used.

There is no loss of generality in assuming that $\sum_1^a \alpha_i = \sum_1^b \beta_j = 0$. To see this, let $\mu_{ij} = \mu' + \alpha_i' + \beta_j'$. Write $\bar{\alpha}' = \Sigma\, \alpha_i'/a$ and $\bar{\beta}' = \Sigma\, \beta_j'/b$. We have $\mu_{ij} = (\mu' + \bar{\alpha}' + \bar{\beta}') + (\alpha_i' - \bar{\alpha}') + (\beta_j' - \bar{\beta}') = \mu + \alpha_i + \beta_j$, where $\Sigma\, \alpha_i = \Sigma\, \beta_j = 0$.

To construct a test of the composite hypothesis $H_0: \beta_1 = \beta_2 = \cdots = \beta_b = 0$ against all alternative hypotheses, we could obtain the corresponding likelihood ratio. However, to gain more insight into such a test, let us reconsider the likelihood ratio test of Section 10.2, namely that of the equality of the means of b distributions. There the important quadratic forms are Q, Q_3, and Q_4, which are related through the equation $Q = Q_4 + Q_3$. That is,

$$abS^2 = \sum_{j=1}^{b} \sum_{i=1}^{a} (\bar{X}_{.j} - \bar{X}_{..})^2 + \sum_{i=1}^{a} \sum_{j=1}^{b} (X_{ij} - \bar{X}_{.j})^2;$$

so we see that the total sum of squares, abS^2, is decomposed into a sum of squares, Q_4, *among* column means and a sum of squares, Q_3, *within* columns. The latter sum of squares, divided by $n = ab$, is the m.l.e. of σ^2, provided that the parameters are in Ω; and we denote it by $\widehat{\sigma^2_\Omega}$. Of course, S^2 is the m.l.e. of σ^2 under ω, here denoted by $\widehat{\sigma^2_\omega}$. So the likelihood ratio $\lambda = (\widehat{\sigma^2_\Omega}/\widehat{\sigma^2_\omega})^{ab/2}$ is a monotone function of the statistic

$$F = \frac{Q_4/(b-1)}{Q_3/[b(a-1)]}$$

upon which the test of the equality of means is based.

To help find a test for $H_0: \beta_1 = \beta_2 = \cdots = \beta_b = 0$, where $\mu_{ij} = \mu + \alpha_i + \beta_j$, return to the decomposition of Example 3, Section 10.1, namely $Q = Q_2 + Q_4 + Q_5$. That is,

$$abS^2 = \sum_{i=1}^{a} \sum_{j=1}^{b} (\bar{X}_{i.} - \bar{X}_{..})^2 + \sum_{i=1}^{a} \sum_{j=1}^{b} (\bar{X}_{.j} - \bar{X}_{..})^2$$

$$+ \sum_{i=1}^{a} \sum_{j=1}^{b} (X_{ij} - \bar{X}_{i.} - \bar{X}_{.j} + \bar{X}_{..})^2;$$

thus the total sum of squares, abS^2, is decomposed into that among *rows* (Q_2), that among *columns* (Q_4), and that *remaining* (Q_5). It is interesting to observe that $\widehat{\sigma_\Omega^2} = Q_5/ab$ is the m.l.e. of σ^2 under Ω and

$$\widehat{\sigma_\omega^2} = \frac{(Q_4 + Q_5)}{ab} = \sum_{i=1}^{a} \sum_{j=1}^{b} \frac{(X_{ij} - \bar{X}_{i.})^2}{ab}$$

is that estimator under ω. A useful monotone function of the likelihood ratio $\lambda = (\widehat{\sigma_\Omega^2}/\widehat{\sigma_\omega^2})^{ab/2}$ is

$$F = \frac{Q_4/(b-1)}{Q_5/[(a-1)(b-1)]},$$

which has, under H_0, an F-distribution with $b - 1$ and $(a - 1)(b - 1)$ degrees of freedom. The hypothesis H_0 is rejected if $F \geq c$, where $\alpha = \Pr(F \geq c; H_0)$.

If we are to compute the power function of the test, we need the distribution of F when H_0 is not true. From Section 10.3 we know, when H_1 is true, that Q_4/σ^2 and Q_5/σ^2 are independent (central or noncentral) chi-square variables. We shall compute the noncentrality parameters of Q_4/σ^2 and Q_5/σ^2 when H_1 is true. We have $E(X_{ij}) = \mu + \alpha_i + \beta_j$, $E(\bar{X}_{i.}) = \mu + \alpha_i$, $E(\bar{X}_{.j}) = \mu + \beta_j$ and $E(\bar{X}_{..}) = \mu$. Accordingly, the noncentrality parameter of Q_4/σ^2 is

$$\frac{a \sum_{j=1}^{b} (\mu + \beta_j - \mu)^2}{\sigma^2} = \frac{a \sum_{j=1}^{b} \beta_j^2}{\sigma^2}$$

and that of Q_5/σ^2 is

$$\frac{\sum_{j=1}^{b} \sum_{i=1}^{a} (\mu + \alpha_i + \beta_j - \mu - \alpha_i - \mu - \beta_j + \mu)^2}{\sigma^2} = 0.$$

Thus, if the hypothesis H_0 is not true, F has a noncentral F-distribution with $b - 1$ and $(a - 1)(b - 1)$ degrees of freedom and noncentrality parameter $a \sum_{j=1}^{b} \beta_j^2/\sigma^2$. The desired probabilities can then be found in tables of the noncentral F-distribution.

A similar argument can be used to construct the F needed to test the equality of row means; that is, this F is essentially the ratio of the sum of squares among rows and Q_5. In particular, this F is defined by

$$F = \frac{Q_2/(a-1)}{Q_5/[(a-1)(b-1)]}$$

and, under $H_0: \alpha_1 = \alpha_2 = \cdots = \alpha_a = 0$, has an F-distribution with $a - 1$ and $(a - 1)(b - 1)$ degrees of freedom.

The analysis-of-variance problem that has just been discussed is usually referred to as a *two-way classification with one observation per cell*. Each combination of i and j determines a cell; thus there is a total of ab cells in this model. Let us now investigate another two-way classification problem, but in this case we take $c > 1$ independent observations per cell.

Let X_{ijk}, $i = 1, 2, \ldots, a$, $j = 1, 2, \ldots, b$, and $k = 1, 2, \ldots, c$, denote $n = abc$ random variables which are independent and which have normal distributions with common, but unknown, variance σ^2. The mean of each X_{ijk}, $k = 1, 2, \ldots, c$, is $\mu_{ij} = \mu + \alpha_i + \beta_j + \gamma_{ij}$, where $\sum_{i=1}^{a} \alpha_i = 0$, $\sum_{j=1}^{b} \beta_j = 0$, $\sum_{i=1}^{a} \gamma_{ij} = 0$, and $\sum_{j=1}^{b} \gamma_{ij} = 0$. For example, take $a = 2, b = 3, \mu = 5, \alpha_1 = 1, \alpha_2 = -1, \beta_1 = 1, \beta_2 = 0, \beta_3 = -1, \gamma_{11} = 1, \gamma_{12} = 1, \gamma_{13} = -2, \gamma_{21} = -1, \gamma_{22} = -1$, and $\gamma_{23} = 2$. Then the means are

$$\mu_{11} = 8, \qquad \mu_{12} = 7, \qquad \mu_{13} = 3,$$
$$\mu_{21} = 4, \qquad \mu_{22} = 3, \qquad \mu_{23} = 5.$$

Note that, if each $\gamma_{ij} = 0$, then

$$\mu_{11} = 7, \qquad \mu_{12} = 6, \qquad \mu_{13} = 5,$$
$$\mu_{21} = 5, \qquad \mu_{22} = 4, \qquad \mu_{23} = 3.$$

That is, if $\gamma_{ij} = 0$, each of the means in the first row is 2 greater than the corresponding mean in the second row. In general, if each $\gamma_{ij} = 0$, the means of row i_1 differ from the corresponding means of row i_2 by a constant. This constant may be different for different choices of i_1 and i_2. A similar statement can be made about the means of columns j_1 and j_2. The parameter γ_{ij} is called the *interaction* associated with cell (i, j). That is, the interaction between the ith level of one classification and the jth level of the other classification is γ_{ij}. One interesting hypothesis to test is that each interaction is equal to zero. This will now be investigated.

From Exercise 10.4 of Section 10.1 we have that

$$\sum_{i=1}^{a} \sum_{j=1}^{b} \sum_{k=1}^{c} (X_{ijk} - \bar{X}_{...})^2 = bc \sum_{i=1}^{a} (\bar{X}_{i..} - \bar{X}_{...})^2 + ac \sum_{j=1}^{b} (\bar{X}_{.j.} - \bar{X}_{...})^2$$
$$+ c \sum_{i=1}^{a} \sum_{j=1}^{b} (\bar{X}_{ij.} - \bar{X}_{i..} - \bar{X}_{.j.} + \bar{X}_{...})^2$$
$$+ \sum_{i=1}^{a} \sum_{j=1}^{b} \sum_{k=1}^{c} (X_{ijk} - \bar{X}_{ij.})^2;$$

that is, the total sum of squares is decomposed into that due to *row* differences, that due to *column* differences, that due to *interaction*, and that *within cells*. The test of

$$H_0 : \gamma_{ij} = 0, \qquad i = 1, 2, \ldots, a, \quad j = 1, 2, \ldots, b,$$

against all possible alternatives is based upon an F with $(a-1)(b-1)$ and $ab(c-1)$ degrees of freedom,

$$F = \frac{\left[c \sum_{i=1}^{a} \sum_{j=1}^{b} (\bar{X}_{ij.} - \bar{X}_{i..} - \bar{X}_{.j.} + \bar{X}_{...})^2\right] \Big/ [(a-1)(b-1)]}{\left[\sum\sum\sum (X_{ijk} - \bar{X}_{ij.})^2\right] \Big/ [ab(c-1)]}.$$

The reader should verify that the noncentrality parameter of this F-distribution is equal to $c \sum_{j=1}^{b} \sum_{i=1}^{a} \gamma_{ij}^2/\sigma^2$. Thus F is central when $H_0 : \gamma_{ij} = 0$, $i = 1, 2, \ldots, a$, $j = 1, 2, \ldots, b$, is true.

EXERCISES

10.19. Show that

$$\sum_{j=1}^{b} \sum_{i=1}^{a} (X_{ij} - \bar{X}_{i.})^2 = \sum_{j=1}^{b} \sum_{i=1}^{a} (X_{ij} - \bar{X}_{i.} - \bar{X}_{.j} + \bar{X}_{..})^2 + a \sum_{j=1}^{b} (\bar{X}_{.j} - \bar{X}_{..})^2.$$

10.20. If at least one $\gamma_{ij} \neq 0$, show that the F, which is used to test that each interaction is equal to zero, has noncentrality parameter equal to $c \sum_{j=1}^{b} \sum_{i=1}^{a} \gamma_{ij}^2/\sigma^2$.

10.21. Using the background of the two-way classification with one observation per cell, show that the maximum likelihood estimators of α_i, β_j, and μ are $\hat{\alpha}_i = \bar{X}_{i.} - \bar{X}_{..}$, $\hat{\beta}_j = \bar{X}_{.j} - \bar{X}_{..}$, and $\hat{\mu} = \bar{X}_{..}$, respectively. Show that these are unbiased estimators of their respective parameters and compute $\text{var}(\hat{\alpha}_i)$, $\text{var}(\hat{\beta}_j)$, and $\text{var}(\hat{\mu})$.

10.22. Prove, using the assumptions of this section, that the linear functions $X_{ij} - \bar{X}_{i.} - \bar{X}_{.j} + \bar{X}_{..}$ and $\bar{X}_{.j} - \bar{X}_{..}$ are uncorrelated.

10.23. Given the following observations associated with a two-way classification with $a = 3$ and $b = 4$, compute the F-statistics used to test the equality of the column means ($\beta_1 = \beta_2 = \beta_3 = \beta_4 = 0$) and the equality of the row means ($\alpha_1 = \alpha_2 = \alpha_3 = 0$), respectively.

Row/Column	1	2	3	4
1	3.1	4.2	2.7	4.9
2	2.7	2.9	1.8	3.0
3	4.0	4.6	3.0	3.9

10.24. With the background of the two-way classification with $c > 1$ observations per cell, show that the maximum likelihood estimators of the parameters are $\hat{\alpha}_i = \overline{X}_{i..} - \overline{X}_{...}$, $\hat{\beta}_j = \overline{X}_{.j.} - \overline{X}_{...}$, $\hat{\gamma}_{ij} = \overline{X}_{ij.} - \overline{X}_{i..} - \overline{X}_{.j.} + \overline{X}_{...}$, and $\hat{\mu} = \overline{X}_{...}$. Show that these are unbiased estimators of the respective parameters. Compute the variance of each estimator.

10.25. Given the following observations in a two-way classification with $a = 3$, $b = 4$, and $c = 2$, compute the F-statistics used to test that all interactions are equal to zero ($\gamma_{ij} = 0$), all column means are equal ($\beta_j = 0$), and all row means are equal ($\alpha_i = 0$), respectively.

Row/Column	1	2	3	4
1	3.1	4.2	2.7	4.9
	2.9	4.9	3.2	4.5
2	2.7	2.9	1.8	3.0
	2.9	2.3	2.4	3.7
3	4.0	4.6	3.0	3.9
	4.4	5.0	2.5	4.2

10.6 A Regression Problem

There is often interest in the relation between two variables, for example, a student's scholastic aptitude test score in mathematics and this same student's grade in calculus. Frequently, one of these variables, say x, is known in advance of the other, and hence there is interest in predicting a future random variable Y. Since Y is a random variable, we cannot predict its future observed value $Y = y$ with certainty. Thus let us first concentrate on the problem of estimating the mean of Y, that is, $E(Y)$. Now $E(Y)$ is usually a function of x; for example, in our illustration with the calculus grade, say Y, we would expect $E(Y)$ to increase with increasing mathematics aptitude score x. Sometimes $E(Y) = \mu(x)$ is assumed to be of a given form, such as linear or quadratic or exponential; that is, $\mu(x)$ could be assumed to be equal to $\alpha + \beta x$ or $\alpha + \beta x + \gamma x^2$ or $\alpha e^{\beta x}$. To estimate $E(Y) = \mu(x)$, or equivalently the parameters α, β, and γ, we observe the random variable Y for each of n possibly different values of x, say $x_1, x_2, \ldots, x_n$, which are not all equal. Once the n independent experiments have been performed, we have n pairs of known numbers $(x_1, y_1), (x_2, y_2), \ldots, (x_n, y_n)$. These pairs are then used to estimate the mean $E(Y)$. Problems like this are often classified under *regression* because $E(Y) = \mu(x)$ is frequently called a regression curve.

Remark. A model for the mean like $\alpha + \beta x + \gamma x^2$, is called a *linear model* because it is linear in the parameters, α, β, and γ. Thus $\alpha e^{\beta x}$ is not a linear model because it is not linear in α and β. Note that, in Sections 10.1 to 10.4, all the means were linear in the parameters and hence linear models.

Let us begin with the case in which $E(Y) = \mu(x)$ is a linear function. The n points are $(x_1, y_1), (x_2, y_2), \ldots, (x_n, y_n)$; so the first problem is that of fitting a straight line to the set of points (see Figure 10.1). In addition to assuming that the mean of Y is a linear function, we assume that, $Y_1, Y_2, \ldots, Y_n$ are independent normal variables with respective means $\alpha + \beta(x_i - \bar{x})$, $i = 1, 2, \ldots, n$, and unknown variance σ^2, where $\bar{x} = \Sigma\, x_i/n$. Their joint p.d.f. is therefore the

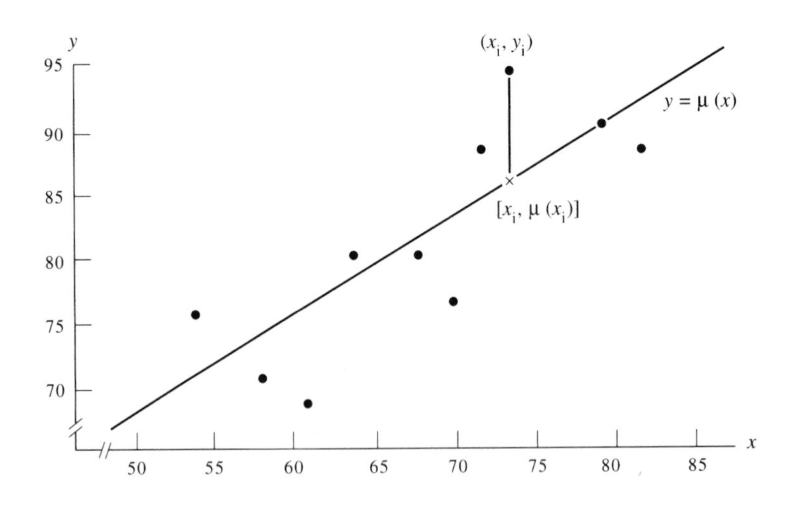

FIGURE 10.1

product of the individual probability density functions; that is, the likelihood function equals

$$L(\alpha, \beta, \sigma^2) = \prod_{i=1}^{n} \frac{1}{\sqrt{2\pi\sigma^2}} \exp\left\{-\frac{[y_i - \alpha - \beta(x_i - \bar{x})]^2}{2\sigma^2}\right\}$$

$$= \left(\frac{1}{2\pi\sigma^2}\right)^{n/2} \exp\left\{-\frac{1}{2\sigma^2} \sum_{i=1}^{n} [y_i - \alpha - \beta(x_i - \bar{x})]^2\right\}$$

To maximize $L(\alpha, \beta, \sigma^2)$, or, equivalently, to minimize

$$-\ln L(\alpha, \beta, \sigma^2) = \frac{n}{2} \ln (2\pi\sigma^2) + \frac{\sum_{i=1}^{n} [y_i - \alpha - \beta(x_i - \bar{x})]^2}{2\sigma^2},$$

we must select α and β to minimize

$$H(\alpha, \beta) = \sum_{i=1}^{n} [y_i - \alpha - \beta(x_i - \bar{x})]^2.$$

Since $|y_i - \alpha - \beta(x_i - \bar{x})| = |y_i - \mu(x_i)|$ is the vertical distance from the point (x_i, y_i) to the line $y = \mu(x)$, we note that $H(\alpha, \beta)$ represents the sum of the squares of those distances. Thus selecting α and β so that the sum of the squares is minimized means that we are fitting the straight line to the data by the *method of least squares.*

To minimize $H(\alpha, \beta)$, we find the two first partial derivatives

$$\frac{\partial H(\alpha, \beta)}{\partial \alpha} = 2 \sum_{i=1}^{n} [y_i - \alpha - \beta(x_i - \bar{x})](-1)$$

and

$$\frac{\partial H(\alpha, \beta)}{\partial \beta} = 2 \sum_{i=1}^{n} [y_i - \alpha - \beta(x_i - \bar{x})][-(x_i - \bar{x})].$$

Setting $\partial H(\alpha, \beta)/\partial \alpha = 0$, we obtain

$$\sum_{i=1}^{n} y_i - n\alpha - \beta \sum_{i=1}^{n} (x_i - \bar{x}) = 0.$$

Since

$$\sum_{i=1}^{n} (x_i - \bar{x}) = 0,$$

we have that

$$\sum_{i=1}^{n} y_i - n\alpha = 0$$

and thus

$$\hat{\alpha} = \bar{Y}.$$

The equation $\partial H(\alpha, \beta)/\partial\beta = 0$ yields, with α replaced by $\bar{y}$,

$$\sum_{i=1}^{n} (y_i - \bar{y})(x_i - \bar{x}) - \beta \sum_{i=1}^{n} (x_i - \bar{x})^2 = 0$$

or, equivalently,

$$\hat{\beta} = \frac{\sum_{i=1}^{n} (Y_i - \bar{Y})(x_i - \bar{x})}{\sum_{i=1}^{n} (x_i - \bar{x})^2} = \frac{\sum_{i=1}^{n} Y_i(x_i - \bar{x})}{\sum_{i=1}^{n} (x_i - \bar{x})^2}.$$

To find the maximum likelihood estimator of σ^2, consider the partial derivative

$$\frac{\partial[-\ln L(\alpha, \beta, \sigma^2)]}{\partial(\sigma^2)} = \frac{n}{2\sigma^2} - \frac{\sum_{i=1}^{n} [y_i - \alpha - \beta(x_i - \bar{x})]^2}{2(\sigma^2)^2}.$$

Setting this equal to zero and replacing α and β by their solutions $\hat{\alpha}$ and $\hat{\beta}$, we obtain

$$\widehat{\sigma^2} = \frac{1}{n} \sum_{i=1}^{n} [Y_i - \hat{\alpha} - \hat{\beta}(x_i - \bar{x})]^2.$$

Of course, due to invariance, $\widehat{\sigma^2} = \hat{\sigma}^2$.

Since $\hat{\alpha}$ is a linear function of independent and normally distributed random variables, $\hat{\alpha}$ has a normal distribution with mean

$$E(\hat{\alpha}) = E\left(\frac{1}{n} \sum_{i=1}^{n} Y_i\right) = \frac{1}{n} \sum_{i=1}^{n} E(Y_i)$$

$$= \frac{1}{n} \sum_{i=1}^{n} [\alpha + \beta(x_i - \bar{x})] = \alpha,$$

and variance

$$\text{var}\,(\hat{\alpha}) = \sum_{i=1}^{n} \left(\frac{1}{n}\right)^2 \text{var}\,(Y_i) = \frac{\sigma^2}{n}.$$

The estimator $\hat{\beta}$ is also a linear function of $Y_1, Y_2, \ldots, Y_n$ and hence has a normal distribution with mean

$$E(\hat{\beta}) = \frac{\sum_{i=1}^{n} (x_i - \bar{x})E(Y_i)}{\sum_{i=1}^{n} (x_i - \bar{x})^2}$$

$$= \frac{\sum_{i=1}^{n} (x_i - \bar{x})[\alpha + \beta(x_i - \bar{x})]}{\sum_{i=1}^{n} (x_i - \bar{x})^2}$$

$$= \frac{\alpha \sum_{i=1}^{n} (x_i - \bar{x}) + \beta \sum_{i=1}^{n} (x_i - \bar{x})^2}{\sum_{i=1}^{n} (x_i - \bar{x})^2} = \beta$$

and variance

$$\text{var}(\hat{\beta}) = \sum_{i=1}^{n} \left[\frac{x_i - \bar{x}}{\sum_{i=1}^{n} (x_i - \bar{x})^2} \right]^2 \text{var}(Y_i)$$

$$= \frac{\sum_{i=1}^{n} (x_i - \bar{x})^2}{\left[\sum_{i=1}^{n} (x_i - \bar{x})^2\right]^2} \sigma^2 = \frac{\sigma^2}{\sum_{i=1}^{n} (x_i - \bar{x})^2}.$$

It can be shown (Exercise 10.27) that

$$\sum_{i=1}^{n} [Y_i - \alpha - \beta(x_i - \bar{x})]^2 = \sum_{i=1}^{n} \{(\hat{\alpha} - \alpha) + (\hat{\beta} - \beta)(x_i - \bar{x}) + [Y_i - \hat{\alpha} - \hat{\beta}(x_i - \bar{x})]\}^2$$

$$= n(\hat{\alpha} - \alpha)^2 + (\hat{\beta} - \beta)^2 \sum_{i=1}^{n} (x_i - \bar{x})^2 + n\widehat{\sigma^2}.$$

or, for brevity,

$$Q = Q_1 + Q_2 + Q_3.$$

Here Q, Q_1, Q_2, and Q_3 are real quadratic forms in the variables

$$Y_i - \alpha - \beta(x_i - \bar{x}), \qquad i = 1, 2, \ldots, n.$$

In this equation, Q represents the sum of the squares of n independent random variables that have normal distributions with means zero and variances σ^2. Thus Q/σ^2 has a chi-square distribution with n degrees of freedom. Each of the random variables $\sqrt{n}(\hat{\alpha} - \alpha)/\sigma$ and $\sqrt{\sum_1^n (x_i - \bar{x})^2}(\hat{\beta} - \beta)/\sigma$ has a normal distribution with zero mean and unit variance; thus each of Q_1/σ^2 and Q_2/σ^2 has a chi-square distribution with 1 degree of freedom. Since Q_3 is nonnegative, we have, in accordance with the theorem of Section 10.1, that Q_1, Q_2, and Q_3 are independent, so that Q_3/σ^2 has a chi-square distribution with $n - 1 - 1 = n - 2$ degrees of freedom. Then each of the random variables

$$T_1 = \frac{[\sqrt{n}(\hat{\alpha} - \alpha)]/\sigma}{\sqrt{Q_3/[\sigma^2(n-2)]}} = \frac{\hat{\alpha} - \alpha}{\sqrt{\hat{\sigma}^2/(n-2)}}$$

and

$$T_2 = \frac{\left[\sqrt{\sum_1^n (x_i - \bar{x})^2}(\hat{\beta} - \beta)\right]\Big/\sigma}{\sqrt{Q_3/[\sigma^2(n-2)]}} = \frac{\hat{\beta} - \beta}{\sqrt{n\hat{\sigma}^2\Big/\left[(n-2)\sum_1^n (x_i - \bar{x})^2\right]}}$$

has a t-distribution with $n - 2$ degrees of freedom. These facts enable us to obtain confidence intervals for α and β. The fact that $n\hat{\sigma}^2/\sigma^2$ has a chi-square distribution with $n - 2$ degrees of freedom provides a means of determining a confidence interval for σ^2. These are some of the statistical inferences about the parameters to which reference was made in the introductory remarks of this section.

Remark. The more discerning reader should quite properly question our constructions of T_1 and T_2 immediately above. We know that the *squares* of the linear forms are independent of $Q_3 = n\hat{\sigma}^2$, but we do not know, at this time, that the linear forms themselves enjoy this independence. This problem arises again in Section 10.7. In Exercise 10.47, a more general problem is proposed, of which the present case is a special instance.

EXERCISES

10.26. Students' scores on the mathematics portion of the ACT examination, x, and on the final examination in first-semester calculus (200 points possible), y, are given.

(a) Calculate the least squares regression line for these data.
(b) Plot the points and the least squares regression line on the same graph.
(c) Find point estimates for α, β, and σ^2.
(d) Find 95 percent confidence intervals for α and β under the usual assumptions.

x	y	x	y
25	138	20	100
20	84	25	143
26	104	26	141
26	112	28	161
28	88	25	124
28	132	31	118
29	90	30	168
32	183		

10.27. Show that

$$\sum_{i=1}^{n} [Y_i - \alpha - \beta(x_i - \bar{x})]^2 = n(\hat{\alpha} - \alpha)^2 + (\hat{\beta} - \beta)^2 \sum_{i=1}^{n} (x_i - \bar{x})^2 + \sum_{i=1}^{n} [Y_i - \hat{\alpha} - \hat{\beta}(x_i - \bar{x})]^2.$$

10.28. Let the independent random variables $Y_1, Y_2, \ldots, Y_n$ have, respectively, the probability density functions $N(\beta x_i, \gamma^2 x_i^2)$, $i = 1, 2, \ldots, n$, where the given numbers $x_1, x_2, \ldots, x_n$ are not all equal and no one is zero. Find the maximum likelihood estimators of β and γ^2.

10.29. Let the independent random variables $Y_1, \ldots, Y_n$ have the joint p.d.f.

$$L(\alpha, \beta, \sigma^2) = \left(\frac{1}{2\pi\sigma^2}\right)^{n/2} \exp\left\{-\frac{1}{2\sigma^2} \sum_{1}^{n} [y_i - \alpha - \beta(x_i - \bar{x})]^2\right\},$$

where the given numbers $x_1, x_2, \ldots, x_n$ are not all equal. Let $H_0 : \beta = 0$ (α and σ^2 unspecified). It is desired to use a likelihood ratio test to test H_0 against all possible alternatives. Find λ and see whether the test can be based on a familiar statistic.

Hint: In the notation of this section show that

$$\sum_{1}^{n} (Y_i - \hat{\alpha})^2 = Q_3 + \hat{\beta}^2 \sum_{1}^{n} (x_i - \bar{x})^2.$$

10.30. Using the notation of Section 10.2, assume that the means μ_j satisfy a linear function of j, namely $\mu_j = c + d[j - (b + 1)/2]$. Let independent random samples of size a be taken from the b normal distributions with common unknown variance σ^2.

(a) Show that the maximum likelihood estimators of c and d are, respectively, $\hat{c} = \bar{X}_{..}$ and

$$\hat{d} = \frac{\sum_{j=1}^{b} [j - (b + 1)/2](\bar{X}_{.j} - \bar{X}_{..})}{\sum_{j=1}^{b} [j - (b + 1)/2]^2}.$$

(b) Show that

$$\sum_{i=1}^{a} \sum_{j=1}^{b} (X_{ij} - \bar{X}_{..})^2 = \sum_{i=1}^{a} \sum_{j=1}^{b} \left[X_{ij} - \bar{X}_{..} - \hat{d}\left(j - \frac{b+1}{2}\right) \right]^2$$

$$+ \hat{d}^2 \sum_{j=1}^{b} a\left(j - \frac{b+1}{2}\right)^2.$$

(c) Argue that the two terms in the right-hand member of part (b), once divided by σ^2, are independent random variables with chi-square distributions provided that $d = 0$.

(d) What F-statistic would be used to test the equality of the means, that is, $H_0 : d = 0$?

10.7 A Test of Independence

Let X and Y have a bivariate normal distribution with means μ_1 and μ_2, positive variances σ_1^2 and σ_2^2, and correlation coefficient ρ. We wish to test the hypothesis that X and Y are independent. Because two jointly normally distributed random variables are independent if and only if $\rho = 0$, we test the hypothesis $H_0 : \rho = 0$ against the hypothesis $H_1 : \rho \neq 0$. A likelihood ratio test will be used. Let (X_1, Y_1), $(X_2, Y_2), \ldots, (X_n, Y_n)$ denote a random sample of size $n > 2$ from the bivariate normal distribution; that is, the joint p.d.f. of these $2n$ random variables is given by

$$f(x_1, y_1)f(x_2, y_2) \cdots f(x_n, y_n).$$

Although it is fairly difficult to show, the statistic that is defined by the likelihood ratio λ is a function of the statistic

$$R = \frac{\sum_{i=1}^{n} (X_i - \bar{X})(Y_i - \bar{Y})}{\sqrt{\sum_{i=1}^{n} (X_i - \bar{X})^2 \sum_{i=1}^{n} (Y_i - \bar{Y})^2}}.$$

This statistic R is called the *correlation coefficient* of the random sample. The likelihood ratio principle, which calls for the rejection of H_0 if $\lambda \leq \lambda_0$, is equivalent to the computed value of $|R| \geq c$. That is, if the absolute value of the correlation coefficient of the sample is too large, we reject the hypothesis that the correlation coefficient of the distribution is equal to zero. To determine a value of c for a satisfactory significance level, it will be necessary to obtain the distribution of R, or a function of R, when H_0 is true. This will now be done.

Let $X_1 = x_1, X_2 = x_2, \ldots, X_n = x_n, n > 2$, where $x_1, x_2, \ldots, x_n$ and $\bar{x} = \sum_1^n x_i/n$ are fixed numbers such that $\sum_1^n (x_i - \bar{x})^2 > 0$. Consider the conditional p.d.f. of $Y_1, Y_2, \ldots, Y_n$, given that $X_1 = x_1$, $X_2 = x_2, \ldots, X_n = x_n$. Because $Y_1, Y_2, \ldots, Y_n$ are independent and, with $\rho = 0$, are also independent of $X_1, X_2, \ldots, X_n$, this conditional p.d.f. is given by

$$\left(\frac{1}{\sqrt{2\pi}\sigma_2}\right)^n \exp\left[-\frac{\sum_1^n (y_i - \mu_2)^2}{2\sigma_2^2}\right].$$

Let R_c be the correlation coefficient, given $X_1 = x_1$, $X_2 = x_2, \ldots, X_n = x_n$, so that

$$\frac{R_c\sqrt{\sum_{i=1}^n (Y_i - \bar{Y})^2}}{\sqrt{\sum_{i=1}^n (x_i - \bar{x})^2}} = \frac{\sum_{i=1}^n (x_i - \bar{x})(Y_i - \bar{Y})}{\sum_{i=1}^n (x_i - \bar{x})^2} = \frac{\sum_{i=1}^n (x_i - \bar{x})Y_i}{\sum_{i=1}^n (x_i - \bar{x})^2}$$

is like $\hat{\beta}$ of Section 10.6 and has mean zero when $\rho = 0$. Thus, referring to T_2 of Section 10.6, we see that

$$\frac{R_c\sqrt{\Sigma(Y_i - \bar{Y})^2}/\sqrt{\Sigma(x_i - \bar{x})^2}}{\sqrt{\dfrac{\sum_{i=1}^n \{Y_i - \bar{Y} - [R_c\sqrt{\Sigma(Y_i - \bar{Y})^2}/\sqrt{\Sigma(x_i - \bar{x})^2}](x_i - \bar{x})\}^2}{(n-2)\Sigma(x_i - \bar{x})^2}}} = \frac{R_c\sqrt{n-2}}{\sqrt{1 - R_c^2}} \tag{1}$$

has, given $X_1 = x_1, \ldots, X_n = x_n$, a conditional t-distribution with $n - 2$ degrees of freedom. Note that the p.d.f., say $g(t)$, of this t-distribution does not depend upon $x_1, x_2, \ldots, x_n$. Now the joint p.d.f. of $X_1, X_2, \ldots, X_n$ and $R\sqrt{n-2}/\sqrt{1-R^2}$, where

$$R = \frac{\sum_1^n (X_i - \bar{X})(Y_i - \bar{Y})}{\sqrt{\sum_1^n (X_i - \bar{X})^2 \sum_1^n (Y_i - \bar{Y})^2}},$$

is the product of $g(t)$ and the joint p.d.f. of $X_1, X_2, \ldots, X_n$. Integration on $x_1, x_2, \ldots, x_n$ yields the marginal p.d.f. of

$R\sqrt{n-2}/\sqrt{1-R^2}$; because $g(t)$ does not depend upon $x_1, x_2, \ldots, x_n$ it is obvious that this marginal p.d.f. is $g(t)$, the conditional p.d.f. of $R_c\sqrt{n-2}/\sqrt{1-R_c^2}$. The change-of-variable technique can now be used to find the p.d.f. of R.

Remarks. Since R has, when $\rho = 0$, a conditional distribution that does not depend upon $x_1, x_2, \ldots, x_n$ (and hence that conditional distribution is, in fact, the marginal distribution of R), we have the remarkable fact that R is independent of $X_1, X_2, \ldots, X_n$. It follows that R is independent of *every function* of $X_1, X_2, \ldots, X_n$ alone, that is, a function that does not depend upon any Y_i. In like manner, R is independent of every function of $Y_1, Y_2, \ldots, Y_n$ alone. Moreover, a careful review of the argument reveals that nowhere did we use the fact that X has a normal marginal distribution. Thus, if X and Y are independent, and if Y has a normal distribution, then R has the same conditional distribution whatever be the distribution of X, subject to the condition $\sum_1^n (x_i - \bar{x})^2 > 0$. Moreover, if $\Pr\left[\sum_1^n (X_i - \bar{X})^2 > 0\right] = 1$, then R has the same marginal distribution whatever be the distribution of X.

If we write $T = R\sqrt{n-2}/\sqrt{1-R^2}$, where T has a t-distribution with $n - 2 > 0$ degrees of freedom, it is easy to show, by the change-of-variable technique (Exercise 10.34), that the p.d.f. of R is given by

$$\begin{aligned} g(r) &= \frac{\Gamma[(n-1)/2]}{\Gamma(\frac{1}{2})\Gamma[(n-2)/2]}(1-r^2)^{(n-4)/2}, \qquad -1 < r < 1, \\ &= 0 \qquad \text{elsewhere.} \end{aligned} \tag{2}$$

We have now solved the problem of the distribution of R, when $\rho = 0$ and $n > 2$, or, perhaps more conveniently, that of $R\sqrt{n-2}/\sqrt{1-R^2}$. The likelihood ratio test of the hypothesis $H_0 : \rho = 0$ against all alternatives $H_1 : \rho \neq 0$ may be based either on the statistic R or on the statistic $R\sqrt{n-2}/\sqrt{1-R^2} = T$, although the latter is easier to use. In either case the significance level of the test is

$$\alpha = \Pr(|R| \geq c_1; H_0) = \Pr(|T| \geq c_2; H_0),$$

where the constants c_1 and c_2 are chosen so as to give the desired value of α.

Remark. It is also possible to obtain an approximate test of size α by using the fact that

$$W = \frac{1}{2}\ln\left(\frac{1+R}{1-R}\right)$$

has an approximate normal distribution with mean $\frac{1}{2}\ln[(1+\rho)/(1-\rho)]$ and variance $1/(n-3)$. We accept this statement without proof. Thus a test of $H_0: \rho = 0$ can be based on the statistic

$$Z = \frac{\frac{1}{2}\ln[(1+R)/(1-R)] - \frac{1}{2}\ln[(1+\rho)/(1-\rho)]}{\sqrt{1/(n-3)}},$$

with $\rho = 0$ so that $\frac{1}{2}\ln[(1+\rho)/(1-\rho)] = 0$. However, using W, we can also test hypotheses like $H_0: \rho = \rho_0$ against $H_1: \rho \neq \rho_0$, where ρ_0 is not necessarily zero. In that case the hypothesized mean of W is

$$\frac{1}{2}\ln\left(\frac{1+\rho_0}{1-\rho_0}\right).$$

EXERCISES

10.31. Show that

$$R = \frac{\sum_1^n (X_i - \bar{X})(Y_i - \bar{Y})}{\sqrt{\sum_1^n (X_i - \bar{X})^2 \sum_1^n (Y_i - \bar{Y})^2}} = \frac{\sum_1^n X_i Y_i - n\bar{X}\bar{Y}}{\sqrt{\left(\sum_1^n X_i^2 - n\bar{X}^2\right)\left(\sum_1^n Y_i^2 - n\bar{Y}^2\right)}}.$$

10.32. A random sample of size $n = 6$ from a bivariate normal distribution yields a value of the correlation coefficient of 0.89. Would we accept or reject, at the 5 percent signficance level, the hypothesis that $\rho = 0$?

10.33. Verify Equation (1) of this section.

10.34. Verify the p.d.f. (2) of this section.

10.8 The Distributions of Certain Quadratic Forms

Remark. It is essential that the reader have the background of the multivariate normal distribution as given in Section 4.10 to understand Sections 10.8 and 10.9.

Let X_i, $i = 1, 2, \ldots, n$, denote independent random variables which are $N(\mu_i, \sigma_i^2)$, $i = 1, 2, \ldots, n$, respectively. Then $Q = \sum_1^n (X_i - \mu_i)^2/\sigma_i^2$ is $\chi^2(n)$. Now Q is a quadratic form in the $X_i - \mu_i$ and Q is seen to be, apart from the coefficient $-\frac{1}{2}$, the random variable which is defined by the exponent on the number e in the joint p.d.f. of $X_1, X_2, \ldots, X_n$. We shall now show that this result can be generalized.

Let $X_1, X_2, \ldots, X_n$ have a multivariate normal distribution with p.d.f.

$$\frac{1}{(2\pi)^{n/2}\sqrt{|\mathbf{V}|}} \exp\left[-\frac{(\mathbf{x}-\boldsymbol{\mu})'\mathbf{V}^{-1}(\mathbf{x}-\boldsymbol{\mu})}{2}\right],$$

where, as usual, the covariance matrix $\mathbf{V}$ is positive definite. We shall show that the random variable Q (a quadratic form in the $X_i - \mu_i$), which is defined by $(\mathbf{x}-\boldsymbol{\mu})'\mathbf{V}^{-1}(\mathbf{x}-\boldsymbol{\mu})$, is $\chi^2(n)$. We have for the m.g.f. $M(t)$ of Q the integral

$$\int_{-\infty}^{\infty} \cdots \int_{-\infty}^{\infty} \frac{1}{(2\pi)^{n/2}\sqrt{|\mathbf{V}|}}$$

$$\times \exp\left[t(\mathbf{x}-\boldsymbol{\mu})'\mathbf{V}^{-1}(\mathbf{x}-\boldsymbol{\mu}) - \frac{(\mathbf{x}-\boldsymbol{\mu})'\mathbf{V}^{-1}(\mathbf{x}-\boldsymbol{\mu})}{2}\right] dx_1 \cdots dx_n$$

$$= \int_{-\infty}^{\infty} \cdots \int_{-\infty}^{\infty} \frac{1}{(2\pi)^{n/2}\sqrt{|\mathbf{V}|}}$$

$$\times \exp\left[-\frac{(\mathbf{x}-\boldsymbol{\mu})'\mathbf{V}^{-1}(\mathbf{x}-\boldsymbol{\mu})(1-2t)}{2}\right] dx_1 \cdots dx_n.$$

With $\mathbf{V}^{-1}$ positive definite, the integral is seen to exist for all real values of $t < \frac{1}{2}$. Moreover, $(1-2t)\mathbf{V}^{-1}$, $t < \frac{1}{2}$, is a positive definite matrix and, since $|(1-2t)\mathbf{V}^{-1}| = (1-2t)^n|\mathbf{V}^{-1}|$, it follows that

$$\frac{1}{(2\pi)^{n/2}\sqrt{|\mathbf{V}|/(1-2t)^n}} \exp\left[-\frac{(\mathbf{x}-\boldsymbol{\mu})'\mathbf{V}^{-1}(\mathbf{x}-\boldsymbol{\mu})(1-2t)}{2}\right]$$

can be treated as a multivariate normal p.d.f. If we multiply our integrand by $(1-2t)^{n/2}$, we have this multivariate p.d.f. Thus the m.g.f. of Q is given by

$$M(t) = \frac{1}{(1-2t)^{n/2}}, \qquad t < \tfrac{1}{2},$$

and Q is $\chi^2(n)$, as we wished to show. This fact is the basis of the chi-square tests that were discussed in Chapter 6.

The remarkable fact that the random variable which is defined by $(\mathbf{x}-\boldsymbol{\mu})'\mathbf{V}^{-1}(\mathbf{x}-\boldsymbol{\mu})$ is $\chi^2(n)$ stimulates a number of questions about quadratic forms in normally distributed variables. We would like to treat this problem in complete generality, but limitations of space forbid this, and we find it necessary to restrict ourselves to some special cases.

Let $X_1, X_2, \ldots, X_n$ denote a random sample of size n from a distribution which is $N(0, \sigma^2)$, $\sigma^2 > 0$. Let $\mathbf{X}' = [X_1, X_2, \ldots, X_n]$ and let $\mathbf{A}$ denote an arbitrary $n \times n$ real symmetric matrix. We shall investigate the distribution of the quadratic form $\mathbf{X}'\mathbf{A}\mathbf{X}$. For instance, we know that $\mathbf{X}'\mathbf{I}_n\mathbf{X}/\sigma^2 = \mathbf{X}'\mathbf{X}/\sigma^2 = \sum_1^n X_i^2/\sigma^2$ is $\chi^2(n)$. First we shall find the m.g.f.of $\mathbf{X}'\mathbf{A}\mathbf{X}/\sigma^2$. Then we shall investigate the conditions that must be imposed upon the real symmetric matrix $\mathbf{A}$ if $\mathbf{X}'\mathbf{A}\mathbf{X}/\sigma^2$ is to have a chi-square distribution. This m.g.f. is given by

$$M(t) = \int_{-\infty}^{\infty} \cdots \int_{-\infty}^{\infty} \left(\frac{1}{\sigma\sqrt{2\pi}}\right)^n \exp\left(\frac{t\mathbf{x}'\mathbf{A}\mathbf{x}}{\sigma^2} - \frac{\mathbf{x}'\mathbf{x}}{2\sigma^2}\right) dx_1 \cdots dx_n$$

$$= \int_{-\infty}^{\infty} \cdots \int_{-\infty}^{\infty} \left(\frac{1}{\sigma\sqrt{2\pi}}\right)^n \exp\left[-\frac{\mathbf{x}'(\mathbf{I} - 2t\mathbf{A})\mathbf{x}}{2\sigma^2}\right] dx_1 \cdots dx_n,$$

where $\mathbf{I} = \mathbf{I}_n$. The matrix $\mathbf{I} - 2t\mathbf{A}$ is positive definite if we take $|t|$ sufficiently small, say $|t| < h$, $h > 0$. Moreover, we can treat

$$\frac{1}{(2\pi)^{n/2}\sqrt{|(\mathbf{I} - 2t\mathbf{A})^{-1}\sigma^2|}} \exp\left[-\frac{\mathbf{x}'(\mathbf{I} - 2t\mathbf{A})\mathbf{x}}{2\sigma^2}\right]$$

as a multivariate normal p.d.f. Now $|(\mathbf{I} - 2t\mathbf{A})^{-1}\sigma^2|^{1/2} = \sigma^n/|\mathbf{I} - 2t\mathbf{A}|^{1/2}$. If we multiply our integrand by $|\mathbf{I} - 2t\mathbf{A}|^{1/2}$, we have this multivariate p.d.f. Hence the m.g.f. of $\mathbf{X}'\mathbf{A}\mathbf{X}/\sigma^2$ is given by

$$M(t) = |\mathbf{I} - 2t\mathbf{A}|^{-1/2}, \qquad |t| < h. \tag{1}$$

It proves useful to express this m.g.f. in a different form. To do this, let $a_1, a_2, \ldots, a_n$ denote the characteristic numbers of $\mathbf{A}$ and let $\mathbf{L}$ denote an $n \times n$ orthogonal matrix such that $\mathbf{L}'\mathbf{A}\mathbf{L} = \text{diag}[a_1, a_2, \ldots, a_n]$. Thus

$$\mathbf{L}'(\mathbf{I} - 2t\mathbf{A})\mathbf{L} = \begin{bmatrix} 1 - 2ta_1 & 0 & \cdots & 0 \\ 0 & 1 - 2ta_2 & \cdots & 0 \\ \vdots & \vdots & & \vdots \\ 0 & 0 & \cdots & 1 - 2ta_n \end{bmatrix}.$$

Then

$$\prod_{i=1}^n (1 - 2ta_i) = |\mathbf{L}'(\mathbf{I} - 2t\mathbf{A})\mathbf{L}| = |\mathbf{I} - 2t\mathbf{A}|.$$

Accordingly, we can write $M(t)$, as given in Equation (1), in the form

$$M(t) = \left[\prod_{i=1}^{n} (1 - 2ta_i)\right]^{-1/2}, \qquad |t| < h. \tag{2}$$

Let r, $0 < r \le n$, denote the rank of the real symmetric matrix $\mathbf{A}$. Then exactly r of the real numbers $a_1, a_2, \ldots, a_n$, say $a_1, \ldots, a_r$, are not zero and exactly $n - r$ of these numbers, say $a_{r+1}, \ldots, a_n$, are zero. Thus we can write the m.g.f. of $\mathbf{X}'\mathbf{A}\mathbf{X}/\sigma^2$ as

$$M(t) = [(1 - 2ta_1)(1 - 2ta_2) \cdots (1 - 2ta_r)]^{-1/2}.$$

Now that we have found, in suitable form, the m.g.f. of our random variable, let us turn to the question of the conditions that must be imposed if $\mathbf{X}'\mathbf{A}\mathbf{X}/\sigma^2$ is to have a chi-square distribution. Assume that $\mathbf{X}'\mathbf{A}\mathbf{X}/\sigma^2$ is $\chi^2(k)$. Then

$$M(t) = [(1 - 2ta_1)(1 - 2ta_2) \cdots (1 - 2ta_r)]^{-1/2} = (1 - 2t)^{-k/2},$$

or, equivalently,

$$(1 - 2ta_1)(1 - 2ta_2) \cdots (1 - 2ta_r) = (1 - 2t)^k, \qquad |t| < h.$$

Because the positive integers r and k are the degrees of these polynomials, and because these polynomials are equal for infinitely many values of t, we have $k = r$, the rank of $\mathbf{A}$. Moreover, the uniqueness of the factorization of a polynomial implies that $a_1 = a_2 = \cdots = a_r = 1$. If each of the nonzero characteristic numbers of a real symmetric matrix is one, the matrix is idempotent, that is, $\mathbf{A}^2 = \mathbf{A}$, and conversely (see Exercise 10.38). Accordingly, if $\mathbf{X}'\mathbf{A}\mathbf{X}/\sigma^2$ has a chi-square distribution, then $\mathbf{A}^2 = \mathbf{A}$ and the random variable is $\chi^2(r)$, where r is the rank of $\mathbf{A}$. Conversely, if $\mathbf{A}$ is of rank r, $0 < r \le n$, and if $\mathbf{A}^2 = \mathbf{A}$, then $\mathbf{A}$ has exactly r characteristic numbers that are equal to one, and the remaining $n - r$ characteristic numbers are equal to zero. Thus the m.g.f. of $\mathbf{X}'\mathbf{A}\mathbf{X}/\sigma^2$ is given by $(1 - 2t)^{-r/2}$, $t < \frac{1}{2}$, and $\mathbf{X}'\mathbf{A}\mathbf{X}/\sigma^2$ is $\chi^2(r)$. This establishes the following theorem.

Theorem 2. *Let Q denote a random variable which is a quadratic form in the observations of a random sample of size n from a distribution which is $N(0, \sigma^2)$. Let $\mathbf{A}$ denote the symmetric matrix of Q and let r, $0 < r \le n$, denote the rank of $\mathbf{A}$. Then Q/σ^2 is $\chi^2(r)$ if and only if $\mathbf{A}^2 = \mathbf{A}$.*

Remark. If the normal distribution in Theorem 2 is $N(\mu, \sigma^2)$, the condition $\mathbf{A}^2 = \mathbf{A}$ remains a necessary and sufficient condition that Q/σ^2 have a chi-square distribution. In general, however, Q/σ^2 is not $\chi^2(r)$ but, instead, Q/σ^2 has a noncentral chi-square distribution if $\mathbf{A}^2 = \mathbf{A}$. The number of degrees of freedom is r, the rank of $\mathbf{A}$, and the noncentrality parameter is $\boldsymbol{\mu}'\mathbf{A}\boldsymbol{\mu}/\sigma^2$, where $\boldsymbol{\mu}' = [\mu, \mu, \ldots, \mu]$. Since $\boldsymbol{\mu}'\mathbf{A}\boldsymbol{\mu} = \mu^2 \sum_{i,j} a_{ij}$, where $\mathbf{A} = [a_{ij}]$, then, if $\mu \neq 0$, the conditions $\mathbf{A}^2 = \mathbf{A}$ and $\sum_{i,j} a_{ij} = 0$ are necessary and sufficient conditions that Q/σ^2 be central $\chi^2(r)$. Moreover, the theorem may be extended to a quadratic form in random variables which have a multivariate normal distribution with positive definite covariance matrix $\mathbf{V}$; here the necessary and sufficient condition that Q have a chi-square distribution is $\mathbf{AVA} = \mathbf{A}$.

EXERCISES

10.35. Let $Q = X_1X_2 - X_3X_4$, where X_1, X_2, X_3, X_4 is a random sample of size 4 from a distribution which is $N(0, \sigma^2)$. Show that Q/σ^2 does not have a chi-square distribution. Find the m.g.f. of Q/σ^2.

10.36. Let $\mathbf{X}' = [X_1, X_2]$ be bivariate normal with matrix of means $\boldsymbol{\mu}' = [\mu_1, \mu_2]$ and positive definite covariance matrix $\mathbf{V}$. Let

$$Q_1 = \frac{X_1^2}{\sigma_1^2(1-\rho^2)} - 2\rho\frac{X_1X_2}{\sigma_1\sigma_2(1-\rho^2)} + \frac{X_2^2}{\sigma_2^2(1-\rho^2)}.$$

Show that Q_1 is $\chi^2(r, \theta)$ and find r and θ. When and only when does Q_1 have a central chi-square distribution?

10.37. Let $\mathbf{X}' = [X_1, X_2, X_3]$ denote a random sample of size 3 from a distribution that is $N(4, 8)$ and let

$$\mathbf{A} = \begin{pmatrix} \frac{1}{2} & 0 & \frac{1}{2} \\ 0 & 1 & 0 \\ \frac{1}{2} & 0 & \frac{1}{2} \end{pmatrix}.$$

Justify the assertion that $\mathbf{X}'\mathbf{A}\mathbf{X}/\sigma^2$ is $\chi^2(2, 6)$.

10.38. Let $\mathbf{A}$ be a real symmetric matrix. Prove that each of the nonzero characteristic numbers of $\mathbf{A}$ is equal to 1 if and only if $\mathbf{A}^2 = \mathbf{A}$.

Hint: Let $\mathbf{L}$ be an orthogonal matrix such that $\mathbf{L}'\mathbf{A}\mathbf{L} =$ diag $[a_1, a_2, \ldots, a_n]$ and note that $\mathbf{A}$ is idempotent if and only if $\mathbf{L}'\mathbf{A}\mathbf{L}$ is idempotent.

10.39. The sum of the elements on the principal diagonal of a square matrix $\mathbf{A}$ is called the trace of $\mathbf{A}$ and is denoted by tr $\mathbf{A}$.

(a) If $\mathbf{B}$ is $n \times m$ and $\mathbf{C}$ is $m \times n$, prove that tr $(\mathbf{BC}) =$ tr $(\mathbf{CB})$.

(b) If $\mathbf{A}$ is a square matrix and if $\mathbf{L}$ is an orthogonal matrix, use the result of part (a) to show that tr $(\mathbf{L}'\mathbf{A}\mathbf{L}) =$ tr $\mathbf{A}$.

(c) If $\mathbf{A}$ is a real symmetric idempotent matrix, use the result of part (b) to prove that the rank of $\mathbf{A}$ is equal to tr $\mathbf{A}$.

10.40. Let $\mathbf{A} = [a_{ij}]$ be a real symmetric matrix. Prove that $\sum_j \sum_i a_{ij}^2$ is equal to the sum of the squares of the characteristic numbers of $\mathbf{A}$.

Hint: If $\mathbf{L}$ is an orthogonal matrix, show that $\sum_j \sum_i a_{ij}^2 = \text{tr}(\mathbf{A}^2) = \text{tr}(\mathbf{L}'\mathbf{A}^2\mathbf{L}) = \text{tr}[(\mathbf{L}'\mathbf{A}\mathbf{L})(\mathbf{L}'\mathbf{A}\mathbf{L})]$.

10.41. Let $\bar{X}$ and S^2 denote, respectively, the mean and the variance of a random sample of size n from a distribution which is $N(0, \sigma^2)$.

(a) If $\mathbf{A}$ denotes the symmetric matrix of $n\bar{X}^2$, show that $\mathbf{A} = (1/n)\mathbf{P}$, where $\mathbf{P}$ is the $n \times n$ matrix, each of whose elements is equal to one.

(b) Demonstrate that $\mathbf{A}$ is idempotent and that the tr $\mathbf{A} = 1$. Thus $n\bar{X}^2/\sigma^2$ is $\chi^2(1)$.

(c) Show that the symmetric matrix $\mathbf{B}$ of nS^2 is $\mathbf{I} - (1/n)\mathbf{P}$.

(d) Demonstrate that $\mathbf{B}$ is idempotent and that tr $\mathbf{B} = n - 1$. Thus nS^2/σ^2 is $\chi^2(n-1)$, as previously proved otherwise.

(e) Show that the product matrix $\mathbf{AB}$ is the zero matrix.

10.9 The Independence of Certain Quadratic Forms

We have previously investigated the independence of linear functions of normally distributed variables (see Exercise 4.132). In this section we shall prove some theorems about the independence of quadratic forms. As we remarked on p. 483, we shall confine our attention to normally distributed variables that constitute a random sample of size n from a distribution that is $N(0, \sigma^2)$.

Let $X_1, X_2, \ldots, X_n$ denote a random sample of size n from a distribution which is $N(0, \sigma^2)$. Let $\mathbf{A}$ and $\mathbf{B}$ denote two real symmetric matrices, each of order n. Let $\mathbf{X}' = [X_1, X_2, \ldots, X_n]$ and consider the two quadratic forms $\mathbf{X}'\mathbf{A}\mathbf{X}$ and $\mathbf{X}'\mathbf{B}\mathbf{X}$. We wish to show that these quadratic forms are independent if and only if $\mathbf{AB} = \mathbf{0}$, the zero matrix. We shall first compute the m.g.f. $M(t_1, t_2)$ of $\mathbf{X}'\mathbf{A}\mathbf{X}/\sigma^2$ and $\mathbf{X}'\mathbf{B}\mathbf{X}/\sigma^2$. We have

$$M(t_1, t_2) = \left(\frac{1}{\sigma\sqrt{2\pi}}\right)^n \int_{-\infty}^{\infty} \cdots \int_{-\infty}^{\infty}$$

$$\exp\left(\frac{t_1\mathbf{x}'\mathbf{A}\mathbf{x}}{\sigma^2} + \frac{t_2\mathbf{x}'\mathbf{B}\mathbf{x}}{\sigma^2} - \frac{\mathbf{x}'\mathbf{x}}{2\sigma^2}\right) dx_1 \cdots dx_n$$

$$= \left(\frac{1}{\sigma\sqrt{2\pi}}\right)^n \int_{-\infty}^{\infty} \cdots \int_{-\infty}^{\infty}$$

$$\exp\left(-\frac{\mathbf{x}'(\mathbf{I} - 2t_1\mathbf{A} - 2t_2\mathbf{B})\mathbf{x}}{2\sigma^2}\right) dx_1 \cdots dx_n.$$

The matrix $\mathbf{I} - 2t_1\mathbf{A} - 2t_2\mathbf{B}$ is positive definite if we take $|t_1|$ and $|t_2|$ sufficiently small, say $|t_1| < h_1$, $|t_2| < h_2$, where $h_1, h_2 > 0$. Then, as on p. 483, we have

$$M(t_1, t_2) = |\mathbf{I} - 2t_1\mathbf{A} - 2t_2\mathbf{B}|^{-1/2}, \qquad |t_1| < h_1, \quad |t_2| < h_2.$$

Let us assume that $\mathbf{X}'\mathbf{A}\mathbf{X}/\sigma^2$ and $\mathbf{X}'\mathbf{B}\mathbf{X}/\sigma^2$ are independent (so that likewise are $\mathbf{X}'\mathbf{A}\mathbf{X}$ and $\mathbf{X}'\mathbf{B}\mathbf{X}$) and prove that $\mathbf{AB} = \mathbf{0}$. Thus we assume that

$$M(t_1, t_2) = M(t_1, 0)M(0, t_2) \tag{1}$$

for all t_1 and t_2 for which $|t_i| < h_i$, $i = 1, 2$. Identity (1) is equivalent to the identity

$$|\mathbf{I} - 2t_1\mathbf{A} - 2t_2\mathbf{B}| = |\mathbf{I} - 2t_1\mathbf{A}||\mathbf{I} - 2t_2\mathbf{B}|, \qquad |t_i| < h_i, \quad i = 1, 2. \tag{2}$$

Let $r > 0$ denote the rank of $\mathbf{A}$ and let $a_1, a_2, \ldots, a_r$ denote the r nonzero characteristic numbers of $\mathbf{A}$. There exists an orthogonal matrix $\mathbf{L}$ such that

$$\mathbf{L}'\mathbf{A}\mathbf{L} = \left[\begin{array}{cccc|c} a_1 & 0 & \cdots & 0 & \\ 0 & a_2 & \cdots & 0 & \mathbf{0} \\ \vdots & \vdots & & \vdots & \\ 0 & 0 & \cdots & a_r & \\ \hline & & \mathbf{0} & & \mathbf{0} \end{array}\right] = \left[\begin{array}{c|c} \mathbf{C}_{11} & \mathbf{0} \\ \hline \mathbf{0} & \mathbf{0} \end{array}\right] = \mathbf{C}$$

for a suitable ordering of $a_1, a_2, \ldots, a_r$. Then $\mathbf{L}'\mathbf{B}\mathbf{L}$ may be written in the identically partitioned form

$$\mathbf{L}'\mathbf{B}\mathbf{L} = \left[\begin{array}{c|c} \mathbf{D}_{11} & \mathbf{D}_{12} \\ \hline \mathbf{D}_{21} & \mathbf{D}_{22} \end{array}\right] = \mathbf{D}.$$

The identity (2) may be written as

$$|\mathbf{L}'||\mathbf{I} - 2t_1\mathbf{A} - 2t_2\mathbf{B}||\mathbf{L}| = |\mathbf{L}'||\mathbf{I} - 2t_1\mathbf{A}||\mathbf{L}||\mathbf{L}'||\mathbf{I} - 2t_2\mathbf{B}||\mathbf{L}|, \tag{2'}$$

or as

$$|\mathbf{I} - 2t_1\mathbf{C} - 2t_2\mathbf{D}| = |\mathbf{I} - 2t_1\mathbf{C}||\mathbf{I} - 2t_2\mathbf{D}|. \tag{3}$$

The coefficient of $(-2t_1)^r$ in the right-hand member of Equation (3) is seen by inspection to be $a_1a_2 \cdots a_r|\mathbf{I} - 2t_2\mathbf{D}|$. It is not so easy to find the coefficient of $(-2t_1)^r$ in the left-hand member of Equation (3).

Conceive of expanding this determinant in terms of minors of order r formed from the first r columns. One term in this expansion is the product of the minor of order r in the upper left-hand corner, namely, $|\mathbf{I}_r - 2t_1\mathbf{C}_{11} - 2t_2\mathbf{D}_{11}|$, and the minor of order $n - r$ in the lower right-hand corner, namely, $|\mathbf{I}_{n-r} - 2t_2\mathbf{D}_{22}|$. Moreover, this product is the only term in the expansion of the determinant that involves $(-2t_1)^r$. Thus the coefficient of $(-2t_1)^r$ in the left-hand member of Equation (3) is $a_1 a_2 \cdots a_r |\mathbf{I}_{n-r} - 2t_2\mathbf{D}_{22}|$. If we equate these coefficients of $(-2t_1)^r$, we have, for all t_2, $|t_2| < h_2$,

$$|\mathbf{I} - 2t_2\mathbf{D}| = |\mathbf{I}_{n-r} - 2t_2\mathbf{D}_{22}|. \tag{4}$$

Equation (4) implies that the nonzero characteristic numbers of the matrices $\mathbf{D}$ and $\mathbf{D}_{22}$ are the same (see Exercise 10.49). Recall that the sum of the squares of the characteristic numbers of a symmetric matrix is equal to the sum of the squares of the elements of that matrix (see Exercise 10.40). Thus the sum of the squares of the elements of matrix $\mathbf{D}$ is equal to the sum of the squares of the elements of $\mathbf{D}_{22}$. Since the elements of the matrix $\mathbf{D}$ are real, it follows that each of the elements of $\mathbf{D}_{11}$, $\mathbf{D}_{12}$, and $\mathbf{D}_{21}$ is zero. Accordingly, we can write $\mathbf{D}$ in the form

$$\mathbf{D} = \mathbf{L}'\mathbf{B}\mathbf{L} = \left[\begin{array}{c|c} \mathbf{0} & \mathbf{0} \\ \hline \mathbf{0} & \mathbf{D}_{22} \end{array}\right].$$

Thus $\mathbf{CD} = \mathbf{L}'\mathbf{A}\mathbf{L}\mathbf{L}'\mathbf{B}\mathbf{L} = \mathbf{0}$ and $\mathbf{L}'\mathbf{A}\mathbf{B}\mathbf{L} = \mathbf{0}$ and $\mathbf{AB} = \mathbf{0}$, as we wished to prove.

To complete the proof of the theorem, we assume that $\mathbf{AB} = \mathbf{0}$. We are to show that $\mathbf{X}'\mathbf{A}\mathbf{X}/\sigma^2$ and $\mathbf{X}'\mathbf{B}\mathbf{X}/\sigma^2$ are independent. We have, for all real values of t_1 and t_2,

$$(\mathbf{I} - 2t_1\mathbf{A})(\mathbf{I} - 2t_2\mathbf{B}) = \mathbf{I} - 2t_1\mathbf{A} - 2t_2\mathbf{B},$$

since $\mathbf{AB} = \mathbf{0}$. Thus

$$|\mathbf{I} - 2t_1\mathbf{A} - 2t_2\mathbf{B}| = |\mathbf{I} - 2t_1\mathbf{A}||\mathbf{I} - 2t_2\mathbf{B}|.$$

Since the m.g.f. of $\mathbf{X}'\mathbf{A}\mathbf{X}/\sigma^2$ and $\mathbf{X}'\mathbf{B}\mathbf{X}/\sigma^2$ is given by

$$M(t_1, t_2) = |\mathbf{I} - 2t_1\mathbf{A} - 2t_2\mathbf{B}|^{-1/2}, \qquad |t_i| < h_i, \quad i = 1, 2,$$

we have

$$M(t_1, t_2) = M(t_1, 0)M(0, t_2),$$

and the proof of the following theorem is complete.

Theorem 3. *Let Q_1 and Q_2 denote random variables which are quadratic forms in the observations of a random sample of size n from a distribution which is $N(0, \sigma^2)$. Let* **A** *and* **B** *denote, respectively, the real symmetric matrices of Q_1 and Q_2. The random variables Q_1 and Q_2 are independent if and only if* $\mathbf{AB} = \mathbf{0}$.

Remark. Theorem 3 remains valid if the random sample is from a distribution which is $N(\mu, \sigma^2)$, whatever be the real value of μ. Moreover, Theorem 2 may be extended to quadratic forms in random variables that have a joint multivariate normal distribution with a positive definite covariance matrix **V**. The necessary and sufficient condition for the independence of two such quadratic forms with symmetric matrices **A** and **B** then becomes $\mathbf{AVB} = \mathbf{0}$. In our Theorem 2, we have $\mathbf{V} = \sigma^2\mathbf{I}$, so that $\mathbf{AVB} = \mathbf{A}\sigma^2\mathbf{IB} = \sigma^2\mathbf{AB} = \mathbf{0}$.

We shall next prove Theorem 1 that was stated in Section 10.1.

Theorem 4. *Let $Q = Q_1 + \cdots + Q_{k-1} + Q_k$, where Q, $Q_1, \ldots, Q_{k-1}, Q_k$ are $k + 1$ random variables that are quadratic forms in the observations of a random sample of size n from a distribution which is $N(0, \sigma^2)$. Let Q/σ^2 be $\chi^2(r)$, let Q_i/σ^2 be $\chi^2(r_i)$, $i = 1, 2, \ldots, k - 1$, and let Q_k be nonnegative. Then the random variables $Q_1, Q_2, \ldots, Q_k$ are independent and, hence, Q_k/σ^2 is $\chi^2(r_k = r - r_1 - \cdots - r_{k-1})$.*

Proof. Take first the case of $k = 2$ and let the real symmetric matrices of Q, Q_1, and Q_2 be denoted, respectively, by $\mathbf{A}$, $\mathbf{A}_1$, $\mathbf{A}_2$. We are given that $Q = Q_1 + Q_2$ or, equivalently, that $\mathbf{A} = \mathbf{A}_1 + \mathbf{A}_2$. We are also given that Q/σ^2 is $\chi^2(r)$ and that Q_1/σ^2 is $\chi^2(r_1)$. In accordance with Theorem 2, p. 484, we have $\mathbf{A}^2 = \mathbf{A}$ and $\mathbf{A}_1^2 = \mathbf{A}_1$. Since $Q_2 \geq 0$, each of the matrices $\mathbf{A}$, $\mathbf{A}_1$, and $\mathbf{A}_2$ is positive semidefinite. Because $\mathbf{A}^2 = \mathbf{A}$, we can find an orthogonal matrix **L** such that

$$\mathbf{L}'\mathbf{AL} = \left[\begin{array}{c|c} \mathbf{I}_r & \mathbf{0} \\ \hline \mathbf{0} & \mathbf{0} \end{array}\right].$$

If then we multiply both members of $\mathbf{A} = \mathbf{A}_1 + \mathbf{A}_2$ on the left by $\mathbf{L}'$ and on the right by **L**, we have

$$\left[\begin{array}{c|c} \mathbf{I}_r & \mathbf{0} \\ \hline \mathbf{0} & \mathbf{0} \end{array}\right] = \mathbf{L}'\mathbf{A}_1\mathbf{L} + \mathbf{L}'\mathbf{A}_2\mathbf{L}.$$

Now each of $\mathbf{A}_1$ and $\mathbf{A}_2$, and hence each of $\mathbf{L}'\mathbf{A}_1\mathbf{L}$ and $\mathbf{L}'\mathbf{A}_2\mathbf{L}$ is positive semidefinite. Recall that, if a real symmetric matrix is positive semidefinite, each element on the principal diagonal is positive or zero. Moreover, if an element on the principal diagonal is zero, then all elements in that row and all elements in that column are zero. Thus $\mathbf{L}'\mathbf{A}\mathbf{L} = \mathbf{L}'\mathbf{A}_1\mathbf{L} + \mathbf{L}'\mathbf{A}_2\mathbf{L}$ can be written as

$$\left[\begin{array}{c|c} \mathbf{I}_r & \mathbf{0} \\ \hline \mathbf{0} & \mathbf{0} \end{array}\right] = \left[\begin{array}{c|c} \mathbf{G}_r & \mathbf{0} \\ \hline \mathbf{0} & \mathbf{0} \end{array}\right] + \left[\begin{array}{c|c} \mathbf{H}_r & \mathbf{0} \\ \hline \mathbf{0} & \mathbf{0} \end{array}\right]. \tag{5}$$

Since $\mathbf{A}_1^2 = \mathbf{A}_1$, we have

$$(\mathbf{L}'\mathbf{A}_1\mathbf{L})^2 = \mathbf{L}'\mathbf{A}_1\mathbf{L} = \left[\begin{array}{c|c} \mathbf{G}_r & \mathbf{0} \\ \hline \mathbf{0} & \mathbf{0} \end{array}\right].$$

If we multiply both members of Equation (5) on the left by the matrix $\mathbf{L}'\mathbf{A}_1\mathbf{L}$, we see that

$$\left[\begin{array}{c|c} \mathbf{G}_r & \mathbf{0} \\ \hline \mathbf{0} & \mathbf{0} \end{array}\right] = \left[\begin{array}{c|c} \mathbf{G}_r & \mathbf{0} \\ \hline \mathbf{0} & \mathbf{0} \end{array}\right] + \left[\begin{array}{c|c} \mathbf{G}_r\mathbf{H}_r & \mathbf{0} \\ \hline \mathbf{0} & \mathbf{0} \end{array}\right],$$

or, equivalently, $\mathbf{L}'\mathbf{A}_1\mathbf{L} = \mathbf{L}'\mathbf{A}_1\mathbf{L} + (\mathbf{L}'\mathbf{A}_1\mathbf{L})(\mathbf{L}'\mathbf{A}_2\mathbf{L})$. Thus $(\mathbf{L}'\mathbf{A}_1\mathbf{L}) \times (\mathbf{L}'\mathbf{A}_2\mathbf{L}) = \mathbf{0}$ and $\mathbf{A}_1\mathbf{A}_2 = \mathbf{0}$. In accordance with Theorem 3, Q_1 and Q_2 are independent. This independence immediately implies that Q_2/σ^2 is $\chi^2(r_2 = r - r_1)$. This completes the proof when $k = 2$. For $k > 2$, the proof may be made by induction. We shall merely indicate how this can be done by using $k = 3$. Take $\mathbf{A} = \mathbf{A}_1 + \mathbf{A}_2 + \mathbf{A}_3$, where $\mathbf{A}^2 = \mathbf{A}$, $\mathbf{A}_1^2 = \mathbf{A}_1$, $\mathbf{A}_2^2 = \mathbf{A}_2$, and $\mathbf{A}_3$ is positive semidefinite. Write $\mathbf{A} = \mathbf{A}_1 + (\mathbf{A}_2 + \mathbf{A}_3) = \mathbf{A}_1 + \mathbf{B}_1$, say. Now $\mathbf{A}^2 = \mathbf{A}$, $\mathbf{A}_1^2 = \mathbf{A}_1$, and $\mathbf{B}_1$ is positive semidefinite. In accordance with the case of $k = 2$, we have $\mathbf{A}_1\mathbf{B}_1 = \mathbf{0}$, so that $\mathbf{B}_1^2 = \mathbf{B}_1$. With $\mathbf{B}_1 = \mathbf{A}_2 + \mathbf{A}_3$, where $\mathbf{B}_1^2 = \mathbf{B}_1$, $\mathbf{A}_2^2 = \mathbf{A}_2$, it follows from the case of $k = 2$ that $\mathbf{A}_2\mathbf{A}_3 = \mathbf{0}$ and $\mathbf{A}_3^2 = \mathbf{A}_3$. If we regroup by writing $\mathbf{A} = \mathbf{A}_2 + (\mathbf{A}_1 + \mathbf{A}_3)$, we obtain $\mathbf{A}_1\mathbf{A}_3 = \mathbf{0}$, and so on.

Remark. In our statement of Theorem 4 we took $X_1, X_2, \ldots, X_n$ to be observations of a random sample from a distribution which is $N(0, \sigma^2)$. We did this because our proof of Theorem 3 was restricted to that case. In fact, if $Q', Q_1', \ldots, Q_k'$ are quadratic forms in any normal variables (including multivariate normal variables), if $Q' = Q_1' + \cdots + Q_k'$, if $Q', Q_1', \ldots, Q_{k-1}'$ are central or noncentral chi-square, and if Q_k' is nonnegative, then $Q_1', \ldots, Q_k'$ are independent and Q_k' is either central or noncentral chi-square.

This section will conclude with a proof of a frequently quoted theorem due to Cochran.

Theorem 5. *Let $X_1, X_2, \ldots, X_n$ denote a random sample from a distribution which is $N(0, \sigma^2)$. Let the sum of the squares of these observations be written in the form*

$$\sum_1^n X_i^2 = Q_1 + Q_2 + \cdots + Q_k,$$

where Q_j is a quadratic form in $X_1, X_2, \ldots, X_n$, with matrix $\mathbf{A}_j$ which has rank r_j, $j = 1, 2, \ldots, k$. The random variables $Q_1, Q_2, \ldots, Q_k$ are independent and Q_j/σ^2 is $\chi^2(r_j)$, $j = 1, 2, \ldots, k$, if and only if $\sum_1^k r_j = n$.

Proof. First assume the two conditions $\sum_1^k r_j = n$ and $\sum_1^n X_i^2 = \sum_1^k Q_j$ to be satisfied. The latter equation implies that $\mathbf{I} = \mathbf{A}_1 + \mathbf{A}_2 + \cdots + \mathbf{A}_k$. Let $\mathbf{B}_i = \mathbf{I} - \mathbf{A}_i$. That is, $\mathbf{B}_i$ is the sum of the matrices $\mathbf{A}_1, \ldots, \mathbf{A}_k$ exclusive of $\mathbf{A}_i$. Let R_i denote the rank of $\mathbf{B}_i$. Since the rank of the sum of several matrices is less than or equal to the sum of the ranks, we have $R_i \le \sum_1^k r_j - r_i = n - r_i$. However, $\mathbf{I} = \mathbf{A}_i + \mathbf{B}_i$, so that $n \le r_i + R_i$ and $n - r_i \le R_i$. Hence $R_i = n - r_i$. The characteristic numbers of $\mathbf{B}_i$ are the roots of the equation $|\mathbf{B}_i - \lambda\mathbf{I}| = 0$. Since $\mathbf{B}_i = \mathbf{I} - \mathbf{A}_i$, this equation can be written as $|\mathbf{I} - \mathbf{A}_i - \lambda\mathbf{I}| = 0$. Thus we have $|\mathbf{A}_i - (1 - \lambda)\mathbf{I}| = 0$. But each root of the last equation is one minus a characteristic number of $\mathbf{A}_i$. Since $\mathbf{B}_i$ has exactly $n - R_i = r_i$ characteristic numbers that are zero, then $\mathbf{A}_i$ has exactly r_i characteristic numbers that are equal to 1. However, r_i is the rank of $\mathbf{A}_i$. Thus each of the r_i nonzero characteristic numbers of $\mathbf{A}_i$ is 1. That is, $\mathbf{A}_i^2 = \mathbf{A}_i$ and thus Q_i/σ^2 is $\chi^2(r_i)$, $i = 1, 2, \ldots, k$. In accordance with Theorem 4, the random variables $Q_1, Q_2, \ldots, Q_k$ are independent.

To complete the proof of Theorem 5, take

$$\sum_1^n X_i^2 = Q_1 + Q_2 + \cdots + Q_k,$$

let $Q_1, Q_2, \ldots, Q_k$ be independent, and let Q_j/σ^2 be $\chi^2(r_j)$, $j = 1, 2, \ldots, k$. Then $\sum_1^k Q_j/\sigma^2$ is $\chi^2\left(\sum_1^k r_j\right)$. But $\sum_1^k Q_j/\sigma^2 = \sum_1^n X_i^2/\sigma^2$ is $\chi^2(n)$. Thus $\sum_1^k r_j = n$ and the proof is complete.

EXERCISES

10.42. Let X_1, X_2, X_3 be a random sample from the normal distribution $N(0, \sigma^2)$. Are the quadratic forms $X_1^2 + 3X_1X_2 + X_2^2 + X_1X_3 + X_3^2$ and $X_1^2 - 2X_1X_2 + \frac{2}{3}X_2^2 - 2X_1X_3 - X_3^2$ independent or dependent?

10.43. Let $X_1, X_2, \ldots, X_n$ denote a random sample of size n from a distribution which is $N(0, \sigma^2)$. Prove that $\sum_1^n X_i^2$ and every quadratic form, which is nonidentically zero in $X_1, X_2, \ldots, X_n$, are dependent.

10.44. Let X_1, X_2, X_3, X_4 denote a random sample of size 4 from a distribution which is $N(0, \sigma^2)$. Let $Y = \sum_1^4 a_i X_i$, where a_1, a_2, a_3, and a_4 are real constants. If Y^2 and $Q = X_1X_2 - X_3X_4$ are independent, determine a_1, a_2, a_3, and a_4.

10.45. Let $\mathbf{A}$ be the real symmetric matrix of a quadratic form Q in the observations of a random sample of size n from a distribution which is $N(0, \sigma^2)$. Given that Q and the mean $\bar{X}$ of the sample are independent. What can be said of the elements of each row (column) of $\mathbf{A}$?

Hint: Are Q and $\bar{X}^2$ independent?

10.46. Let $\mathbf{A}_1, \mathbf{A}_2, \ldots, \mathbf{A}_k$ be the matrices of $k > 2$ quadratic forms $Q_1, Q_2, \ldots, Q_k$ in the observations of a random sample of size n from a distribution which is $N(0, \sigma^2)$. Prove that the pairwise independence of these forms implies that they are mutually independent.

Hint: Show that $\mathbf{A}_i\mathbf{A}_j = \mathbf{0}$, $i \neq j$, permits $E[\exp(t_1Q_1 + t_2Q_2 + \cdots + t_kQ_k)]$ to be written as a product of the moment-generating functions of $Q_1, Q_2, \ldots, Q_k$.

10.47. Let $\mathbf{X}' = [X_1, X_2, \ldots, X_n]$, where $X_1, X_2, \ldots, X_n$ are observations of a random sample from a distribution which is $N(0, \sigma^2)$. Let $\mathbf{b}' = [b_1, b_2, \ldots, b_n]$ be a real nonzero matrix, and let $\mathbf{A}$ be a real symmetric matrix of order n. Prove that the linear form $\mathbf{b}'\mathbf{X}$ and the quadratic form $\mathbf{X}'\mathbf{A}\mathbf{X}$ are independent if and only if $\mathbf{b}'\mathbf{A} = \mathbf{0}$. Use this fact to prove that $\mathbf{b}'\mathbf{X}$ and $\mathbf{X}'\mathbf{A}\mathbf{X}$ are independent if and only if the two quadratic forms, $(\mathbf{b}'\mathbf{X})^2 = \mathbf{X}'\mathbf{b}\mathbf{b}'\mathbf{X}$ and $\mathbf{X}'\mathbf{A}\mathbf{X}$, are independent.

10.48. Let Q_1 and Q_2 be two nonnegative quadratic forms in the observations of a random sample from a distribution which is $N(0, \sigma^2)$. Show that another quadratic form Q is independent of $Q_1 + Q_2$ if and only if Q is independent of each of Q_1 and Q_2.

Hint: Consider the orthogonal transformation that diagonalizes the matrix of $Q_1 + Q_2$. After this transformation, what are the forms of the matrices of Q, Q_1, and Q_2 if Q and $Q_1 + Q_2$ are independent?

10.49. Prove that Equation (4) of this section implies that the nonzero characteristic numbers of the matrices $\mathbf{D}$ and $\mathbf{D}_{22}$ are the same.

Hint: Let $\lambda = 1/(2t_2)$, $t_2 \neq 0$, and show that Equation (4) is equivalent to $|\mathbf{D} - \lambda\mathbf{I}| = (-\lambda)^r|\mathbf{D}_{22} - \lambda\mathbf{I}_{n-r}|$.

10.50. Here Q_1 and Q_2 are quadratic forms in observations of a random sample from $N(0, 1)$. If Q_1 and Q_2 are independent and if $Q_1 + Q_2$ has a chi-square distribution, prove that Q_1 and Q_2 are chi-square variables.

10.51. Often in regression the mean of the random variable Y is a linear function of p-values $x_1, x_2, \ldots, x_p$, say $\beta_1 x_1 + \beta_2 x_2 + \cdots + \beta_p x_p$, where $\boldsymbol{\beta}' = (\beta_1, \beta_2, \ldots, \beta_p)$ are the *regression coefficients*. Suppose that n values, $\mathbf{Y}' = (Y_1, Y_2, \ldots, Y_n)$, are observed for the x-values in $\mathbf{X} = (x_{ij})$, where $\mathbf{X}$ is an $n \times p$ *design matrix* and its ith row is associated with Y_i, $i = 1, 2, \ldots, n$. Assume that $\mathbf{Y}$ is multivariate normal with mean $\mathbf{X}\boldsymbol{\beta}$ and covariance matrix $\sigma^2\mathbf{I}$, where $\mathbf{I}$ is the $n \times n$ identity matrix.

(a) Note that $Y_1, Y_2, \ldots, Y_n$ are independent. Why?

(b) Since $\mathbf{Y}$ should approximately equal its mean $\mathbf{X}\boldsymbol{\beta}$, we estimate $\boldsymbol{\beta}$ by solving the *normal equations* $\mathbf{X}'\mathbf{Y} = \mathbf{X}'\mathbf{X}\boldsymbol{\beta}$ for $\boldsymbol{\beta}$. Assuming that $\mathbf{X}'\mathbf{X}$ is nonsingular, solve the equations to get $\hat{\boldsymbol{\beta}} = (\mathbf{X}'\mathbf{X})^{-1}\mathbf{X}'\mathbf{Y}$. Show that $\hat{\boldsymbol{\beta}}$ has a multivariate normal distribution with mean $\boldsymbol{\beta}$ and covariance matrix $\sigma^2(\mathbf{X}'\mathbf{X})^{-1}$.

(c) Show that

$$(\mathbf{Y} - \mathbf{X}\boldsymbol{\beta})'(\mathbf{Y} - \mathbf{X}\boldsymbol{\beta}) = (\hat{\boldsymbol{\beta}} - \boldsymbol{\beta})'(\mathbf{X}'\mathbf{X})(\hat{\boldsymbol{\beta}} - \boldsymbol{\beta}) + (\mathbf{Y} - \mathbf{X}\hat{\boldsymbol{\beta}})'(\mathbf{Y} - \mathbf{X}\hat{\boldsymbol{\beta}}),$$

say $Q = Q_1 + Q_2$ for convenience.

(d) Show that Q_1/σ^2 is $\chi^2(p)$.

(e) Show that Q_1 and Q_2 are independent.

(f) Argue that Q_2/σ^2 is $\chi^2(n - p)$.

(g) Find c so that cQ_1/Q_2 has an F-distribution.

(h) The fact that a value d can be found so that $\Pr(cQ_1/Q_2 \le d) = 1 - \alpha$ could be used to find a $100(1 - \alpha)$ percent confidence ellipsoid for $\boldsymbol{\beta}$. Explain.

(i) If the coefficient matrix $\boldsymbol{\beta}$ has the prior distribution that is multivariate normal with mean matrix $\boldsymbol{\beta}_0$ and covariance matrix Σ_0, what is the posterior distribution of $\boldsymbol{\beta}$, given $\hat{\boldsymbol{\beta}}$?

10.52. Say that G.P.A. (Y) is thought to be a linear function of a "coded" high school rank (x_2) and a "coded" American College Testing score (x_3), namely, $\beta_1 + \beta_2 x_2 + \beta_3 x_3$. Note that all x_1 values equal 1. We observe the following five points:

x_1	x_2	x_3	Y
1	1	2	3
1	4	3	6
1	2	2	4
1	4	2	4
1	3	2	4

(a) Compute $\mathbf{X}'\mathbf{X}$ and $\hat{\boldsymbol{\beta}} = (\mathbf{X}'\mathbf{X})^{-1}\mathbf{X}'\mathbf{Y}$.

(b) Compute a 95 percent confidence ellipsoid for $\boldsymbol{\beta}' = (\beta_1, \beta_2, \beta_3)$.

ADDITIONAL EXERCISES

10.53. Let μ_1, μ_2, μ_3 be, respectively, the means of three normal distributions with a common but unknown variance σ^2. In order to test, at the $\alpha = 5$ percent significance level, the hypothesis $H_0 : \mu_1 = \mu_2 = \mu_3$ against all possible alternative hypotheses, we take an independent random sample of size 4 from each of these distributions. Determine whether we accept or reject H_0 if the observed values from these three distributions are, respectively,

X_1: 5 9 6 8

X_2: 11 13 10 12

X_3: 10 6 9 9

10.54. The driver of a diesel-powered automobile decided to test the quality of three types of diesel fuel sold in the area based on mpg. Test the null hypothesis that the three means are equal using the following data. Make the usual assumptions and take $\alpha = 0.05$.

Brand A: 38.7 39.2 40.1 38.9

Brand B: 41.9 42.3 41.3

Brand C: 40.8 41.2 39.5 38.9 40.3

10.55. We wish to compare compressive strengths of concrete corresponding to $a = 3$ different drying methods (treatments). Concrete is mixed in batches that are just large enough to produce three cylinders. Although care is taken to achieve uniformity, we expect some variability among the $b = 5$ batches used to obtain the following compressive strengths. (There is little reason to suspect interaction and hence only one observation is taken in each cell.)

	Batch				
Treatment	B_1	B_2	B_3	B_4	B_5
A_1	52	47	44	51	42
A_2	60	55	49	52	43
A_3	56	48	45	44	38

(a) Use the 5 percent significance level and test $H_A : \alpha_1 = \alpha_2 = \alpha_3 = 0$ against all alternatives.

(b) Use the 5 percent significance level and test $H_B : \beta_1 = \beta_2 = \beta_3 = \beta_4 = \beta_5 = 0$ against all alternatives.

10.56. With $a = 3$ and $b = 4$, find μ, α_i, β_j, and γ_{ij}, if μ_{ij}, $i = 1, 2, 3$ and $j = 1, 2, 3, 4$, are given by

$$\begin{array}{cccc} 6 & 7 & 7 & 12 \\ 10 & 3 & 11 & 8 \\ 8 & 5 & 9 & 10 \end{array}$$

10.57. Two experiments gave the following results:

n	$\bar{x}$	$\bar{y}$	s_x	s_y	r
100	10	20	5	8	0.70
200	12	22	6	10	0.80

Calculate r for the combined sample.

10.58. Consider the following matrices: $\mathbf{Y}$ is $n \times 1$, $\boldsymbol{\beta}$ is $p \times 1$, $\mathbf{X}$ is $n \times p$ and of rank p. Let $\mathbf{Y}$ be $N(\mathbf{X}\boldsymbol{\beta}, \sigma^2\mathbf{I})$. Discuss the joint p.d.f. of $\hat{\boldsymbol{\beta}} = (\mathbf{X}'\mathbf{X})^{-1}\mathbf{X}'\mathbf{Y}$ and $\mathbf{Y}'[\mathbf{I} - \mathbf{X}(\mathbf{X}'\mathbf{X})^{-1}\mathbf{X}']\mathbf{Y}/\sigma^2$.

10.59. Fit $y = a + x$ to the data

x	0	1	2
y	1	3	4

by the method of least squares.

10.60. Fit by the method of least squares the plane $z = a + bx + cy$ to the five points (x, y, z): $(-1, -2, 5)$, $(0, -2, 4)$, $(0, 0, 4)$, $(1, 0, 2)$, $(2, 1, 0)$.

10.61. Let the 4×1 matrix $\mathbf{Y}$ be multivariate normal $N(\mathbf{X}\boldsymbol{\beta}, \sigma^2\mathbf{I})$, where the 4×3 design matrix equals

$$\mathbf{X} = \begin{bmatrix} 1 & 1 & 2 \\ 1 & -1 & 2 \\ 1 & 0 & -3 \\ 1 & 0 & -1 \end{bmatrix}$$

and $\boldsymbol{\beta}$ is the 3×1 regression coefficient matrix.

(a) Find the mean matrix and the covariance matrix of $\hat{\boldsymbol{\beta}} = (\mathbf{X}'\mathbf{X})^{-1}\mathbf{X}'\mathbf{Y}$.

(b) If we observe $\mathbf{Y}'$ to be equal to $(6, 1, 11, 3)$, compute $\hat{\boldsymbol{\beta}}$.

10.62. Let the independent normal random variables $Y_1, Y_2, \ldots, Y_n$ have, respectively, the probability density functions $N(\mu, \gamma^2 x_i^2)$, $i = 1, 2, \ldots, n$, where the given $x_1, x_2, \ldots, x_n$ are not all equal and no one of which is zero. Discuss the test of the hypothesis $H_0: \gamma = 1$, μ unspecified, against all alternatives $H_1: \gamma \neq 1$, μ unspecified.

10.63. Let $Y_1, Y_2, \ldots, Y_n$ be n independent normal variables with common unknown variance σ^2. Let Y_i have mean βx_i, $i = 1, 2, \ldots, n$, where $x_1, x_2, \ldots, x_n$ are known but not all the same and β is an unknown constant. Find the likelihood ratio test for $H_0: \beta = 0$ against all alternatives. Show that this likelihood ratio test can be based on a statistic that has a well-known distribution.

10.64. Consider the multivariate normal p.d.f. $f(\mathbf{x}; \boldsymbol{\mu}, \boldsymbol{\Sigma})$ where the known parameters equal either $\boldsymbol{\mu}_1, \boldsymbol{\Sigma}_1$ or $\boldsymbol{\mu}_2, \boldsymbol{\Sigma}_2$, respectively.

(a) If $\boldsymbol{\Sigma}_1 = \boldsymbol{\Sigma}_2$ is known to equal $\boldsymbol{\Sigma}$, classify $\mathbf{X}$ as being in the second of these distributions if

$$\frac{f(\mathbf{x}; \boldsymbol{\mu}_1, \boldsymbol{\Sigma})}{f(\mathbf{x}; \boldsymbol{\mu}_2, \boldsymbol{\Sigma})} \le k;$$

otherwise, $\mathbf{X}$ is classified as being from the first distribution. Show that this rule is based upon a linear function of $\mathbf{X}$ and determine its distribution. This allows us to compute the probabilities of misclassification.

(b) If $\boldsymbol{\Sigma}_1$ and $\boldsymbol{\Sigma}_2$ are different but known, show that

$$\frac{f(\mathbf{x}; \boldsymbol{\mu}_1, \boldsymbol{\Sigma}_1)}{f(\mathbf{x}; \boldsymbol{\mu}_2, \boldsymbol{\Sigma}_2)} \le k$$

can be based upon a second degree polynomial in $\mathbf{X}$. When either $\boldsymbol{\Sigma}_1$ or $\boldsymbol{\Sigma}_2$ is the correct covariance matrix, does this expression have a chi-square distribution?

CHAPTER 11

Nonparametric Methods

11.1 Confidence Intervals for Distribution Quantiles

We shall first define the concept of a quantile of a distribution of a random variable of the continuous type. Let X be a random variable of the continuous type with p.d.f. $f(x)$ and distribution function $F(x)$. Let p denote a positive proper fraction and assume that the equation $F(x) = p$ has a unique solution for x. This unique root is denoted by the symbol ξ_p and is called the *quantile* (of the distribution) *of order* p. Thus $\Pr(X \le \xi_p) = F(\xi_p) = p$. For example, the quantile of order $\frac{1}{2}$ is the median of the distribution and $\Pr(X \le \xi_{0.5}) = F(\xi_{0.5}) = \frac{1}{2}$.

In Chapter 6 we computed the probability that a certain random interval includes a special point. Frequently, this special point was a parameter of the distribution of probability under consideration. Thus we are led to the notion of an interval estimate of a parameter. If the parameter happens to be a quantile of the distribution, and if we work with certain functions of the order statistics, it will be seen that this method of statistical inference is applicable to all distributions of the continuous type. We call these methods *distribution-free* or *nonparametric* methods of inference.

To obtain a distribution-free confidence interval for ξ_p, the quantile of order p, of a distribution of the continuous type with distribution function $F(x)$, take a random sample $X_1, X_2, \ldots, X_n$ of size n from that distribution. Let $Y_1 < Y_2 < \cdots < Y_n$ be the order statistics of the sample. Take $Y_i < Y_j$ and consider the event $Y_i < \xi_p < Y_j$. For the ith order statistic Y_i to be less than ξ_p it must be true that at least i of the X values are less than ξ_p. Moreover, for the jth order statistic to be greater than ξ_p, fewer than j of the X values are less than ξ_p. That is, if we say that we have a "success" when an individual X value is less than ξ_p, then, in the n independent trials, there must be at least i successes but fewer than j successes for the event $Y_i < \xi_p < Y_j$ to occur. But since the probability of success on each trial is $\Pr(X < \xi_p) = F(\xi_p) = p$, the probability of this event is

$$\Pr(Y_i < \xi_p < Y_j) = \sum_{w=i}^{j-1} \frac{n!}{w!\,(n-w)!} p^w(1-p)^{n-w},$$

the probability of having at least i, but less than j, successes. When particular values of n, i, and j are specified, this probability can be computed. By this procedure, suppose it has been found that $\gamma = \Pr(Y_i < \xi_p < Y_j)$. Then the probability is γ that the random interval (Y_i, Y_j) includes the quantile of order p. If the experimental values of Y_i and Y_j are, respectively, y_i and y_j, the interval (y_i, y_j) serves as a 100γ percent confidence interval for ξ_p, the quantile of order p. An illustrative example follows.

Example 1. Let $Y_1 < Y_2 < Y_3 < Y_4$ be the order statistics of a random sample of size 4 from a distribution of the continuous type. The probability that the random interval (Y_1, Y_4) includes the median $\xi_{0.5}$ of the distribution will be computed. We have

$$\Pr(Y_1 < \xi_{0.5} < Y_4) = \sum_{w=1}^{3} \frac{4!}{w!\,(4-w)!}\left(\frac{1}{2}\right)^4 = 0.875.$$

If Y_1 and Y_4 are observed to be $y_1 = 2.8$ and $y_4 = 4.2$, respectively, the interval $(2.8, 4.2)$ is an 87.5 percent confidence interval for the median $\xi_{0.5}$ of the distribution.

For samples of fairly large size, we can approximate the binomial probabilities with those associated with normal distributions, as illustrated in the next example.

Example 2. Let the following numbers represent the values of the order statistics of $n = 27$ observations obtained in a random sample from a certain distribution of the continuous type:

61, 69, 71, 74, 79, 80, 83, 84, 86, 87, 92, 93, 96, 100,

104, 105, 113, 121, 122, 129, 141, 143, 156, 164, 191, 217, 276.

Say that we are interested in estimating the 25th percentile $\xi_{0.25}$ (that is, the quantile of order 0.25) of the distribution. Since $(n + 1)p = 28(\frac{1}{4}) = 7$, the seventh order statistic, $y_7 = 83$, could serve as a point estimate of $\xi_{0.25}$. To get a confidence interval for $\xi_{0.25}$, consider two order statistics, one less than y_7 and the other greater, for illustration, y_4 and y_{10}. What is the confidence coefficient associated with the interval (y_4, y_{10})? Of course, before the sample is drawn, we know that

$$\gamma = \Pr(Y_4 < \xi_{0.25} < Y_{10}) = \sum_{w=4}^{9} \binom{27}{w} (0.25)^w (0.75)^{27-w}.$$

That is,

$$\gamma = \Pr(3.5 < W < 9.5),$$

where W is $b(27, \frac{1}{4})$ with mean $\frac{27}{4} = 6.75$ and variance $\frac{81}{16}$. Hence γ is approximately equal to

$$\Phi\left(\frac{9.5 - 6.75}{\frac{9}{4}}\right) - \Phi\left(\frac{3.5 - 6.75}{\frac{9}{4}}\right) = \Phi\left(\frac{11}{9}\right) - \Phi\left(-\frac{13}{9}\right) = 0.814.$$

Thus $(y_4 = 74, y_{10} = 87)$ serves as an approximate 81.4 percent confidence interval for $\xi_{0.25}$. It should be noted that we could choose other intervals also, for illustration, $(y_3 = 71, y_{11} = 92)$, and these would have different confidence coefficients. The persons involved in the study must select the desired confidence coefficient, and then the appropriate order statistics, Y_i and Y_j, are taken in such a way that i and j are fairly symmetrically located about $(n + 1)p$.

EXERCISES

11.1. Let Y_n denote the nth order statistic of a random sample of size n from a distribution of the continuous type. Find the smallest value of n for which $\Pr(\xi_{0.9} < Y_n) \geq 0.75$.

11.2. Let $Y_1 < Y_2 < Y_3 < Y_4 < Y_5$ denote the order statistics of a random sample of size 5 from a distribution of the continuous type. Compute:

(a) $\Pr(Y_1 < \xi_{0.5} < Y_5)$.
(b) $\Pr(Y_1 < \xi_{0.25} < Y_3)$.
(c) $\Pr(Y_4 < \xi_{0.80} < Y_5)$.

11.3. Compute $\Pr(Y_3 < \xi_{0.5} < Y_7)$ if $Y_1 < \cdots < Y_9$ are the order statistics of a random sample of size 9 from a distribution of the continuous type.

11.4. Find the smallest value of n for which $\Pr(Y_1 < \xi_{0.5} < Y_n) \geq 0.99$, where $Y_1 < \cdots < Y_n$ are the order statistics of a random sample of size n from a distribution of the continuous type.

11.5. Let $Y_1 < Y_2$ denote the order statistics of a random sample of size 2 from a distribution which is $N(\mu, \sigma^2)$, where σ^2 is known.
(a) Show that $\Pr(Y_1 < \mu < Y_2) = \frac{1}{2}$ and compute the expected value of the random length $Y_2 - Y_1$.
(b) If $\bar{X}$ is the mean of this sample, find the constant c such that $\Pr(\bar{X} - c\sigma < \mu < \bar{X} + c\sigma) = \frac{1}{2}$, and compare the length of this random interval with the expected value of that of part (a).

11.6. Let $Y_1 < Y_2 < \cdots < Y_{25}$ be the order statistics of a random sample of size $n = 25$ from a distribution of the continuous type. Compute approximately:
(a) $\Pr(Y_8 < \xi_{0.5} < Y_{18})$.
(b) $\Pr(Y_2 < \xi_{0.2} < Y_9)$.
(c) $\Pr(Y_{18} < \xi_{0.8} < Y_{23})$.

11.7. Let $Y_1 < Y_2 < \cdots < Y_{100}$ be the order statistics of a random sample of size $n = 100$ from a distribution of the continuous type. Find $i < j$ so that $\Pr(Y_i < \xi_{0.2} < Y_j)$ is about equal to 0.95.

11.8. Let $\xi_{1/4}$ be the 25th percentile of a distribution of the continuous type. Let $Y_1 < Y_2 < \cdots < Y_{48}$ be the order statistics of a random sample of size $n = 48$ from this distribution.
(a) In terms of "binomial probabilities," to what is $\Pr(Y_9 < \xi_{1/4} < Y_{16})$ equal?
(b) How would you approximate this answer with "normal probabilities"?
(c) Find i such that $\Pr(Y_{12-i} < \xi_{1/4} < Y_{13+i})$ is as close as possible to 0.95 (using the normal approximation).

11.2 Tolerance Limits for Distributions

We propose now to investigate a problem that has something of the same flavor as that treated in Section 11.1. Specifically, can we compute the probability that a certain random interval includes (or *covers*) a preassigned percentage of the probability for the distribution under consideration? And, by appropriate selection of the random interval, can we be led to an additional distribution-free method of statistical inference?

Let X be a random variable with distribution function $F(x)$ of the continuous type. The random variable $Z = F(X)$ is an important random variable, and its distribution is given in Example 1, Section 4.1. It is our purpose now to make an interpretation. Since $Z = F(X)$ has the p.d.f.

$$\begin{aligned} h(z) &= 1, \qquad 0 < z < 1, \\ &= 0 \qquad \text{elsewhere,} \end{aligned}$$

then, if $0 < p < 1$, we have

$$\Pr[F(X) \le p] = \int_0^p dz = p.$$

Now $F(x) = \Pr(X \le x)$. Since $\Pr(X = x) = 0$, then $F(x)$ is the fractional part of the probability for the distribution of X that is between $-\infty$ and x. If $F(x) \le p$, then no more than $100p$ percent of the probability for the distribution of X is between $-\infty$ and x. But recall $\Pr[F(X) \le p] = p$. That is, the probability that the random variable $Z = F(X)$ is less than or equal to p is precisely the probability that the random interval $(-\infty, X)$ contains no more than $100p$ percent of the probability for the distribution. For example, the probability that the random interval $(-\infty, X)$ contains no more than 70 percent of the probability for the distribution is 0.70; and the probability that the random interval $(-\infty, X)$ contains more than 70 percent of the probability for the distribution is $1 - 0.70 = 0.30$.

We now consider certain functions of the order statistics. Let $X_1, X_2, \ldots, X_n$ denote a random sample of size n from a distribution that has a positive and continuous p.d.f. $f(x)$ if and only if $a < x < b$; and let $F(x)$ denote the associated distribution function. Consider the random variables $F(X_1), F(X_2), \ldots, F(X_n)$. These random variables are independent and each, in accordance with Example 1, Section 4.1, has a uniform distribution on the interval $(0, 1)$. Thus $F(X_1), F(X_2), \ldots, F(X_n)$ is a random sample of size n from a uniform distribution on the interval $(0, 1)$. Consider the order statistics of this random sample $F(X_1), F(X_2), \ldots, F(X_n)$. Let Z_1 be the smallest of these $F(X_i)$, Z_2 the next $F(X_i)$ in order of magnitude, $\ldots$, and Z_n the largest $F(X_i)$. If $Y_1, Y_2, \ldots, Y_n$ are the order statistics of the initial random sample $X_1, X_2, \ldots, X_n$, the fact that $F(x)$ is a nondecreasing (here, strictly increasing) function of x implies that $Z_1 = F(Y_1)$, $Z_2 = F(Y_2), \ldots, Z_n = F(Y_n)$. Thus the joint p.d.f. of $Z_1, Z_2, \ldots, Z_n$ is given by

$$
\begin{aligned}
h(z_1, z_2, \ldots, z_n) &= n!, \qquad 0 < z_1 < z_2 < \cdots < z_n < 1, \\
&= 0 \qquad \text{elsewhere.}
\end{aligned}
$$

This proves a special case of the following theorem.

Theorem 1. *Let $Y_1, Y_2, \ldots, Y_n$ denote the order statistics of a random sample of size n from a distribution of the continuous type that has p.d.f. $f(x)$ and distribution function $F(x)$. The joint p.d.f. of the random variables $Z_i = F(Y_i)$, $i = 1, 2, \ldots, n$, is*

$$
\begin{aligned}
h(z_1, z_2, \ldots, z_n) &= n!, \qquad 0 < z_1 < z_2 < \cdots < z_n < 1, \\
&= 0 \qquad \text{elsewhere.}
\end{aligned}
$$

Because the distribution function of $Z = F(X)$ is given by z, $0 < z < 1$, the marginal p.d.f. of $Z_k = F(Y_k)$ is the following beta p.d.f.:

$$
\begin{aligned}
h_k(z_k) &= \frac{n!}{(k-1)!\,(n-k)!} z_k^{k-1}(1 - z_k)^{n-k}, \qquad 0 < z_k < 1, \\
&= 0 \qquad \text{elsewhere.}
\end{aligned} \tag{1}
$$

Moreover, the joint p.d.f. of $Z_i = F(Y_i)$ and $Z_j = F(Y_j)$ is, with $i < j$, given by

$$
\begin{aligned}
h_{ij}(z_i, z_j) &= \frac{n!\, z_i^{i-1}(z_j - z_i)^{j-i-1}(1 - z_j)^{n-j}}{(i-1)!\,(j-i-1)!\,(n-j)!}, \qquad 0 < z_i < z_j < 1, \\
&= 0 \qquad \text{elsewhere.}
\end{aligned} \tag{2}
$$

Consider the difference $Z_j - Z_i = F(Y_j) - F(Y_i)$, $i < j$. Now $F(y_j) = \Pr(X \le y_j)$ and $F(y_i) = \Pr(X \le y_i)$. Since $\Pr(X = y_i) = \Pr(X = y_j) = 0$, then the difference $F(y_j) - F(y_i)$ is that fractional part of the probability for the distribution of X that is between y_i and y_j. Let p denote a positive proper fraction. If $F(y_j) - F(y_i) \ge p$, then at least $100p$ percent of the probability for the distribution of X is between y_i and y_j. Let it be given that $\gamma = \Pr[F(Y_j) - F(Y_i) \ge p]$. Then the random interval (Y_i, Y_j) has probability γ of containing at least $100p$ percent of the probability for the distribution of X. If now y_i and y_j denote, respectively, experimental values of Y_i and Y_j, the interval (y_i, y_j) either does or does not contain at least $100p$ percent of the probability for the distribution of X. However, we refer to the interval (y_i, y_j) as a 100γ percent *tolerance interval* for $100p$ percent of the probability for the distribution of X. In like vein, y_i and y_j are called 100γ percent *tolerance limits* for $100p$ percent of the probability for the distribution of X.

One way to compute the probability $\gamma = \Pr[F(Y_j) - F(Y_i) \geq p]$ is to use Equation (2), which gives the joint p.d.f. of $Z_i = F(Y_i)$ and $Z_j = F(Y_j)$. The required probability is then given by

$$\gamma = \Pr(Z_j - Z_i \geq p) = \int_0^{1-p} \int_{p+z_i}^{1} h_{ij}(z_i, z_j)\, dz_j\, dz_i.$$

Sometimes, this is a rather tedious computation. For this reason and for the reason that *coverages* are important in distribution-free statistical inference, we choose to introduce at this time the concept of a coverage.

Consider the random variables $W_1 = F(Y_1) = Z_1$, $W_2 = F(Y_2) - F(Y_1) = Z_2 - Z_1$, $W_3 = F(Y_3) - F(Y_2) = Z_3 - Z_2, \ldots, W_n = F(Y_n) - F(Y_{n-1}) = Z_n - Z_{n-1}$. The random variable W_1 is called a *coverage* of the random interval $\{x : -\infty < x < Y_1\}$ and the random variable W_i, $i = 2, 3, \ldots, n$, is called a *coverage* of the random interval $\{x : Y_{i-1} < x < Y_i\}$. We shall find the joint p.d.f. of the n coverages $W_1, W_2, \ldots, W_n$. First we note that the inverse functions of the associated transformation are given by

$$\begin{aligned} z_1 &= w_1, \\ z_2 &= w_1 + w_2, \\ z_3 &= w_1 + w_2 + w_3, \\ &\vdots \\ z_n &= w_1 + w_2 + w_3 + \cdots + w_n. \end{aligned}$$

We also note that the Jacobian is equal to 1 and that the space of positive probability density is

$$\{(w_1, w_2, \ldots, w_n) : 0 < w_i,\ i = 1, 2, \ldots, n,\ w_1 + \cdots + w_n < 1\}.$$

Since the joint p.d.f. of $Z_1, Z_2, \ldots, Z_n$ is $n!$, $0 < z_1 < z_2 < \cdots < z_n < 1$, zero elsewhere, the joint p.d.f. of the n coverages is

$$\begin{aligned} k(w_1, \ldots, w_n) &= n!, \qquad 0 < w_i, \quad i = 1, \ldots, n, \quad w_1 + \cdots + w_n < 1, \\ &= 0 \qquad \text{elsewhere.} \end{aligned}$$

A reexamination of Example 1 of Section 4.5 reveals that this is a Dirichlet p.d.f. with $k = n$ and $\alpha_1 = \alpha_2 = \cdots = \alpha_{n+1} = 1$.

Because the p.d.f. $k(w_1, \ldots, w_n)$ is symmetric in $w_1, w_2, \ldots, w_n$, it is evident that the distribution of every sum of r, $r < n$, of these coverages $W_1, \ldots, W_n$ is exactly the same for each fixed value of r. For instance, if $i < j$ and $r = j - i$, the distribution of $Z_j - Z_i = F(Y_j) - F(Y_i) = W_{i+1} + W_{i+2} + \cdots + W_j$ is exactly the same as that of $Z_{j-i} = F(Y_{j-i}) = W_1 + W_2 + \cdots + W_{j-i}$. But we know that the p.d.f. of Z_{j-i} is the beta p.d.f. of the form

$$h_{j-i}(v) = \frac{\Gamma(n+1)}{\Gamma(j-i)\Gamma(n-j+i+1)} v^{j-i-1}(1-v)^{n-j+i}, \qquad 0 < v < 1,$$

$$= 0 \qquad \text{elsewhere.}$$

Consequently, $F(Y_j) - F(Y_i)$ has this p.d.f. and

$$\Pr[F(Y_j) - F(Y_i) \geq p] = \int_p^1 h_{j-i}(v)\,dv.$$

Example 1. Let $Y_1 < Y_2 < \cdots < Y_6$ be the order statistics of a random sample of size 6 from a distribution of the continuous type. We want to use the observed interval (y_1, y_6) as a tolerance interval for 80 percent of the distribution. Then

$$\gamma = \Pr[F(Y_6) - F(Y_1) \geq 0.8]$$

$$= 1 - \int_0^{0.8} 30v^4(1-v)\,dv,$$

because the integrand is the p.d.f. of $F(Y_6) - F(Y_1)$. Accordingly,

$$\gamma = 1 - 6(0.8)^5 + 5(0.8)^6 = 0.34,$$

approximately. That is, the observed values of Y_1 and Y_6 will define a 34 percent tolerance interval for 80 percent of the probability for the distribution.

Example 2. Each of the coverages W_i, $i = 1, 2, \ldots, n$, has the beta p.d.f.

$$k_1(w) = n(1-w)^{n-1}, \qquad 0 < w < 1,$$

$$= 0 \qquad \text{elsewhere,}$$

because $W_1 = Z_1 = F(Y_1)$ has this p.d.f. Accordingly, the mathematical expectation of each W_i is

$$\int_0^1 nw(1-w)^{n-1}\,dw = \frac{1}{n+1}.$$

Now the coverage W_i can be thought of as the area under the graph of the p.d.f. $f(x)$, above the x-axis, and between the lines $x = Y_{i-1}$ and $x = Y_i$. (We take $Y_0 = -\infty$.) Thus the expected value of each of these random areas W_i, $i = 1, 2, \ldots, n$, is $1/(n+1)$. That is, the order statistics partition the probability for the distribution into $n + 1$ parts, and the expected value of each of these parts is $1/(n+1)$. More generally, the expected value of $F(Y_j) - F(Y_i)$, $i < j$, is $(j-i)/(n+1)$, since $F(Y_j) - F(Y_i)$ is the sum of $j - i$ of these coverages. This result provides a reason for calling Y_k, where $(n+1)p = k$, the $(100p)$th *percentile of the sample*, since

$$E[F(Y_k)] = \frac{k}{n+1} = \frac{(n+1)p}{n+1} = p.$$

EXERCISES

11.9. Let Y_1 and Y_n be, respectively, the first and nth order statistics of a random sample of size n from a distribution of the continuous type having distribution function $F(x)$. Find the smallest value of n such that $\Pr[F(Y_n) - F(Y_1) \geq 0.5]$ is at least 0.95.

11.10. Let Y_2 and Y_{n-1} denote the second and the $(n-1)$st order statistics of a random sample of size n from a distribution of the continuous type having distribution function $F(x)$. Compute $\Pr[F(Y_{n-1}) - F(Y_2) \geq p]$, where $0 < p < 1$.

11.11. Let $Y_1 < Y_2 < \cdots < Y_{48}$ be the order statistics of a random sample of size 48 from a distribution of the continuous type. We want to use the observed interval (y_4, y_{45}) as a 100γ percent tolerance interval for 75 percent of the distribution.

(a) To what is γ equal?

(b) Approximate the integral in part (a) by noting that it can be written as a partial sum of a binomial p.d.f., which in turn can be approximated by probabilities associated with a normal distribution.

11.12. Let $Y_1 < Y_2 < \cdots < Y_n$ be the order statistics of a random sample of size n from a distribution of the continuous type having distribution function $F(x)$.

(a) What is the distribution of $U = 1 - F(Y_j)$?

(b) Determine the distribution of $V = F(Y_n) - F(Y_j) + F(Y_i) - F(Y_1)$, where $i < j$.

11.13. Let $Y_1 < Y_2 < \cdots < Y_{10}$ be the order statistics of a random sample from a continuous-type distribution with distribution function $F(x)$. What is the joint distribution of $V_1 = F(Y_4) - F(Y_2)$ and $V_2 = F(Y_{10}) - F(Y_6)$?

11.3 The Sign Test

Some of the chi-square tests of Section 6.6 are illustrative of the type of tests that we investigate in the remainder of this chapter. Recall, in that section, we tested the hypothesis that the distribution of a certain random variable X is a specified distribution. We did this in the following manner. The space of X was partitioned into k mutually disjoint sets $A_1, A_2, \ldots, A_k$. The probability p_{i0} that $X \in A_i$ was computed under the assumption that the specified distribution is the correct distribution, $i = 1, 2, \ldots, k$. The original hypothesis was then replaced by the less restrictive hypothesis

$$H_0 : \Pr(X \in A_i) = p_{i0}, \qquad i = 1, 2, \ldots, k;$$

and a chi-square test, based upon a statistic that was denoted by Q_{k-1}, was used to test the hypothesis H_0 against all alternative hypotheses.

There is a certain subjective element in the use of this test, namely the choice of k and of $A_1, A_2, \ldots, A_k$. But it is important to note that the limiting distribution of Q_{k-1}, under H_0, is $\chi^2(k-1)$; that is, the distribution of Q_{k-1} is *free* of $p_{10}, p_{20}, \ldots, p_{k0}$ and, accordingly, of the specified distribution of X. Here, and elsewhere, "under H_0" means when H_0 is true. A test of a hypothesis H_0 based upon a statistic whose distribution, under H_0, does not depend upon the specified distribution or any parameters of that distribution is called a *distribution-free* or a *nonparametric test.*

Next, let $F(x)$ be the unknown distribution function of the random variable X. Let there be given two numbers ξ and p_0, where $0 < p_0 < 1$. We wish to test the hypothesis $H_0 : F(\xi) = p_0$, that is, the hypothesis that $\xi = \xi_{p_0}$, the quantile of order p_0 of the distribution of X. We could use the statistic Q_{k-1}, with $k = 2$, to test H_0 against all alternatives. Suppose, however, that we are interested only in the alternative hypothesis, which is $H_1 : F(\xi) > p_0$. One procedure is to base the test of H_0 against H_1 upon the random variable Y, which is the number of observations less than or equal to ξ in a random sample of size n from the distribution. The statistic Y can be thought of as the number of "successes" throughout n independent trials. Then, if H_0 is true, Y is $b[n, p_0 = F(\xi)]$; whereas if H_0 is false, Y is $b[n, p = F(\xi)]$ whatever be the distribution function $F(x)$. We reject H_0 and accept H_1 if and only if the observed value $y \geq c$, where c is an integer selected such that $\Pr(Y \geq c; H_0)$ is some reasonable significance level α. The power function of the test is given by

$$K(p) = \sum_{y=c}^{n} \binom{n}{y} p^y (1-p)^{n-y}, \qquad p_0 \leq p < 1,$$

where $p = F(\xi)$. In certain instances, we may wish to approximate $K(p)$ by using an approximation to the binomial distribution.

Suppose that the alternative hypothesis to $H_0 : F(\xi) = p_0$ is $H_1 : F(\xi) < p_0$. Then the critical region is a set $\{y : y \leq c_1\}$. Finally, if the alternative hypothesis is $H_1 : F(\xi) \neq p_0$, the critical region is a set $\{y : y \leq c_2 \text{ or } c_3 \leq y\}$.

Frequently, $p_0 = \frac{1}{2}$ and, in that case, the hypothesis is that the given number ξ is a median of the distribution. In the following example, this value of p_0 is used.

Example 1. Let $X_1, X_2, \ldots, X_{10}$ be a random sample of size 10 from a distribution with distribution function $F(x)$. We wish to test the hypothesis $H_0 : F(72) = \frac{1}{2}$ against the alternative hypothesis $H_1 : F(72) > \frac{1}{2}$. Let Y be the number of sample items that are less than or equal to 72. Let the observed value of Y be y, and let the test be defined by the critical region $\{y : y \geq 8\}$. The power function of the test is given by

$$K(p) = \sum_{y=8}^{10} \binom{10}{y} p^y (1-p)^{10-y}, \qquad \frac{1}{2} \leq p < 1,$$

where $p = F(72)$. In particular, the significance level is

$$\alpha = K\left(\frac{1}{2}\right) = \left[\binom{10}{8} + \binom{10}{9} + \binom{10}{10}\right]\left(\frac{1}{2}\right)^{10} = \frac{7}{128}.$$

In many places in the literature, the test that we have just described is called the *sign test*. The reason for this terminology is that the test is based upon a statistic Y that is equal to the number of nonpositive signs in the sequence $X_1 - \xi, X_2 - \xi, \ldots, X_n - \xi$. In the next section a distribution-free test, which considers both the sign and the magnitude of each deviation $X_i - \xi$, is studied.

EXERCISES

11.14. Suggest a chi-square test of the hypothesis which states that a distribution is one of the beta type, with parameters $\alpha = 2$ and $\beta = 2$. Further, suppose that the test is to be based upon a random sample of size 100. In the solution, give k, define $A_1, A_2, \ldots, A_k$, and compute each

p_{i0}. If possible, compare your proposal with those of other students. Are any of them the same?

11.15. Let $X_1, X_2, \ldots, X_{48}$ be a random sample of size 48 from a distribution that has the distribution function $F(x)$. To test $H_0: F(41) = \frac{1}{4}$ against $H_1: F(41) < \frac{1}{4}$, use the statistic Y, which is the number of sample observations less than or equal to 41. If the observed value of Y is $y \le 7$, reject H_0 and accept H_1. If $p = F(41)$, find the power function $K(p)$, $0 < p \le \frac{1}{4}$, of the test. Approximate $\alpha = K(\frac{1}{4})$.

11.16. Let $X_1, X_2, \ldots, X_{100}$ be a random sample of size 100 from a distribution that has distribution function $F(x)$. To test $H_0: F(90) - F(60) = \frac{4}{5}$ against $H_1: F(90) - F(60) > \frac{4}{5}$, use the statistic Y, which is the number of sample observations less than or equal to 90 but greater than 60. If the observed value of Y, say y, is such that $y \ge c$, reject H_0. Find c so that $\alpha = 0.05$, approximately.

11.17. Let $X_1, X_2, \ldots, X_n$ be a random sample from some continuous-type distribution. We wish to consider only *unbiased* estimators of $\Pr(X \le c)$, where c is a fixed constant.

(a) What would you use as an unbiased estimator if you had no additional assumptions about the distribution?

(b) What would you use as an unbiased estimator if you knew the distribution was normal with unknown mean μ and variance $\sigma^2 = 1$?

11.4 A Test of Wilcoxon

Suppose that $X_1, X_2, \ldots, X_n$ is a random sample from a distribution with distribution function $F(x)$. We have considered a test of the hypothesis $F(\xi) = \frac{1}{2}$, ξ given, which is based upon the signs of the deviations $X_1 - \xi, X_2 - \xi, \ldots, X_n - \xi$. In this section a statistic is studied that takes into account not only these signs, but also the magnitudes of the deviations.

To find such a statistic that is distribution-free, we must make two additional assumptions:

1. $F(x)$ is the distribution function of a continuous type of random variable X.
2. The p.d.f. $f(x)$ of X has a graph that is symmetric about the vertical axis through $\xi_{0.5}$, the median (which we assume to be unique) of the distribution.

Thus

$$F(\xi_{0.5} - x) = 1 - F(\xi_{0.5} + x)$$

and

$$f(\xi_{0.5} - x) = f(\xi_{0.5} + x),$$

for all x. Moreover, the probability that any two observations of a random sample are equal is zero, and in our discussion we shall assume that no two are equal.

The problem is to test the hypothesis that the median $\xi_{0.5}$ of the distribution is equal to a fixed number, say ξ. Thus we may, in all cases and without loss of generality, take $\xi = 0$. The reason for this is that if $\xi \neq 0$, then the fixed ξ can be subtracted from each sample observation and the resulting variables can be used to test the hypothesis that their underlying distribution is symmetric about zero. Hence our conditions on $F(x)$ and $f(x)$ become $F(-x) = 1 - F(x)$ and $f(-x) = f(x)$, respectively.

To test the hypothesis $H_0 : F(0) = \frac{1}{2}$, we proceed by first ranking $X_1, X_2, \ldots, X_n$ according to magnitude, disregarding their algebraic signs. Let R_i be the rank of $|X_i|$ among $|X_1|, |X_2|, \ldots, |X_n|$, $i = 1, 2, \ldots, n$. For example, if $n = 3$ and if we have $|X_2| < |X_3| < |X_1|$, then $R_1 = 3$, $R_2 = 1$, and $R_3 = 2$. Thus $R_1, R_2, \ldots, R_n$ is an arrangement of the first n positive integers $1, 2, \ldots, n$. Further, let Z_i, $i = 1, 2, \ldots, n$, be defined by

$$\begin{aligned} Z_i &= -1, \qquad \text{if } X_i < 0, \\ &= 1, \qquad \text{if } X_i > 0. \end{aligned}$$

If we recall that $\Pr(X_i = 0) = 0$, we see that it does not change the probabilities whether we associate $Z_i = 1$ or $Z_i = -1$ with the outcome $X_i = 0$.

The statistic $W = \sum_{i=1}^{n} Z_i R_i$ is the Wilcoxon statistic. Note that in computing this statistic we simply associate the sign of each X_i with the rank of its absolute value and sum the resulting n products.

If the alternative to the hypothesis $H_0 : \xi_{0.5} = 0$ is $H_1 : \xi_{0.5} > 0$, we reject H_0 if the observed value of W is an element of the set $\{w : w \geq c\}$. This is due to the fact that large positive values of W indicate that most of the large deviations from zero are positive. For alternatives $\xi_{0.5} < 0$ and $\xi_{0.5} \neq 0$ the critical regions are, respectively, the sets $\{w : w \leq c_1\}$ and $\{w : w \leq c_2 \text{ or } w \geq c_3\}$. To compute probabilities like $\Pr(W \geq c; H_0)$, we need to determine the distribution of W, under H_0.

To help us find the distribution of W, when $H_0 : F(0) = \frac{1}{2}$ is true, we note the following facts:

1. The assumption that $f(x) = f(-x)$ ensures that $\Pr(X_i < 0) = \Pr(X_i > 0) = \frac{1}{2}$, $i = 1, 2, \ldots, n$.
2. Now $Z_i = -1$ if $X_i < 0$ and $Z_i = 1$ if $X_i > 0$, $i = 1, 2, \ldots, n$. Hence we have $\Pr(Z_i = -1) = \Pr(Z_i = 1) = \frac{1}{2}$, $i = 1, 2, \ldots, n$. Moreover, $Z_1, Z_2, \ldots, Z_n$ are independent because $X_1, X_2, \ldots, X_n$ are independent.
3. The assumption that $f(x) = f(-x)$ also assures that the rank R_i of $|X_i|$ does not depend upon the sign Z_i of X_i. More generally, $R_1, R_2, \ldots, R_n$ are independent of $Z_1, Z_2, \ldots, Z_n$.
4. A sum W is made up of the numbers $1, 2, \ldots, n$, each number with either a positive or a negative sign.

The preceding observations enable us to say that $W = \sum_1^n Z_i R_i$ has the same distribution as the random variable $V = \sum_1^n V_i$, where $V_1, V_2, \ldots, V_n$ are independent and

$$\Pr(V_i = i) = \Pr(V_i = -i) = \tfrac{1}{2},$$

$i = 1, 2, \ldots, n$. That $V_1, V_2, \ldots, V_n$ are independent follows from the fact that $Z_1, Z_2, \ldots, Z_n$ have that property; that is, the numbers $1, 2, \ldots, n$ always appear in a sum W and those numbers receive their algebraic signs by independent assignment. Thus each of $V_1, V_2, \ldots, V_n$ is like one and only one of $Z_1R_1, Z_2R_2, \ldots, Z_nR_n$.

Since W and V have the same distribution, the m.g.f. of W is that of V,

$$M(t) = E\left[\exp\left(t\sum_1^n V_i\right)\right] = \prod_{i=1}^n E(e^{tV_i})$$

$$= \prod_{i=1}^n \left(\frac{e^{-it} + e^{it}}{2}\right).$$

We can express $M(t)$ as the sum of terms of the form $(a_j/2^n)e^{b_jt}$. When $M(t)$ is written in this manner, we can determine by inspection the p.d.f. of the discrete-type random variable W. For example, the smallest value of W is found from the term $(1/2^n)e^{-t}e^{-2t}\cdots e^{-nt} = (1/2^n)e^{-n(n+1)t/2}$ and it is $-n(n+1)/2$. The probability of this value of W is the coefficient $1/2^n$. To make these statements more concrete, take $n = 3$. Then

$$M(t) = \left(\frac{e^{-t} + e^{t}}{2}\right)\left(\frac{e^{-2t} + e^{2t}}{2}\right)\left(\frac{e^{-3t} + e^{3t}}{2}\right)$$

$$= (\tfrac{1}{8})(e^{-6t} + e^{-4t} + e^{-2t} + 2 + e^{2t} + e^{4t} + e^{6t}).$$

Thus the p.d.f. of W, for $n = 3$, is given by

$$\begin{aligned} g(w) &= \tfrac{1}{8}, \qquad w = -6, -4, -2, 2, 4, 6, \\ &= \tfrac{2}{8}, \qquad w = 0, \\ &= 0 \qquad \text{elsewhere.} \end{aligned}$$

The mean and the variance of W are more easily computed directly than by working with the m.g.f. $M(t)$. Because $V = \sum_1^n V_i$ and $W = \sum_1^n Z_i R_i$ have the same distribution, they have the same mean and the same variance. When the hypothesis $H_0 : F(0) = \frac{1}{2}$ is true, it is easy to determine the values of these two characteristics of the distribution of W. Since $E(V_i) = 0$, $i = 1, 2, \ldots, n$, we have

$$\mu_W = E(W) = \sum_1^n E(V_i) = 0.$$

The variance of V_i is $(-i)^2(\frac{1}{2}) + (i)^2(\frac{1}{2}) = i^2$. Thus the variance of W is

$$\sigma_w^2 = \sum_1^n i^2 = \frac{n(n+1)(2n+1)}{6}.$$

For large values of n, the determination of the exact distribution of W becomes tedious. Accordingly, one looks for an approximating distribution. Although W is distributed as the sum of n random variables that are independent, our form of the central limit theorem cannot be applied because the n random variables do not have identical distributions. However, a more general theorem, due to Liapounov, states that if U_i has mean μ_i and variance σ_i^2, $i = 1, 2, \ldots, n$, if $U_1, U_2, \ldots, U_n$ are independent, if $E(|U_i - \mu_i|^3)$ is finite for every i, and if

$$\lim_{n\to\infty} \frac{\sum_{i=1}^{n} E(|U_i - \mu_i|^3)}{\left(\sum_{i=1}^{n} \sigma_i^2\right)^{3/2}} = 0,$$

then

$$\frac{\sum_{i=1}^{n} U_i - \sum_{i=1}^{n} \mu_i}{\sqrt{\sum_{i=1}^{n} \sigma_i^2}}$$

has a limiting distribution that is $N(0, 1)$. For our variables $V_1, V_2, \ldots, V_n$ we have

$$E(|V_i - \mu_i|^3) = i^3(\tfrac{1}{2}) + i^3(\tfrac{1}{2}) = i^3;$$

and it is known that

$$\sum_{i=1}^{n} i^3 = \frac{n^2(n+1)^2}{4}.$$

Now

$$\lim_{n\to\infty} \frac{n^2(n+1)^2/4}{[n(n+1)(2n+1)/6]^{3/2}} = 0$$

because the numerator is of order n^4 and the denominator is of order $n^{9/2}$. Thus

$$\frac{W}{\sqrt{n(n+1)(2n+1)/6}}$$

is approximately $N(0, 1)$ when H_0 is true. This allows us to approximate probabilities like $\Pr(W \geq c; H_0)$ when the sample size n is large.

Example 1. Let $\xi_{0.5}$ be the median of a symmetric distribution that is of the continuous type. To test, with $\alpha = 0.01$, the hypothesis $H_0 : \xi_{0.5} = 75$ against $H_1 : \xi_{0.5} > 75$, we observed a random sample of size $n = 18$. Let it be given that the deviations of these 18 values from 75 are the following numbers:

$$1.5, -0.5, 1.6, 0.4, 2.3, -0.8, 3.2, 0.9, 2.9,$$

$$0.3, 1.8, -0.1, 1.2, 2.5, 0.6, -0.7, 1.9, 1.3.$$

The experimental value of the Wilcoxon statistic is equal to

$$w = 11 - 4 + 12 + 3 + 15 - 7 + 18 + 8 + 17 + 2 + 13 - 1$$
$$+ 9 + 16 + 5 - 6 + 14 + 10 = 135.$$

Since, with $n = 18$ so that $\sqrt{n(n+1)(2n+1)/6} = 45.92$, we have that

$$0.01 = \Pr\left(\frac{W}{45.92} \geq 2.326\right) = \Pr(W \geq 106.8).$$

Because $w = 135 > 106.8$, we reject H_0 at the approximate 0.01 significance level. The p-value associated with $135/45.92 = 2.94$ is about $1 - 0.998 = 0.002$ since $\Phi(2.94) = 0.998$.

There are many modifications and generalizations of the Wilcoxon statistic. One generalization is the following: Let $c_1 \le c_2 \le \cdots \le c_n$ be nonnegative numbers. Then, in the Wilcoxon statistic, replace the ranks $1, 2, \ldots, n$ by $c_1, c_2, \ldots, c_n$, respectively. For example, if $n = 3$ and if we have $|X_2| < |X_3| < |X_1|$, then $R_1 = 3$ is replaced by c_3, $R_2 = 1$ by c_1, and $R_3 = 2$ by c_2. In this example, the generalized statistic is given by $Z_1c_3 + Z_2c_1 + Z_3c_2$. Similar to the Wilcoxon statistic, this generalized statistic is distributed under H_0, as the sum of n independent random variables, the ith of which takes each of the values $c_i \neq 0$ and $-c_i$ with probability $\frac{1}{2}$; if $c_i = 0$, that variable takes the value $c_i = 0$ with probability 1. Some special cases of this statistic are proposed in the Exercises.

EXERCISES

11.18. The observed values of a random sample of size 10 from a distribution that is symmetric about $\xi_{0.5}$ are 10.2, 14.1, 9.2, 11.3, 7.2, 9.8, 6.5, 11.8, 8.7, 10.8. Use Wilcoxon's statistic to test the hypothesis $H_0 : \xi_{0.5} = 8$ against $H_1 : \xi_{0.5} > 8$ if $\alpha = 0.05$. Even though n is small, use the normal approximation and find the p-value.

11.19. Find the distribution of W for $n = 4$ and $n = 5$.
Hint: Multiply the moment-generating function of W, with $n = 3$, by $(e^{-4t} + e^{4t})/2$ to get that of W, with $n = 4$.

11.20. Let $X_1, X_2, \ldots, X_n$ be independent. If the p.d.f. of X_i is uniform over the interval $(-2^{1-i}, 2^{1-i})$, $i = 1, 2, 3, \ldots,$ show that Liapounov's condition is not satisfied. The sum $\sum_{i=1}^{n} X_i$ does not have an approximate normal distribution because the first random variables in the sum tend to dominate it.

11.21. If $n = 4$ and, in the notation of the text, $c_1 = 1, c_2 = 2, c_3 = c_4 = 3$, find the distribution of the generalization of the Wilcoxon statistic, say W_g. For a general n, find the mean and the variance of W_g if $c_i = i$, $i \le n/2$, and $c_i = [n/2] + 1$, $i > n/2$, where $[z]$ is the greatest integer function. Does Liapounov's condition hold here?

11.22. A modification of Wilcoxon's statistic that is frequently used is achieved by replacing R_i by $R_i - 1$; that is, use the modification

$W_m = \sum_{1}^{n} Z_i(R_i - 1)$. Show that $W_m/\sqrt{(n-1)n(2n-1)/6}$ has a limiting distribution that is $N(0, 1)$.

11.23. If, in the discussion of the generalization of the Wilcoxon statistic, we let $c_1 = c_2 = \cdots = c_n = 1$, show that we obtain a statistic equivalent to that used in the sign test.

11.24. If $c_1, c_2, \ldots, c_n$ are selected so that $i/(n+1) = \int_0^{c_i} \sqrt{2/\pi}\, e^{-x^2/2}\, dx$, $i = 1, 2, \ldots, n$, the generalized Wilcoxon W_g is an example of a *normal scores statistic*. If $n = 9$, compute the mean and the variance of this W_g.

11.25. If $c_i = 2^i$, $i = 1, 2, \ldots, n$, the corresponding W_g is called the *binary statistic*. Find the mean and the variance of this W_g. Is Liapounov's condition satisfied?

11.26. In the definition of Wilcoxon's statistic, let W_1 be the sum of the ranks of those observations of the sample that are positive and let W_2 be the sum of the ranks of those observations that are negative. Then $W = W_1 - W_2$.
(a) Show that $W = 2W_1 - n(n+1)/2$ and $W = n(n+1)/2 - 2W_2$.
(b) Compute the mean and the variance of each of W_1 and W_2.

11.27. Let $X_1, X_2, \ldots, X_{2n}$ be a random sample of size $2n$ from a continuous-type distribution that is symmetric about zero. Modify the Wilcoxon statistic by replacing the scores (ranks) $1, 2, \ldots, 2n$ by the scores consisting of n ones and n twos. Call this statistic W.
(a) Find the variance of W.
(b) Argue that $E(e^{tW}) = \left(\dfrac{e^{-t} + e^{t}}{2}\right)^n \left(\dfrac{e^{-2t} + e^{2t}}{2}\right)^n$.
(c) Evaluate $\lim_{n\to\infty} E(e^{tW/\sqrt{n}})$. What is the limiting distribution of $W/\sqrt{n}$?

11.5 The Equality of Two Distributions

In Sections 11.3 and 11.4, some tests of hypotheses about one distribution were investigated. In this section, as in the next section, various tests of the equality of two distributions are studied. By the equality of two distributions, we mean that the two distribution functions, say F and G, have $F(z) = G(z)$ for all values of z.

The first test that we discuss is a natural extension of the chi-square test. Let X and Y be independent variables with distribution functions $F(x)$ and $G(y)$, respectively. We wish to test the hypothesis that $F(z) = G(z)$, for all z. Let us partition the real line into k mutually disjoint sets $A_1, A_2, \ldots, A_k$. Define

$$p_{i1} = \Pr(X \in A_i), \qquad i = 1, 2, \ldots, k,$$

and

$$p_{i2} = \Pr(Y \in A_i), \qquad i = 1, 2, \ldots, k.$$

If $F(z) = G(z)$, for all z, then $p_{i1} = p_{i2}$, $i = 1, 2, \ldots, k$. Accordingly, the hypothesis that $F(z) = G(z)$, for all z, is replaced by the less restrictive hypothesis

$$H_0: p_{i1} = p_{i2}, \qquad i = 1, 2, \ldots, k.$$

But this is exactly the problem of testing the equality of two multinomial distributions that was considered in Example 3, Section 6.6, and the reader is referred to that example for the details.

Some statisticians prefer a procedure which eliminates some of the subjectivity of selecting the partitions. For a fixed positive integer k, proceed as follows. Consider a random sample of size m from the distribution of X and an independent random sample of size n from the distribution of Y. Let the experimental values be denoted by $x_1, x_2, \ldots, x_m$ and $y_1, y_2, \ldots, y_n$. Then combine the two samples into one sample of size $m + n$ and order the $m + n$ values (not their absolute values) in ascending order of magnitude. These ordered items are then partitioned into k parts in such a way that each part has the same number of items. (If the sample sizes are such that this is impossible, a partition with approximately the same number of items in each group suffices.) In effect, then, the partition $A_1, A_2, \ldots, A_k$ is determined by the experimental values themselves. This does not alter the fact that the statistic, discussed in Example 3, Section 6.6, has a limiting distribution that is $\chi^2(k - 1)$. Accordingly, the procedures used in that example may be used here.

Among the tests of this type there is one that is frequently used. It is essentially a test of the equality of the medians of two distributions. To simplify the discussion, we assume that $m + n$.

the size of the combined sample, is an even number, say $m + n = 2h$, where h is a positive integer. We take $k = 2$ and then the combined sample of size $m + n = 2h$, which has been ordered, is separated into two parts, a "lower half" and an "upper half," each containing $h = (m + n)/2$ of the experimental values of X and Y. The statistic, suggested by Example 3, Section 6.6, could be used because it has, when H_0 is true, a limiting distribution that is $\chi^2(1)$. However, it is more interesting to find the exact distribution of another statistic which enables us to test the hypothesis H_0 against the alternative $H_1 : F(z) \geq G(z)$ or against the alternative $H_1 : F(z) \leq G(z)$ as opposed to merely $F(z) \neq G(z)$. [Here, and in the sequel, alternatives $F(z) \geq G(z)$ and $F(z) \leq G(z)$ and $F(z) \neq G(z)$ mean that strict inequality holds on some set of positive probability measure.] This other statistic is V, which is the number of observed values of X that are in the lower half of the combined sample. If the observed value of V is quite large, one might suspect that the median of the distribution of X is smaller than that of the distribution of Y. Thus the critical region of this test of the hypothesis $H_0 : F(z) = G(z)$, for all z, against $H_1 : F(z) \geq G(z)$ is of the form $V \geq c$. Because our combined sample is of even size, there is no unique median of the sample. However, one can arbitrarily insert a number between the hth and $(h + 1)$st ordered items and call it the median of the sample. On this account, a test of the sort just described is called a *median test*. Incidentally, if the alternative hypothesis is $H_1 : F(z) \leq G(z)$, the critical region is of the form $V \leq c$.

The distribution of V is quite easy to find if the distribution functions $F(x)$ and $G(y)$ are of the continuous type and if $F(z) = G(z)$, for all z. We shall now show that V has a hypergeometric p.d.f. Let $m + n = 2h$, h a positive integer. To compute $\Pr(V = v)$, we need the probability that exactly v of $X_1, X_2, \ldots, X_m$ are in the lower half of the ordered combined sample. Under our assumptions, the probability is zero that any two of the $2h$ random variables are equal. The smallest h of the $m + n = 2h$ items can be selected in any one of $\binom{2h}{h}$ ways. Each of these ways has the same probability. Of these $\binom{2h}{h}$ ways, we need to count the number of those in which exactly v of the m values of X (and hence $h - v$ of the n values of Y)

appear in the lower h items. But this is $\binom{m}{v}\binom{n}{h-v}$. Thus the p.d.f. of V is the hypergeometric p.d.f.

$$k(v) = \Pr(V = v) = \frac{\binom{m}{v}\binom{n}{h-v}}{\binom{m+n}{h}}, \qquad v = 0, 1, 2, \dots, m,$$

$$= 0 \qquad \text{elsewhere,}$$

where $m + n = 2h$.

The reader may be momentarily puzzled by the meaning of $\binom{n}{h-v}$ for $v = 0, 1, 2, \dots, m$. For example, let $m = 17$, $n = 3$, so that $h = 10$. Then we have $\binom{3}{10-v}$, $v = 0, 1, \dots, 17$. However, we take $\binom{n}{h-v}$ to be zero if $h - v$ is negative or if $h - v > n$.

If $m + n$ is an odd number, say $m + n = 2h + 1$, it is left to the reader to show that the p.d.f. $k(v)$ gives the probability that exactly v of the m values of X are among the lower h of the combined $2h + 1$ values; that is, exactly v of the m values of X are less than the median of the combined sample.

If the distribution functions $F(x)$ and $G(y)$ are of the continuous type, there is another rather simple test of the hypothesis that $F(z) = G(z)$, for all z. This test is based upon the notion of *runs* of values of X and of values of Y. We shall now explain what we mean by runs. Let us again combine the sample of m values of X and the sample of n values of Y into one collection of $m + n$ ordered items arranged in ascending order of magnitude. With $m = 7$ and $n = 8$ we might find that the 15 ordered items were in the arrangement

$$\underline{x}\ \underline{yyy}\ \underline{xx}\ \underline{y}\ \underline{x}\ \underline{yy}\ \underline{xxx}\ \underline{yy}.$$

Note that in this ordering we have underscored the groups of successive values of the random variable X and those of the random variable Y. If we read from left to right, we would say that we have a *run* of one value of X, followed by a *run* of three values of Y, followed by a *run* of two values of X, and so on. In our example, there is a total of eight runs. Three are runs of length 1; three are runs of length 2; and two are runs of length 3. Note that the total number of runs is always one more than the number of unlike adjacent symbols.

Of what can runs be suggestive? Suppose that with $m = 7$ and $n = 8$ we have the following ordering:

$$\underline{xxxxx}\ \underline{y}\ \underline{xx}\ \underline{yyyyyyy}.$$

To us, this strongly suggests that $F(z) > G(z)$. For if, in fact, $F(z) = G(z)$ for all z, we would anticipate a greater number of runs. And if the first run of five values of X were interchanged with the last run of seven values of Y, this would suggest that $F(z) < G(z)$. But runs can be suggestive of other things. For example, with $m = 7$ and $n = 8$, consider the runs.

$$\underline{yyyy}\ \underline{xxxxxxx}\ \underline{yyyy}.$$

This suggests to us that the medians of the distributions of X and Y may very well be about the same, but that the "spread" (measured possibly by the standard deviation) of the distribution of X is considerably less than that of the distribution of Y.

Let the random variable R equal the number of runs in the combined sample, once the combined sample has been ordered. Because our random variables X and Y are of the continuous type, we may assume that no two of these sample items are equal. We wish to find the p.d.f. of R. To find this distribution, when $F(z) = G(z)$, we shall suppose that all arrangements of the m values of X and the n values of Y have equal probabilities. We shall show that

$$\Pr(R=2k+1) = \left\{\binom{m-1}{k}\binom{n-1}{k-1} + \binom{m-1}{k-1}\binom{n-1}{k}\right\}\Big/\binom{m+n}{m}$$

$$\Pr(R = 2k) = 2\binom{m-1}{k-1}\binom{n-1}{k-1}\Big/\binom{m+n}{m} \tag{1}$$

when $2k$ and $2k + 1$ are elements of the space of R.

To prove formulas (1), note that we can select the m positions for the m values of X from the $m + n$ positions in any one of $\binom{m+n}{m}$ ways. Since each of these choices yields one arrangement, the probability of each arrangement is equal to $1\Big/\binom{m+n}{m}$. The problem is now to determine how many of these arrangements yield $R = r$, where r is an integer in the space of R. First, let $r = 2k + 1$, where k is a positive integer. This means that there must be $k + 1$ runs of the ordered values of X and k runs of the ordered values of Y or vice versa. Consider first the number of ways of obtaining $k + 1$ runs of the m values of X. We can form $k+1$ of these runs by inserting k "dividers" into the $m-1$ spaces between the values of X, with no more than one divider per space. This can be done in any one of $\binom{m-1}{k}$ ways. Similarly, we can construct k runs of the n values of Y by inserting $k-1$ dividers into the $n-1$ spaces between the values of Y, with no more than one divider per space. This can be done in any one of $\binom{n-1}{k-1}$ ways. The joint operation can be performed in any one of $\binom{m-1}{k}\binom{n-1}{k-1}$ ways. These two sets of runs can be placed together to form $r = 2k+1$ runs. But we could also have k runs of the values of X and $k+1$ runs of the values of Y. An argument similar to the preceding shows that this can be affected in any one of

$$\binom{m-1}{k-1}\binom{n-1}{k}$$

ways. Thus

$$\Pr(R = 2k+1) = \frac{\binom{m-1}{k}\binom{n-1}{k-1} + \binom{m-1}{k-1}\binom{n-1}{k}}{\binom{m+n}{m}},$$

which is the first of formulas (1).

If $r = 2k$, where k is a positive integer, we see that the ordered values of X and the ordered values of Y must each be separated into k runs. These operations can be performed in any one of $\binom{m-1}{k-1}$ and $\binom{n-1}{k-1}$ ways, respectively. These two sets of runs can be placed together to form $r = 2k$ runs. But we may begin with either a run of values of X or a run of values of Y. Accordingly, the probability of $2k$ runs is

$$\Pr(R = 2k) = \frac{2\binom{m-1}{k-1}\binom{n-1}{k-1}}{\binom{m+n}{m}},$$

which is the second of formulas (1).

If the critical region of this *run test* of the hypothesis

$H_0: F(z) = G(z)$ for all z is of the form $R \le c$, it is easy to compute $\alpha = \Pr(R \le c; H_0)$, provided that m and n are small. Although it is not easy to show, the distribution of R can be approximated, with large sample sizes m and n, by a normal distribution with mean

$$\mu = E(R) = 2\frac{mn}{m+n} + 1$$

and variance

$$\sigma^2 = \frac{(\mu - 1)(\mu - 2)}{m + n - 1}.$$

The run test may also be used to test for *randomness*. That is, it can be used as a check to see if it is reasonable to treat $X_1, X_2, \ldots, X_s$ as a random sample of size s from some continuous distribution. To facilitate the discussion, take s to be even. We are given the s values of X to be $x_1, x_2, \ldots, x_s$, which are not ordered by magnitude but by the order in which they were observed. However, there are $s/2$ of these values, each of which is smaller than the remaining $s/2$ values. Thus we have a "lower half" and an "upper half" of these values. In the sequence $x_1, x_2, \ldots, x_s$, replace each value X that is in the lower half by the letter L and each value in the upper half by the letter U. Then, for example, with $s = 10$, a sequence such as

$$L\ L\ L\ L\ U\ L\ U\ U\ U\ U$$

may suggest a trend toward increasing values of X; that is, these values of X may not reasonably be looked upon as being the observations of a random sample. If trend is the only alternative to randomness, we can make a test based upon R and reject the hypothesis of randomness if $R \le c$. To make this test, we would use the p.d.f. of R with $m = n = s/2$. On the other hand if, with $s = 10$, we find a sequence such as

$$L\ U\ L\ U\ L\ U\ L\ U\ L\ U,$$

our suspicions are aroused that there may be a nonrandom effect which is cyclic even though $R = 10$. Accordingly, to test for a trend or a cyclic effect, we could use a critical region of the form $R \le c_1$ or $R \ge c_2$.

If the sample size s is odd, the number of sample items in the "upper half" and the number in the "lower half" will differ by one. Then, for example, we could use the p.d.f. of R with $m = (s - 1)/2$ and $n = (s + 1)/2$, or vice versa.

EXERCISES

11.28. Let 3.1, 5.6, 4.7, 3.8, 4.2, 3.0, 5.1, 3.9, 4.8 and 5.3, 4.0, 4.9, 6.2, 3.7, 5.0, 6.5, 4.5, 5.5, 5.9, 4.4, 5.8 be observed independent samples of sizes $m = 9$ and $n = 12$ from two distributions. With $k = 3$, use a chi-square test to test, with $\alpha = 0.05$ approximately, the equality of the two distributions.

11.29. In the median test, with $m = 9$ and $n = 7$, find the p.d.f. of the random variable V, the number of values of X in the lower half of the combined sample. In particular, what are the values of the probabilities $\Pr(V = 0)$ and $\Pr(V = 9)$?

11.30. In the notation of the text, use the median test and the data given in Exercise 11.28 to test, with $\alpha = 0.05$, approximately, the hypothesis of the equality of the two distributions against the alternative hypothesis that $F(z) \geq G(z)$. If the exact probabilities are too difficult to determine for $m = 9$ and $n = 12$, approximate these probabilities.

11.31. Using the notation of this section, let U be the number of observed values of X in the smallest d items of the combined sample of $m + n$ items. Argue that

$$\Pr(U = u) = \binom{m}{u}\binom{n}{d-u}\bigg/\binom{m+n}{d}, \qquad u = 0, 1, \ldots, m.$$

The statistic U could be used to test the equality of the $(100p)$th percentiles, where $(m + n)p = d$, of the distributions of X and Y.

11.32. In the discussion of the run test, let the random variables R_1 and R_2 be, respectively, the number of runs of the values of X and the number of runs of the values of Y. Then $R = R_1 + R_2$. Let the pair (r_1, r_2) of integers be in the space of (R_1, R_2); then $|r_1 - r_2| \leq 1$. Show that the joint p.d.f. of R_1 and R_2 is $2\binom{m-1}{r_1-1}\binom{n-1}{r_2-1}\bigg/\binom{m+n}{m}$ if $r_1 = r_2$; that this joint p.d.f. is $\binom{m-1}{r_1-1}\binom{n-1}{r_2-1}\bigg/\binom{m+n}{m}$ if $|r_1 - r_2| = 1$; and is zero elsewhere. Show that the marginal p.d.f. of R_1 is $\binom{m-1}{r_1-1}\binom{n+1}{r_1}\bigg/\binom{m+n}{m}$, $r_1 = 1, \ldots, m$, and is zero elsewhere. Find $E(R_1)$. In a similar manner, find $E(R_2)$. Compute $E(R) = E(R_1) + E(R_2)$.

11.6 The Mann–Whitney–Wilcoxon Test

We return to the problem of testing the equality of two distributions of the continuous type. Let X and Y be independent random variables of the continuous type. Let $F(x)$ and $G(y)$ denote, respectively, the distribution functions of X and Y and let $X_1, X_2, \ldots, X_m$ and $Y_1, Y_2, \ldots, Y_n$ denote independent samples from these distributions. We shall discuss the Mann–Whitney–Wilcoxon test of the hypothesis $H_0 : F(z) = G(z)$ for all values of z.

Let us define

$$\begin{aligned} Z_{ij} &= 1, \qquad X_i < Y_j, \\ &= 0, \qquad X_i > Y_j, \end{aligned}$$

and consider the statistic

$$U = \sum_{j=1}^{n} \sum_{i=1}^{m} Z_{ij}.$$

We note that

$$\sum_{i=1}^{m} Z_{ij}$$

counts the number of values of X that are less than $Y_j, j = 1, 2, \ldots, n$. Thus U is the sum of these n counts. For example, with $m = 4$ and $n = 3$, consider the observations

$$x_2 < y_3 < x_1 < x_4 < y_1 < x_3 < y_2.$$

There are three values of x that are less than y_1; there are four values of x that are less than y_2; and there is one value of x that is less than y_3. Thus the experimental value of U is $u = 3 + 4 + 1 = 8$.

Clearly, the smallest value which U can take is zero, and the largest value is mn. Thus the space of U is $\{u : u = 0, 1, 2, \ldots, mn\}$. If U is large, the values of Y tend to be larger than the values of X, and this suggests that $F(z) \geq G(z)$ for all z. On the other hand, a small value of U suggests that $F(z) \leq G(z)$ for all z. Thus, if we test the hypothesis $H_0 : F(z) = G(z)$ for all z against the alternative hypothesis $H_1 : F(z) \geq G(z)$ for all z, the critical region is of the form $U \geq c_1$. If the alternative hypothesis is $H_1 : F(z) \leq G(z)$ for all z, the critical region is of the form $U \leq c_2$. To determine the size of a critical region, we need the distribution of U when H_0 is true.

If u belongs to the space of U, let us denote $\Pr(U = u)$ by the symbol $h(u; m, n)$. This notation focuses attention on the sample sizes m and n. To determine the probability $h(u; m, n)$, we first note that we have $m + n$ positions to be filled by m values of X and n values of Y. We can fill m positions with the values of X in any one of $\binom{m+n}{m}$ ways. Once this has been done, the remaining n positions can be filled with the values of Y. When H_0 is true, each of these arrangements has the same probability, $1\Big/\binom{m+n}{m}$. The final right-hand position of an arrangement may be either a value of X or a value of Y. This position can be filled in any one of $m + n$ ways, m of which are favorable to X and n of which are favorable to Y. Accordingly, the probability that an arrangement ends with a value of X is $m/(m + n)$ and the probability that an arrangement terminates with a value of Y is $n/(m + n)$.

Now U can equal u in two mutually exclusive and exhaustive ways: (1) The final right-hand position (the largest of the $m + n$ values) in the arrangement may be a value of X and the remaining $(m - 1)$ values of X and the n values of Y can be arranged so as to have $U = u$. The probability that $U = u$, given an arrangement that terminates with a value of X, is given by $h(u; m - 1, n)$. Or (2) the largest value in the arrangement can be a value of Y. This value of Y is greater than m values of X. If we are to have $U = u$, the sum of $n - 1$ counts of the m values of X with respect to the remaining $n - 1$ values of Y must be $u - m$. Thus the probability that $U = u$, given an arrangement that terminates in a value of Y, is given by $h(u - m; m, n - 1)$. Accordingly, the probability that $U = u$ is

$$h(u; m, n) = \left(\frac{m}{m+n}\right)h(u; m-1, n) + \left(\frac{n}{m+n}\right)h(u-m; m, n-1).$$

We impose the following reasonable restrictions upon the function $h(u; m, n)$:

$$\begin{aligned} h(u; 0, n) &= 1, \qquad u = 0, \\ &= 0, \qquad u > 0, \quad n \geq 1, \end{aligned}$$

and

$$\begin{aligned} h(u; m, 0) &= 1, \qquad u = 0, \\ &= 0, \qquad u > 0, \quad m \geq 1, \end{aligned}$$

and

$$h(u; m, n) = 0, \qquad u < 0, \quad m \geq 0, \quad n \geq 0.$$

Then it is easy, for small values m and n, to compute these probabilities. For example, if $m = n = 1$, we have

$$h(0; 1, 1) = \tfrac{1}{2}h(0; 0, 1) + \tfrac{1}{2}h(-1; 1, 0) = \tfrac{1}{2} \cdot 1 + \tfrac{1}{2} \cdot 0 = \tfrac{1}{2},$$

$$h(1; 1, 1) = \tfrac{1}{2}h(1; 0, 1) + \tfrac{1}{2}h(0; 1, 0) = \tfrac{1}{2} \cdot 0 + \tfrac{1}{2} \cdot 1 = \tfrac{1}{2};$$

and if $m = 1$, $n = 2$, we have

$$h(0; 1, 2) = \tfrac{1}{3}h(0; 0, 2) + \tfrac{2}{3}h(-1; 1, 1) = \tfrac{1}{3} \cdot 1 + \tfrac{2}{3} \cdot 0 = \tfrac{1}{3},$$

$$h(1; 1, 2) = \tfrac{1}{3}h(1; 0, 2) + \tfrac{2}{3}h(0; 1, 1) \quad = \tfrac{1}{3} \cdot 0 + \tfrac{2}{3} \cdot \tfrac{1}{2} = \tfrac{1}{3},$$

$$h(2; 1, 2) = \tfrac{1}{3}h(2; 0, 2) + \tfrac{2}{3}h(1; 1, 1) \quad = \tfrac{1}{3} \cdot 0 + \tfrac{2}{3} \cdot \tfrac{1}{2} = \tfrac{1}{3}.$$

In Exercise 11.33 the reader is to determine the distribution of U when $m = 2$, $n = 1$; $m = 2$, $n = 2$; $m = 1$, $n = 3$; and $m = 3$, $n = 1$.

For large values of m and n, it is desirable to use an approximate distribution of U. Consider the mean and the variance of U when the hypothesis $H_0 : F(z) = G(z)$, for all values of z, is true. Since $U = \sum_{j=1}^{n} \sum_{i=1}^{m} Z_{ij}$, then

$$E(U) = \sum_{i=1}^{m} \sum_{j=1}^{n} E(Z_{ij}).$$

But

$$E(Z_{ij}) = (1) \Pr(X_i < Y_j) + (0) \Pr(X_i > Y_j) = \tfrac{1}{2}$$

because, when H_0 is true, $\Pr(X_i < Y_j) = \Pr(X_i > Y_j) = \frac{1}{2}$. Thus

$$E(U) = \sum_{i=1}^{m} \sum_{j=1}^{n} \left(\frac{1}{2}\right) = \frac{mn}{2}.$$

To compute the variance of U, we first find

$$E(U^2) = \sum_{k=1}^{n} \sum_{h=1}^{m} \sum_{j=1}^{n} \sum_{i=1}^{m} E(Z_{ij}Z_{hk})$$

$$= \sum_{j=1}^{n} \sum_{i=1}^{m} E(Z_{ij}^2) + \sum_{\substack{k=1 \\ k \neq j}}^{n} \sum_{j=1}^{n} \sum_{i=1}^{m} E(Z_{ij}Z_{ik})$$

$$+ \sum_{j=1}^{n} \sum_{\substack{h=1 \\ h \neq i}}^{m} \sum_{i=1}^{m} E(Z_{ij}Z_{hj}) + \sum_{\substack{k=1 \\ k \neq j}}^{n} \sum_{j=1}^{n} \sum_{\substack{h=1 \\ h \neq i}}^{m} \sum_{i=1}^{m} E(Z_{ij}Z_{hk}).$$

Note that there are mn terms in the first of these sums, $mn(n - 1)$ in the second, $mn(m - 1)$ in the third, and $mn(m - 1)(n - 1)$ in the fourth. When H_0 is true, we know that X_i, X_h, Y_j, and Y_k, $i \neq h, j \neq k$, are independent and have the same distribution of the continuous type. Thus $\Pr(X_i < Y_j) = \frac{1}{2}$. Moreover, $\Pr(X_i < Y_j, X_i < Y_k) = \frac{1}{3}$ because this is the probability that a designated one of three items is less than each of the other two. Similarly, $\Pr(X_i < Y_j, X_h < Y_j) = \frac{1}{3}$. Finally, $\Pr(X_i < Y_j, X_h < Y_k) = \Pr(X_i < Y_j)\Pr(X_h < Y_k) = \frac{1}{4}$. Hence we have

$$E(Z_{ij}^2) = (1)^2 \Pr(X_i < Y_j) = \tfrac{1}{2},$$

$$E(Z_{ij}Z_{ik}) = (1)(1) \Pr(X_i < Y_j, X_i < Y_k) = \tfrac{1}{3}, \qquad j \neq k,$$

$$E(Z_{ij}Z_{hj}) = (1)(1) \Pr(X_i < Y_j, X_h < Y_j) = \tfrac{1}{3}, \qquad i \neq h,$$

and

$$E(Z_{ij}Z_{hk}) = (1)(1) \Pr(X_i < Y_j, X_h < Y_k) = \tfrac{1}{4}, \qquad i \neq h, \quad j \neq k.$$

Thus

$$E(U^2) = \frac{mn}{2} + \frac{mn(n-1)}{3} + \frac{mn(m-1)}{3} + \frac{mn(m-1)(n-1)}{4}$$

and

$$\sigma_U^2 = mn\left[\frac{1}{2} + \frac{n-1}{3} + \frac{m-1}{3} + \frac{(m-1)(n-1)}{4} - \frac{mn}{4}\right]$$

$$= \frac{mn(m+n+1)}{12}.$$

Although it is fairly difficult to prove, it is true, when $F(z) = G(z)$ for all z, that

$$\frac{U - \dfrac{mn}{2}}{\sqrt{\dfrac{mn(m+n+1)}{12}}}$$

has, if each of m and n is large, an approximate distribution that is $N(0, 1)$. This fact enables us to compute, approximately, various significance levels.

Prior to the introduction of the statistic U in the statistical literature, it had been suggested that a test of H_0: $F(z) = G(z)$, for all z, be based upon the following statistic, say T (not Student's t). Let T be the sum of the ranks of $Y_1, Y_2, \ldots, Y_n$ among the $m + n$ items $X_1, \ldots, X_m, Y_1, \ldots, Y_n$, once this combined sample has been ordered. In Exercise 11.35 the reader is asked to show that

$$U = T - \frac{n(n+1)}{2}.$$

This formula provides another method of computing U and it shows that a test of H_0 based on U is equivalent to a test based on T. A generalization of T is considered in Section 11.8.

***Example* 1.** With the assumptions and the notation of this section, let $m = 10$ and $n = 9$. Let the observed values of X be as given in the first row and the observed values of Y as in the second row of the following display:

$$4.3, 5.9, 4.9, 3.1, 5.3, 6.4, 6.2, 3.8, 7.5, 5.8,$$

$$5.5, 7.9, 6.8, 9.0, 5.6, 6.3, 8.5, 4.6, 7.1.$$

Since, in the combined sample, the ranks of the values of y are 4, 7, 8, 12, 14, 15, 17, 18, 19, we have the experimental value of T to be equal to $t = 114$. Thus $u = 114 - 45 = 69$. If $F(z) = G(z)$ for all z, then, approximately,

$$0.05 = \Pr\left(\frac{U - 45}{12.247} \geq 1.645\right) = \Pr(U \geq 65.146).$$

Accordingly, at the 0.05 significance level, we reject the hypothesis $H_0 : F(z) = G(z)$, for all z, and accept the alternative hypothesis $H_1 : F(z) \geq G(z)$, for all z.

EXERCISES

11.33. Compute the distribution of U in each of the following cases: (a) $m = 2$, $n = 1$; (b) $m = 2$, $n = 2$; (c) $m = 1$, $n = 3$; (d) $m = 3$, $n = 1$.

11.34. Suppose that the hypothesis $H_0 : F(z) = G(z)$, for all z, is not true. Let $p = \Pr(X_i < Y_j)$. Show that U/mn is an unbiased estimator of p and that it converges in probability to p as $m \to \infty$ and $n \to \infty$.

11.35. Show that $U = T - [n(n + 1)]/2$.

Hint: Let $Y_{(1)} < Y_{(2)} < \cdots < Y_{(n)}$ be the order statistics of the random sample $Y_1, Y_2, \ldots, Y_n$. If R_i is the rank of $Y_{(i)}$ in the combined ordered sample, note that $Y_{(i)}$ is greater than $R_i - i$ values of X.

11.36. In Example 1 of this section assume that the values came from two normal distributions with means μ_1 and μ_2, respectively, and with common variance σ^2. Calculate the Student's t which is used to test the hypothesis $H_0 : \mu_1 = \mu_2$. If the alternative hypothesis is $H_1 : \mu_1 < \mu_2$, do we accept or reject H_0 at the 0.05 significance level?

11.7 Distributions Under Alternative Hypotheses

In this section we discuss certain problems that are related to a nonparametric test when the hypothesis H_0 is not true. Let X and Y be independent random variables of the continuous type with distribution functions $F(x)$ and $G(y)$, respectively, and probability density functions $f(x)$ and $g(y)$. Let $X_1, X_2, \ldots, X_m$ and $Y_1, Y_2, \ldots, Y_n$ denote independent random samples from these distributions. Consider the hypothesis $H_0 : F(z) = G(z)$ for all values of z. It has been seen that the test of this hypothesis may be based upon the statistic U, which, when the hypothesis H_0 is true, has a distribution that does not depend upon $F(z) = G(z)$. Or this test can be based upon the statistic $T = U + n(n + 1)/2$, where T is the sum of the ranks of $Y_1, Y_2, \ldots, Y_n$ in the combined sample. To elicit some information about the distribution of T when the alternative hypothesis is true, let us consider the joint distribution of the ranks of these values of Y.

Let $Y_{(1)} < Y_{(2)} < \cdots < Y_{(n)}$ be the order statistics of the sample $Y_1, Y_2, \ldots, Y_n$. Order the combined sample, and let R_i be the rank of $Y_{(i)}$, $i = 1, 2, \ldots, n$. Thus there are $i - 1$ values of Y and $R_i - i$ values of X that are less than $Y_{(i)}$. Moreover, there are $R_i - R_{i-1} - 1$ values of X between $Y_{(i-1)}$ and $Y_{(i)}$. If it is given that $Y_{(1)} = y_1 < Y_{(2)} = y_2 < \cdots < Y_{(n)} = y_n$, then the conditional probability

$$\Pr(R_1 = r_1, R_2 = r_2, \ldots, R_n = r_n | y_1 < y_2 < \cdots < y_n), \tag{1}$$

where $r_1 < r_2 < \cdots < r_n \le m + n$ are positive integers, can be computed by using the multinomial p.d.f. in the following manner. Define the following sets: $A_1 = \{x : -\infty < x < y_1\}$, $A_i = \{x : y_{i-1} < x < y_i\}$, $i = 2, \ldots, n$, $A_{n+1} = \{x : y_n < x < \infty\}$. The conditional probabilities of these sets are, respectively, $p_1 = F(y_1)$, $p_2 = F(y_2) - F(y_1), \ldots, p_n = F(y_n) - F(y_{n-1})$, $p_{n+1} = 1 - F(y_n)$. Then the conditional probability of display (1) is given by

$$\frac{m!\, p_1^{r_1 - 1} p_2^{r_2 - r_1 - 1} \cdots p_n^{r_n - r_{n-1} - 1} p_{n+1}^{m+n-r_n}}{(r_1 - 1)!\,(r_2 - r_1 - 1)! \cdots (r_n - r_{n-1} - 1)!\,(m + n - r_n)!}.$$

To find the unconditional probability $\Pr(R_1 = r_1, R_2 = r_2, \ldots, R_n = r_n)$, which we denote simply by $\Pr(r_1, \ldots, r_n)$, we multiply the conditional probability by the joint p.d.f. of $Y_{(1)} < Y_{(2)} < \cdots < Y_{(n)}$, namely $n!\, g(y_1)g(y_2)\cdots g(y_n)$, and then integrate on $y_1, y_2, \ldots, y_n$. That is,

$$\Pr(r_1, r_2, \ldots, r_n) = \int_{-\infty}^{\infty} \cdots \int_{-\infty}^{y_3} \int_{-\infty}^{y_2} \Pr(r_1, \ldots, r_n | y_1 < \cdots < y_n) n!$$

$$\times\, g(y_1) \cdots g(y_n)\, dy_1 \cdots dy_n,$$

where $\Pr(r_1, \ldots, r_n | y_1 < \cdots < y_n)$ denotes the conditional probability in display (1).

Now that we have the joint distribution of $R_1, R_2, \ldots, R_n$, we can find, theoretically, the distributions of functions of $R_1, R_2, \ldots, R_n$ and, in particular, the distribution of $T = \sum_1^n R_i$. From the latter we can find that of $U = T - n(n+1)/2$. To point out the extremely tedious computational problems of distribution theory that we encounter, we give an example. In this example we use the assumptions of this section.

Example 1. Suppose that an hypothesis H_0 is not true but that in fact $f(x) = 1, 0 < x < 1$, zero elsewhere, and $g(y) = 2y, 0 < y < 1$, zero elsewhere. Let $m = 3$ and $n = 2$. Note that the space of U is the set $\{u\colon u = 0, 1, \ldots, 6\}$. Consider $\Pr(U = 5)$. This event $U = 5$ occurs when and only when $R_1 = 3$, $R_2 = 5$, since in this section $R_1 < R_2$ are the ranks of $Y_{(1)} < Y_{(2)}$ in the combined sample and $U = R_1 + R_2 - 3$. Because $F(x) = x, 0 < x \le 1$, we have

$$\begin{aligned}\Pr(U = 5) &= \Pr(R_1 = 3, R_2 = 5)\\ &= \int_0^1 \int_0^{y_2} \frac{3!\, y_1^2(y_2 - y_1)}{2!\,1!} 2!\,(2y_1)(2y_2)\, dy_1\, dy_2\\ &= 24 \int_0^1 \left(\frac{y_2^6}{4} - \frac{y_2^6}{5}\right) dy_2 = \tfrac{6}{35}.\end{aligned}$$

Consider next $\Pr(U = 4)$. The event $U = 4$ occurs if $R_1 = 2$, $R_2 = 5$ or if $R_1 = 3$, $R_2 = 4$. Thus

$$\Pr(U = 4) = \Pr(R_1 = 2, R_2 = 5) + \Pr(R_1 = 3, R_2 = 4);$$

the computation of each of these probabilities is similar to that of $\Pr(R_1 = 3, R_2 = 5)$. This procedure may be continued until we have computed $\Pr(U = u)$ for each $u \in \{u\colon u = 0, 1, \ldots, 6\}$.

In the preceding example the probability density functions and the sample sizes m and n were selected so as to provide relatively simple integrations. The reader can discover for himself or herself how tedious, and even difficult, the computations become if the sample sizes are large or if the probability density functions are not of a simple functional form.

EXERCISES

11.37. Let the probability density functions of X and Y be those given in Example 1 of this section. Further, let the sample sizes be $m = 5$ and $n = 3$. If $R_1 < R_2 < R_3$ are the ranks of $Y_{(1)} < Y_{(2)} < Y_{(3)}$ in the combined sample, compute $\Pr(R_1 = 2, R_2 = 6, R_3 = 8)$.

11.38. Let $X_1, X_2, \ldots, X_m$ be a random sample of size m from a distribution of the continuous type with distribution function $F(x)$ and p.d.f. $F'(x) = f(x)$. Let $Y_1, Y_2, \ldots, Y_n$ be a random sample from a distribution with distribution function $G(y) = [F(y)]^\theta$, $0 < \theta$. If $\theta \neq 1$, this distribution is called a *Lehmann alternative*. With $\theta = 2$, show that

$$\Pr(r_1, r_2, \ldots, r_n) = \frac{2^n r_1(r_2 + 1)(r_3 + 2)\cdots(r_n + n - 1)}{\binom{m+n}{m}(m + n + 1)(m + n + 2)\cdots(m + 2n)}.$$

11.39. Let X_1, X_2, X_3 be a random sample from a continuous-type distribution with distribution function $F(x)$ and p.d.f. $f(x) = F'(x)$. Let Y_1, Y_2 be a random sample of size $n = 2$ from a distribution with distribution function $G(y) = [F(y)]^2$. In the combined sample of 5, determine the probability that the Y values have ranks 1 and 3; that is, the order is $y\ x\ y\ x\ x$.

11.40. To generalize the results of Exercise 11.38, let $G(y) = h[F(y)]$, where $h(z)$ is a differentiable function such that $h(0) = 0$, $h(1) = 1$, and $h'(z) > 0$, $0 < z < 1$. Show that

$$\Pr(r_1, r_2, \ldots, r_n) = \frac{E[h'(V_{r_1})h'(V_{r_2})\cdots h'(V_{r_n})]}{\binom{m+n}{m}},$$

where $V_1 < V_2 < \cdots < V_{m+n}$ are the order statistics of a random sample of size $m + n$ from the uniform distribution over the interval $(0, 1)$.

11.8 Linear Rank Statistics

In this section we consider a type of distribution-free statistic that is, among other things, a generalization of the Mann–Whitney–Wilcoxon statistic. Let $V_1, V_2, \ldots, V_N$ be a random sample of size N from a distribution of the continuous type. Let R_i be the rank of V_i among $V_1, V_2, \ldots, V_N$, $i = 1, 2, \ldots, N$; and let $c(i)$

be a scoring function defined on the first N positive integers—that is, let $c(1), c(2), \ldots, c(N)$ be some appropriately selected constants. If $a_1, a_2, \ldots, a_N$ are constants, then a statistic of the form

$$L = \sum_{i=1}^{N} a_i c(R_i)$$

is called a *linear rank statistic.*

To see that this type of statistic is actually a generalization of both the Mann–Whitney–Wilcoxon statistic and also that statistic associated with the median test, let $N = m + n$ and

$$V_1 = X_1, \ldots, V_m = X_m, V_{m+1} = Y_1, \ldots, V_N = Y_n.$$

These two special statistics result from the following respective assignments for $c(i)$ and $a_1, a_2, \ldots, a_N$:

1. Take $c(i) = i$, $a_1 = \cdots = a_m = 0$ and $a_{m+1} = \cdots = a_N = 1$, so that

$$L = \sum_{i=1}^{N} a_i c(R_i) = \sum_{i=m+1}^{m+n} R_i,$$

which is the sum of the ranks of $Y_1, Y_2, \ldots, Y_n$ among the $m + n$ observations (a statistic denoted by T in Section 11.6).

2. Take $c(i) = 1$, provided that $i \le (m + n)/2$, zero otherwise. If $a_1 = \cdots = a_m = 1$ and $a_{m+1} = \cdots = a_N = 0$, then

$$L = \sum_{i=1}^{N} a_i c(R_i) = \sum_{i=1}^{m} c(R_i),$$

which is equal to the number of the m values of X that are in the lower half of the combined sample of $m + n$ observations (a statistic used in the median test of Section 11.5).

To determine the mean and the variance of L, we make some observations about the joint and marginal distributions of the ranks $R_1, R_2, \ldots, R_N$. Clearly, from the results of Section 4.6 on the distribution of order statistics of a random sample, we observe that each permutation of the ranks has the same probability,

$$\Pr(R_1 = r_1, R_2 = r_2, \ldots, R_N = r_N) = \frac{1}{N!},$$

where $r_1, r_2, \ldots, r_N$ is any permutation of the first N positive integers. This implies that the marginal p.d.f. of R_i is

$$g_i(r_i) = \frac{1}{N}, \qquad r_i = 1, 2, \ldots, N,$$

zero elsewhere, because the number of permutations in which $R_i = r_i$ is $(N - 1)!$ so that

$$\sum_{\text{all } (r_1, \ldots, r_{i-1}, r_{i+1}, \ldots, r_N)} \sum \cdots \sum \frac{1}{N!} = \frac{(N-1)!}{N!} = \frac{1}{N}.$$

In a similar manner, the joint marginal p.d.f. of R_i and R_j, $i \neq j$, is

$$g_{ij}(r_i, r_j) = \frac{1}{N(N-1)}, \qquad r_i \neq r_j,$$

zero elsewhere. That is, the $(n - 2)$-fold summation

$$\sum \cdots \sum \frac{1}{N!} = \frac{(N-2)!}{N!} = \frac{1}{N(N-1)},$$

where the summation is over all permutations in which $R_i = r_i$ and $R_j = r_j$.

Among other things these properties of the distribution of R_1, $R_2, \ldots, R_N$ imply that

$$E[c(R_i)] = \sum_{r_i=1}^{N} c(r_i)\left(\frac{1}{N}\right) = \frac{c(1) + \cdots + c(N)}{N}.$$

If, for convenience, we let $c(k) = c_k$, then

$$E[c(R_i)] = \sum_{k=1}^{N} \left(\frac{c_k}{N}\right) = \bar{c},$$

say, for all $i = 1, 2, \ldots, N$. In addition, we have that

$$\sigma^2_{c(R_i)} = \sum_{r_i=1}^{N} [c(r_i) - \bar{c}]^2 \left(\frac{1}{N}\right) = \sum_{k=1}^{N} \frac{(c_k - \bar{c})^2}{N},$$

for all $i = 1, 2, \ldots, N$.

A simple expression for the covariance of $c(R_i)$ and $c(R_j)$, $i \neq j$, is a little more difficult to determine. That covariance is

$$E\{[c(R_i) - \bar{c}][c(R_j) - \bar{c}]\} = \sum_{k \neq h} \sum \frac{(c_k - \bar{c})(c_h - \bar{c})}{N(N-1)}.$$

However, since

$$0 = \left[\sum_{k=1}^{N} (c_k - \bar{c})\right]^2 = \sum_{k=1}^{N} (c_k - \bar{c})^2 + \sum_{k \neq h} \sum (c_k - \bar{c})(c_h - \bar{c}),$$

the covariance can be written simply as

$$E\{[c(R_i) - \bar{c}][c(R_j) - \bar{c}]\} = -\sum_{k=1}^{N} \frac{(c_k - \bar{c})^2}{N(N-1)}.$$

With these results, we first observe that the mean of L is

$$\mu_L = E\left[\sum_{i=1}^{N} a_i c(R_i)\right] = \sum_{i=1}^{N} a_i E$$

$$[c(R_i)] = \sum_{i=1}^{N} a_i \bar{c} = N\bar{a}\bar{c},$$

where $\bar{a} = (\Sigma\, a_i)/N$. Second, note that the variance of L is

$$\begin{aligned}
\sigma_L^2 &= \sum_{i=1}^{N} a_i^2 \sigma_{c(R_i)}^2 + \sum_{i \neq j}\sum a_i a_j E\{[c(R_i) - \bar{c}][c(R_j) - \bar{c}]\} \\
&= \sum_{i=1}^{N} a_i^2 \sum_{k=1}^{N} \frac{(c_k - \bar{c})^2}{N} + \sum_{i \neq j}\sum a_i a_j \left[-\sum_{k=1}^{N} \frac{(c_k - \bar{c})^2}{N(N-1)}\right] \\
&= \left[\sum_{k=1}^{N} \frac{(c_k - \bar{c})^2}{N(N-1)}\right]\left[(N-1)\sum_{i=1}^{N} a_i^2 - \sum_{i \neq j}\sum a_i a_j\right].
\end{aligned}$$

However, we can determine a substitute for the second factor by observing that

$$\begin{aligned}
N\sum_{i=1}^{N} (a_i - \bar{a})^2 &= N\sum_{i=1}^{N} a_i^2 - N^2\bar{a}^2 \\
&= N\sum_{i=1}^{N} a_i^2 - \left(\sum_{i=1}^{N} a_i\right)^2 \\
&= N\sum_{i=1}^{N} a_i^2 - \left[\sum_{i=1}^{N} a_i^2 + \sum_{i \neq j}\sum a_i a_j\right] \\
&= (N-1)\sum_{i=1}^{N} a_i^2 - \sum_{i \neq j}\sum a_i a_j.
\end{aligned}$$

So, making this substitution in σ_L^2, we finally have that

$$\begin{aligned}
\sigma_L^2 &= \left[\sum_{k=1}^{N} \frac{(c_k - \bar{c})^2}{N(N-1)}\right]\left[N\sum_{i=1}^{N} (a_i - \bar{a})^2\right] \\
&= \frac{1}{N-1}\sum_{i=1}^{N} (a_i - \bar{a})^2 \sum_{k=1}^{N} (c_k - \bar{c})^2.
\end{aligned}$$

In the special case in which $N = m + n$ and

$$L = \sum_{i=m+1}^{N} c(R_i),$$

the reader is asked to show that (Exercise 11.41)

$$\mu_L = n\bar{c}, \qquad \sigma_L^2 = \frac{mn}{N(N-1)} \sum_{k=1}^{N} (c_k - \bar{c})^2.$$

A further simplification when $c_k = c(k) = k$ yields

$$\mu_L = \frac{n(m+n+1)}{2}, \qquad \sigma_L^2 = \frac{mn(m+n+1)}{12};$$

these latter are, respectively, the mean and the variance of the statistic T as defined in Section 11.6.

As in the case of the Mann–Whitney–Wilcoxon statistic, the determination of the exact distribution of a linear rank statistic L can be very difficult. However, for many selections of the constants $a_1, a_2, \ldots, a_N$ and the scores $c(1), c(2), \ldots, c(N)$, the ratio $(L - \mu_L)/\sigma_L$ has, for large N, an approximate distribution that is $N(0, 1)$. This approximation is better if the scores $c(k) = c_k$ are like an ideal sample from a normal distribution, in particular, symmetric and without extreme values. For example, use of normal scores defined by

$$\frac{k}{N+1} = \int_{-\infty}^{c_k} \frac{1}{\sqrt{2\pi}} \exp\left(-\frac{w^2}{2}\right) dw$$

makes the approximation better. However, even with the use of ranks, $c(k) = k$, the approximation is reasonably good, provided that N is large enough, say around 30 or greater.

In addition to being a generalization of statistics such as those of Mann, Whitney, and Wilcoxon, we give two additional applications of linear rank statistics in the following illustrations.

Example 1. Let $X_1, X_2, \ldots, X_n$ denote n random variables. However, suppose that we question whether they are observations of a random sample due either to possible lack of independence or to the fact that $X_1, X_2, \ldots, X_n$ might not have the same distributions. In particular, say we suspect a trend toward larger and larger values in the sequence $X_1, X_2, \ldots, X_n$. If $R_i = \text{rank}\,(X_i)$, a statistic that could be used to test the alternative (trend) hypothesis is $L = \sum_{i=1}^{n} iR_i$. Under the assumption (H_0) that the n random variables are actually observations of a random sample from a distribution of the continuous type, the reader is asked to show that (Exercise 11.42)

$$\mu_L = \frac{n(n+1)^2}{4}, \qquad \sigma_L^2 = \frac{n^2(n+1)^2(n-1)}{144}.$$

The critical region of the test is of the form $L \geq d$, and the constant d can be determined either by using the normal approximation or referring to a tabulated distribution of L so that $\Pr(L \geq d; H_0)$ is approximately equal to a desired significance level α.

Example 2. Let $(X_1, Y_1), (X_2, Y_2), \ldots, (X_n, Y_n)$ be a random sample from a bivariate distribution of the continuous type. Let R_i be the rank of X_i among $X_1, X_2, \ldots, X_n$ and Q_i be the rank of Y_i among $Y_1, Y_2, \ldots, Y_n$. If X and Y have a large positive correlation coefficient, we would anticipate that R_i and Q_i would tend to be large or small together. In particular, the correlation coefficient of $(R_1, Q_1), (R_2, Q_2), \ldots, (R_n, Q_n)$, namely the Spearman rank correlation coefficient,

$$\frac{\sum_{i=1}^{n} (R_i - \bar{R})(Q_i - \bar{Q})}{\sqrt{\sum_{i=1}^{n} (R_i - \bar{R})^2 \sum_{i=1}^{n} (Q_i - \bar{Q})^2}},$$

would tend to be large. Since $R_1, R_2, \ldots, R_n$ and $Q_1, Q_2, \ldots, Q_n$ are permutations of $1, 2, \ldots, n$, this correlation coefficient can be shown (Exercise 11.43) to equal

$$\frac{\sum_{i=1}^{n} R_i Q_i - n(n+1)^2/4}{n(n^2-1)/12},$$

which in turn equals

$$1 - \frac{6 \sum_{i=1}^{n} (R_i - Q_i)^2}{n(n^2-1)}.$$

From the first of these two additional expressions for Spearman's statistic, it is clear that $\sum_{i=1}^{n} R_i Q_i$ is an equivalent statistic for the purpose of testing the independence of X and Y, say H_0. However, note that if H_0 is true, then the distribution of $\sum_{i=1}^{n} Q_i R_i$, which is not a linear rank statistic, and $L = \sum_{i=1}^{n} iR_i$ are the same. The reason for this is that the ranks $R_1, R_2, \ldots, R_n$ and the ranks $Q_1, Q_2, \ldots, Q_n$ are independent because of the independence of X and Y. Hence, under H_0, pairing $R_1, R_2, \ldots, R_n$ at random with $1, 2, \ldots, n$ is distributionally equivalent to pairing those ranks with $Q_1, Q_2, \ldots, Q_n$, which is simply a permutation of $1, 2, \ldots, n$. The mean and the variance of L is given in Example 1.

EXERCISES

11.41. Use the notation of this section.

(a) Show that the mean and the variance of $L = \sum_{i=m+1}^{N} c(R_i)$ are equal to the expressions in the text.

(b) In the special case in which $L = \sum_{i=m+1}^{N} R_i$, show that μ_L and σ_L^2 are those of T considered in Section 11.6.
Hint: Recall that

$$\sum_{k=1}^{N} k^2 = \frac{N(N+1)(2N+1)}{6}.$$

11.42. If $X_1, X_2, \ldots, X_n$ is a random sample from a distribution of the continuous type and if $R_i = \text{rank}\,(X_i)$, show that the mean and the variance of $L = \Sigma\, iR_i$ are $n(n+1)^2/4$ and $n^2(n+1)^2(n-1)/144$, respectively.

11.43. Verify that the two additional expressions, given in Example 2, for the Spearman rank correlation coefficient are equivalent to the first one.

Hint: $\sum R_i^2 = n(n+1)(2n+1)/6$ and $\Sigma\,(R_i - Q_i)^2/2 = \Sigma\,(R_i^2 + Q_i^2)/2 - \Sigma\, R_iQ_i$.

11.44. Let $X_1, X_2, \ldots, X_6$ be a random sample of size $n = 6$ from a distribution of the continuous type. Let $R_i = \text{rank}\,(X_i)$ and take $a_1 = a_6 = 9$, $a_2 = a_5 = 4$, $a_3 = a_4 = 1$. Find the mean and the variance of $L = \sum_{i=1}^{6} a_iR_i$, a statistic that could be used to detect a parabolic trend in $X_1, X_2, \ldots, X_6$.

11.45. Let R_i be the rank of X_i, $i = 1, 2, \ldots, 9$. The statistic $W = (R_1 + R_2 + R_3) + 2(R_4 + R_5 + R_6) + 3(R_7 + R_8 + R_9)$ is used to test a trend in the data. If, in fact, $X_1, X_2, \ldots, X_9$ are observations of a random sample from a continuous-type distribution, what are the mean and the variance of W?

11.46. Let X_1, X_2, X_3, X_4, X_5 be a random sample of size $n = 5$ from a continuous-type distribution. Let R_i be the rank of X_i, $i = 1, 2, 3, 4, 5$.

(a) Compute the mean and the variance of $L = R_5 - R_1$.
(b) Find the distribution of L.

11.47. In the notation of this section show that the covariance of the two linear rank statistics, $L_1 = \sum_{i=1}^{N} a_ic(R_i)$ and $L_2 = \sum_{i=1}^{N} b_id(R_i)$, is equal to

$$\sum_{i=1}^{N} (a_i - \bar{a})(b_i - \bar{b}) \sum_{k=1}^{N} (c_k - \bar{c})(d_k - \bar{d})/(N-1),$$

where, for convenience, $d_k = d(k)$.

11.9 Adaptive Nonparametric Methods

Frequently, an investigator is tempted to evaluate several test statistics associated with a single hypothesis and then use the one statistic that best supports his or her position, usually rejection. Obviously, this type of procedure changes the actual significance level of the test from the nominal α that is used. However, there is a way in which the investigator can first look at the data and then select a test statistic without changing this significance level. For illustration, suppose there are three possible test statistics W_1, W_2, W_3 of the hypothesis H_0 with respective critical regions C_1, C_2, C_3 such that $\Pr(W_i \in C_i; H_0) = \alpha$, $i = 1, 2, 3$. Moreover, suppose that a statistic Q, based upon the same data, selects one and only one of the statistics W_1, W_2, W_3, and that W is then used to test H_0. For example, we choose to use the test statistic W_i if $Q \in D_i$, $i = 1, 2, 3$, where the events defined by D_1, D_2, and D_3 are mutually exclusive and exhaustive. Now if Q and each W_i are independent when H_0 is true, then the probability of rejection, using the entire procedure (selecting and testing), is, under H_0,

$$\begin{aligned}
&\Pr(Q \in D_1, W_1 \in C_1) + \Pr(Q \in D_2, W_2 \in C_2) + \Pr(Q \in D_3, W_3 \in C_3) \\
&\quad = \Pr(Q \in D_1)\Pr(W_1 \in C_1) + \Pr(Q \in D_2)\Pr(W_2 \in C_2) \\
&\qquad + \Pr(Q \in D_3)\Pr(W_3 \in C_3) \\
&\quad = \alpha[\Pr(Q \in D_1) + \Pr(Q \in D_2) + \Pr(Q \in D_3)] = \alpha.
\end{aligned}$$

That is, the procedure of selecting W_i using an independent statistic Q and then constructing a test of significance level α with the statistic W_i has overall significance level α.

Of course, the important element in this procedure is the ability to be able to find a selector Q that is independent of each test statistic W. This can frequently be done by using the fact that the complete sufficient statistics for the parameters, given by H_0, are independent of every statistic whose distribution is free of those parameters. For illustration, if independent random samples of sizes m and n arise from two normal distributions with respective means μ_1 and μ_2 and common variance σ^2, then the complete sufficient statistics $\bar{X}$, $\bar{Y}$, and

$$V = \sum_1^m (X_i - \bar{X})^2 + \sum_1^n (Y_i - \bar{Y})^2$$

for μ_1, μ_2, and σ^2 are independent of every statistic whose distribution is free of μ_1, μ_2, and σ^2 such as

$$\frac{\sum_1^m (X_i - \bar{X})^2}{\sum_1^n (Y_i - \bar{Y})^2}, \quad \frac{\sum_1^m |X_i - \text{median}\,(X_i)|}{\sum_1^n |Y_i - \text{median}\,(Y_i)|}, \quad \frac{\text{range}\,(X_1, X_2, \ldots, X_m)}{\text{range}\,(Y_1, Y_2, \ldots, Y_n)}.$$

Thus, in general, we would hope to be able to find a selector Q that is a function of the complete sufficient statistics for the parameters, under H_0, so that it is independent of the test statistics.

It is particularly interesting to note that it is relatively easy to use this technique in *nonparametric* methods by using the independence result based upon complete sufficient statistics for *parameters*. How can we use an argument depending on parameters in nonparametric methods? Although this does sound strange, it is due to the unfortunate choice of a name in describing this broad area of nonparametric methods. Most statisticians would prefer to describe the subject as being *distribution-free*, since the test statistics have distributions that do not depend on the underlying distribution of the continuous type, described by either the distribution function F or the p.d.f. f. In addition, the latter name provides the clue for our application here because we have many test statistics whose distributions are free of the unknown (infinite vector) "parameter" F (or f). We now must find complete sufficient statistics for the distribution function F of the continuous type. In many instances, this is easy to do.

In Exercise 7.50, Section 7.7, it is shown that the order statistics $Y_1 < Y_2 < \cdots < Y_n$ of a random sample of size n from a distribution of the continuous type with p.d.f. $F'(x) = f(x)$ are sufficient statistics for the "parameter" f (or F). Moreover, if the family of distributions contains all probability density functions of the continuous type, the family of joint probability density functions of $Y_1, Y_2, \ldots, Y_n$ is also complete. We accept this latter fact without proof, as it is beyond the level of this text; but doing so, we can now say that the order statistics $Y_1, Y_2, \ldots, Y_n$ are complete sufficient statistics for the parameters f (or F).

Accordingly, our selector Q will be based upon those complete sufficient statistics, the order statistics under H_0. This allows us to independently choose a distribution-free test appropriate for this type of underlying distribution, and thus increase our power. Although it is well known that distribution-free tests hold the significance level α

for all underlying distributions of the continuous type, they have often been criticized because their powers are sometimes low. The independent selection of the distribution-free test to be used can help correct this. So selecting—or adapting the test to the data—provides a new dimension to nonparametric tests, which usually improves the power of the overall test.

A statistical test that maintains the significance level close to a desired significance level α for a wide variety of underlying distributions with good (not necessarily the best for any one type of distribution) power for all these distributions is described as being *robust*. As an illustration, the T (Student's t) used to test the equality of the means of two normal distributions is quite robust *provided* that the underlying distributions are rather close to normal ones with common variance. However, if the class of distributions includes those that are not too close to normal ones, such as the Cauchy distribution, the test based upon T is *not* robust; the significance level is not maintained and the power of the T-test is low with Cauchy distributions. As a matter of fact, the test based on the Mann–Whitney–Wilcoxon statistic (Section 11.6) is a much more robust test than that based upon T if the class of distributions is fairly wide (in particular, if long-tailed distributions such as the Cauchy are included).

An illustration of this *adaptive distribution-free procedure* that is robust is provided by considering a test of the equality of two distributions of the continuous type. From the discussion in Section 11.8, we know that we could construct many linear rank statistics by changing the scoring function. However, we concentrate on three such statistics mentioned explicitly in that section: that based on normal scores, say L_1; that of Mann–Whitney–Wilcoxon, say L_2; and that of the median test, say L_3. Moreover, respective critical regions C_1, C_2, and C_3 are selected so that, under the equality of the two distributions, we have

$$\alpha = \Pr(L_1 \in C_1) = \Pr(L_2 \in C_2) = \Pr(L_3 \in C_3).$$

Of course, we would like to use the test given by $L_1 \in C_1$ if the tails of the distributions are like or shorter than those of the normal distributions. With distributions having somewhat longer tails, $L_2 \in C_2$ provides an excellent test. And with distributions having very long tails, the test based on $L_3 \in C_3$ is quite satisfactory.

In order to select the appropriate test in an independent manner we let $V_1 < V_2 < \cdots < V_N$, where $N = m + n$, be the order statistics of the combined sample, which is of size N. Recall that if the two distributions are equal and thus have the same distribution function F, these order statistics are the complete sufficient statistics for the parameter F. Hence every statistic based on $V_1, V_2, \ldots, V_N$ is independent of L_1, L_2, and L_3, since the latter statistics have distributions that do not depend upon F. In particular, the kurtosis (Exercise 1.102, Section 1.9) of the combined sample,

$$K = \frac{\frac{1}{N}\sum_{i=1}^{N}(V_i - \bar{V})^4}{\left[\frac{1}{N}\sum_{i=1}^{N}(V_i - \bar{V})^2\right]^2},$$

is independent of L_1, L_2, and L_3. From Exercise 3.64, Section 3.4, we know that the kurtosis of the normal distribution is 3; hence if the two distributions were equal and normal we would expect K to be about 3. Of course, a longer-tailed distribution has a bigger kurtosis. Thus one simple way of defining the independent selection procedure would be to let

$$D_1 = \{k : k \le 3\}, \qquad D_2 = \{k : 3 < k \le 8\}, \qquad D_3 = \{k : 8 < k\}.$$

These choices are not necessarily the best way of selecting the appropriate test, but they are reasonable and illustrative of the adaptive procedure. From the independence of K and (L_1, L_2, L_3), we know that the overall test has significance level α. Since a more appropriate test has been selected, the power will be relatively good throughout a wide range of distributions. Accordingly, this distribution-free adaptive test is robust.

EXERCISES

11.48. Let $F(x)$ be a distribution function of a distribution of the continuous type which is symmetric about its median ξ. We wish to test $H_0 : \xi = 0$ against $H_1 : \xi > 0$. Use the fact that the $2n$ values, X_i and $-X_i$, $i = 1, 2, \ldots, n$, after ordering, are complete sufficient statistics for F, provided that H_0 is true. Then construct an adaptive distribution-free test based upon Wilcoxon's statistic and two of its modifications given in Exercises 11.23 and 11.24.

11.49. Suppose that the hypothesis H_0 concerns the independence of two random variables X and Y. That is, we wish to test $H_0 : F(x, y) = F_1(x)F_2(y)$, where F, F_1, and F_2 are the respective joint and marginal distribution functions of the continuous type, against all alternatives. Let (X_1, Y_1), $(X_2, Y_2), \ldots, (X_n, Y_n)$ be a random sample from the joint distribution. Under H_0, the order statistics of $X_1, X_2, \ldots, X_n$ and the order statistics of $Y_1, Y_2, \ldots, Y_n$ are, respectively, complete sufficient statistics for F_1 and F_2. Use Spearman's statistic (Example 2, Section 11.8) and at least two modifications of it to create an adaptive distribution-free test of H_0.

Hint: Instead of ranks, use normal and median scores (Section 11.8) to obtain two additional correlation coefficients. The one associated with the median scores is frequently called the *quadrant test*.

ADDITIONAL EXERCISES

11.50. Let $Y_1 < Y_2 < \cdots < Y_5$ be the order statistics of a random sample of size $n = 5$ from a distribution of the continuous type with distribution function F. Compute $\Pr\{F(Y_2) + [1 - F(Y_4)] \ge \frac{1}{2}\}$.

11.51. Let $Y_1 < Y_2 < \cdots < Y_8$ be the order statistics of a random sample of size $n = 8$ from a distribution of the continuous type with median ξ. Compute $\Pr(Y_2 < \xi < Y_7)$.

11.52. Let $X_1, X_2, \ldots, X_8$ be a random sample of size $n = 8$ from a symmetric distribution of the continuous type with distributional median equal to zero. Modify the regular one-sample Wilcoxon W by replacing the ranks 1, 2, 3, 4, 5, 6, 7, 8 by the scores 1, 1, 2, 2, 2, 2, 3, 3, to obtain W_g. Compute the mean and variance of W_g.

11.53. Let $Y_1 < Y_2 < Y_3 < Y_4$ be the order statistics of a random sample of size $n = 4$ from a continuous-type distribution with distribution function $F(x)$ and unknown 75th percentile $\xi_{0.75}$.
(a) What is $\Pr(Y_3 \le \xi_{0.75} < Y_4)$?
(b) What is the p.d.f. of $V = F(Y_4) - F(Y_3)$?

11.54. Let X_1, X_2, X_3, X_4, X_5 be a random sample from a continuous-type distribution that is symmetric about zero. If we modify the one-sample Wilcoxon by replacing the ranks 1, 2, 3, 4, 5 by the scores 1, 1, 1, 3, 4, what is the m.g.f. of this new statistic?

11.55. Let X_1, X_2 and Y_1, Y_2 be independent random samples, each of size $n = 2$, from distributions with respective probability density functions $f(x) = 1$, $0 < x < 1$, zero elsewhere, and $g(y) = 3y^2$, $0 < y < 1$, zero elsewhere. Compute the probability that the ranks of the Y-values in the combined sample of size 4 are 2 and 4.

11.56. Let $X_1, X_2, \ldots, X_6$ be a random sample of size $n = 6$ from a continuous-type distribution with distribution function $F(x)$. Let

$R_i = \text{rank}(X_i)$ and consider the scores $c(1) = c(2) = 1$, $c(3) = 2$, $c(4) = 3$, $c(5) = c(6) = 4$.
(a) What are the mean and the variance of $L = c(R_4) + c(R_5) + c(R_6)$?
(b) Why are L and the sample range $R = \max(X_i) - \min(X_i)$ independent?

11.57. Let X_1, X_2, X_3, X_4, X_5 be a random sample from a distribution with p.d.f. $f(x) = e^{-x}$, $0 < x < \infty$, zero elsewhere. Find the probability that both X_3 and X_5 are less than X_1, X_2, and X_4. Is this answer the same for every underlying distribution of the continuous type?

11.58. Let $X_1, X_2, \ldots, X_5$ be a random sample of size $n = 5$ from a distribution of the continuous type. Let $R_i = \text{rank}(X_i)$ among $X_1, X_2, \ldots, X_5$. Find the mean and variance of

$$L = R_1 + 2(R_2 + R_3 + R_4) + 3R_5.$$

11.59. Let X_1, X_2, X_3 denote a random sample of size $n = 3$ from a continuous-type distribution with distribution function F. It is well known that the order statistics $Y_1 < Y_2 < Y_3$ are complete sufficient statistics for F.
(a) Let the statistic U be defined as follows:

$$U = \begin{cases} 1, & \text{if } X_1 \text{ is the sample median } Y_2 \\ 0, & \text{otherwise.} \end{cases}$$

Find the distribution of U.
(b) Argue that U and (Y_1, Y_2, Y_3) are independent.
(c) Argue that U and $\bar{X}$ are independent.

11.60. Let X have a p.d.f. $f(x)$ of the continuous type that is symmetric about zero; that is, $f(x) = f(-x)$ for all real x. Show that the joint m.g.f.

$$E\{\exp[t_1|X| + t_2 \text{ sign}(X)]\} = \left[2\int_0^\infty e^{t_1 x} f(x)\,dx\right]\left[\frac{e^{-t_2} + e^{t_2}}{2}\right],$$

and thus $|X|$ and sign (X) are independent.

11.61. Let $X - \theta \stackrel{d}{=} \theta - X$ mean that $X - \theta$ has the same distribution as $\theta - X$; thus X has a symmetric distribution about θ. Say that Y and X are independent random variables and Y has a distribution which is also symmetric about θ. Show that $X - Y$ has a distribution that is symmetric about zero.

Hint: Write $X - Y = X - \theta - (Y - \theta) \stackrel{d}{=} \theta - X - (\theta - Y) = Y - X$.

11.62. Let $X_1, X_2, \ldots, X_n$ be a random sample from a distribution that is symmetric about θ. Let W and T be two statistics enjoying the following properties, respectively:

$$W(x_1, \ldots, x_n) = W(-x_1, \ldots, -x_n),$$

$$W(x_1 + h, \ldots, x_n + h) = W(x_1, \ldots, x_n),$$

so that W is an even location invariant statistic like S^2 or the range;

$$T(x_1, \ldots, x_n) = -T(-x_1, \ldots, -x_n),$$

$$T(x_1 + h, \ldots, x_n + h) = T(x_1, \ldots, x_n) + h,$$

so that T is an odd location statistic like $\bar{X}$ or the median. Show that

$$[T(X_1, \ldots, X_n) - \theta, \quad W(X_1, \ldots, X_n)]$$

$$\stackrel{d}{=} [\theta - T(X_1, \ldots, X_n), \quad W(X_1, \ldots, X_n)].$$

Hint: Write the left-hand member as

$$[T(X_1 - \theta, \ldots, X_n - \theta), \quad W(X_1 - \theta, \ldots, X_n - \theta)]$$

using the properties of T and W. Then use the fact that substitute $X_i - \theta \stackrel{d}{=} \theta - X_i$, $i = 1, 2, \ldots, n$.

11.63. The result of Exercise 11.62 implies that T has a conditional distribution that is symmetric about θ, given $W = w$. Of course, T has an unconditional symmetric distribution about θ. Moreover, it also implies that if appropriate expectations exist, $E(T|W = w) = \theta$ and $\text{cov}(T, W) = 0$. Suppose that $T_1, T_2, \ldots, T_k$ and $W_1, W_2, \ldots, W_k$ represent k such T and W statistics, so that $W_1 + \cdots + W_k = 1$.

(a) Show that $E\left[\sum_{i=1}^{k} W_i T_i\right] = \theta$.

(b) Let $T_1 = \bar{X}$ and $T_2 = m$, the sample median. Let $W_1 = 1$ if $K \le 4$ and zero otherwise, where K is the sample kurtosis, and let $W_2 = 1 - W_1$. Consider $T = \sum_{i=1}^{2} W_i T_i$. Is its expectation equal to θ? If so, note that T is an adaptive unbiased estimator which equals $\bar{X}$ for certain X-values and m for others.

APPENDIX A

References

[1] Anderson, T. W., *An Introduction to Multivariate Statistical Analysis*, 2nd ed., John Wiley & Sons, Inc., New York, 1984.

[2] Basu, D., "On Statistics Independent of a Complete Sufficient Statistic," *Sankhyā*, **15**, 377 (1955).

[3] Box, G. E. P., and Muller, M. A., "A Note on the Generation of Random Normal Deviates," *Ann. Math. Stat.*, **29**, 610 (1958).

[4] Carpenter, O., "Note on the Extension of Craig's Theorem to Noncentral Variates," *Ann. Math. Stat.*, **21**, 455 (1950).

[5] Casella, G., and Berger, R. L., *Statistical Inference*, Wadsworth & Brooks/Cole, Belmont, Calif., 1990.

[6] Cochran, W. G., "The Distribution of Quadratic Forms in a Normal System, with Applications to the Analysis of Covariance," *Proc. Cambridge Phil. Soc.*, **30**, 178 (1934).

[7] Craig, A. T., "Bilinear Forms in Normally Correlated Variables," *Ann. Math. Stat.*, **18**, 565 (1947).

[8] Craig, A. T., "Note on the Independence of Certain Quadratic Forms," *Ann. Math. Stat.*, **14**, 195 (1943).

[9] Curtiss, J. H., "A Note on the Theory of Moment Generating Functions," *Ann. Math. Stat.*, **13**, 430 (1942).

[10] De Groot, M. H., *Probability and Statistics*, 2nd ed., Addison-Wesley Publishing Co., Reading, Mass., 1986.

[11] Fisher, R. A., "On the Mathematical Foundation of Theoretical Statistics," *Phil. Trans. Royal Soc. London*, Series A, **222**, 309 (1921).
[12] Graybill, F. A., *Theory and Application of the Linear Model*, Wadsworth Publishing Co., Inc., Belmont, Calif., 1976.
[13] Hájek, J., and Šidák, Z., *Theory of Rank Tests*, Academia, Prague, 1967.
[14] Hogg, R. V., "Adaptive Robust Procedures: A Partial Review and Some Suggestions for Future Applications and Theory," *J. Amer. Stat. Assoc.*, **69**, 909 (1974).
[15] Hogg, R. V., "Testing the Equality of Means of Rectangular Populations," *Ann. Math. Stat.*, **24**, 691 (1953).
[16] Hogg, R. V., and Craig, A. T., "On the Decomposition of Certain Chi-Square Variables," *Ann. Math. Stat.*, **29**, 608 (1958).
[17] Hogg, R. V., and Craig, A. T., "Sufficient Statistics in Elementary Distribution Theory," *Sankhyā*, **17**, 209 (1956).
[18] Huber, P., "Robust Statistics: A Review," *Ann. Math. Stat.*, **43**, 1041 (1972).
[19] Johnson, N. L., and Kotz, S., *Continuous Univariate Distributions*, Vols. 1 and 2, Houghton Mifflin Company, Boston, 1970.
[20] Koopman, B. O., "On Distributions Admitting a Sufficient Statistic," *Trans. Amer. Math. Soc.*, **39**, 399 (1936).
[21] Lancaster, H. O., "Traces and Cumulants of Quadratic Forms in Normal Variables," *J. Royal Stat. Soc.*, Series B, **16**, 247 (1954).
[22] Lehmann, E. L., *Testing Statistical Hypotheses*, 2nd ed., John Wiley & Sons, Inc., New York, 1986.
[23] Lehmann, E. L., *Theory of Point Estimation*, Wadsworth & Brooks/Cole, Pacific Grove, Calif., 1983.
[24] Lehmann, E. L., and Scheffé, H., "Completeness, Similar Regions, and Unbiased Estimation," *Sankhyā*, **10**, 305 (1950).
[25] Lévy, P., *Théorie de l'addition des variables aléatoires*, Gauthier-Villars, Paris, 1937.
[26] Mann, H. B., and Whitney, D. R., "On a Test of Whether One of Two Random Variables Is Stochastically Larger Than the Other," *Ann. Math. Stat.*, **18**, 50 (1947).
[27] Neymann, J., "Su un teorema concernente le cosiddette statistiche sufficienti," *Giornale dell' Istituto degli Attuari*, **6**, 320 (1935).
[28] Neyman, J., and Pearson, E. S., "On the Problem of the Most Efficient Tests of Statistical Hypotheses," *Phil. Trans. Royal Soc. London*, Series A, **231**, 289 (1933).
[29] Pearson, K., "On the Criterion That a Given System of Deviations from the Probable in the Case of a Correlated System of Variables Is Such That It Can Be Reasonably Supposed to Have Arisen from Random Sampling," *Phil. Mag.*, Series 5, **50**, 157 (1900).
[30] Pitman, E. J. G., "Sufficient Statistics and Intrinsic Accuracy," *Proc. Cambridge Phil. Soc.*, **32**, 567 (1936).

[31] Rao, C. R., *Linear Statistical Inference and Its Applications*, John Wiley & Sons, Inc., New York, 1965.

[32] Scheffé, H., *The Analysis of Variance*, John Wiley & Sons, Inc., New York, 1959.

[33] Wald, A., *Sequential Analysis*, John Wiley & Sons, Inc., New York, 1947.

[34] Wilcoxon, F., "Individual Comparisons by Ranking Methods," *Biometrics Bull.*, **1**, 80 (1945).

APPENDIX B

Tables

TABLE I

The Poisson Distribution

$$\Pr(X \leq x) = \sum_{w=0}^{x} \frac{\mu^{w} e^{-\mu}}{w!}$$

	$\mu = E(X)$											
x	0.5	1.0	1.5	2.0	3.0	4.0	5.0	6.0	7.0	8.0	9.0	10.0
0	0.607	0.368	0.223	0.135	0.050	0.018	0.007	0.002	0.001	0.000	0.000	0.000
1	0.910	0.736	0.558	0.406	0.199	0.092	0.040	0.017	0.007	0.003	0.001	0.000
2	0.986	0.920	0.809	0.677	0.423	0.238	0.125	0.062	0.030	0.014	0.006	0.003
3	0.998	0.981	0.934	0.857	0.647	0.433	0.265	0.151	0.082	0.042	0.021	0.010
4	1.000	0.996	0.981	0.947	0.815	0.629	0.440	0.285	0.173	0.100	0.055	0.029
5		0.999	0.996	0.983	0.916	0.785	0.616	0.446	0.301	0.191	0.116	0.067
6		1.000	0.999	0.995	0.966	0.889	0.762	0.606	0.450	0.313	0.207	0.130
7			1.000	0.999	0.988	0.949	0.867	0.744	0.599	0.453	0.324	0.220
8				1.000	0.996	0.979	0.932	0.847	0.729	0.593	0.456	0.333
9					0.999	0.992	0.968	0.916	0.830	0.717	0.587	0.458
10					1.000	0.997	0.986	0.957	0.901	0.816	0.706	0.583
11						0.999	0.995	0.980	0.947	0.888	0.803	0.697
12						1.000	0.998	0.991	0.973	0.936	0.876	0.792
13							0.999	0.996	0.987	0.966	0.926	0.864
14							1.000	0.999	0.994	0.983	0.959	0.917
15								0.999	0.998	0.992	0.978	0.951
16								1.000	0.999	0.996	0.989	0.973
17									1.000	0.998	0.995	0.986
18										0.999	0.998	0.993
19										1.000	0.999	0.997
20											1.000	0.998
21												0.999
22												1.000

TABLE II

*The Chi-Square Distribution**

$$\Pr(X \le x) = \int_0^x \frac{1}{\Gamma(r/2)2^{r/2}} w^{r/2-1} e^{-w/2}\, dw$$

	Pr (X ≤ x)					
r	0.01	0.025	0.050	0.95	0.975	0.99
1	0.000	0.001	0.004	3.84	5.02	6.63
2	0.020	0.051	0.103	5.99	7.38	9.21
3	0.115	0.216	0.352	7.81	9.35	11.3
4	0.297	0.484	0.711	9.49	11.1	13.3
5	0.554	0.831	1.15	11.1	12.8	15.1
6	0.872	1.24	1.64	12.6	14.4	16.8
7	1.24	1.69	2.17	14.1	16.0	18.5
8	1.65	2.18	2.73	15.5	17.5	20.1
9	2.09	2.70	3.33	16.9	19.0	21.7
10	2.56	3.25	3.94	18.3	20.5	23.2
11	3.05	3.82	4.57	19.7	21.9	24.7
12	3.57	4.40	5.23	21.0	23.3	26.2
13	4.11	5.01	5.89	22.4	24.7	27.7
14	4.66	5.63	6.57	23.7	26.1	29.1
15	5.23	6.26	7.26	25.0	27.5	30.6
16	5.81	6.91	7.96	26.3	28.8	32.0
17	6.41	7.56	8.67	27.6	30.2	33.4
18	7.01	8.23	9.39	28.9	31.5	34.8
19	7.63	8.91	10.1	30.1	32.9	36.2
20	8.26	9.59	10.9	31.4	34.2	37.6
21	8.90	10.3	11.6	32.7	35.5	38.9
22	9.54	11.0	12.3	33.9	36.8	40.3
23	10.2	11.7	13.1	35.2	38.1	41.6
24	10.9	12.4	13.8	36.4	39.4	43.0
25	11.5	13.1	14.6	37.7	40.6	44.3
26	12.2	13.8	15.4	38.9	41.9	45.6
27	12.9	14.6	16.2	40.1	43.2	47.0
28	13.6	15.3	16.9	41.3	44.5	48.3
29	14.3	16.0	17.7	42.6	45.7	49.6
30	15.0	16.8	18.5	43.8	47.0	50.9

*This table is abridged and adapted from "Tables of Percentage Points of the Incomplete Beta Function and of the Chi-Square Distribution," *Biometrika*, **32** (1941). It is published here with the kind permission of Professor E. S. Pearson on behalf of the author, Catherine M. Thompson, and of the Biometrika Trustees.

TABLE III

The Normal Distribution

$$\Pr(X \leq x) = \Phi(x) = \int_{-\infty}^{x} \frac{1}{\sqrt{2\pi}} e^{-w^2/2}\, dw$$

$$[\Phi(-x) = 1 - \Phi(x)]$$

x	$\Phi(x)$	x	$\Phi(x)$	x	$\Phi(x)$
0.00	0.500	1.10	0.864	2.05	0.980
0.05	0.520	1.15	0.875	2.10	0.982
0.10	0.540	1.20	0.885	2.15	0.984
0.15	0.560	1.25	0.894	2.20	0.986
0.20	0.579	1.282	0.900	2.25	0.988
0.25	0.599	1.30	0.903	2.30	0.989
0.30	0.618	1.35	0.911	2.326	0.990
0.35	0.637	1.40	0.919	2.35	0.991
0.40	0.655	1.45	0.926	2.40	0.992
0.45	0.674	1.50	0.933	2.45	0.993
0.50	0.691	1.55	0.939	2.50	0.994
0.55	0.709	1.60	0.945	2.55	0.995
0.60	0.726	1.645	0.950	2.576	0.995
0.65	0.742	1.65	0.951	2.60	0.995
0.70	0.758	1.70	0.955	2.65	0.996
0.75	0.773	1.75	0.960	2.70	0.997
0.80	0.788	1.80	0.964	2.75	0.997
0.85	0.802	1.85	0.968	2.80	0.997
0.90	0.816	1.90	0.971	2.85	0.998
0.95	0.829	1.95	0.974	2.90	0.998
1.00	0.841	1.960	0.975	2.95	0.998
1.05	0.853	2.00	0.977	3.00	0.999

TABLE IV

*The t-Distribution**

$$\Pr(T \le t) = \int_{-\infty}^{t} \frac{\Gamma[(r+1)/2]}{\sqrt{\pi r}\,\Gamma(r/2)(1+w^2/r)^{(r+1)/2}}\,dw$$

$$[\Pr(T \le -t) = 1 - \Pr(T \le t)]$$

	Pr (T ≤ t)				
r	0.90	0.95	0.975	0.99	0.995
1	3.078	6.314	12.706	31.821	63.657
2	1.886	2.920	4.303	6.965	9.925
3	1.638	2.353	3.182	4.541	5.841
4	1.533	2.132	2.776	3.747	4.604
5	1.476	2.015	2.571	3.365	4.032
6	1.440	1.943	2.447	3.143	3.707
7	1.415	1.895	2.365	2.998	3.499
8	1.397	1.860	2.306	2.896	3.355
9	1.383	1.833	2.262	2.821	3.250
10	1.372	1.812	2.228	2.764	3.169
11	1.363	1.796	2.201	2.718	3.106
12	1.356	1.782	2.179	2.681	3.055
13	1.350	1.771	2.160	2.650	3.012
14	1.345	1.761	2.145	2.624	2.977
15	1.341	1.753	2.131	2.602	2.947
16	1.337	1.746	2.120	2.583	2.921
17	1.333	1.740	2.110	2.567	2.898
18	1.330	1.734	2.101	2.552	2.878
19	1.328	1.729	2.093	2.539	2.861
20	1.325	1.725	2.086	2.528	2.845
21	1.323	1.721	2.080	2.518	2.831
22	1.321	1.717	2.074	2.508	2.819
23	1.319	1.714	2.069	2.500	2.807
24	1.318	1.711	2.064	2.492	2.797
25	1.316	1.708	2.060	2.485	2.787
26	1.315	1.706	2.056	2.479	2.779
27	1.314	1.703	2.052	2.473	2.771
28	1.313	1.701	2.048	2.467	2.763
29	1.311	1.699	2.045	2.462	2.756
30	1.310	1.697	2.042	2.457	2.750

*This table is abridged from Table III of Fisher and Yates; *Statistical Tables for Biological, Agricultural, and Medical Research*, published by Oliver and Boyd, Ltd., Edinburgh, by permission of the authors and publishers.

TABLE V

*The F-Distribution**

$$\Pr(F \le b) = \int_0^b \frac{\Gamma[(r_1 + r_2)/2](r_1/r_2)^{r_1/2} w^{r_1/2 - 1}}{\Gamma(r_1/2)\Gamma(r_2/2)(1 + r_1 w/r_2)^{(r_1 + r_2)/2}}\, dw$$

$\Pr(F \le b)$	r_2	r_1 = 1	2	3	4	5	6	7	8	9	10	12	15
0.95	1	161	200	216	225	230	234	237	239	241	242	244	246
0.975		648	800	864	900	922	937	948	957	963	969	977	985
0.99		4052	4999	5403	5625	5764	5859	5928	5982	6023	6056	6106	6157
0.95	2	18.5	19.0	19.2	19.2	19.3	19.3	19.4	19.4	19.4	19.4	19.4	19.4
0.975		38.5	39.0	39.2	39.2	39.3	39.3	39.4	39.4	39.4	39.4	39.4	39.4
0.99		98.5	99.0	99.2	99.2	99.3	99.3	99.4	99.4	99.4	99.4	99.4	99.4
0.95	3	10.1	9.55	9.28	9.12	9.01	8.94	8.89	8.85	8.81	8.79	8.74	8.70
0.975		17.4	16.0	15.4	15.1	14.9	14.7	14.6	14.5	14.5	14.4	14.3	14.3
0.99		34.1	30.8	29.5	28.7	28.2	27.9	27.7	27.5	27.3	27.2	27.1	26.9
0.95	4	7.71	6.94	6.59	6.39	6.26	6.16	6.09	6.04	6.00	5.96	5.91	5.86
0.975		12.2	10.6	9.98	9.60	9.36	9.20	9.07	8.98	8.90	8.84	8.75	8.66
0.99		21.2	18.0	16.7	16.0	15.5	15.2	15.0	14.8	14.7	14.5	14.4	14.2
0.95	5	6.61	5.79	5.41	5.19	5.05	4.95	4.88	4.82	4.77	4.74	4.68	4.62
0.975		10.0	8.43	7.76	7.39	7.15	6.98	6.85	6.76	6.68	6.62	6.52	6.43
0.99		16.3	13.3	12.1	11.4	11.0	10.7	10.5	10.3	10.2	10.1	9.89	9.72

$\Pr(F \le b)$	r_2	r_1: 1	2	3	4	5	6	7	8	9	10	12	15
0.95	6	5.99	5.14	4.76	4.53	4.39	4.28	4.21	4.15	4.10	4.06	4.00	3.94
0.975		8.81	7.26	6.60	6.23	5.99	5.82	5.70	5.60	5.52	5.46	5.37	5.27
0.99		13.7	10.9	9.78	9.15	8.75	8.47	8.26	8.10	7.98	7.87	7.72	7.56
0.95	7	5.59	4.74	4.35	4.12	3.97	3.87	3.79	3.73	3.68	3.64	3.57	3.51
0.975		8.07	6.54	5.89	5.52	5.29	5.12	4.99	4.90	4.82	4.76	4.67	4.57
0.99		12.2	9.55	8.45	7.85	7.46	7.19	6.99	6.84	6.72	6.62	6.47	6.31
0.95	8	5.32	4.46	4.07	3.84	3.69	3.58	3.50	3.44	3.39	3.35	3.28	3.22
0.975		7.57	6.06	5.42	5.05	4.82	4.65	4.53	4.43	4.36	4.30	4.20	4.10
0.99		11.3	8.65	7.59	7.01	6.63	6.37	6.18	6.03	5.91	5.81	5.67	5.52
0.95	9	5.12	4.26	3.86	3.63	3.48	3.37	3.29	3.23	3.18	3.14	3.07	3.01
0.975		7.21	5.71	5.08	4.72	4.48	4.32	4.20	4.10	4.03	3.96	3.87	3.77
0.99		10.6	8.02	6.99	6.42	6.06	5.80	5.61	5.47	5.35	5.26	5.11	4.96
0.95	10	4.96	4.10	3.71	3.48	3.33	3.22	3.14	3.07	3.02	2.98	2.91	2.85
0.975		6.94	5.46	4.83	4.47	4.24	4.07	3.95	3.85	3.78	3.72	3.62	3.52
0.99		10.0	7.56	6.55	5.99	5.64	5.39	5.20	5.06	4.94	4.85	4.71	4.56
0.95	12	4.75	3.89	3.49	3.26	3.11	3.00	2.91	2.85	2.80	2.75	2.69	2.62
0.975		6.55	5.10	4.47	4.12	3.89	3.73	3.61	3.51	3.44	3.37	3.28	3.18
0.99		9.33	6.93	5.95	5.41	5.06	4.82	4.64	4.50	4.39	4.30	4.16	4.01
0.95	15	4.54	3.68	3.29	3.06	2.90	2.79	2.71	2.64	2.59	2.54	2.48	2.40
0.975		6.20	4.77	4.15	3.80	3.58	3.41	3.29	3.20	3.12	3.06	2.96	2.86
0.99		8.68	6.36	5.42	4.89	4.56	4.32	4.14	4.00	3.89	3.80	3.67	3.52

*This table is abridged and adapted from "Tables of Percentage Points of the Inverted Beta Distribution," *Biometrika*, **33** (1943). It is published here with the kind permission of Professor E. S. Pearson on behalf of the authors, Maxine Merrington and Catherine M. Thompson, and of the *Biometrika* Trustees.

APPENDIX C

Answers to Selected Exercises

CHAPTER 1

1.1 (a) $\{x : x = 0, 1, 2, 3, 4\}$; $\{x : x = 2\}$.
(b) $\{x : 0 < x < 3\}$; $\{x : 1 \le x < 2\}$.

1.2 (a) $\{x : 0 < x < \frac{5}{8}\}$.

1.7 (a) $\{x : 0 < x < 3\}$.
(b) $\{(x, y) : 0 < x^2 + y^2 < 4\}$.

1.8 (a) $\{x : x = 2\}$.
(b) Null set.
(c) $\{(x, y) : x^2 + y^2 = 0\}$.

1.9 $\frac{80}{81}$; 1.

1.10 $\frac{11}{16}$; 0; 1.

1.11 $\frac{8}{3}$; 0; $\pi/2$.

1.12 $\frac{1}{2}$; 0; $\frac{2}{9}$.

1.13 $\frac{1}{6}$; 0.

1.15 10.

1.18 $\frac{1}{4}$; $\frac{1}{13}$; $\frac{1}{52}$; $\frac{4}{13}$.

1.19 $\frac{31}{32}$; $\frac{3}{64}$; $\frac{1}{32}$; $\frac{63}{64}$.

1.20 0.3.

1.21 e^{-4}; $1 - e^{-4}$; 1.

1.22 $\frac{1}{2}$.

1.26 (a) $\binom{6}{4}\Big/\binom{16}{4}$.
(b) $\binom{10}{4}\Big/\binom{16}{4}$.

1.27 $1 - \binom{990}{5}\Big/\binom{1000}{5}$.

1.29 (b) $1 - \binom{10}{3}\Big/\binom{20}{3}$.

1.34 (a) $\frac{1}{7}$. (b) $\frac{5}{56}$.
(c) $\left[\binom{3}{x}\Big/\binom{8}{x}\right][5/(8 - x)]$.

1.37 $\frac{1}{3}$.

1.38 $\frac{9}{20}$; $\frac{2}{3}$.

1.39 $\frac{5}{14}$.

1.40 $\frac{3}{7}$, $\frac{4}{7}$.

1.42 (a) 0.18. (b) 0.72.
(c) 0.88.

1.45 0.1029 for (a), (b), (c), (d). (e) 0.4116.
1.46 $\frac{1}{4}$, $\frac{3}{4}$.
1.47 $\frac{9}{13}$, $\frac{1}{13}$, $\frac{1}{13}$, $\frac{1}{13}$, $\frac{1}{13}$.
1.48 (a) $\frac{1}{2}$. (b) $\frac{1}{21}$.
1.49 $\frac{1}{5}$, $\frac{1}{5}$, $\frac{1}{5}$.
1.51 (a) $\dfrac{\binom{13}{x}\binom{39}{5-x}}{\binom{52}{5}}$, $x=0, 1, \ldots, 5$.
(b) $\dfrac{\binom{39}{5}+\binom{13}{1}\binom{39}{4}}{\binom{52}{5}}$.
1.54 (a) $\frac{1}{10}$, $x=1, 2, \ldots, 9$. (b) $\frac{4}{10}$.
1.56 $\frac{6}{36}$, $x=0$; $\dfrac{12-2x}{36}$, $x=1, 2, 3, 4, 5$.
1.59 $\frac{3}{4}$.
1.61 $\frac{5}{8}$; $\frac{7}{8}$; $\frac{3}{8}$.
1.63 $e^{-2}-e^{-3}$.
1.64 $\frac{1}{27}$, 1; $\frac{2}{9}$, $\frac{25}{36}$.
1.66 (a) 1. (b) $\frac{2}{3}$. (c) 2.
1.69 (a) 0, $x<0$; $1-(1-x)^3$, $0 \le x < 1$; 1, $1 \le x$; $1-\sqrt[3]{\frac{3}{4}}$, $1-\sqrt[3]{\frac{1}{2}}$.
1.71 (a) $\frac{1}{4}$. (b) 0. (c) $\frac{1}{4}$. (d) 0.
1.72 0, $y<0$; y^2, $0 \le y < 1$; 1, $1 \le y$, $2y$, $0<y<1$; 0 elsewhere.
1.74 $\frac{1}{2}$; $\frac{1}{4}$.
1.76 0, $x<0$; $1-e^{-x/2}$, $0 \le x$. $\frac{1}{2}e^{-x/2}$, $0<x$; 0 elsewhere.
1.79 $1/3\sqrt{y}$, $0<y<1$; $1/6\sqrt{y}$, $1<y<4$; 0 elsewhere.
1.80 2; 86.4; -160.8.
1.81 3; 11; 27.
1.83 (a) $\frac{3}{4}$. (b) $\frac{1}{4}$, $\frac{1}{2}$.
1.85 $7.80.
1.88 $\frac{7}{3}$.
1.89 (a) 1.5, 0.75. (b) 0.5, 0.05. (c) 2; does not exist.
1.90 $e^t/(2-e^t)$, $t<\ln 2$; 2; 2.
1.99 10; 0; 2; -30.
1.101 $-\dfrac{2\sqrt{2}}{5}$, $\dfrac{2\sqrt{2}}{5}$.
1.103 $1/2p$; $\frac{3}{2}$; $\frac{5}{2}$; 5; 50.
1.105 $\frac{31}{12}$, $\frac{167}{144}$.
1.110 $\frac{5}{8}$, $\frac{37}{192}$.
1.114 0.84.

CHAPTER 2

2.1 $\frac{15}{64}$; 0; $\frac{1}{2}$; $\frac{1}{2}$.
2.2 $\frac{1}{4}$.
2.6 ze^{-z}, $0<z<\infty$; 0 elsewhere.
2.7 $-\ln z$, $0<z<1$; 0 elsewhere.
2.10 $5x_2^4$, $0<x_2<1$; 0 elsewhere.
2.11 $(3x_1+2)/(6x_1+3)$; $(6x_1^2+6x_1+1)/(2)(6x_1+3)^2$.
2.13 $3x_2/4$; $3x_2^2/80$.
2.18 (b) $1/e$.
2.20 (a) 1. (b) -1. (c) 0.
2.21 (a) $7/\sqrt{804}$.
2.31 $\frac{5}{81}$.
2.32 $\frac{7}{8}$.
2.36 $\frac{1}{2}$.
2.38 (a) $\frac{1}{6}$, 0.
2.39 $1-(1-y)^{12}$, $0 \le y < 1$; $12(1-y)^{11}$, $0<y<1$.
2.40 $g(y)=[y^3-(y-1)^3]/6^3$, $y=1, 2, 3, 4, 5, 6$.
2.42 $b_2=\sigma_1(\rho_{12}-\rho_{13}\rho_{23})/[\sigma_2(1-\rho_{23}^2)]$; $b_3=\sigma_1(\rho_{13}-\rho_{12}\rho_{23})/[\sigma_3(1-\rho_{23}^2)]$.

CHAPTER 3

3.1 $\frac{40}{81}$.
3.4 $\frac{147}{512}$.
3.6 5.
3.8 $\frac{3}{16}$.
3.10 $\frac{65}{81}$.
3.13 $(\frac{1}{3})(\frac{2}{3})^{x-3}$, $x=3, 4, 5, \ldots$.
3.14 $\frac{5}{72}$.
3.17 $\frac{1}{6}$.
3.18 $\frac{24}{625}$.
3.20 $\frac{11}{6}$; $x_1/2$; $\frac{11}{6}$.
3.21 $\frac{25}{4}$.
3.22 0.09.
3.25 $4^x e^{-4}/x!$, $x=0, 1, 2, \ldots$.
3.26 0.84.
3.31 2.
3.33 (a) $\exp[-2+e^{t_2}(1+e^{t_1})]$.
(b) $\mu_1=1$, $\mu_2=2$,
$\sigma_1^2=1$, $\sigma_2^2=2$,
$\rho=\sqrt{2}/2$.
(c) $y/2$.
3.34 0.05.
3.35 0.831, 12.8.
3.36 0.90.
3.37 $\chi^2(4)$.
3.39 $3e^{-3y}$, $0<y<\infty$.
3.40 2, 0.95.
3.45 $\frac{11}{16}$.
3.46 $\chi^2(2)$.
3.49 0.067; 0.685.
3.51 71.3, 189.7.
3.52 $\sqrt{\ln 2/\pi}$.
3.57 0.774.
3.58 $\sqrt{2/\pi}$; $(\pi-2)/\pi$.
3.59 0.90.
3.60 0.477.
3.61 0.461.
3.62 $N(0, 1)$.
3.63 0.433.
3.64 0; 3.
3.69 $N(0, 2)$.
3.70 (a) 0.574.
(b) 0.735.
3.71 (a) 0.264. (b) 0.440.
(c) 0.433. (d) 0.642.
3.73 $\rho=\frac{4}{5}$.
3.74 (38.2, 43.4).

CHAPTER 4

4.2 $\frac{405}{1024}$.
4.3 0.405.
4.6 $\frac{16}{15}$.
4.7 $\frac{1}{8}$.
4.9 $(n+1)/2$; $(n^2-1)/12$.
4.10 $a+b\bar{x}$; $b^2s_x^2$.
4.11 $\chi^2(2)$.
4.14 $\frac{1}{2}$, $0<y<1$;
$1/2y^2$, $1<y<\infty$.
4.15 y^{15}, $0\le y<1$; $15y^{14}$,
$0<y<1$.
4.16 $\frac{4}{7}$.
4.17 $\frac{1}{3}$, $y=3, 5, 7$.
4.19 $(\frac{1}{2})^{\sqrt[3]{y}}$, $y=1, 8, 27, \ldots$.
4.20

y_1	$g_1(y_1)$
1	$\frac{1}{36}$
2	$\frac{4}{36}$
3	$\frac{6}{36}$
4	$\frac{4}{36}$
6	$\frac{12}{36}$
9	$\frac{9}{36}$

4.25 $\frac{1}{27}$, $0<y<27$.
4.32 $y_1 e^{-y_1}$, $0<y_1<\infty$.
4.34 $(2y_1)(4y_2^3)$, $0<y_1<1$,
$0<y_2<1$.
4.35 $\alpha/(\alpha+\beta)$;
$\alpha\beta/[(\alpha+\beta+1)(\alpha+\beta)^2]$.
4.36 (a) 20. (b) 1260. (c) 495.
4.37 $\frac{10}{243}$.
4.40 0.05.
4.43 1/4.74, 3.33.
4.48 $(1/\sqrt{2\pi})^3 y_1^2 e^{-y_1^2/2} \sin y_3$,
$0\le y_1<\infty$, $0\le y_2<2\pi$,
$0\le y_3\le\pi$.

4.49 $y_2 y_3^2 e^{-y_3}$, $0<y_1<1$, $0<y_2<1$, $0<y_3<\infty$.
4.53 $1/(2\sqrt{y})$, $0<y<1$.
4.54 $e^{-y_1/2}/(2\pi\sqrt{y_1-y_2^2})$, $-\sqrt{y_1}<y_2<\sqrt{y_1}$, $0<y_1<\infty$.
4.56 $1-(1-e^{-3})^4$.
4.57 $\frac{1}{8}$.
4.62 $\frac{5}{16}$.
4.63 $48z_1z_2^3z_3^5$, $0<z_1<1$, $0<z_2<1$, $0<z_3<1$.
4.64 $\frac{7}{12}$.
4.69 $\frac{1}{4}$.
4.70 $6uv(u+v)$, $0<u<v<1$.
4.75

y	$g(y)$
2	$\frac{1}{36}$
3	$\frac{2}{36}$
4	$\frac{3}{36}$
5	$\frac{4}{36}$
6	$\frac{5}{36}$
7	$\frac{6}{36}$
8	$\frac{5}{36}$
9	$\frac{4}{36}$
10	$\frac{3}{36}$
11	$\frac{2}{36}$
12	$\frac{1}{36}$

4.76 0.24.
4.79 0.159.
4.82 0.159.
4.88 0.818.
4.91 (b) -1 or 1. (c) $Z_i=\sigma_i Y_i+\mu_i$.
4.92 $\sum_1^n a_ib_i=0$.
4.94 6.41.
4.95 $n=16$.
4.97 $(n-1)\sigma^2/n$; $2(n-1)\sigma^4/n^2$.
4.98 0.90.
4.100 0.945.
4.102 0.618.
4.103 0.78.
4.104 $\frac{8}{3}$; $\frac{2}{9}$.
4.105 7.
4.107 2.5; 0.25.
4.109 -5; $60-12\sqrt{6}$.
4.110 $\sigma_1/\sqrt{\sigma_1^2+\sigma_2^2}$.
4.113 0.265.
4.115 22.5, $\frac{261}{4}$.
4.116 $r_2>4$.
4.118 $\mu_2\sigma_1/\sqrt{\sigma_1^2\sigma_2^2+\mu_1^2\sigma_2^2+\mu_2^2\sigma_1^2}$.
4.121 $5/\sqrt{39}$.
4.125 $e^{\mu+\sigma^2/2}$; $e^{2\mu+\sigma^2}(e^{\sigma^2}-1)$.

CHAPTER 5

5.1 Degenerate at μ.
5.2 Gamma ($\alpha=1$, $\beta=1$).
5.3 Gamma ($\alpha=1$, $\beta=1$).
5.4 Gamma ($\alpha=2$; $\beta=1$).
5.13 0.682.
5.14 (b) 0.815.
5.17 Degenerate at $\mu_2+(\sigma_2/\sigma_1)(x-\mu_1)$.
5.18 (b) $N(0, 1)$.
5.19 (b) $N(0, 1)$.
5.21 0.954.
5.23 0.840.
5.26 0.08.
5.28 0.267.
5.29 0.682.
5.35 $N(0, 1)$.

CHAPTER 6

6.1 (a) $\bar{X}$. (b) $-n/\ln(X_1X_2\cdots X_n)$. (c) $\bar{X}$. (d) The median. (e) The first order statistic.
6.2 The first order statistic Y_1, $\sum_1^n (X_i-Y_1)/n$.

6.4 $\frac{4}{25}, \frac{11}{25}, \frac{7}{25}$.
6.5 $Y_1 = \min(X_i)$; $n/\ln[(X_1X_2\cdots X_n)/Y_1^n]$.
6.7 (b) $\bar{X}/(1-\bar{X})$. (d) $\bar{X}$. (e) $\bar{X}-1$.
6.9 $1-e^{-2/\bar{X}}$.
6.10 Multiply by $n/(n-1)$.
6.12 $(Y_1+Y_n)/2$, $(Y_n-Y_1)/2$; $E[(Y_n-Y_1)/2] = \rho(n-1)/(n+1)$.
6.14 (77.28, 85.12).
6.15 24 or 25.
6.16 (3.7, 5.7).
6.17 160.
6.23 $(5\bar{x}/6, 5\bar{x}/4)$.
6.25 1692.
6.26 3.19 to 3.61.
6.28 3.92 to 31.50.
6.30 $(-3.6, 2.0)$.
6.35 135 or 136.
6.38 $\frac{1}{4}+\frac{3}{4}\ln\frac{3}{4}$; $\frac{7}{16}+\frac{9}{8}\ln\frac{3}{4}$.
6.39 $\frac{11}{64}$; $(31)3^8/4^9$.
6.42 $n = 19$ or 20.
6.43 $K(\frac{1}{2}) = 0.062$; $K(\frac{1}{12}) = 0.920$.
6.44 $n \approx 73$, $c \approx 42$.
6.46 (a) Reject. (b) p-value ≈ 0.005.
6.49 (c) p-value ≈ 0.005.
6.51 23.3.
6.52 2.91.
6.53 $q_3 = \frac{176}{21} > 7.81$, reject H_0.
6.55 $b \le 8$ or $32 \le b$.
6.56 $q_3 = \frac{22}{9} < 11.3$, accept H_0.
6.57 $6.4 < 9.49$, accept H_0.
6.59 $\hat{p} = (X_1 + X_2/2)/(X_1+X_2+X_3)$.

CHAPTER 7

7.4 $\frac{1}{3}, \frac{2}{3}$.
7.5 $\delta_1(y)$.
7.6 $b = 0$; does not exist.
7.7 Does not exist.
7.17 $\prod_{i=1}^{n} [X_i(1-X_i)]$.
7.19 $60y_3^2(y_5-y_3)/\theta^5$; $6y_5/5$; $\theta^2/7$; $\theta^2/35$.
7.20 $(1/\theta^2)e^{-y_1/\theta}$, $0 < y_2 < y_1 < \infty$; $y_1/2$; $\theta^2/2$.
7.22 $\Sigma X_i^2/n$; $\Sigma X_i/n$; $(n+1)Y_n/n$
7.24 X; X.
7.25 Y_1/n.
7.27 $Y_1 - 1/n$.
7.29 $Y_1 = \sum_1^n X_i$; $Y_1/4n$; yes.
7.37 $\bar{x}$.
7.40 $\bar{X}^2 - 1/n$.
7.43 $\left(\frac{n-1}{n}\right)^Y\left(1+\frac{Y}{n-1}\right)$.
7.51 $\frac{Y_1+Y_n}{2}$, $\frac{(n+1)(Y_n-Y_1)}{2(n-1)}$.
7.55 Y_1, $\Sigma(Y_i - Y_1)/n$.

CHAPTER 8

8.2 $[y\tau^2 + \mu\sigma^2/n]/(\tau^2+\sigma^2/n)$.
8.3 $\beta(y+\alpha)/(n\beta+1)$.
8.9 $\sqrt[6]{2}$ if $y_4 < 1$, $\sqrt[6]{2}\, y_4$ if $1 \le y_4$.
8.13 θ^2/n; $\theta^2/n(n+2)$.
8.15 (a) $4/\theta^2$.
8.17 (d) $\text{var}(\hat{\theta}) = \frac{1}{nI(\theta)} = \frac{\theta^2}{5n}$.
8.22 2.17; 2.44.
8.25 2.20.

CHAPTER 9

9.4 $\sum_1^{10} x_i^2 \ge 18.3$; yes; yes.

9.6 $3\sum_{1}^{10} x_i^2 + 2\sum_{1}^{10} x_i \geq c.$

9.7 95 or 96; 76.7.

9.9 38 or 39; 15.

9.10 0.08; 0.875.

9.11 $(1-\theta)^9(1+9\theta)$.

9.12 $1,\ 0<\theta\leq\frac{1}{2};\ 1/(16\theta^4),$ $\frac{1}{2}<\theta<1;\ 1-15/(16\theta^4),$ $1\leq\theta.$

9.14 53 or 54, 5.6.

9.17 Reject H_0 if $\overline{x}\geq 77.564$.

9.18 26 or 27; reject H_0 if $\overline{x}\leq 24$.

9.19 220 or 221; reject H_0 if $y\geq 17$.

9.23 $t=3>2.262$, reject H_0.

9.24 $|t|=2.27>2.145$, reject H_0.

9.37 $c_0(n)=(14.4)\times(n\ln 1.5-\ln 9.5)$; $c_1(n)=(14.4)\times(n\ln 1.5+\ln 18)$.

9.38 $c_0(n)=(0.05n-\ln 8)/\ln 3.5$; $c_1(n)=(0.05n-\ln 4.5)/\ln 3.5$.

9.41 (b) $c=0.18$; 0.64, (c) $c=0.5$; 0.16; 0.84. (d) $c=0.23$; 0.06; 0.68.

9.44 $(9y-20x)/30\leq c$.

CHAPTER 10

10.9 6.39.

10.12 $r+\theta,\ 2r+4\theta$.

10.13 $r_2(\theta+r_1)/[r_1(r_2-2)]$.

10.23 7.00, 9.98.

10.25 4.79, 22.82, 30.73.

10.26 (a) $4.483x+6.483$.

10.28 $\hat{\beta}=\sum(X_i/nc_i)$, $\sum[(X_i-\hat{\beta}c_i)^2/nc_i^2]$.

10.32 Reject H_0.

10.44 $a_i=0,\ i=1, 2, 3, 4.$

10.45 $\sum_{j=1}^{n} a_{ij}=0,\ i=1, 2, \ldots, n.$

CHAPTER 11

11.2 (a) $\frac{15}{16}$. (b) 675/1024; (c) $(0.8)^4$.

11.4 8.

11.6 0.954; 0.92; 0.788.

11.9 8.

11.12 (a) Beta $(n-j+1, j)$. (b) Beta $(n-j+i-1,\ j-i+2)$.

11.15 0.067.

11.18 Reject H_0.

11.25 0; $4(4^n-1)/3$; no.

11.37 $\frac{2}{99}$.

11.44 98; $\frac{686}{3}$.

Index

Statistical Inference

Confidence Intervals for Means: Normal Assumptions

$\bar{x} \pm a\sigma/\sqrt{n}$, where $\Phi(a) = 1 - \alpha/2$, for μ with σ^2 known

$\bar{x} \pm bs/\sqrt{n-1}$, where $\Pr(T \le b) = 1 - \alpha/2$, for μ with σ^2 unknown

$\bar{x}_1 - \bar{x}_2 \pm b\sqrt{\dfrac{n_1 S_1^2 + n_2 S_2^2}{n_1 + n_2 - 2}\left(\dfrac{1}{n_1} + \dfrac{1}{n_2}\right)}$ for $\mu_1 - \mu_2$

Approximate Confidence Intervals for Binomial Parameters

$\dfrac{y}{n} \pm a\sqrt{\dfrac{(y/n)(1 - y/n)}{n}}$, where $\Phi(a) = 1 - \alpha/2$, for p

$\dfrac{y_1}{n_1} - \dfrac{y_2}{n_2} \pm a\sqrt{\dfrac{(y_1/n_1)(1 - y_1/n_1)}{n_1} + \dfrac{(y_2/n_2)(1 - y_2/n_2)}{n_2}}$, for $p_1 - p_2$

One-Sided Tests of Hypotheses: Normal Assumptions

$H_0 : \mu = \mu_0$ against $H_1 : \mu > \mu_0$. Reject H_0 if

$$\frac{\bar{X} - \mu_0}{S/\sqrt{n-1}} \ge c, \qquad \text{where } \Pr(T \ge c) = \alpha$$

$H_0 : \mu_1 = \mu_2$ against $H_1 : \mu_1 > \mu_2$. Reject if

$$\frac{\bar{X}_1 - \bar{X}_2}{\sqrt{\dfrac{n_1 S_1^2 + n_2 S_2^2}{n_1 + n_2 - 2}\left(\dfrac{1}{n_1} + \dfrac{1}{n_2}\right)}} \ge d, \qquad \text{where } \Pr(T \ge d) = \alpha$$

One-Sided Test of Hypothesis About p

$H_0 : p = p_0$ against $H_1 : p > p_0$. Reject if

$$\frac{Y/n - p_0}{\sqrt{p_0(1 - p_0)/n}} \ge k, \qquad \text{where } \Phi(k) = 1 - \alpha$$

Chi-Square Test

Reject null hypothesis concerning probabilities if

$$\sum_{\text{all cells}} \frac{(Obs_i - Exp_i)^2}{Exp_i} \ge h, \text{ where } h \text{ is the } 100(1 - \alpha)$$

percentile of $\chi^2(r)$, where r is the difference of the dimension of the total parameter space and that of the parameter space under the null hypothesis